T0329091

Polycomb Group Proteins

Translational Epigenetics Series

Trygve O. Tollefsbol, Series Editor

Transgenerational Epigenetics
Edited by Trygve O. Tollefsbol, 2014

Personalized Epigenetics
Edited by Trygve O. Tollefsbol, 2015

Epigenetic Technological Applications
Edited by Y. George Zheng, 2015

Epigenetic Cancer Therapy
Edited by Steven G. Gray, 2015

DNA Methylation and Complex Human Disease
By Michel Neidhart, 2015

Epigenomics in Health and Disease
Edited by Mario F. Fraga and Agustin F. Fernández, 2015

DNA Biomarkers and Diagnostics
Edited by José Luis García-Giménez, 2015

Drug Discovery in Cancer Epigenetics
Edited by Gerda Egger and Paola Arimondo, 2015

Medical Epigenetics
Edited by Trygve O. Tollefsbol, 2016

Chromatin Signaling
Edited by Olivier Binda and Martin Fernandez-Zapico, 2016

Polycomb Group Proteins

Edited by

Vincenzo Pirrotta
Department of Molecular Biology and Biochemistry
Rutgers University, Piscataway, NY,
United States

AMSTERDAM • BOSTON • HEIDELBERG • LONDON
NEW YORK • OXFORD • PARIS • SAN DIEGO
SAN FRANCISCO • SINGAPORE • SYDNEY • TOKYO
Academic Press is an imprint of Elsevier

Academic Press is an imprint of Elsevier
125 London Wall, London EC2Y 5AS, United Kingdom
525 B Street, Suite 1800, San Diego, CA 92101-4495, United States
50 Hampshire Street, 5th Floor, Cambridge, MA 02139, United States
The Boulevard, Langford Lane, Kidlington, Oxford OX5 1GB, United Kingdom

Notices

Knowledge and best practice in this field are constantly changing. As new research and experience broaden our understanding, changes in research methods, professional practices, or medical treatment may become necessary.

Practitioners and researchers must always rely on their own experience and knowledge in evaluating and using any information, methods, compounds, or experiments described herein. In using such information or methods they should be mindful of their own safety and the safety of others, including parties for whom they have a professional responsibility.

To the fullest extent of the law, neither the Publisher nor the authors, contributors, or editors, assume any liability for any injury and/or damage to persons or property as a matter of products liability, negligence or otherwise, or from any use or operation of any methods, products, instructions, or ideas contained in the material herein.

Library of Congress Cataloging-in-Publication Data
A catalog record for this book is available from the Library of Congress

British Library Cataloguing-in-Publication Data
A catalogue record for this book is available from the British Library

ISBN: 978-0-12-809737-3

Publisher: Sara Tenney
Acquisition Editor: Sara Tenney
Editorial Project Manager: Fenton Coulthurst
Production Project Manager: Julia Haynes
Designer: Matt Limbert

Typeset by TNQ Books and Journals

Contents

List of Contributors xiii
Editor's Biography xv
Acknowledgments xvii

1. Introduction to Polycomb Group Mechanisms

V. Pirrotta

References 3

2. The Role of RAWUL and SAM in Polycomb Repression Complex 1 Assembly and Function

C.A. Kim

Introduction 5
Polycomb Group 6
Polycomb Group RAWUL 7
Structural Basis for RAWUL-Binding Selectivity 9
RAWUL Heterodimerization as a Template for Additional Interactions 11
Internal Tandem Duplications in BCOR PUFD Found in Tumors 12
Polycomb Group SAMs 14
Ph SAM Polymerization Is Required for Function 16
Ph SAM Polymerization Is Required for Subnuclear Organization of PRC1 17
What Is the Stoichiometry of SAM-Polymerized PRC1? 18
SAM Polymer Regulation 20
How Does Ph SAM Polymerization Affect PRC1 H2Aub Activity? 21
Scm SAM—Dependent Repression 22
Summary 23
List of Acronyms and Abbreviations 24
Acknowledgments 25
References 25

3. The Chromodomain of Polycomb: Methylation Reader and Beyond

S. Qin, L. Li and J. Min

Introduction 33
Polycomb Chromodomain Specifically Recognizes H3K27me3 35
Structural Basis for the Polycomb–H3K27me3 Interaction Specificity 38
Mammalian Polycomb Homologs Bind Differentially to Methylated Histone H3 41
Cross Talk Between Histone Methylation and Other Posttranslational Modifications 44
Putative Nonhistone Targets of Polycomb Group Chromodomains 45
Noncoding RNA: Noncanonical Partners of Polycomb Group Chromodomains 46
Chemical Probes for CBX7 Chromodomain 48
Polycomb Homologs From Yeast and Plant: Evolutionarily Conserved Biological Significance of Chromodomain 50
Conclusion 51
List of Acronyms and Abbreviations 52
Glossary 52
References 53

4. Unraveling the Roles of Canonical and Noncanonical PRC1 Complexes

E.M. Conway and A.P. Bracken

Background 57
The Mechanism of Action of Canonical Polycomb Repressive Complexes 1 59
The Mechanism of Action of Noncanonical Polycomb Repressive Complexes 1 63
The Biological Importance of Canonical Polycomb Repressive Complex 1, Noncanonical Polycomb Repressive Complex 1, and H2A Monoubiquitination 65
Perspectives 72
List of Acronyms and Abbreviations 74
References 75

5. Structure and Biochemistry of the Polycomb Repressive Complex 1 Ubiquitin Ligase Module

A.G. Cochran

Introduction 81
Ubiquitin Conjugation and Deconjugation 82
PRC1 Is a RING E3 Ligase 85

The Active E3 Ligase Is a Heterodimer of Bml1 and Ring1 89
Biochemical and Structural Studies of the RING Domains of the Bml1-Ring1B Heterodimer 89
Recognition of E2 Enzymes by Bml1-Ring1B 91
Interaction of Bml1-Ring1B-UbcH5c With Substrate 93
Structure of the PRC1 Ubiquitin Ligase Module Bound to Nucleosome 94
Variant PRC1 Complexes and the Mechanism of Ubiquitin Transfer 96
Which PRC1 Contributes Most to H2A Ubiquitination? 99
Is H2A Ubiquitination Really Important? 100
What Does uH2A Do? 102
What's Next for PRC1 and uH2A? 104
References 104

6. Cooperative Recruitment of Polycomb Complexes by Polycomb Response Elements

Y.B. Schwartz

Introduction 111
Polycomb Group Protein Complexes 112
Polycomb Response Elements of *Drosophila* 112
Polycomb Response Elements as Cellular Memory Modules 114
Sequence-Specific DNA-Binding Proteins Implicated in PRE Function 115
PREs as DNA Platforms for Cooperative Recruitment of PcG Complexes 118
Parallels Between Polycomb Targeting in *Drosophila* and Mammals 121
Conclusion 122
List of Acronyms and Abbreviations 123
Glossary 124
Acknowledgments 124
References 124

7. Polycomb Function and Nuclear Organization

F. Bantignies and G. Cavalli

Introduction 131
Polycomb Complexes and Their Action on Chromatin 132
Polycomb Domains 135
Polycomb and Chromatin Compaction 137
Polycomb Group Target Loci Form Dynamic Multilooped Three-Dimensional Structures 138
Polycomb-Repressed Domains Form a Subset of Topologically Associating Domains 141

Long-Range Chromosomal Interactions and Three-Dimensional Gene Networks 144
Potential Role for Noncoding RNA in Polycomb Group–Dependent Three-Dimensional Organization 148
Polycomb and Three-Dimensional Genomics in Cancer and Other Diseases 151
Concluding Remarks 151
List of Acronyms and Abbreviations 152
Acknowledgments 153
References 153

8. Molecular Architecture of the Polycomb Repressive Complex 2

C.S. Huang, E. Nogales and C. Ciferri

Introduction 165
PRC2 Electron Microscopy Studies 169
PRC2 X-ray Crystallography Studies 171
EED Recognizes a Number of Histone Motifs 171
Drosophila Nurf55 (RbAp48) + Histone H3 or Suz12 173
Architecture of a Ternary Ezh2–EED–Suz12 Complex 174
Mechanism of H3K27M Inhibition 179
Mechanism of H3K27me3 Activation 180
Summary and Outlook 182
List of Acronyms and Abbreviations 184
References 185

9. Polycomb Repressive Complex 2 Structure and Function

D. Holoch and R. Margueron

Introduction: Discovery of PRC2 192
Discovery of the ESC-E(Z) Complex 192
PRC2: A Histone Methyltransferase Activity Required for PcG-Mediated Silencing 193
PRC2 Evolutionary Conservation 195
The PRC2 Core Complex 197
Complex Architecture and Requirement of the Subunits for Enzymatic Activity 197
Control of PRC2 Activity by the Chromatin Context 199
Functional Roles of H3K27me3, H3K27me2, and H3K27me1 202
The Alternative PRC2 Catalytic Subunit EZH1 204
PRC2 Cofactors 206
AEBP2 207
PCL 208
JARID2 209

Recently Identified PRC2 Cofactors 211
PRC2 Within the Polycomb Machinery 212
Concluding Remarks: On the Deterministic or Responsive Role of PRC2 in Transcriptional Regulation 214
List of Acronyms and Abbreviations 216
Acknowledgments 216
References 216

10. Regulation of PRC2 Activity

N. Liu and B. Zhu

Polycomb Repressive Complex 2 and Its Enzymatic Activity 225
Activity of Polycomb Repressive Complex 2 226
Structure of Polycomb Repressive Complex 2 227
Solo EZH2 229
Role of H3K27 Methylation 230
Embryonic Ectoderm Development Facilitates the Propagation of H3K27 Methylation 233
Polycomb Repressive Complex 2 Is Stimulated by Dense Chromatin 234
Cross Talk Among Histone Modifications 236
H2A K119 Ubiquitination Stimulates Polycomb Repressive Complex 2 Activity 237
H3K4ME3 and H3K36ME2/3 Inhibit Polycomb Repressive Complex 2 Activity 239
H3S28 Phosphorylation Antagonizes Polycomb Silencing 240
Accessory Components Modulate Polycomb Repressive Complex 2 Activity 241
AEBP2 242
PCL Proteins (PHF1, MTF2, PHF19) 243
JARID2 244
EZH1-Containing Polycomb Repressive Complex 2 246
H3K27M Inhibits Polycomb Repressive Complex 2 Activity and Leads to Pediatric Glioblastoma 247
Conclusion 250
List of Acronyms and Abbreviations 250
Glossary 250
Acknowledgments 251
References 251

11. Activating Mutations of the EZH2 Histone Methyltransferase in Cancer

R.G. Kruger, A.P. Graves and M.T. McCabe

Introduction to Chromatin and EZH2 259
Amplification and Overexpression of EZH2 in Cancer 262
Regulation of Normal B-Cell Differentiation by EZH2 262
Mutation and Biochemical Activity of EZH2 264

Discovery and Incidence of EZH2 Tyrosine 641 Mutations 264
Biochemical Activity of Y641 EZH2 Mutants 264
Discovery of Additional Gain-of-Function EZH2 Mutations 266
Altered Substrate Specificity of A677G and A687V EZH2 Mutants 267
Structural Rationale for Altered Substrate Specificity in EZH2 Mutants 268
Y641 Mutations Have Dual Effects on Substrate Preference 271
The A677G Mutation Optimizes Y641 Positioning for All Three Methylation Reactions 272
EZH2 A687 Coordinates a Water Molecule Required for Substrate Monomethylation 272
Cellular Activity of EZH2 Mutants 274
Loss-of-Function EZH2 Mutations Commonly Occur in Myeloid Malignancies 275
Discovery of EZH2 Inhibitors 275
Mechanistic and Phenotypic Effects of EZH2 Inhibitors in Cancer Cells 278
Conclusions 280
List of Acronyms and Abbreviations 281
References 282

12. PcG Proteins in *Caenorhabditis elegans*

B. Tursun

Introduction 289
PRC1 292
PRC1 in Other Species 292
No Obvious PRC1 in *Caenorhabditis elegans*? 294
SOR-1 and SAM Domain Containing SOP-2: Worm-Specific PRC1-Like Proteins? 295
PRC2 296
Identification of PRC2 Subunits in *Caenorhabditis elegans* 296
PRC2 Is Important for Germline Development and X Chromosome Repression 296
Transgenerational Inheritance of PRC2-Mediated Repression 298
H3K9me2 Compensates for the Loss of H3K27me3 on Sperm Chromosomes 299
Transmission of H3K27me3 in Dividing Cells During Embryonic Development 300
PRC2 Safeguards Germ Cells From Somatic Differentiation 301
PRC2 Restricts Plasticity of Embryonic Cells but is Dispensable for Somatic Differentiation in Worms 302
PRC2 Function in Somatic Cells of *Caenorhabditis elegans* 303
Antagonizing PRC2 Activity 304
Noncoding RNA-Mediated H3K27 Methylation 306
PcG Recruitment—Noncoding RNAs or PREs? 306

Conclusions 307
List of Acronyms and Abbreviation 309
References 309

13. Global Functions of PRC2 Complexes
V. Pirrotta

Introduction 317
Targeted Silencing Functions 318
Global Functions of PRC2 319
Genomic Distribution of H3K27 Methylation 321
Role of Global H3K27 Methylation 323
The Role of UTX: H3K27 Demethylation or Not? 324
H3K27 Acetylation 326
Roaming Activities 328
The Accessibility Hypothesis 329
Recruitment of PRC2 by a PRC1 Type of Complex 332
Does H2A Ubiquitylation Play a Role in Global PRC2 Activity? 333
Polycomb Repressive Activities 336
Evolutionary Aspects of PRC2 Function 338
References 342

Index 349

List of Contributors

Frédéric Bantignies, Institute of Human Genetics, CNRS UPR 1142, Montpellier, France; University of Montpellier, Montpellier, France

Adrian P. Bracken, Trinity College Dublin, Dublin 2, Ireland

Giacomo Cavalli, Institute of Human Genetics, CNRS UPR 1142, Montpellier, France; University of Montpellier, Montpellier, France

Claudio Ciferri, Genentech, Inc., South San Francisco, CA, United States

Andrea G. Cochran, Genentech, Inc., South San Francisco, CA, United States

Eric M. Conway, Trinity College Dublin, Dublin 2, Ireland

Alan P. Graves, GlaxoSmithKline, Collegeville, PA, United States

Daniel Holoch, PSL Research University, Paris, France; INSERM U934, CNRS UMR3215, Paris, France

Christine S. Huang, Genentech, Inc., South San Francisco, CA, United States

Chongwoo A. Kim, Midwestern University, Glendale, AZ, United States

Ryan G. Kruger, GlaxoSmithKline, Collegeville, PA, United States

Li Li, University of Toronto, Toronto, ON, Canada

Nan Liu, Institute of Biophysics, Chinese Academy of Sciences, Beijing, China

Raphaël Margueron, PSL Research University, Paris, France; INSERM U934, CNRS UMR3215, Paris, France

Michael T. McCabe, GlaxoSmithKline, Collegeville, PA, United States

Jinrong Min, University of Toronto, Toronto, ON, Canada

Eva Nogales, University of California, Berkeley, CA, United States; Lawrence Berkeley National Laboratory, Berkeley, CA, United States; Howard Hughes Medical Institute, UC Berkeley, Berkeley, CA, United States

Vincenzo Pirrotta, Rutgers University, Piscataway, NJ, United States

Su Qin, Southern University of Science and Technology, Shenzhen, Guangdong, China

Yuri B. Schwartz, Umeå University, Umeå, Sweden

Baris Tursun, Max-Delbrück-Center for Molecular Medicine, Berlin, Germany

Bing Zhu, Institute of Biophysics, Chinese Academy of Sciences, Beijing, China; University of Chinese Academy of Sciences, Beijing, China

Editor's Biography

Vincenzo Pirrotta was born in Palermo, Sicily, in 1942. He attended Harvard University as an undergraduate, graduate student, and postdoctoral fellow, studying first physical chemistry and then molecular biology with Matthew Meselson and Mark Ptashne. After a year as visiting scientist at the Karolinska Institutet in Stockholm, he was appointed assistant professor at the Biozentrum of the University of Basel in 1972. In 1977, he became group leader at the newly opened European Molecular Biology Laboratory in Heidelberg, where he began to work on the molecular genetics of *Drosophila*. In 1995, he became full professor at the Baylor College of Medicine in Houston, Texas. In 2002 he moved again to be professor in the Department of Zoology of the University of Geneva and in 2004 he became distinguished professor in the Department of Molecular Biology and Biochemistry of Rutgers University in Piscataway, New Jersey. His work has dealt with multiple aspects of genome regulation from bacteriophage repressors to *Drosophila* developmental gene expression, chromatin insulators, epigenetics, and Polycomb mechanisms.

Acknowledgments

My thanks, as always, go to my wife Donna McCabe for her unfailing help and infinite patience.

Chapter 1

Introduction to Polycomb Group Mechanisms

V. Pirrotta
Rutgers University, Piscataway, NJ, United States

Chapter Outline

References 3

Polycomb (*Pc*) mutations were first described by Ed Lewis in 1978 [1], and the gene was characterized as a repressor of the *Drosophila* homeotic genes, which are key developmental genes that determine the anterior–posterior identity of embryonic domains. The name, as usual in *Drosophila*, comes from a common phenotype of the mutation. In this case, the dominant phenotype of flies heterozygous for a *Pc* loss of function mutation is the appearance of a sex comb—a row of thick bristles usually found only on the anterior legs of male flies, on the second and sometimes third legs. This is due to a decreased repression of homeotic genes owing to reduced levels of Pc protein. Pc mechanisms are now known to control not just homeotic genes but as a general chromatin-modifying mechanism that generates a repressive state and controls the expression of most key genes that control growth and differentiation in metazoans.

The molecular study of Pc mechanisms began with the cloning and sequencing of the *Pc* gene [2], which first revealed the existence of the chromodomain, a structural motif found also in the heterochromatin protein HP1. The genetic analyses, followed by biochemical studies showed that a whole group of genes/proteins was involved and was necessary for the repression of homeotic genes. The genes/proteins belonging to this group are, for the most part, not structurally related but are involved in a common repressive mechanism and are often collectively referred to as Pc Group or PcG genes/proteins. The main PcG proteins are in fact components of two different types of complexes: one type built around a core with an enzymatic activity that ubiquitylates histone H2A at lysine 119 and the other type with a methyltransferase activity that methylates histone H3 at lysine 27. A major

Polycomb Group Proteins. http://dx.doi.org/10.1016/B978-0-12-809737-3.00001-5

revolution in our understanding of PcG complexes has come from the discovery of the variety of such complexes and their roles. Complexes of the first type are often called Polycomb repressive complexes one or PRC1-like. Some PRC1 complexes include a Pc-related component that contains a chromodomain and is able to recognize trimethylated histone H3 lysine 27 (H3K27me3). Complexes of the second type are called PRC2. Some PRC2 complexes can bind to histone H2A ubiquitylated at lysine 119. The mutual relationship between at least some PRC1 and some PRC2 complexes may account for the epigenetic features of the repressed states they generate.

Despite enormous progress in understanding PcG mechanisms in the past 25 years, some of the basic questions: "how are they recruited?" and "how do they repress?" are still not clearly answered, and there is considerable divergence of opinion in the field, as may be discerned by reading some of the chapters in this volume. This is a healthy variety that demonstrates the vigor and rapid pace of research. The implementation of PcG mechanisms varies in different organisms with different developmental strategies, and it may vary within one organism from one gene to another, depending on the regulatory needs of the target gene. In flies, for example, the domains of expression of homeotic genes are set by PcG mechanisms early in development and tend to be maintained in progeny cells. This is not necessarily true for all other PcG target genes. In mammals, PcG mechanisms are used for tactical purposes in a rather dynamic way, and they often give way either to specific activators or to other silencing mechanisms such as DNA methylation for more long-term shutting off of genes or parts of the genome during development. There may be a blurring of the distinction between heterochromatic silencing and PcG silencing in some situations. Unlike the Drosophila Pc protein, the mammalian CBX homologs often interact with trimethylated lysine nine of histone H3 as well or better than that with H3K27me3. Therefore, if histone methyl-lysines can indeed recruit PcG complexes, we must expect a significant presence of canonical PRC1 complexes in heterochromatin but whether they play a role there is not known.

Current research is beginning to suggest new roles for PcG proteins beyond the traditional mechanisms of chromatin repression. Increased evidence is being reported for a degree of binding of PRC1 complexes to enhancers and promoters in the absence of H3K27me3 or of PRC2 [3], in some cases even facilitating rather than repressing transcription [4]. Most intriguing but unfortunately not discussed in the present volume is the fact that some PcG activities may take place outside of the nucleus. These activities have occasionally surfaced in the vast PcG literature but have not been pursued in detail—a role for BMI1, a PRC1 component, in normal mitochondrial function [5]. PRC2 has been reported and has been shown to shuttle between the nucleus and the cytoplasm where it plays a role in the cytoskeletal response to extracellular signaling [6]. A recent report has argued that PRC2, targeted by a long noncoding RNA, binds to and methylates Wnt/β-catenin, increasing its stability and promoting Wnt signaling [7]. Clearly, PcG mechanisms have not ceased to surprise us.

The chapters of the present volume are roughly divided between those dealing primarily with PRC1 and those focusing on PRC2. Present chapter and Chapter 2 deal with structural features of PRC1 complexes, including the chromodomain; Chapter 3 describes the complex relationships among different types of PRC1 complexes, including those that lack a chromodomain component; Chapter 4 delves into the biochemistry of the ubiquityl transferase activity of PRC1 complexes; Chapter 5 considers Pc response elements in *Drosophila* and how to make the recruitment of PcG complexes dependent on the cooperation of multiple elements. Chapter 6 examines the role of PcG in nuclear organization and how this affects PcG repression. With Chapter 7 begins the discussion of the PRC2 complex, seen here from the three-dimensional structure of its components. Chapter 8 takes up the functional aspects of PRC2 and the role of its components. Chapter 9 discusses the remarkable feedback and feedforward mechanisms that govern PRC2 enzymatic activities depending on the chromatin context. In Chapter 10 we see how mutations that deregulate the activity of the PRC2 complex can lead to cancer. In Chapter 11 other uses of the PRC2 complex come to the fore in the roundworm *Caenorhabditis elegans*, another model organism in which many genomic and physiological features have been dissected. The question of how other organisms use Pc complexes is taken up also in Chapters 8 and 12 to try to understand how these delicate mechanisms have evolved. Chapter 12 focuses on genome-wide activities that have been little understood until recently, as opposed to targeted activities that are more widely known. The division of the chapters into those dealing with PRC1 and those dealing with PRC2 is only approximate. In many cases, discussion of one necessitates discussion of the other, making a neat division impossible and undesirable.

REFERENCES

[1] Lewis EB. A gene complex controlling segmentation in *Drosophila*. Nature 1978;276: 565–70.

[2] Paro R, Hogness DS. Polycomb protein shares a homologous domain with a heterochromatin-associated protein of *Drosophila*. Proc Natl Acad Sci USA 1991;88:263–7.

[3] Kloet S, et al. The dynamic interactome and genomic targets of Polycomb complexes during stem-cell differentiation. Nat Struct Mol Biol 2016;23:682–90.

[4] Schaaf CA, et al. The *Drosophila Enhancer of split* gene complex: architecture and coordinate regulation by Notch, Cohesin, and Polycomb group proteins. G3 2013;3:1785–94.

[5] Liu J, et al. Bmi1 regulates mitochondrial function and the DNA damage response pathway. Nature 2009;459:387–92.

[6] Su I-H, et al. Polycomb group protein Ezh2 controls actin polymerization and cell signaling. Cell 2005;121:425–36.

[7] Zhu P, et al. lnc-β-Catm elicits Ezh2-dependent β-catenin stabilization and sustains liver CSC self-renewal. Nat Struct Mol Biol 2016;23:631–9.

Chapter 2

The Role of RAWUL and SAM in Polycomb Repression Complex 1 Assembly and Function

C.A. Kim
Midwestern University, Glendale, AZ, United States

Chapter Outline

Introduction **5**
Polycomb Group **6**
Polycomb Group RAWUL **7**
Structural Basis for RAWUL-Binding Selectivity 9
RAWUL Heterodimerization as a Template for Additional Interactions 11
Internal Tandem Duplications in BCOR PUFD Found in Tumors 12
Polycomb Group SAMs **14**
Ph SAM Polymerization Is Required for Function 16
Ph SAM Polymerization Is Required for Subnuclear Organization of PRC1 17
What Is the Stoichiometry of SAM-Polymerized PRC1? 18
SAM Polymer Regulation 20
How Does Ph SAM Polymerization Affect PRC1 H2Aub Activity? 21
Scm SAM–Dependent Repression 22
Summary **23**
List of Acronyms and Abbreviations **24**
Acknowledgments **25**
References **25**

INTRODUCTION

Genome architecture has emerged as a critical component of gene regulation [1]. Three-dimensional perspective of genomes obtained in part from advancing technologies that probe long distance chromosomal contacts provides a far richer and complex view of gene regulation compared to earlier, more linear paradigms. Moreover, determining how a cell is able to faithfully inherit a specific gene expression program upon cell division must now encompass an approach that not only considers the precise reestablishment of the modifications to DNA and histones but also the shape of the genome. The reestablishment of the marks

Polycomb Group Proteins. http://dx.doi.org/10.1016/B978-0-12-809737-3.00002-7

and chromosome structure are, not surprisingly, intimately related [2]. What are the forces that determine the assembly of these large chromosome structures? The major driving force that is fundamental to gene regulation, and ultimately to the formation of these large assemblies, is the collection of noncovalent interactions [3]. The focus here is on two protein–protein interaction domains, RING finger and tryptophan-asparpate 40-associated ubiquitin-like (RAWUL) and Sterile Alpha Motif (SAM), which underlie the assembly of the Polycomb group (PcG) complex called Polycomb repression complex 1 (PRC1), and consequently, the higher order chromatin state mediated by PRC1.

POLYCOMB GROUP

Preservation of chromatin states is requisite for heritable gene silencing. The PcG has long been known for such a role in *Drosophila* development, including for the repression of a family of developmental regulatory genes called the homeotic (HOX) genes. The PcG silences the HOX genes in cells in which their expression is not required and maintains that repression for the lifetime of the fly. In addition to the role in development, PcG proteins maintain pluripotency of mammalian stem cells, control cell proliferation and are the targets of cancer therapeutics [4,5].

All PcG proteins function within multiprotein assemblies [6,7]. The first such PcG complex, isolated from *Drosophila* embryos by Robert Kingston's laboratory, was named Polycomb repressive complex 1 (PRC1) [8]. It consists of four core components observed at stoichiometric levels [9,10]: Polycomb (Pc) (the original member of the PcG), Posterior sex combs (Psc), Polyhomeotic (Ph), and Sex combs extra (Sce; also called, among others, dRING1, where RING stands for 'really interesting new gene'). In vitro, PRC1 can compact nucleosomal arrays into states that are inaccessible to chromatin remodeling enzymes [8,10–12]. PRC1 also houses ubiquitinated histone H2A (H2Aub) ligase activity [13,14], a modification associated with transcription repression, although recent studies suggest that PRC1 H2Aub activity is not required for gene repression [15,16]. The RING finger domain of RING1B, the human ortholog of Sce, has been shown to house the residues directly responsible for the catalytic activity that is stimulated by its dimerization with the RING finger domain of a PcG RING Finger (PCGF) protein [17,18], a mammalian ortholog of Psc. While the number of homologs for the PRC1 components in the *Drosophila* genome is limited, they have made a sizable expansion through evolution (Table 2.1). Various permutations of mammalian PRC1 complexes that utilize particular paralogs of each of the four core components have been isolated (Fig. 2.1). How many of all the different PRC1 permutations exist in cells and how a particular, functionally distinct, PRC1 is able to assemble in both flies and mammals remain to be determined.

Networks of noncovalent interactions determine the selective assembly of different PRC1 permutations. These interactions will not only regulate

TABLE 2.1 Canonical PRC1 Proteins in *Drosophila* and Human

D. melanogaster	Pc	Sce	Psc, Su(z)2	Ph (distal and proximal)
Human	CBX2, 4, 6, 7, 8	RING1A, RING1B	PCGF1, 2, 3, 4, 5, 6	PHC1, 2, 3

PRC1-mediated repression but are also likely to influence the organization of the 3D genome. The folding of the genome has many parallels to the protein-folding problem [1]. Following this parallel, the noncovalent interactions that mediate assembly of distinct PRC1s would be analogous to the thermodynamic principles that underlie the formation of secondary and tertiary protein structures, or in the case of the genome, the formation of topologically associated chromatin domains. For example, protein–protein interactions likely influence the clustering of genomic elements separated by long distances, an important feature of PcG-mediated repression in which members of PRC1 play a key role [2,19–22].

Dissecting the functions of each of the structured domains within PRC1 provides some hints as to how its members contribute to long-range interactions (Fig. 2.1). The chromodomain in Polycomb (Pc) and the chromobox (CBX) proteins allow association with methylated histones. While the preferred modification that the chromodomain binds to has generally been considered to be trimethylation of lysine 27 of histone H3 (H3K27me3), chromo domains of the different CBX proteins have varying affinities for different histone modifications and some with higher affinity than to H3K27me3 [23]. The chromodomain of Cbx7 can even bind RNA [23]. The zinc-binding phenylalanine cysteine serine (FCS) domain of Polyhomeotic homolog 1 (PHC1) is also capable of binding nucleic acids [24]. The FCS may also be involved in protein–protein interactions as the FCS-containing regions within PcG proteins, Sex comb on midleg-related gene containing four malignant brain tumor (MBT) domains (Sfmbt) and Sex comb on midleg (Scm), can directly interact [25]. The protein–protein interaction domains that are most likely to contribute to selective assembly of the different PRC1s and allowing PRC1s to have such influence in genome architecture are the RAWUL and SAM. These domains are discussed in greater detail in the following sections.

POLYCOMB GROUP RAWUL

The RAWUL (previously referred to as the helix-loop-helix (HLH), Ub fold, C-RING1B for the RING1B RAWUL) domain is a ubiquitin fold, protein interaction domain present in all PcG Sce/RING and Psc/PCGF proteins. More

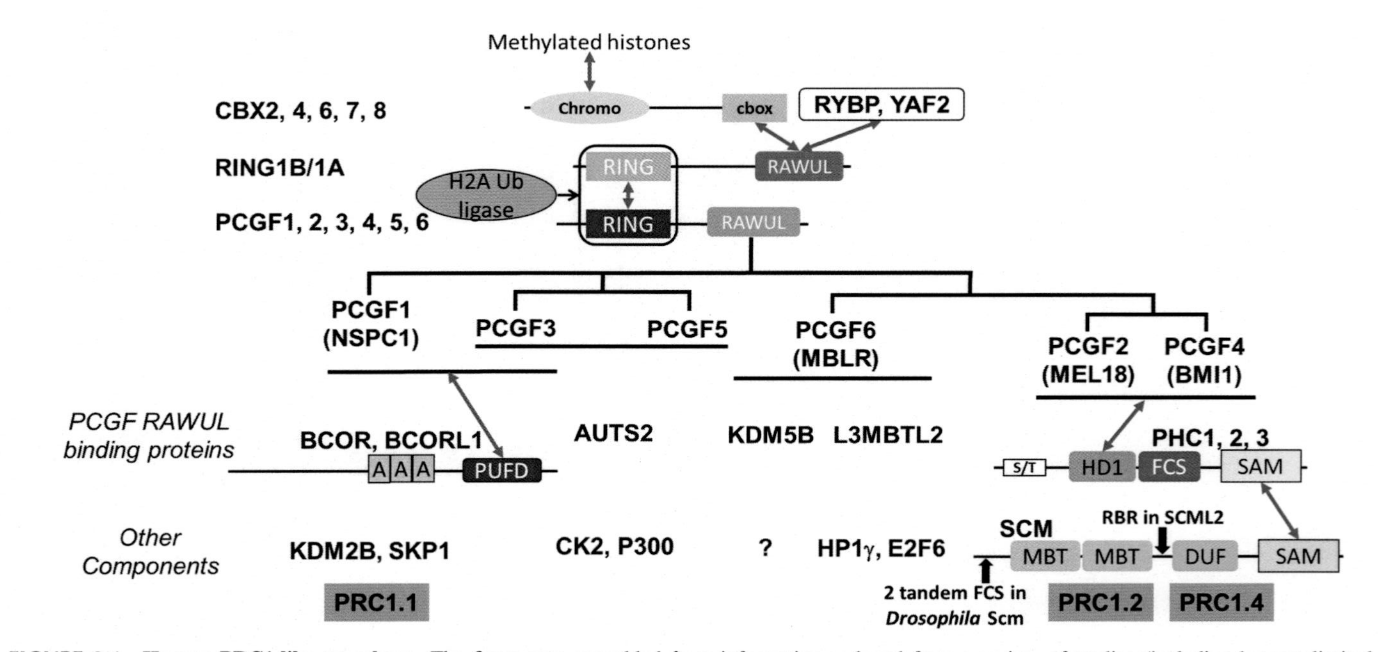

FIGURE 2.1 Human PRC1-like complexes. The figure was assembled from information gathered from a variety of studies (including but not limited to [8,10,28,42,98–101]). *Blue arrows* indicate direct interactions. Domain structures for selected proteins are shown. PRC1.1, 1.2, and 1.4 are designations used by Gao et al. [28] stemming from the central role played by the particular PCGF protein in the assembly of the larger complex.

than any other noncovalent interaction domain of the PRC1 proteins, the RAWUL contributes to the assembly of functionally distinct PRC1s that stem from incorporating different paralogs of the different PRC1 core members. Biophysical studies of the RING1B and PCGF1 RAWULs have revealed that despite both proteins utilizing the same mode of binding for their interactions, the individual RAWULs bind in a highly selective fashion, binding with high affinity to only a select group of proteins. Not only do these interactions serve in assembling the distinct PRC1s but also the formation of these 1:1 interactions leads to greater diversity of higher order assemblies whereby the initial RAWUL heterodimerization with its primary partner then allows interaction with another protein which neither the RAWUL nor its primary partner can bind on its own.

The investigation of PRC1 permutations has emerged as a very active topic of research in the field [7,26,27]. Understanding the molecular basis for how the different PRC1 paralogs assemble to form distinct complexes has important functional ramifications. For example, different PRC1s can have much overlap in the binding locations throughout the genome, yet they are frequently observed to regulate different sets of genes [28–31]. Moreover, only a subset of possible complexes has been observed to form in cells, and these complexes differ in their H2A ubiquitylation abilities. For example, a PRC1 complex housing RING1B/PCGF1 and the histone demethylase lysine demethylase 2B (KDM2B) was observed to be the major PRC1 responsible for H2Aub activity in embryonic stem cells, while another PRC1 present in the same cells, which contains PCGF2 [32] (along with RING1B, CBX7, and PHC1 [30]), has little role in H2A ubiquitylation. This is analogous to the situation in *Drosophila* embryos where a PRC1 complex lacking Ph or Pc but housing dKdm2, the *Drosophila* ortholog of KDM2B, is largely responsible for the H2Aub activity rather than the canonical PRC1 containing all four core members [33]. This study from Peter Verrijzer's laboratory showed that reducing levels of dKdm2 resulted in a marked reduction of H2A ubiquitylation equivalent to that observed when Sce levels are reduced. The PCGF and CBX/RING1B YY1–Binding Protein (RYBP)/YY1-associated factor (YAF) proteins have been identified as critical determinants in the assembly of functionally distinct PRC1s [28,30,34,35]. The interaction between these proteins is mediated by the RAWUL suggesting a key role for the RAWUL in determining which particular PRC1 forms.

Structural Basis for RAWUL-Binding Selectivity

Structural studies have revealed how the RAWUL contributes to the assembly of functionally distinct PRC1s [36–39]. The RING1B RAWUL is capable of binding the cbox domain to all five CBX proteins [40] as well as to RYBP and its closely related paralog YAF2 [37,41]. While there is sufficient sequence similarity among the CBX cbox domains for their grouping as an identifiable domain, they share little sequence identity with the stretch of residues within RYBP/YAF2 that bind the RAWUL [37]. In addition, cbox and RYBP are unfolded in the absence of the RING1B RAWUL [37,40]. While the lack of any structural features in the absence of the RAWUL could suggest a tendency for nonspecific interactions, the RING1B RAWUL binds with high affinity to only the cbox domains and RYBP/YAF2. Structures of the RING1B RAWUL as well as the PCGF1 RAWUL [37,38] in complex with their binding partners have identified formation of an intermolecular beta sheet as the major RAWUL-binding selectivity determinant. The RAWUL-binding partner forms an antiparallel beta sheet that augments the major beta sheet of the RAWUL ubiquitin fold (Fig. 2.2). The combined, intermolecular beta sheet surrounds the central helix of the ubiquitin fold of the RAWUL. The selectivity stems from precise complementary contacts made by the side chains of the residues of the augmenting beta sheet to the residues of the RAWUL central helix.

This selectivity of the RAWUL interactions reduces the number of different PRC1 permutations in the following manner. There are a total of 252 possible different permutations of PRC1 accounting for seven CBX/RYBP/YAF, two RING, six PCGF, and three PHC proteins. The RAWULs of PCGF2 and PCGF4 are the only ones able to bind the PHC proteins [37]. Thus, the number of canonical PRC1s, those that include the original four members, is reduced to 84. Consistent with this is the absence of PHC paralogs in the purifications of the PRC1 complex that houses PCGF1. Instead of PHC, these

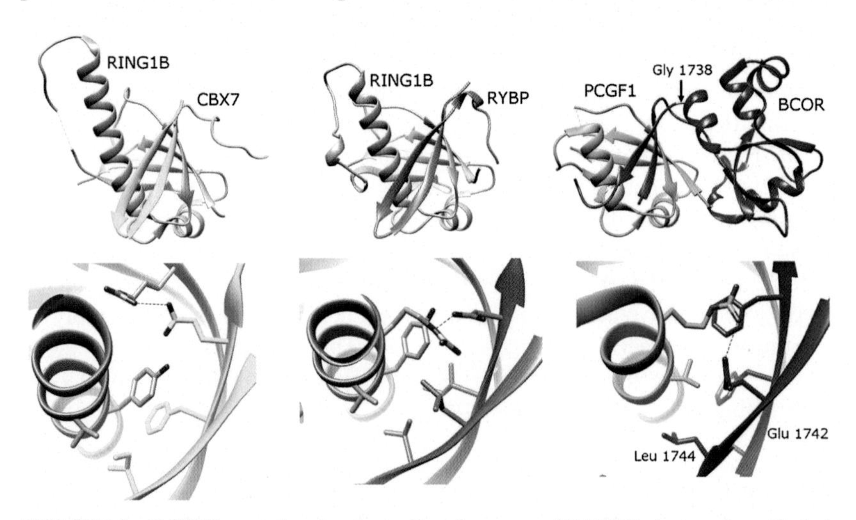

FIGURE 2.2 RAWUL complex structures. Top: Structures of RAWULs in complex with their binding partners [37,38]. Bottom: A close-up view of the augmenting beta sheet structure that provides the binding selectivity of the interaction. Key residues in the PCGF1–BCOR interaction that appear likely to be affected by the ITDs found in CCSK and CNS-PNET are highlighted. The figures were prepared from the following coordinates: PDBIDs 3GS2, 3IXS, and 4HPL for RING1B/CBX7, RING1B/RYBP, and PCGF1/BCOR, respectively.

PCGF1 PRC1s include BCL6 corepressor (BCOR) and KDM2B [28,42,43]. The other PCGF RAWULs exhibit their own selective binding, thereby serving a key role in assembling a particular PRC1 (Fig. 2.1). While the RAWUL-binding selectivity does reduce the allowable number of PRC1 permutations, there still remain approximately 196 different PRC1 permutations that are possible. It is currently not known how many of the 196 actually form in cells nor what the functional consequences are for the different permutations.

RAWUL Heterodimerization as a Template for Additional Interactions

The RAWUL also has a potential role in determining what other proteins, other than the primary RAWUL-binding partner, are included in the larger complex. The RAWUL of PCGF1 is involved in such an assembly. Structure and binding studies involving PCGF1, BCOR (or its close paralog BCL6 corepressor-like 1 (BCORL1)), and KDM2B revealed that the dimer between PCGF1 and BCOR (or BCORL1) can assemble with KDM2B which neither PCGF1 nor BCOR/BCORL1 can perform alone. The inability of either the RAWUL or its binding partner alone to associate with the third protein may stem from the conformational changes involved in the RAWUL interaction. While the RING1B RAWUL alone is folded, it is not in a single conformation [40]. Its binding partners are also disordered in the absence of the RAWUL. The approximate 30 residues of the CBX cbox proteins and the similarly sized region within RYBP are both unfolded prior to their association with RING1B RAWUL [37,40]. The approximate 115 residue PCGF1 ubiquitin fold discriminator (PUFD) domain of BCOR that binds the PCGF1 RAWUL is an independently folded domain, yet, the augmenting beta sheet region of the BCOR PUFD is unstructured in the absence of the RAWUL (Kim, unpublished). On dimerization, a conformational tightening occurs to both the RAWUL and its binding partner whereby the heterodimer exists in a single conformation [37,40]. The structural differences between the unbound and dimer states along with a new binding template created when the two proteins unite would then allow association with the third component. Thus assembly of the larger complex does not occur through a series of independent 1:1 protein—protein interactions. Rather, assembly involving the RAWUL occurs in an ordered manner, initiated by a heterodimer formation followed by recruitment of additional proteins (Fig. 2.3A). A consequence for this ordered assembly may be to allow for proper spatial and temporal considerations when resetting the genome after cell division. As discussed below, the polymerization of PRC1s with SAMs may similarly assemble in a sequence that is required to reestablish the genome state (Fig. 2.3B). It will be of interest to determine if RAWULs other than the one from PCGF1 can similarly act as a binding template on dimerization with their primary binding partner.

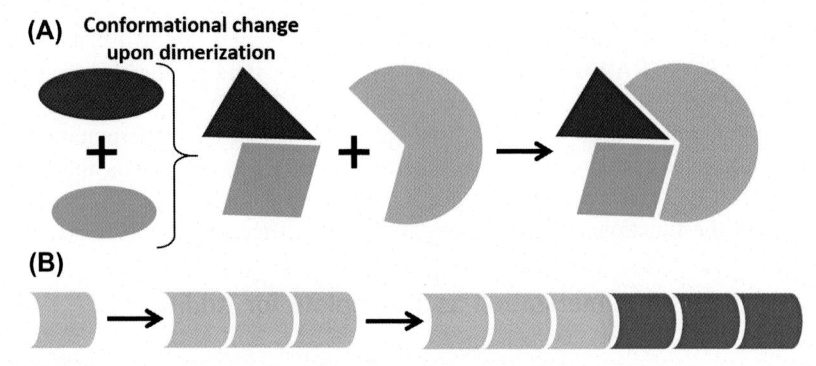

FIGURE 2.3 Hierarchical assembly of an epigenetic complex. (A) Biophysical and binding studies of the RAWUL suggest a model whereby the individual proteins (ovals), such as the RAWUL, are unable to bind a third component without first interacting with its primary binding partner. (B) Potential sequence of assembly of homo- and hetero-SAM polymers.

In the context of epigenetic regulatory proteins, there are precedents for the need of such a mode of assembly. Take for example, the assembly of the histone octamer as demonstrated by Roger Kornberg's early work identifying the octamer's suboligomeric states [44]. During either transcription or replication, histone octamer disassembly occurs through eviction of the H2A-H2B dimer while retaining the (H3-H4)2 tetramer on the DNA. These suboligomeric states, as identified by Kornberg, serve as a foundation for inheriting the identical chromatin state in that a vast majority of posttranslational modifications (80%) required to reestablish the gene expression program occur on the tails of H3 and H4. Reestablishing the three-dimensional structure of the genome must extend beyond the histone octamer to the additional layers of genome assembly. Could the RAWUL interactions contribute to mediating such larger assemblies? It is tempting to speculate that, like histones, noncovalent interactions that occur in a particular order along with the posttranslational modifications that occur on these proteins may serve in an analogous fashion in reestablishing the three-dimensional genome.

Internal Tandem Duplications in BCOR PUFD Found in Tumors

The potential harmful consequence from disrupting the PCGF1 RAWUL interaction with the BCOR PUFD was recently revealed by the identification of internal tandem duplications (ITDs) that occur in-frame within the *BCOR* gene in two different types of pediatric tumors: clear cell sarcoma of the kidney [45,46] and primitive neuroectodermal tumors of the central nervous system (CNS-PNET) [47]. All the varying ITDs found in the different tumors found in both kidney and brain were shown to have in common the insertion of ~20–40 duplicated amino acid residues within the BCOR PUFD sequence (Table 2.2). Interestingly, none of the ITDs are predicted to disrupt the

TABLE 2.2 CCSK and CNS-PNET With Internal Tandem Duplications (ITDs) Within the BCOR PUFD

CCSK: Ueno-Yokohata et al. [45]	CCSK: Roy et al. [46]	CNS-PNET: Sturm et al. [47]*	Length of ITD (Residues)	ITD Insertion Site (After Indicated Amino Acid)
CCSK2, 13, 15	504T	MB_15–0048, CNS-PNET_15–0302, EPN_15–0019, MB_15–0049, CNS-PNET_15-0223	36, 38	Leu1737
		GBM_15-0010	20	Ser1740
CCSK17, 19	382T, 385T	MB_15–0047, CNS-PNET_15–0319, EPN_15–0027, EPN_15–0025, GBM_15-0003	30	Val1741
		ETMR_15-0011, GBM_15-0001	32, 42	Trp1743
		CNS-PNET_15-0029	21	His1745
CCSK1, 3, 5, 6, 8, 10	380T, 474T, 501T		31	Asp1752
CCSK4, 7, 9, 11, 14, 16, 18, 20	347T, 383T, 384T, 499T, 624T		32	Trp1755 (last BCOR residue)

All ITDs are predicted to affect the PCGF1 augmenting beta sheet of BCOR PUFD. The C-terminal augmenting beta strand begins after Gly1738 and includes PCGF1-contacting residues Glu1742 and Leu1744 (Fig. 2.2, right). * The authors note one additional tumor with an alteration not specifically noted and thus not listed in the table. The ITD for this alteration is inserted after Phe1637, which is part of the N-terminal augmenting beta strand of the BCOR PUFD. Like the C-terminal beta strand, the N-terminal strand makes key contacts with PCGF1.

hydrophobic core of the protein and thus would allow the stable three-dimensional folding of the PUFD core. Rather, all the ITDs, in one way or another, alter the PCGF1 RAWUL augmenting the beta sheet of the BCOR PUFD. As this beta sheet makes key contacts with PCGF1 (Table 2.2, Fig. 2.2, right), the ITDs would most likely disrupt binding to PCGF1 RAWUL, and consequently, disrupt KDM2B binding. This structural analysis highlights the key molecular event that may lead to disease, which in turn could help in the development of targeted therapies to treat these diseases.

POLYCOMB GROUP SAMs

Canonical PRC1s house a polyhomeotic protein (Ph in *Drosophila*, PHCs in humans). An important and distinguishing feature of these PRC1s compared to the other permutations is a protein interaction module found at the C-terminus of the PHC proteins called SAM. Although the *Drosophila* Ph SAM was identified as being able to self-associate in vitro as a polymer in 2002, it is only recently that evidence has begun to emerge in support of the important functional role of the SAM polymer structure. Recent evidence has revealed that SAMs are required for clustering PRC1 complexes, likely providing a structural template from which a higher order chromatin architecture is assembled and one that is essential for repression mediated by canonical PRC1s.

SAMs are ~7 kDa, independently folded protein domains. Their original identification within yeast proteins that play an important role in mating and their predicted structure consisting of alpha helices led to their name designation [**S**terile **A**lpha **M**otif, they have also been referred to as Scm, Ph, L(3) mbt (SPM) in PcG literature] [48]. SAMs are quite prevalent throughout eukaryotic genomes and are even found in bacteria [49–51]. They are involved in a myriad of noncovalent interactions mostly consisting of protein–protein interactions though some SAMs can bind RNA [52–55] and even membranes in vitro [56,57]. While some SAMs are monomeric or form limited oligomeric states, many SAMs, including several SAMs of PcG proteins (Table 2.3), possess an intriguing ability to self-associate forming an open-ended, left-handed helical polymer [58–60] (Fig. 2.4). Thus, SAM polymerization appears to be a key feature in influencing PcG function and chromatin structure.

The first SAM polymer to be identified was an e26 transformation-specific (Ets) family gene regulatory protein called translocation ets leukemia (TEL) [61]. The left-handed helical polymer architecture for TEL SAM has been observed for all subsequent polymeric SAMs whose structures have been determined [58–66]. While most assemble utilizing six SAMs per turn of the polymer helix, this number can vary [60,66]. Interestingly, the SAM from PHC3, a human ortholog of Ph, can form polymers using either five or six SAMs per turn [60] perhaps reflective of an ability to serve as a scaffold for variable chromatin architectures. These SAM polymers assemble through head-to-tail mediated

TABLE 2.3 PcG SAM Proteins

D. melanogaster	Human
Ph	PHC1, 2, 3
Scm	SCMH1, SCML1, SCML2, SCML4
Sfmbt	SFMBT1, SFMBT2
L(3)mbt	L3MBTL1, L3MBTL3, L3MBT4

While all SAMs from the listed proteins scored relatively high in predicting their polymeric state [96], Sfmbt SAM was shown to be a monomer [87]. In vitro evidence for SAM polymers has been obtained for Ph [58], Scm [59], PHC3 [77], L3MBT3 [97], and L3MBTL4 [97]. Human homolog L3MBTL2 does not contain a SAM.

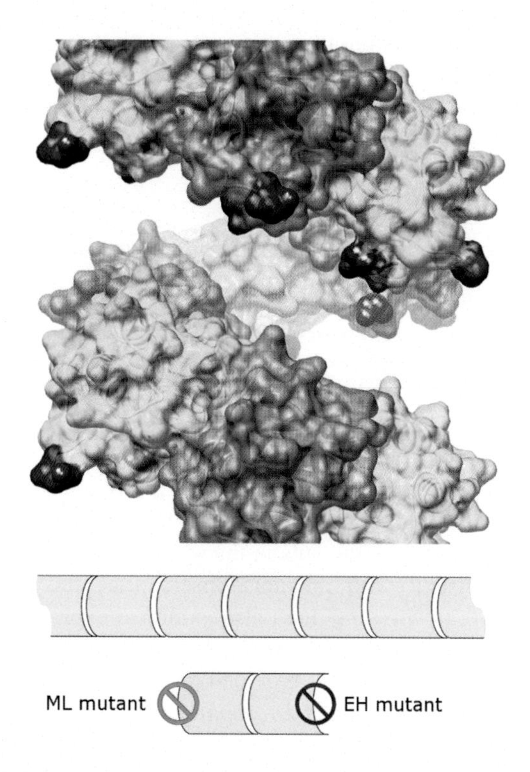

FIGURE 2.4 Ph SAM polymer structure. Top: Three-dimensional structure representation of the Ph SAM polymer (PDBID: 1KW4). Each SAM in the polymer is shown in alternating colors. The N- and C-termini of the SAMs are indicated by their blue and red colors, respectively. Bottom: Illustration of the head-to-tail arrangement observed for all SAM polymers. A mutation on either the ML- or EH-binding surfaces renders the SAM monomeric. However, the remaining native binding surface allows interaction with the opposite surface mutant SAM.

contacts via two different binding surfaces named for their location in the structure first observed for TEL SAM. The residues that make up the mid-loop (ML) surface typically occur near the middle of the SAM sequence while the residues that constitute the second binding surface are on or near the C-terminal helix of the SAM, thus designating the end-helix (EH) surface (Fig. 2.4). The ML surface of one SAM interacts with the EH surface of another, allowing the formation of extended polymer chains. A mutation on either the ML or EH surfaces disrupts polymerization. The opposite binding surface of either of these mutants, however, is native and remains capable of binding the opposite mutant SAM allowing measurement of the native SAM—SAM affinity. For the PcG SAMs from Ph and Scm, these were measured to be quite strong (190 and 50 nM K_D, respectively [58,59]), reflecting the likelihood that these SAM polymers form in vivo. The SAMs of all PcG proteins are found at the C-terminus of their respective sequences. Having both the N- and C-termini point outward from the polymer helix axis of the SAMs (Fig. 2.4, blue and red highlighted residues) makes it feasible to both form the helical polymer while sterically accommodating the other parts of the protein if they were to extend away from either termini of the SAM (see below). Together, these structural features gave strong support for allowing polymerization to occur in vivo while suggesting the likely important role SAM polymerization plays in protein function.

Ph SAM Polymerization Is Required for Function

Even though the conservation of the helical polymer architecture in diverse proteins strongly suggested that SAM polymerization serves a common and important functional purpose, evidence of PcG SAM polymerization being important for function was slow to emerge. In 2012, evidence was finally provided showing that Ph SAM polymerization was required for Ph-mediated repression [67]. Ph constructs with SAM mutations at either the ML or EH surfaces were unable to repress transcription of a reporter gene compared to the wild-type version. Furthermore, transgenic flies that overexpressed these same SAM polymer-deficient mutants in the wing disc resulted in overgrowth of the tissue, consistent with previous observations for Ph mutant flies [68—70]. These overgrown wing discs also exhibited expression of the HOX gene *Abdominal-B* (*Abd-B*), which is normally repressed in this tissue. Recent work from Jürg Müller's laboratory provided further support for the important role of Ph SAM polymerization in *Drosophila* [71]. A mutant *ph* transgene with the SAM deletion failed to rescue the phenotype of flies that do not express any *ph*. The SAM EH mutant *ph* transgene, while slightly better than the SAM deletion transgene, still exhibited severe deficiencies in its rescue ability including the misexpression of *Abd-B*. In mice, the expression of polymer-deficient Phc2, a homolog of Ph, exhibited skeletal transformations and derepression of genes, including HOX genes [21]. Together, these functional studies have quite convincingly settled the importance of SAM polymerization for PcG-mediated repression.

Ph SAM Polymerization Is Required for Subnuclear Organization of PRC1

With the importance of polymerization to protein function firmly established, a key issue that is beginning to be addressed is the precise role of SAM polymerization in gene repression. Given the architecture of the SAM polymer, it seems likely that it could direct the formation of a specific, higher order chromatin state. Precisely what this structure is, however, far from being determined. While a highly speculative chromatin state model was proposed for the TEL SAM polymer [61], Ph being a part of a multiprotein complex predicts a far greater complexity in the chromatin state than TEL. Recent work led by Haruhiko Koseki [21] and Nicole Francis [72] has begun to dissect the role of SAM polymerization in mediating these repressive higher order chromatin structures.

The Koseki group investigated the role of Phc2 SAM polymerization in nuclear clustering. PcG proteins play a role in creating the properly folded 3D chromosomes in part by mediating the assembly of nuclear clusters generically referred to as PcG bodies [73]. Their formation is dependent on PcG proteins and house many repressed genes in the clusters including those that are separated by long distances [19,74]. The Koseki group revealed that polymerization of Phc2 via its SAM is required for the formation of these nuclear clusters and for repressing the genes that reside within them. A telling experiment involved the use of a Phc2 construct lacking the Homology domain 1 (HD1) domain while also housing a polymerization-deficient EH surface mutation on its SAM. The HD1 domain binds directly to the RAWUL of either PCGF2 or PCGF4 [38] thereby representing the protein—protein interaction that connects the PHC proteins to PRC1. Thus the deletion of HD1 and the SAM EH mutation leaves only the SAM ML-binding surface to assemble with PRC1 and influence PRC1-mediated nuclear clustering. This mutant was indeed able to associate with endogenous PRC1 components and disrupt PRC1-mediated clustering. This Phc2 construct with only a viable SAM ML surface acted in a dominant negative fashion to disrupt condensation of distant HOX genes into a single locus and consequently their derepression. Such a dominant negative role for polymer-deficient SAM mutant has also been observed for *Drosophila* Ph SAM mutants [67,72]. This result suggests that Phc2 SAM polymerization, and not the protein interaction domain directly connecting Phc2 to PRC1, plays the major role in organizing nuclear clusters. The possible mode of repression that occurs within the SAM polymer—mediated clusters was suggested by a substantial reduction of RNA polymerase II at Phc2 SAM—mediated repression sites leading the authors to propose that repression may occur via the exclusion of RNA polymerase II.

Recent work from the laboratories of Robert Kingston and Xiaowei Zhuang and led by Nicole Francis' group provided greater detail of the Ph polymer—mediated chromatin state [72]. Utilizing stochastic optical reconstruction microscopy (STORM) to image *Drosophila* S2 cells, a more precise measurement of the size of the PcG bodies was possible. PcG bodies were measured to actually have a wide variance in size distribution ranging from 30 nm (resolution limit) to greater than 700 nm. Thus, what had previously been generically classified as "PcG bodies" appears to be quite heterogeneous. Furthermore, previously identified single clusters were likely to be many clusters that were unable to be resolved due to resolution limits of the microscopy methods. As was observed by the Koseki group, formation of the nuclear clusters was dependent on Ph SAM polymerization. Further probing the chromatin state using circularized chromatin conformation capture coupled with next-gneration sequencing (4C-seq) methods provided an even closer "view" of Ph SAM—mediated repression of *Abd-B*. The 4C-seq results revealed that the expression of the Ph-ML mutant increased interactions between the *Abd-B* regulatory regions called infra-abdominal (*iab*) elements and the *Abd-B* promoter likely leading to the expression of *Abd-B*, consistent with earlier studies showing Ph SAM depolymerization leading to *Abd-B* derepression [67,71]. In contrast, Ph SAM polymer—mediated clustering precluded interactions between the *iab* elements and *Abd-B*, thus maintaining the repression of *Abd-B*. This model of what has recently been described as a *Drosophila* chromosomal topological domain [2] is the most resolved perspective of the effect SAM polymerization has on chromatin structure revealing a significant change to the genome structure that occurs when Ph SAM polymerization is altered. The influence of Ph, and thus in part due to Ph SAM polymerization, was further revealed in a separate study investigating the structural features of different chromatin states using high-resolution three-dimensional STORM [75]. The chromatin structure that is highly influenced by the PcG was measured to be the most densely packed, even compared to other repressed chromatin domains. On knockdown of Ph, the size of the PcG domain increased to a size similar to that of active chromatin states while also exhibiting greater overlap with neighboring chromatin domains. With the increased emphasis and improvement on microscopy and chromatin conformation capture methods, the chromatin states mediated by PcG SAM polymerization are likely to be clarified even further in the coming years.

What Is the Stoichiometry of SAM-Polymerized PRC1?

Ph SAM polymerization affecting genome architecture raises an intriguing and important question of how many Ph, and by extension PRC1, assemble as a consequence of Ph SAM polymerization. If SAM polymerization occurs unimpeded, PRC1 oligomerization would depend on the concentration of Ph. However, this does not appear to be the case, at least for Ph. Built into its own amino acid sequence is a means by which Ph limits the length of polymers mediated by its SAM. The sequence that links the *Drosophila* Ph FCS domain and the SAM, or the SAM linker, can influence the length of SAM polymers that can form in vitro [67]. Unlike the isolated SAM that forms open-ended polymers, a construct housing the Ph SAM and its linker is limited in its ability to polymerize and is only able to oligomerize to about five to six units based on sedimentation velocity analytical ultracentrifugation data. The sequence N-terminal to Scm SAM may also function to limit the extent of Scm polymerization. Rather than localizing to specific chromosome locations as observed for the wild-type full-length Scm, the isolated Scm SAM alone is shown to occupy long extended regions on polytene chromosomes consistent with its polymerization [76]. Interestingly, the SAM polymer limiting property of the linker is not conserved for human PHC SAM. In contrast to the *Drosophila* Ph SAM linker, the linker to PHC3 SAM allows open-ended polymerization [77]. While it is possible that PHC3 could polymerize via its SAM indefinitely, a more plausible explanation is that an alternative mode exists by which PHC3 SAM polymerization is controlled. The mode by which a stretch of about 100 unstructured amino acids allows the SAM to either polymerize indefinitely or form only a short oligomer is unclear. What is evident is that the molecular basis for this control stems from the amino acid content in the linker sequences. Scrambling the sequence but maintaining the amino acid content had the identical effect on SAM polymerization [77].

The ability of the SAM linker to influence polymerization raises the possibility that other parts of the protein or even the larger multiprotein assembly could also affect polymerization. Extending polymerization beyond just several PRC1 units presents a challenge when considering the potential steric clashes stemming from PRC1 residing on the periphery of the SAM helical architecture. The red sphere shown in Fig. 2.5 is representative of the volume

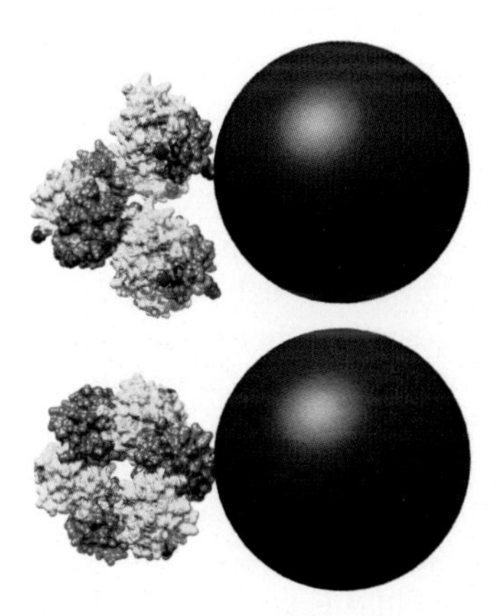

FIGURE 2.5 PRC1 around Ph SAM. Two views are shown of the calculated volume of PRC1 surrounding the Ph SAM polymer. The sphere represents the volume of PRC1 calculated using the partial specific volume of the four core components of PRC1 (Psc, Pc, Sce, and Ph minus the SAM sequence).

occupied by one *Drosophila* PRC1 attached to the N-terminus of one Ph SAM. It thus includes the Ph sequence outside of the SAM, one Psc, one Pc, and one Sce. It is important to note that the volume corresponds to a sequence that is packed into a three-dimensional structure. The sequence analysis of PRC1 components, especially so for Ph and Psc, predicts significant regions of unstructured residues, which would expand the volume occupied by PRC1. In addition, the unstructured Ph SAM linker is kept close to the SAM helix as suggested by its ability to limit polymerization on its own, and thus would keep the other PRC1 components close to the SAM polymer (as in Fig. 2.5) and not allow the utilization of the infinite space that extends away from the SAM polymer. This crude analysis suggests that oligomerizing PRC1 via Ph SAM polymerization is likely to be limited to just several PRC1 units. As alluded to earlier, however, longer PRC1 polymers may be possible if not all PRC1 components are included for every Ph or the unstructured residues within PRC1 proteins play a specific role in promoting polymerization similar to what is observed for the PHC3 SAM linker [77].

SAM Polymer Regulation

A broadly applicable question important for all proteins that have polymeric SAMs is how polymerization is regulated. Jürg Müller's group has recently identified a posttranslational modification of Ph as a potential means of triggering the availability of a functional form of Ph [71]. Ph and its mouse ortholog, Phc3, are modified by the addition of an O-linked β-N-Acetylglucosamine (O-GlcNAc) [78–80]. The covalent attachment of O-GlcNAc occurs on the hydroxyl group of either Ser or Thr residues in the Ser/Thr-rich region (S/T region, Fig. 2.1) immediately N-terminal to the HD1 of Ph. The modification can reversibly transition Ph from a nonproductive disordered aggregate to one that is functional and able to mediate repression via ordered Ph SAM polymerization. The formation of the disordered aggregates was suggested to be dependent on the SAM as its deletion resulted in increased solubility compared to wild type when expressed in insect cells. Also, attaching the S/T region to just the SAM resulted in the same O-GlcNAc–dependent aggregation behavior as full-length Ph. The SAM-dependent disordered aggregation, however, appears independent of SAM polymerization as Ph with EH and ML mutations exhibited the identical behavior as wild-type Ph. In other words, Ph appears to require the SAM to form disordered aggregates when Ph is not modified by O-GlcNAc, but the aggregation is not mediated through SAM polymerization. Mueller's group extended their studies to the human orthologs and showed that the human PHC2 and PHC3 exhibit similar behavior as Ph even though the stretches of S/T residues in the human proteins are not immediately adjacent to the conserved domains of Ph but rather are several hundred residues away. Why Ph modified by O-GlcNAc is able to repress transcription while the nonmodified aggregate form does not remains a mystery. This is so because even the nonmodified aggregated Ph is still observed to be in complex with other PRC1 components and is found at gene loci where O-GlcNAc–modified Ph is also observed [78].

Another potential mode for regulating Ph SAM polymerization stems from its ability to bind other SAMs. The Scm SAM can self-associate and polymerize while also being able to bind Ph SAM utilizing the equivalent ML and EH surfaces of Scm SAM [59,81–83]. Affinity measurements of all the different SAM–SAM combinations between the ML- and EH-binding surfaces of both Ph and Scm SAMs and the structure of the Scm SAM–Ph SAM complex led to a model whereby the two individual SAM polymers connect at a single point [59] forming a potential boundary between Ph and Scm SAM polymer–mediated complexes. This model would be consistent with observations that Scm is present at less than the stoichiometric amounts of the other PRC1 components [9]. If such a copolymer does occur, it will be of interest to determine where and how such connections occur on chromatin. Another heterotypic SAM–SAM interaction that is likely to occur is between the three mammalian Ph orthologs. The sequence identity between any combination of SAMs of PHC1, 2, and 3 is all above 75%. Moreover, all the residues that make up the ML- and EH-binding surfaces between the three SAMs are completely conserved [60]. Thus, in contrast to the situation where Scm and Ph SAMs form a single-junction copolymer, it may be possible that PHC1, 2, and 3 are interspersed along a single PRC1 polymer chain. It has been previously observed that Phc1 and 2 function synergistically [84] and colocalize in cells [21], and, like Phc2, polymer-deficient SAM mutations on Phc1 disrupt nuclear clustering [21].

It will be of interest to determine if O-GlcNAc or some undiscovered SAM polymer regulation is at all correlated with cell division. It has been speculated that SAM polymerization would propagate repression through recruitment of chromatin [21]. In such a scenario, a prepolymerized Ph would nucleate a chromatin state with subsequent SAM polymerization correlating with the reestablishment of the gene repression pattern. Future studies that address these and other hypotheses regarding SAM polymerization will be quite informative in determining the role of the genome structure in establishing, regulating, and maintaining the gene expression program.

How Does Ph SAM Polymerization Affect PRC1 H2Aub Activity?

Another question that has yet to be addressed is how SAM polymerization affects other PRC1 functions. Phc2 SAM polymerization–dependent repression, while significant, affected only a limited number of PcG-regulated genes [21]. In addition, there are some Ph-bound sites in the *Drosophila* genome that require Ph SAM polymerization and others that do not [72]. These results suggest that SAM polymerization, and its contribution to the genome architecture, represents only a portion of the repertoire of repressive mechanisms of PRC1s that house Ph proteins. For example, it remains to be determined whether Ph SAM polymerization can affect PRC1 H2Aub activity. A PRC1 composed of RING1B, RING1A, PCGF4, and PHC2 was isolated and shown to have H2Aub activity [13]. Subsequent studies, however, have shown that the PHC component is not necessary for activity [17,85]. Moreover, Ph is not present nor is it needed for the PRC1-like complex that carries out the H2Aub function in *Drosophila* [33]. The same study reported that the recombinant canonical PRC1 containing the four core components was defective in its H2Aub activity in vitro. Clearly, additional studies will be required to determine if SAM polymerization has any effect on this and possibly other repressive functions of PRC1.

Scm SAM—Dependent Repression

As noted above, Scm is another SAM-containing PcG protein that can influence PRC1 function via its SAM interaction with Ph SAM. Despite this interaction, Scm is not considered a core member of PRC1. The reason for its exclusion stems largely from Scm being in substoichiometric amounts in both *Drosophila* [9] and human [86] PRC1 purifications. In addition, the coimmunoprecipitation (co-IP) signal is weak with Ph in *Drosophila* embryo extracts [83]. Similar weak co-IP signals have been observed with recombinant proteins expressed in Sf9 cells [25], though others have shown a more robust PRC1 interaction with Scm that is dependent on Ph [83]. This discrepancy may point to a means of regulating the order of assembly of different polymeric states after cell division (Fig. 2.3B). Scm shows similar tenuous associations with other well-defined PcG complexes. For example, Scm can associate with PcG proteins Sfmbt and Pho [76,87], the core members of the PcG complex called Pleiohomeotic (Pho) repression complex (PhoRC). Scm, however, is absent in purifications of PhoRC [25,88]. Nevertheless, Scm and its SAM play a vital role in PcG-mediated repression not just through its association with PRC1 but with other PcG complexes as well.

The importance of Scm SAM was first revealed in studies where an Scm with its SAM deleted failed to repress *Drosophila* reporter genes in an in vivo transcription assay [89]. In the same study, Scm function showed dependence on Ph as reducing levels of Ph corresponded to reduced Scm repressive ability. In a study led by Jeffrey Simon, Scm mutants with altered ML- or EH-binding surfaces failed to rescue the lethal phenotype of either a null or hypomorphic *Scm* mutant flies [83]. In humans, SCML2B, the Scm homolog of SCML2 that lacks the SAM, hinders ability to bind chromatin and recruit members of PRC1 [90,91]. A recent study from Jürg Müller's laboratory revealed the molecular details by which Scm SAM is able to recruit PRC1 [87]. In addition to their previous observation of Scm and Sfmbt interacting via their N-terminal regions [25], a direct interaction between their SAMs was observed [87]. Biochemical and structural analysis revealed Scm SAM playing a bridging role, utilizing both its ML and EH surfaces to bind the EH surface of monomeric Sfmbt SAM and the ML surface of Ph SAM, respectively. This network of SAM interactions provides the path by which PRC1-mediated repression is brought to PhoRC bound PREs. What remains to be determined is the length of Scm SAM and Ph SAM–mediated polymerization within this assembly. It seems reasonable to expect that the ability of these SAMs to polymerize is regulated and in turn could allow for slightly altered repressive mechanisms that are adjusted to different sites of PRC1 repressive activity.

Scm SAM may also influence the function of an entirely different PcG complex called PRC2. Using a combination of cross-linking and tandem affinity purification (BioTAP-XL) [92] allowing identification of more transiently associated proteins in large chromatin associated assemblies, Scm was identified within a complex that included a core member of PRC2 [76]. The precise protein regions responsible for this association has not yet been identified. The consequence of this association, however, was demonstrated when knocking down *Scm* resulted in the loss of PRC2-associated signals on polytene chromosomes. The SAM of Scm appears to play a role in PRC2 localization as a prior study had shown compromised ability of a PRC2 component to associate with chromatin in *Drosophila* expressing an Scm SAM mutant [93].

Aside from the potential molecular role Scm plays in recruiting PRC1 and PRC2 to chromatin [91,93], Danny Reinberg's group revealed how SCML2 assists PRC1-mediated repression in mammals [94]. SCML2 binds USP7 [91,95], a protein deubiquitinase whose targets include PRC1 component PCGF4 [94]. A region within SCML2 spanning the MBT and Domain of unknown function 3588 (DUF) domains was necessary and sufficient to directly bind USP7. Thus SCML2 functions to bridge USP7 with PRC1. Consequently, PRC1 is stabilized leading to the maintenance of H2Aub activity. This molecular function, however, may be unique for the mammalian SCML2 as it is the only Scm homolog that houses an RNA-binding region between the MBT and DUF domains, which is requisite for the interaction with USP7.

SUMMARY

Noncovalent interactions constitute the most influential components that determine the structure of the genome. The myriad of protein–protein interaction domains that constitute PRC1 are emerging as key architectural pillars in creating a particular chromatin state. Our current understanding of PRC1 complexes suggests a hierarchical regulation of their assembly (Fig. 2.3). For the RAWULs, there is an initial protein–protein interaction with its primary binding partner followed by secondary assembly with proteins that recognize the primary complex. Adding to the complexity of PRC1 is its polymerization through SAM interactions. As with all multistep biochemical processes, the sequence of steps required to reestablish the chromatin state offers opportunities to regulate the process via protein expression and their posttranslational modifications. We are merely beginning to understand how the structures of these proteins affect this assembly pathway and the consequent genome state. Continued characterization of PRC1 including ongoing structure determination efforts using both X-ray crystallography and cryo-electron microscopy, in combination with the ever advancing field of probing the 3D genome, should provide a much clearer perspective of genome structures and their effect on gene regulation.

LIST OF ACRONYMS AND ABBREVIATIONS

4C-seq Circularized chromatin conformation capture coupled with next-gneration sequencing
Abd-B *Abdominal-B*
BCOR BCL6 corepressor
BCORL1 BCL6 corepressor-like 1
cbox chromo-box domain
CBX Chromodomain-containing Pc orthologues
CCSK Clear cell sarcoma of the kidney
CNS-PNET primitive neuroectodermal tumors of the central nervous system
DUF Domain of unknown function 3588
EH End-helix
Ets E26 transformation-specific
FCS Phenylalanine cysteine serine domain
H2Aub Ubiquitinated histone H2A
H3K27me3 Histone trimethylated on lysine 27
HD1 Homology domain 1
HLH Helix-loop-helix cbox
HOX Homeotic
iab *infra-abdominal*
ITD Internal tandem duplication
KDM2B Lysine demethylase 2B
MBT Malignant brain tumor
ML Mid-loop
O-GlcNAc O-linked β-N-acetylglucosamine
Pc Polycomb
PcG Polycomb group
PCGF Polycomb group ring finger
Ph Polyhomeotic
PHC Polyhomeotic homolog
Pho Pleiohomeotic
PhoRC Pleiohomeotic (Pho) repression complex
PRC1 Polycomb repression complex 1
PRC2 Polycomb repression complex 2
Psc Posterior sex combs
PUFD PCGF1 ubiquitin fold discriminator

RAWUL RING finger- and tryptophan-asparpate 40-associated ubiquitin like
RBR RNA binding region
RING Really interesting new gene
RYBP RING1B YY1 binding protein
SAM Sterile Alpha Motif
Sce Sex combs extra
Scm Sex comb on midleg
SCML2 Sex comb on midleg-like 2
Sfmbt Sex comb on midleg-related gene containing four mbt domains
SKP1 S-phase kinase-associated protein 1
SPM Scm, Ph, L(3)mbt
STORM Stochastic optical reconstruction microscopy
TEL Translocation ets leukemia
USP7 Ubiquitin-specific protease 7
YAF2 YY1 associated factor 2

ACKNOWLEDGMENTS

I would like to thank Drs. Vivian Bardwell and Micah Gearhart for their insights and comments regarding the *BCOR* ITDs. I would also like to thank Dr. Nicole Francis for comments on the manuscript and sharing of data prior to publication.

This work was supported by the NIH (R01GM114338).

REFERENCES

[1] Sexton T, Cavalli G. The role of chromosome domains in shaping the functional genome. Cell March 12, 2015;160(6):1049–59.

[2] Sexton T, Yaffe E, Kenigsberg E, Bantignies F, Leblanc B, Hoichman M, et al. Three-dimensional folding and functional organization principles of the Drosophila genome. Cell February 03, 2012;148(3):458–72.

[3] Ptashne M. Binding reactions: epigenetic switches, signal transduction and cancer. Curr Biol March 24, 2009;19(6):R234–41.

[4] Steffen PA, Ringrose L. What are memories made of? How Polycomb and Trithorax proteins mediate epigenetic memory. Nat Rev Mol Cell Biol May 2014;15(5):340–56.

[5] Helin K, Dhanak D. Chromatin proteins and modifications as drug targets. Nature October 24, 2013;502(7472):480–8.

[6] Simon JA, Kingston RE. Occupying chromatin: Polycomb mechanisms for getting to genomic targets, stopping transcriptional traffic, and staying put. Mol Cell March 07, 2013;49(5):808–24.

[7] Schwartz YB, Pirrotta V. A new world of Polycombs: unexpected partnerships and emerging functions. Nat Rev Genet December 2013;14(12):853–64.

[8] Shao Z, Raible F, Mollaaghababa R, Guyon JR, Wu CT, Bender W, et al. Stabilization of chromatin structure by PRC1, a Polycomb complex. Cell July 09, 1999;98(1):37–46.

[9] Saurin AJ, Shao Z, Erdjument-Bromage H, Tempst P, Kingston RE. A Drosophila Polycomb group complex includes Zeste and dTAFII proteins. Nature August 09, 2001;412(6847):655–60.

[10] Francis NJ, Saurin AJ, Shao Z, Kingston RE. Reconstitution of a functional core polycomb repressive complex. Mol Cell September 2001;8(3):545–56.

[11] Francis NJ, Kingston RE, Woodcock CL. Chromatin compaction by a Polycomb group protein complex. Science November 26, 2004;306(5701):1574–7.

[12] Grau DJ, Chapman BA, Garlick JD, Borowsky M, Francis NJ, Kingston RE. Compaction of chromatin by diverse Polycomb group proteins requires localized regions of high charge. Genes Dev October 15, 2011;25(20):2210–21.

[13] Wang H, Wang L, Erdjument-Bromage H, Vidal M, Tempst P, Jones RS, et al. Role of histone H2A ubiquitination in Polycomb silencing. Nature October 14, 2004;431(7010):873–8.

[14] de Napoles M, Mermoud JE, Wakao R, Tang YA, Endoh M, Appanah R, et al. Polycomb group proteins Ring1A/B link ubiquitylation of histone H2A to heritable gene silencing and X inactivation. Dev Cell November 2004;7(5):663–76.

[15] Pengelly AR, Kalb R, Finkl K, Muller J. Transcriptional repression by PRC1 in the absence of H2A monoubiquitylation. Genes Dev July 15, 2015;29(14):1487–92.

[16] Illingworth RS, Moffat M, Mann AR, Read D, Hunter CJ, Pradeepa MM, et al. The E3 ubiquitin ligase activity of RING1B is not essential for early mouse development. Genes Dev September 15, 2015;29(18):1897–902.

[17] Cao R, Tsukada Y, Zhang Y. Role of Bmi-1 and Ring1A in H2A ubiquitylation and Hox gene silencing. Mol Cell December 22, 2005;20(6):845–54.

[18] Buchwald G, van der Stoop P, Weichenrieder O, Perrakis A, van Lohuizen M, Sixma TK. Structure and E3-ligase activity of the ring-ring complex of polycomb proteins Bmi1 and Ring1b. EMBO J June 07, 2006;25(11):2465–74.

[19] Bantignies F, Roure V, Comet I, Leblanc B, Schuettengruber B, Bonnet J, et al. Polycomb-dependent regulatory contacts between distant Hox loci in *Drosophila*. Cell January 21, 2011;144(2):214–26.

[20] Rosa S, De Lucia F, Mylne JS, Zhu D, Ohmido N, Pendle A, et al. Physical clustering of FLC alleles during Polycomb-mediated epigenetic silencing in vernalization. Genes Dev September 01, 2013;27(17):1845–50.

[21] Isono K, Endo TA, Ku M, Yamada D, Suzuki R, Sharif J, et al. SAM domain polymerization links subnuclear clustering of PRC1 to gene silencing. Dev Cell September 30, 2013;26(6):565–77.

[22] Schoenfelder S, Sugar R, Dimond A, Javierre BM, Armstrong H, Mifsud B, et al. Polycomb repressive complex PRC1 spatially constrains the mouse embryonic stem cell genome. Nat Genet October 2015;47(10):1179–86.

[23] Bernstein E, Duncan EM, Masui O, Gil J, Heard E, Allis CD. Mouse Polycomb proteins bind differentially to methylated histone H3 and RNA and are enriched in facultative heterochromatin. Mol Cell Biol April 2006;26(7):2560–9.

[24] Wang R, Ilangovan U, Leal BZ, Robinson AK, Amann BT, Tong CV, et al. Identification of nucleic acid binding residues in the FCS domain of the polycomb group protein polyhomeotic. Biochemistry June 07, 2011;50(22):4998–5007.

[25] Grimm C, Matos R, Ly-Hartig N, Steuerwald U, Lindner D, Rybin V, et al. Molecular recognition of histone lysine methylation by the Polycomb group repressor dSfmbt. EMBO J July 08, 2009;28(13):1965–77.

[26] Gil J, O'Loghlen A. PRC1 complex diversity: where is it taking us? Trends Cell Biol November 2014;24(11):632–41.

[27] Koppens M, van Lohuizen M. Context-dependent actions of Polycomb repressors in cancer. Oncogene March 17, 2016;35(11):1341–52.

[28] Gao Z, Zhang J, Bonasio R, Strino F, Sawai A, Parisi F, et al. PCGF homologs, CBX proteins, and RYBP define functionally distinct PRC1 family complexes. Mol Cell February 10, 2012;45(3):344–56.

[29] Maertens GN, El Messaoudi-Aubert S, Racek T, Stock JK, Nicholls J, Rodriguez-Niedenfuhr M, et al. Several distinct polycomb complexes regulate and co-localize on the INK4a tumor suppressor locus. PLoS One July 28, 2009;4(7):e6380.

[30] Morey L, Pascual G, Cozzuto L, Roma G, Wutz A, Benitah SA, et al. Nonoverlapping functions of the Polycomb group Cbx family of proteins in embryonic stem cells. Cell Stem Cell January 06, 2012;10(1):47–62.

[31] Pemberton H, Anderton E, Patel H, Brookes S, Chandler H, Palermo R, et al. Genome-wide co-localization of Polycomb orthologs and their effects on gene expression in human fibroblasts. Genome Biol February 03, 2014;15(2). http://dx.doi.org/10.1186/gb-2014-15-2-r23. :R23.

[32] Wu X, Johansen JV, Helin K. Fbxl10/Kdm2b recruits polycomb repressive complex 1 to CpG islands and regulates H2A ubiquitylation. Mol Cell 2013;49:1134–46.

[33] Lagarou A, Mohd-Sarip A, Moshkin YM, Chalkley GE, Bezstarosti K, Demmers JA, et al. dKDM2 couples histone H2A ubiquitylation to histone H3 demethylation during Polycomb group silencing. Genes Dev October 15, 2008;22(20):2799–810.

[34] O'Loghlen A, Munoz-Cabello AM, Gaspar-Maia A, Wu HA, Banito A, Kunowska N, et al. MicroRNA regulation of Cbx7 mediates a switch of Polycomb orthologs during ESC differentiation. Cell Stem Cell January 06, 2012;10(1):33–46.

[35] Klauke K, Radulovic V, Broekhuis M, Weersing E, Zwart E, Olthof S, et al. Polycomb Cbx family members mediate the balance between haematopoietic stem cell self-renewal and differentiation. Nat Cell Biol April 2013;15(4):353–62.

[36] Bezsonova I, Walker JR, Bacik JP, Duan S, Dhe-Paganon S, Arrowsmith CH. Ring1B contains a ubiquitin-like docking module for interaction with Cbx proteins. Biochemistry November 10, 2009;48(44):10542–8.

[37] Wang R, Taylor AB, Leal BZ, Chadwell LV, Ilangovan U, Robinson AK, et al. Polycomb group targeting through different binding partners of RING1B C-terminal domain. Structure August 11, 2010;18(8):966–75.

[38] Junco SE, Wang R, Gaipa JC, Taylor AB, Schirf V, Gearhart MD, et al. Structure of the Polycomb group protein PCGF1 in complex with BCOR reveals basis for binding selectivity of PCGF homologs. Structure April 02, 2013;21(4):665–71.

[39] Wong SJ, Gearhart MD, Taylor AB, Nanyes DR, Ha DJ, Robinson AK, et al. KDM2B Recruitment of the Polycomb Group Complex, PRC1.1, Requires Cooperation between PCGF1 and BCORL1. Structure October 4, 2016;24(10):1795–801.

[40] Wang R, Ilangovan U, Robinson AK, Schirf V, Schwarz PM, Lafer EM, et al. Structural transitions of the RING1B C-terminal region upon binding the polycomb cbox domain. Biochemistry August 05, 2008;47(31):8007–15.

[41] Garcia E, Marcos-Gutierrez C, Del Mar Lorente M, Moreno JC, Vidal M. RYBP, a new repressor protein that interacts with components of the mammalian Polycomb complex, and with the transcription factor YY1. EMBO J June 15, 1999;18(12):3404–18.

[42] Gearhart MD, Corcoran CM, Wamstad JA, Bardwell VJ. Polycomb group and SCF ubiquitin ligases are found in a novel BCOR complex that is recruited to BCL6 targets. Mol Cell Biol September 2006;26(18):6880–9.

[43] Sanchez C, Sanchez I, Demmers JA, Rodriguez P, Strouboulis J, Vidal M. Proteomics analysis of Ring1B/Rnf2 interactors identifies a novel complex with the Fbxl10/Jhdm1B histone demethylase and the Bcl6 interacting corepressor. Mol Cell Proteomics May 2007;6(5):820–34.

[44] Kornberg RD, Thomas JO. Chromatin structure; oligomers of the histones. Science May 24, 1974;184(4139):865–8.

[45] Ueno-Yokohata H, Okita H, Nakasato K, Akimoto S, Hata J, Koshinaga T, et al. Consistent in-frame internal tandem duplications of BCOR characterize clear cell sarcoma of the kidney. Nat Genet August 2015;47(8):861–3.

[46] Roy A, Kumar V, Zorman B, Fang E, Haines KM, Doddapaneni H, et al. Recurrent internal tandem duplications of BCOR in clear cell sarcoma of the kidney. Nat Commun November 17, 2015;6:8891.

[47] Sturm D, Orr BA, Toprak UH, Hovestadt V, Jones DT, Capper D, et al. New brain tumor entities emerge from molecular classification of CNS-PNETs. Cell February 25, 2016;164(5):1060–72.

[48] Ponting CP. SAM: a novel motif in yeast sterile and Drosophila polyhomeotic proteins. Protein Sci September 1995;4(9):1928–30.

[49] Bonin I, Muhlberger R, Bourenkov GP, Huber R, Bacher A, Richter G, et al. Structural basis for the interaction of *Escherichia coli* NusA with protein N of phage lambda. Proc Natl Acad Sci USA September 21, 2004;101(38):13762–7.

[50] Tosi T, Cioci G, Jouravleva K, Dian C, Terradot L. Structures of the tumor necrosis factor alpha inducing protein Tipalpha: a novel virulence factor from *Helicobacter pylori*. FEBS Lett May 19, 2009;583(10):1581–5.

[51] Hernandez JA, Phillips AH, Erbil WK, Zhao D, Demuez M, Zeymer C, et al. A sterile alpha-motif domain in NafY targets apo-NifDK for iron-molybdenum cofactor delivery via a tethered domain. J Biol Chem February 25, 2011;286(8):6321–8.

[52] Green JB, Gardner CD, Wharton RP, Aggarwal AK. RNA recognition via the SAM domain of Smaug. Mol Cell June 2003;11(6):1537–48.

[53] Aviv T, Lin Z, Lau S, Rendl LM, Sicheri F, Smibert CA. The RNA-binding SAM domain of Smaug defines a new family of post-transcriptional regulators. Nat Struct Biol August 2003;10(8):614–21.

[54] Oberstrass FC, Lee A, Stefl R, Janis M, Chanfreau G, Allain FH. Shape-specific recognition in the structure of the Vts1p SAM domain with RNA. Nat Struct Mol Biol February 2006;13(2):160–7.

[55] Johnson PE, Donaldson LW. RNA recognition by the Vts1p SAM domain. Nat Struct Mol Biol February 2006;13(2):177–8.

[56] Barrera FN, Poveda JA, Gonzalez-Ros JM, Neira JL. Binding of the C-terminal sterile alpha motif (SAM) domain of human p73 to lipid membranes. J Biol Chem November 21, 2003;278(47):46878–85.

[57] Bhunia A, Domadia PN, Mohanram H, Bhattacharjya S. NMR structural studies of the Ste11 SAM domain in the dodecyl phosphocholine micelle. Proteins February 01, 2009;74(2):328–43.

[58] Kim CA, Gingery M, Pilpa RM, Bowie JU. The SAM domain of polyhomeotic forms a helical polymer. Nat Struct Biol June 2002;9(6):453–7.

[59] Kim CA, Sawaya MR, Cascio D, Kim W, Bowie JU. Structural organization of a sex-comb-on-midleg/polyhomeotic copolymer. J Biol Chem July 29, 2005;280(30):27769–75.

[60] Nanyes DR, Junco SE, Taylor AB, Robinson AK, Patterson NL, Shivarajpur A, et al. Multiple polymer architectures of human polyhomeotic homolog 3 sterile alpha motif. Proteins October 2014;82(10):2823–30.

[61] Kim CA, Phillips ML, Kim W, Gingery M, Tran HH, Robinson MA, et al. Polymerization of the SAM domain of TEL in leukemogenesis and transcriptional repression. EMBO J August 01, 2001;20(15):4173–82.

[62] Qiao F, Song H, Kim CA, Sawaya MR, Hunter JB, Gingery M, et al. Derepression by depolymerization; structural insights into the regulation of Yan by Mae. Cell July 23, 2004;118(2):163–73.

[63] Baron MK, Boeckers TM, Vaida B, Faham S, Gingery M, Sawaya MR, et al. An architectural framework that may lie at the core of the postsynaptic density. Science January 27, 2006;311(5760):531–5.

[64] Harada BT, Knight MJ, Imai S, Qiao F, Ramachander R, Sawaya MR, et al. Regulation of enzyme localization by polymerization: polymer formation by the SAM domain of diacylglycerol kinase delta1. Structure March 2008;16(3):380–7.

[65] Di Pietro SM, Cascio D, Feliciano D, Bowie JU, Payne GS. Regulation of clathrin adaptor function in endocytosis: novel role for the SAM domain. EMBO J March 17, 2010;29(6):1033–44.

[66] Leettola CN, Knight MJ, Cascio D, Hoffman S, Bowie JU. Characterization of the SAM domain of the PKD-related protein ANKS6 and its interaction with ANKS3. BMC Struct Biol July 07, 2014;14:17.

[67] Robinson AK, Leal BZ, Chadwell LV, Wang R, Ilangovan U, Kaur Y, et al. The growth-suppressive function of the polycomb group protein polyhomeotic is mediated by polymerization of its sterile alpha motif (SAM) domain. J Biol Chem March 16, 2012;287(12):8702–13.

[68] Martinez AM, Schuettengruber B, Sakr S, Janic A, Gonzalez C, Cavalli G. Polyhomeotic has a tumor suppressor activity mediated by repression of Notch signaling. Nat Genet October 2009;41(10):1076–82.

[69] Classen AK, Bunker BD, Harvey KF, Vaccari T, Bilder D. A tumor suppressor activity of Drosophila Polycomb genes mediated by JAK-STAT signaling. Nat Genet October 2009;41(10):1150–5.

[70] Feng S, Huang J, Wang J. Loss of the Polycomb group gene polyhomeotic induces non-autonomous cell overproliferation. EMBO Rep February 01, 2011;12(2):157–63.

[71] Gambetta MC, Muller J. O-GlcNAcylation prevents aggregation of the Polycomb group repressor polyhomeotic. Dev Cell December 08, 2014;31(5):629–39.

[72] Wani AH, Boettiger AN, Schorderet P, Ergun A, Munger C, Sadreyev RI, et al. Chromatin topology is coupled to Polycomb group protein subnuclear organization. Nat Commun January 13, 2016;7:10291.

[73] Pirrotta V, Li HB. A view of nuclear Polycomb bodies. Curr Opin Genet Dev April 2012;22(2):101–9.

[74] Eskeland R, Leeb M, Grimes GR, Kress C, Boyle S, Sproul D, et al. Ring1B compacts chromatin structure and represses gene expression independent of histone ubiquitination. Mol Cell May 14, 2010;38(3):452–64.

[75] Boettiger AN, Bintu B, Moffitt JR, Wang S, Beliveau BJ, Fudenberg G, et al. Super-resolution imaging reveals distinct chromatin folding for different epigenetic states. Nature January 21, 2016;529(7586):418–22.

[76] Kang H, McElroy KA, Jung YL, Alekseyenko AA, Zee BM, Park PJ, et al. Sex comb on midleg (Scm) is a functional link between PcG-repressive complexes in Drosophila. Genes Dev June 01, 2015;29(11):1136–50.

[77] Robinson AK, Leal BZ, Nanyes DR, Kaur Y, Ilangovan U, Schirf V, et al. Human polyhomeotic homolog 3 (PHC3) sterile alpha motif (SAM) linker allows open-ended polymerization of PHC3 SAM. Biochemistry July 10, 2012;51(27):5379–86.

[78] Gambetta MC, Oktaba K, Muller J. Essential role of the glycosyltransferase sxc/Ogt in polycomb repression. Science July 03, 2009;325(5936):93–6.

[79] Chalkley RJ, Thalhammer A, Schoepfer R, Burlingame AL. Identification of protein O-GlcNAcylation sites using electron transfer dissociation mass spectrometry on native peptides. Proc Natl Acad Sci USA June 02, 2009;106(22):8894–9.

[80] Myers SA, Panning B, Burlingame AL. Polycomb repressive complex 2 is necessary for the normal site-specific O-GlcNAc distribution in mouse embryonic stem cells. Proc Natl Acad Sci USA June 07, 2011;108(23):9490–5.

[81] Peterson AJ, Kyba M, Bornemann D, Morgan K, Brock HW, Simon J. A domain shared by the Polycomb group proteins Scm and ph mediates heterotypic and homotypic interactions. Mol Cell Biol November 1997;17(11):6683−92.

[82] Kyba M, Brock HW. The SAM domain of polyhomeotic, RAE28, and scm mediates specific interactions through conserved residues. Dev Genet 1998;22(1):74−84.

[83] Peterson AJ, Mallin DR, Francis NJ, Ketel CS, Stamm J, Voeller RK, et al. Requirement for sex comb on midleg protein interactions in Drosophila polycomb group repression. Genetics July 2004;167(3):1225−39.

[84] Isono K, Fujimura Y, Shinga J, Yamaki M, O-Wang J, Takihara Y, et al. Mammalian polyhomeotic homologues phc2 and phc1 act in synergy to mediate polycomb repression of Hox genes. Mol Cell Biol August 2005;25(15):6694−706.

[85] Wei J, Zhai L, Xu J, Wang H. Role of Bmi1 in H2A ubiquitylation and Hox gene silencing. J Biol Chem August 11, 2006;281(32):22537−44.

[86] Levine SS, Weiss A, Erdjument-Bromage H, Shao Z, Tempst P, Kingston RE. The core of the polycomb repressive complex is compositionally and functionally conserved in flies and humans. Mol Cell Biol September 2002;22(17):6070−8.

[87] Frey F, Sheahan T, Finkl K, Stoehr G, Mann M, Benda C, et al. Molecular basis of PRC1 targeting to Polycomb response elements by PhoRC. Genes Dev May 01, 2016;30(9):1116−27.

[88] Klymenko T, Papp B, Fischle W, Kocher T, Schelder M, Fritsch C, et al. A Polycomb group protein complex with sequence-specific DNA-binding and selective methyl-lysine-binding activities. Genes Dev May 01, 2006;20(9):1110−22.

[89] Roseman RR, Morgan K, Mallin DR, Roberson R, Parnell TJ, Bornemann DJ, et al. Long-range repression by multiple polycomb group (PcG) proteins targeted by fusion to a defined DNA-binding domain in Drosophila. Genetics May 2001;158(1):291−307.

[90] Lecona E, Rojas LA, Bonasio R, Johnston A, Fernandez-Capetillo O, Reinberg D. Polycomb protein SCML2 regulates the cell cycle by binding and modulating CDK/CYCLIN/p21 complexes. PLoS Biol December 2013;11(12):e1001737.

[91] Bonasio R, Lecona E, Narendra V, Voigt P, Parisi F, Kluger Y, et al. Interactions with RNA direct the Polycomb group protein SCML2 to chromatin where it represses target genes. Elife July 01, 2014;3:e02637.

[92] Alekseyenko AA, McElroy KA, Kang H, Zee BM, Kharchenko PV, Kuroda MI. BioTAP-xl: cross-linking/tandem affinity purification to study DNA targets, RNA, and protein components of chromatin-associated complexes. Curr Protoc Mol Biol January 05, 2015;109:21.30.1−21.30.32.

[93] Wang L, Jahren N, Miller EL, Ketel CS, Mallin DR, Simon JA. Comparative analysis of chromatin binding by Sex Comb on Midleg (SCM) and other polycomb group repressors at a Drosophila Hox gene. Mol Cell Biol June 2010;30(11):2584−93.

[94] Lecona E, Narendra V, Reinberg D. USP7 cooperates with SCML2 to regulate the activity of PRC1. Mol Cell Biol April 2015;35(7):1157−68.

[95] Sowa ME, Bennett EJ, Gygi SP, Harper JW. Defining the human deubiquitinating enzyme interaction landscape. Cell July 23, 2009;138(2):389−403.

[96] Meruelo AD, Bowie JU. Identifying polymer-forming SAM domains. Proteins January 2009;74(1):1−5.

[97] Knight MJ, Leettola C, Gingery M, Li H, Bowie JU. A human sterile alpha motif domain polymerizome. Protein Sci October 2011;20(10):1697−706.

[98] Trojer P, Cao AR, Gao Z, Li Y, Zhang J, Xu X, et al. L3MBTL2 protein acts in concert with PcG protein-mediated monoubiquitination of H2A to establish a repressive chromatin structure. Mol Cell May 20, 2011;42(4):438–50.

[99] Vandamme J, Volkel P, Rosnoblet C, Le Faou P, Angrand PO. Interaction proteomics analysis of polycomb proteins defines distinct PRC1 complexes in mammalian cells. Mol Cell Proteomics April 2011;10(4):M110.002642.

[100] Gao Z, Lee P, Stafford JM, von Schimmelmann M, Schaefer A, Reinberg D. An AUTS2-Polycomb complex activates gene expression in the CNS. Nature December 18, 2014;516(7531):349–54.

[101] Lee MG, Norman J, Shilatifard A, Shiekhattar R. Physical and functional association of a trimethyl H3K4 demethylase and Ring6a/MBLR, a polycomb-like protein. Cell March 09, 2007;128(5):877–87.

Chapter 3

The Chromodomain of Polycomb: Methylation Reader and Beyond

S. Qin[1], L. Li[2], J. Min[2]
[1]Southern University of Science and Technology, Shenzhen, Guangdong, China; [2]University of Toronto, Toronto, ON, Canada

Chapter Outline

Introduction	33
Polycomb Chromodomain Specifically Recognizes H3K27me3	35
Structural Basis for the Polycomb–H3K27me3 Interaction Specificity	38
Mammalian Polycomb Homologs Bind Differentially to Methylated Histone H3	41
Cross Talk Between Histone Methylation and Other Posttranslational Modifications	44
Putative Nonhistone Targets of Polycomb Group Chromodomains	45
Noncoding RNA: Noncanonical Partners of Polycomb Group Chromodomains	46
Chemical Probes for CBX7 Chromodomain	48
Polycomb Homologs From Yeast and Plant: Evolutionarily Conserved Biological Significance of Chromodomain	50
Conclusion	51
List of Acronyms and Abbreviations	52
Glossary	52
References	53

INTRODUCTION

The *Polycomb* (*Pc*) gene was first identified in 1947 through classic genetic studies using the model organism *Drosophila melanogaster*. The gene name, *Polycomb*, was coined from the most prominent phenotypical characteristic associated with its deficiency: the formation of ectopic sex combs on the second and third legs of adult male flies [1]. Decades of research effort revealed that Pc is a key developmental regulator required to maintain homeotic gene (*Hox*) repression of the Bithorax complex (BX-C) [2]. Meanwhile, a novel group of proteins emerged, which were found to possess

Polycomb Group Proteins. http://dx.doi.org/10.1016/B978-0-12-809737-3.00003-9

transcriptionally repressive activities during development similar to the Pc protein, and thereby are collectively termed the Pc group (PcG) proteins [3]. Genetic and biochemical studies of the PcG members in fruit flies and mammals converged to elucidate the canonical molecular mechanisms of PcG-mediated repressive effects on transcription, and two multiprotein Pc repressive complexes PRC1 and PRC2 have been recruited to the PcG-targeted genes and collaborated to mediate the transcriptional repressive activities.

In *Drosophila*, the PRC1 core complex consists of Pc, Polyhomeotic (Ph), Posterior sex combs, and Sex combs extra (also known as RING), while the PRC2 core complex comprises Enhancer of zeste (E(z)), suppressor of zeste, and extra sex combs (Esc) [4]. PRC1 and PRC2 are often recruited to the same target locations and functionally act in a sequential manner, as proposed in the "canonical model" of PcG repression. Namely, the PRC2 complex methylates lysine 27 of histone H3 (H3K27me3) of its target genes. This methylation event usually initiates on a genomic region called the nucleation site, followed by the spreading of the methylation mark over adjacent nucleosomes. The robust deposition of H3K27me3 triggers the recruitment of the PRC1 complex to the same chromosomal target sites. Subsequently, the PRC1 complex modifies the chromosomal target sites by ubiquitinating lysine 119 of histone H2A. These histone posttranslational modifications (PTMs) adjacent to the target genes result in other complementary factors being recruited, leading to transcription repression and chromatin remodeling at the target sites [5]. In this model, the chromodomain of Pc as a component of the PRC1 complex serves as a critical bridge linking the two important histone PTM events by specifically recognizing the H3K27me3 mark generated by the PRC2 complex [6–8].

The chromodomain was first identified by comparison of the Pc sequence with that of HP1 (heterochromatin-associated protein one) in *Drosophila* [9]. HP1 is a conserved protein that plays important roles in chromatin packaging and gene silencing. Since both the Pc and HP1 proteins participate in cellular processes leading to changes in the organization of the chromatin, this shared domain has been named as "chromo domain" (*chr*omatin *o*rganization *mo*difier). Together with previous genetic studies, the molecular similarity between Pc and HP1 supports the suggestion of a common mechanism used for generating heterochromatin and repressing homeotic genes. Indeed, analogous to the recognition of the H3K27me3 mark by the Pc chromodomain, the HP1 chromodomain recognizes another gene silencing mark trimethylated lysine 9 of histone H3 (H3K9me3) generated by histone H3K9 methyltransferase SUV39H1 [10,11]. As a matter of fact, our understanding of the Pc chromodomain was largely promoted by the study of HP1, thanks to their similarity.

In this chapter, we will summarize our current knowledge on the chromodomain of Pc and its mammalian homologs with an emphasis on their mechanistic role as a histone methylation reader from a structural point of view. We will also discuss the cross talk of the methyl-lysine mark with other histone PTMs and the recent progress in chemical probe development for these domains. Our discussion on Pc will be in parallel with the HP1 protein due to their high similarity.

POLYCOMB CHROMODOMAIN SPECIFICALLY RECOGNIZES H3K27me3

On the basis of elegant studies using *Drosophila* transgenic cell lines as well as transient tissue culture cells, it was revealed that the chromodomain of Pc is absolutely required for the binding of Pc protein to chromatin, which is important for its functions [12]. Specifically, mutations of the Pc chromodomain, including deletion as well as point mutations, abolish the chromosomal binding capability of the Pc protein, whereas carboxy-terminal truncations of the Pc protein do not affect its chromosomal binding ability [12]. Interestingly, a similar phenomenon was also observed for HP1. The HP1 chromodomain, like that of the Pc protein, has chromosome-binding activities, but it binds at distinct chromosomal sites. Analogously, point mutations in the HP1 chromodomain effectively nullify the ability of HP1 to promote gene silencing [13]. Furthermore, in the fission yeast *Schizosaccharomyces pombe*, the correct localization of Swi6 (the HP1 equivalent) depends on Clr4, a homolog of histone methyltransferase SUV39H1 that specifically methylates H3K9me3 [14,15]. These results led scientists to propose the HP1 chromodomain as a reader of the H3K9me3 mark 10 years after the identification of the chromodomain [10,11]. A point mutation in the chromodomain, which disrupts the gene silencing activity of HP1 in *Drosophila*, also abolishes its methyl-lysine–binding activity [10]. Genetic and biochemical analyses in *S. pombe* showed that the methyltransferase activity of Clr4 is necessary for the targeted localization of Swi6 at centromeric heterochromatin and for gene silencing [10]. These results provide a stepwise model for the formation of transcriptionally silent heterochromatin: SUV39H1/Clr4 places a "methyl mark" on histone H3, which is then recognized by HP1/Swi6 through its chromodomain [10]. In particular, the association of HP1 with methylated mononucleosomes could be completely disrupted by the addition of excess H3K9me3 peptide, suggesting that HP1 recognizes H3K9me3 in the context of mononucleosome [10]. Functionally, the interaction of HP1 with H3K9me3 is essential for the epigenetic control of heterochromatin assembly in vivo [10,16]. Therefore, the finding that the HP1 chromodomain specifically binds to H3K9me3 is a major breakthrough in the field of chromatin biology. Soon after that, the Pc chromodomain was reported to bind H3K27me3, a mark of repressed homeotic genes generated by a multiprotein complex containing E(z) and Esc, or their human counterparts EZH2 and EED [6–8].

The amino acid sequences surrounding lysine 9 and lysine 27 in the H3 tail are very similar. In particular, they share a consensus sequence ARKS (Ala-Arg-Lys-Ser) (Fig. 3.1A). However, quantitative measurements showed a

(A)

H3_K9	ARTKQTARKSTG
H3_K27	QLATKAARKSAP
H3t_K27	QLATKVARKSAP
H1.4_K26	TPVKKKARKSAG
G9a_K185	PKVHRARKTMS
SETDB1_K1170	STRGFALKSTH
DNMT3A_K47	QEPSTTARKVGR

(B)

β1 β2 η1 β3 η2 α1

				pI
dPc	23	DLVYAAEKIIQKRVKKGVVEYRVKWKGWNQRYNTWEPEVNILDRRLIDIYEQTN	76	9.40
CBX2	9	EQVFAAECILSKRLRKGKLEYLVKWRGWSSKHNSWEPEENILDPRLLLAFQKKE	62	9.31
CBX4	8	EHVFAVESIEKKRIRKGRVEYLVKWRGWSPKYNTWEPEENILDPRLLIAFQNRE	61	9.23
CBX6	8	ERVFAAESIIKRRIRKGRIEYLVKWKGWAIKYSTWEPEENILDSRLIAAFEQKE	61	9.52
CBX7	8	EQVFAVESIRKKRVRKGKVEYLVKWKGWPPKYSTWEPEEHILDPRLVMAYEEKE	61	9.02
CBX8	8	ERVFAAEALLKRRIRKGRMEYLVKWKGWSQKYSTWEPEENILDARLLAAFEERE	61	9.23
dHP1	21	EEEYAVEKIIDRRVRKGKVEYYLKWKGYPETENTWEPENNLDCQDLIQQYEASR	74	4.73
CBX1	18	EEEYVVEKVLDRRVVKGKVEYLLKWKGFSDEDNTWEPEENLDCPDLIAEFLQSQ	71	4.28
CBX3	27	PEEFVVEKVLDRRVVNGKVEYFLKWKGFTDADNTWEPEENLDCPELIEAFLNSQ	80	4.28
CBX5	17	EEEYVVEKVLDRRVVKGQVEYLLKWKGFSEEHNTWEPEKNLDCPELISEFMKKY	70	4.92

FIGURE 3.1 Sequence alignments. (A) ARKS (Ala-Arg-Lys-Ser)-like motif-containing histone and nonhistone targets of chromodomain; the methyl-lysine site is highlighted in yellow and the ARKS-like motif is underlined. The residues of particular notice are shown in red. *H3t*, testis-specific histone H3 variant. (B) Chromodomains of Pc and HP1 proteins from *Drosophila* and human; the secondary structure elements of Pc are shown on top and the theoretical isoelectric points of each protein are shown on the right. The residues involved in the methyl-lysine interactions are highlighted in yellow. The "hydrophobic clasp" of Pc proteins and the "polar clasp" of HP1 proteins are highlighted in cyan and magenta, respectively.

strong preference of the chromodomain of *Drosophila* Pc protein for H3K27me3, whereas the chromodomain of *Drosophila* HP1 protein binds preferentially to H3K9me3 (Table 3.1) [17]. Specifically, the Pc protein's dissociation constant (K_d) is about 5 μM for the H3K27me3 peptide, but becomes 25-fold weaker or about 125 μM for the H3K9me3 peptide. In striking contrast, the HP1 chromodomain bound to the H3K9me3 peptide with an affinity of about 4 μM, but bound 16-fold weaker to the H3K27me3 peptide with an affinity of about 64 μM [17]. Furthermore, the binding affinities of the Pc protein to H3K27me1/2 peptides were about five times weaker than its binding affinity to the H3K27me3 peptide, but were still much stronger than its binding affinity to the H3K9me3 peptide. On the other hand, the binding affinities of HP1 to H3K9me1/2 peptides decreased about 15-fold and 2-fold, respectively, when compared with its binding affinity to H3K9me3 peptide. Collectively, these results indicate that Pc prefers H3K27me3, whereas HP1 prefers both H3K9me3 and H3K9me2 [17].

TABLE 3.1 Dissociation Constants (μM) of Polycomb and HP1 Chromodomains to H3K9me and H3K27me Peptides [17,25,26]

		K9me1	K9me2	K9me3	K27me1	K27me2	K27me3
Fruit fly	Pc	>1000	>1000	125 ± 28	20 ± 3	28 ± 4	5 ± 1
Mouse	Cbx2	382 ± 29	396 ± 36	41 ± 6	>500	143 ± 4	44 ± 5
	Cbx4	>500	261 ± 21	49 ± 9	>500	>500	150 ± 20
	Cbx6	>500	>500	>500	>500	>500	330 ± 120
	Cbx7	267 ± 48	79 ± 12	12 ± 3	>500	136 ± 23	22 ± 5
	Cbx8	>500	>500	>500	>500	>500	165 ± 20
Human	CBX2			>500			185 ± 20
	CBX4			70 ± 7			205 ± 20
	CBX6			>500			>500
	CBX7			55 ± 5			110 ± 17
	CBX8			>500			>500
Fruit fly	HP1	46 ± 9	7 ± 2	4 ± 1			64 ± 7
Mouse	Cbx5	10	7	2 ± 1	>500	286	204
Human	CBX1			5 ± 2			NB
	CBX3			15 ± 8			NB
	CBX5			30 ± 5			NB

STRUCTURAL BASIS FOR THE POLYCOMB–H3K27me3 INTERACTION SPECIFICITY

The chromodomain shares structural similarities with the Tudor, PWWP (Pro-Trp-Trp-Pro), and malignant brain tumor (MBT) domains. These domains are collectively referred to as the "Royal family" and are believed to originate from a common ancestor [18]. The shared structural features of the Royal family members include an antiparallel β-barrel-like fold formed by four to five β-strands, though the canonical chromodomain deviates slightly by harboring only three β-strands and requiring the binding ligand to complete the β-barrel fold through forming an extra β-strand [19]. Functionally, most of the Royal family members are found to be able to recognize methyl-lysine or methyl-arginine residues through a conserved aromatic cage located at the bottom of the β-barrel [19]. Nevertheless, interactions with residues flanking methyl-lysine or methyl-arginine are highly diverse. Even for the chromodomains, their structures and ligand-binding abilities exhibit variability. Based on their structures and mechanisms of target recognition, the chromodomains with high sequence homology to the HP1 chromodomain are referred to as canonical, while the remaining members that harbor variable insertions and bind distinct ligands are referred to as noncanonical [20].

The Pc chromodomain consists of three antiparallel β-strands followed by an α-helix. Two short 3_{10} helices are inserted in the linkers between β2 and β3, and between β3 and α1 (Fig. 3.1B). This structure is highly similar to that of the HP1 chromodomain and is recognized as a canonical chromodomain (Fig. 3.2A and B). The histone H3 peptide with a trimethylated lysine 27 is bound as a β-strand between β1 and the loop connecting β3 and α1. In many respects the interaction between Pc and the H3K27me3 peptide is similar to that between HP1 and H3K9me3 peptide, particularly for the ARKS motif interactions [17,21]. First, methylated lysine 27 is bound in a hydrophobic pocket formed by three aromatic residues, Y26, W47, and W50 (Fig. 3.2C). Distinct from lysine acetylation that neutralizes the positive charge carried by the epsilon amino group of the lysine side chain, and from serine/threonine phosphorylation that creates a negative charge, lysine or arginine methylation does not change the net charge of the modified residue, but redistributes the charge through methyl groups. Quantitative binding energy study on HP1 confirmed that the positive charge of the methyl-lysine is important for binding, as substitution of the methyl-lysine residue in the H3K9 peptide with a neutral trimethyl-lysine analog, the tert-butylnorleucine, results in 30-fold weaker binding to HP1 [22]. In HP1, the aromatic cage recognizing methylated lysine 9 is further assisted by a glutamic acid E52, which may form a hydrogen bond with dimethylated lysine 9 through an ordered water molecule (Fig. 3.2D). The position corresponding to E52 of HP1 in Pc is occupied by a tyrosine Tyr54, which lacks the ability to form a corresponding hydrogen bond (Fig. 3.2C). This feature may explain why the Pc chromodomain prefers the

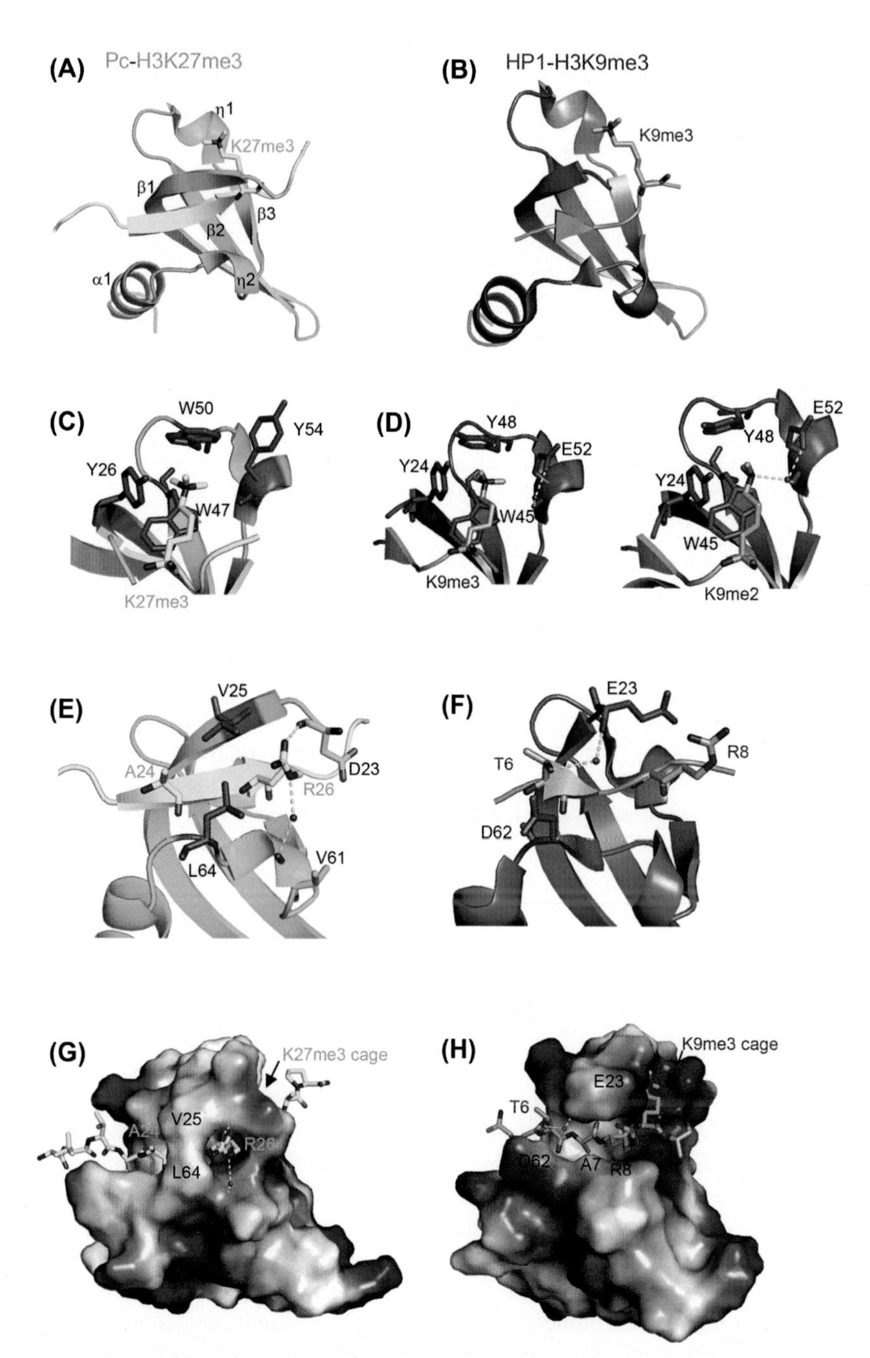

FIGURE 3.2 Structural basis for specific recognition of H3K27me3 by Pc chromodomain compared with the recognition of H3K9me3 by HP1. (A) and (B), overall structures of Pc (PDB code: 1PFB) and HP1 (PDB code: 1KNE) complexes, with methyl-lysine shown in stick mode; (C) and (D), aromatic cages recognizing tri- or di-methylated lysine; (E) and (F), distinct "clasps" of Pc and HP1; (G) and (H), electronic potential surfaces of Pc and HP1 bound with their ligands.

H3K27me3 to the H3K27me2 peptide, whereas the HP1 chromodomain binds to H3K9me2/3 peptides with a similar affinity.

Furthermore, the main-chain carbonyl and amino groups of the histone peptide residues preceding the methyl-lysine are involved in β-sheet-like hydrogen bonding with the chromodomain residues located at the N-terminus and residues located in the loop connecting β3 and α1, thus completing a β-barrel. Also, the hydroxyl group of S28 of histone H3 makes hydrogen bonds to Glu 58 and Asn 62 of the Pc chromodomain. This interaction is conserved in all Pc and HP1 homologs, and contributes to the establishment of a serine phosphorylation–mediated molecular switch (see below). Moreover, A25 of histone H3 is buried in a shallow hydrophobic pocket surrounded by Ala 28, Trp 47, Ile 63, and Leu 68. The size of this pocket is not large enough to accommodate any amino acid side chain except an alanine side chain, and thus this feature enables the Pc/HP1 chromodomains to distinguish the H3K9me3 and H3K27me3 methylation marks from other histone methylation marks.

However, it is still unclear why the Pc and HP1 show specific binding preferences for H3K27me3 or H3K9me3, respectively, because the intermolecular interactions identified above in Pc-H3K27me3 are mostly conserved in the HP1–H3K9me3 interaction. Two research groups provided distinct explanations to this puzzle. Wolfgang Fischle et al. proposed that the peptide-binding grooves of Pc and HP1 show distinct features. The most striking difference is the extent of peptide–protein interactions in these two complexes [17]. A total of six residues (Q5 to S10) of the H3K9me3 peptide were observed to interact with the HP1 chromodomain. In the structure of Pc complex with H3K27me3, a total of nine residues corresponding to the sequence stretch from L20 to S28 are involved in intermolecular interactions, and the Pc chromodomain recognizes an extended surface encompassing n-4 through n-7 residues (where n corresponds to the methyl-lysine). On the other hand, HP1 appears to be more discriminating for the n-3 position (T6 vs. A24). Indeed, mutation of T6 to an alanine (corresponding to residue A24) reduced the peptide-binding affinity of HP1 six-fold. In contrast, when A24 is changed to threonine, the peptide-binding affinity of Pc did not change significantly [17].

Alternatively, Jinrong Min et al. proposed that a potential chromodomain dimer could account for the binding specificity of Pc [21]. There are several evidences to support this dimer model. Firstly, the residues involved in the protein–protein interaction to form a dimer are specifically conserved in the Pc family of proteins. Secondly, there are very few solvent molecules at the interface of the two Pc complexes. Thirdly, dynamic light scattering in solution shows that the chromodomain of Pc has an apparent molecular mass of 14.2 kD, which is close to twice the calculated mass of a monomer, 6.8 kD. Fourthly and more importantly, earlier studies have found that a mammalian Pc chromodomain forms oligomers in solution [23]. The idea that the Pc chromodomain mediates self-association in vivo is also supported by the observation that the chimeric Pc-HP1 protein, in which the HP1 chromodomain is replaced by the Pc chromodomain, has the ability to target endogenous Pc protein to ectopic sites in heterochromatin [13]. The Pc chromodomain dimer juxtaposes the two H3-binding clefts in an antiparallel fashion and results in histone–histone interactions involving L20, T22, and A24 of histone H3. Thus both the histone H3 sequence at positions 20, 22, and 24 and the dimerization of the Pc chromodomain are key determinants for the recognition of the H3K27me3 code. Most recently, a similar dimer of Rhino (an HP1 variant in *Drosophila*) chromodomain was observed and found to be important for both H3K9me3 binding and in vivo function [24]. Taken together, it seems that both an extended recognition groove that binds five additional residues preceding the ARKS motif and dimerization of the Pc chromodomain contribute to the specific Pc–H3K27me3 interaction.

MAMMALIAN POLYCOMB HOMOLOGS BIND DIFFERENTIALLY TO METHYLATED HISTONE H3

Both PRC1 and PRC2 are conserved in mammals, and mammals have several homologs for each subunit of these complexes. Specifically, Pc protein is encoded by a single gene in *Drosophila*, while its mammalian homologs have expanded into five family members known as Chromobox 2 (CBX2), CBX4, CBX6, CBX7, and CBX8. Three other CBX proteins, CBX1, CBX3, and CBX5, are known as variants of HP1 protein, named as HP1β, HP1γ, and HP1α, respectively (Fig. 3.1B). All of these proteins have a highly conserved chromodomain at the N-terminus.

Despite a high degree of conservation with their *Drosophila* homologs, the mouse Cbx chromodomains display significant differences in binding preferences. Not all Pc chromodomains bind preferentially to H3K27me3 (Table 3.1); rather, Cbx2 and Cbx7 display strong binding affinities toward both H3K9me3 and H3K27me3, and Cbx4 prefers H3K9me3. Cbx7, in particular, displays strong binding affinity for both H3K9me3 and H3K27me3 with a dissociation constant in the low micromolar range. Neither Cbx6 nor Cbx8 binds significantly to H3K27me3 or H3K9me3 [25].

Similar results were obtained for human CBX proteins [26]. Most human Pc homologs have a wide range of binding affinities toward both H3K27me3 and H3K9me3 without a distinct selectivity for one. CBX4 and CBX7 bound to both methylated marks but behaved more like an HP1 protein with modest binding to H3K9me3 and two- to three-fold weaker binding to H3K27me3. CBX2 was the only Pc-like chromodomain with a clear preference for the H3K27me3 mark; however, the affinity was rather weak. Furthermore, CBX2, CBX4, and CBX7 were insensitive to mutation at the n-3 position (T6 or A24). Finally, both CBX6 and CBX8 bound to H3K9me3 or H3K27me3 very weakly, with K_d >500 μM. Distinct from Pc homologs, the human HP1 homologs (CBX1, -3, -5) showed significant preference for H3K9me3 peptides and were sensitive to mutation of T6 to alanine, as observed with their *Drosophila* homologs, confirming that these chromodomains can distinguish H3K9me3 and H3K27me3 sequences via the third residue preceding the methyl-lysine [26].

A "clasp theory" was proposed to interpret the loss of substrate selectivity for mammalian Pc homologs [26]. For instance, in the CBX7–H3K27me3 complex structure, a "hydrophobic clasp" formed by Val-10 and Leu-49 of CBX7 interacts with the n-3 position A24 of the H3K27me3 peptide, and this interaction is highly conserved in all Pc-class complex structures (Figs. 3.1B and 3.2E). A peptide–array assay, screening for amino acid substitution of the residues surrounding K27me3 using all 20 kinds of amino acids, revealed that many other amino acids including threonine and especially hydrophobic residues at the n-3 position of the peptide are favorable for binding. Thus, the hydrophobic clasp is an important structural contributor to the lack of selectivity of human Pc homologs toward the methyl-lysine marks. In the HP1 subfamily of chromodomains, the "clasp" residues are two negatively charged residues called "polar fingers," which are conserved in the HP1 homologs and determine that only threonine and valine can be tolerated at the n-3 position of the peptide (Figs. 3.1B and 3.2F). However, the "clasp" residues cannot be simply swapped between the HP1 and Pc-classes to switch binding selectivity but are context dependent within each chromodomain. Another significant difference between the Pc and HP1 chromodomains is their surface charge. The HP1 chromodomains have theoretical isoelectric points of less than five, whereas the Pc chromodomains have theoretical isoelectric points of more than nine, which means that HP1 chromodomains are negatively charged and Pc chromodomains are positively charged under physiological conditions (Figs. 3.1B and 3.2G and H). Considering that the histone tail is also positively charged, this feature may be responsible for the weaker binding affinities of the Pc chromodomains. Comparison of the CBX6 complexes with either H3K9me3 or H3K27me3 peptide further reveals structural plasticity of the Pc-like chromodomains (Fig. 3.3A). When bound to H3K27me3, Arg52 of CBX6 forms a salt bridge with Asp50 of CBX6 to avoid electrostatic exclusion with K23 of histone H3. When bound to H3K9me3, Arg52 of CBX6 forms hydrogen bonds with Q5 of histone H3. Of note, this arginine–aspartic acid pair is conserved in the Pc chromodomains but absent in the HP1 chromodomains, which may also contribute to the loss of selectivity of mammalian Pc homologs toward H3K9me3 and H3K27me3. In vivo, different CBX proteins associate with nucleosomes in different subnuclear regions in both ES cells and fibroblasts as revealed by the bimolecular fluorescence complementation analysis [27]. Furthermore, neither the chromodomains nor H3 K27 trimethylation is required for chromatin association by CBX proteins in cells, and

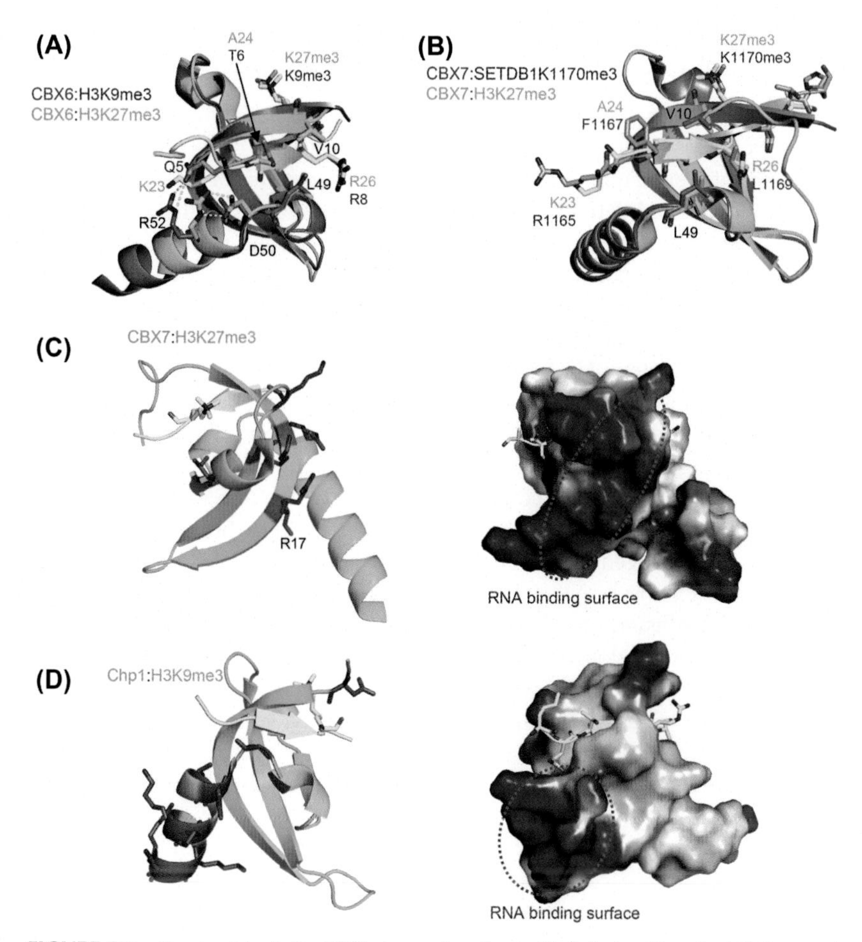

FIGURE 3.3 Structural basis for CBX chromodomains to bind distinct ligands. (A) Structural comparison of CBX6 binding to H3K9me3 (PDB code: 3GV6) and H3K27me3 (PDB code: 3I90). Note the conformation difference of the D50-R52 pair of CBX6. (B) Structural comparison of CBX7 binding to H3K27me3 (PDB code: 4X3K) and a peptide from SETDB1 (PDB code: 4X3S). A "hydrophobic clasp" formed by V10 and L49 of CBX7 interacts with two hydrophobic residues F1167 and L1169 of SETDB1. (C) Left, residues of CBX7 perturbed by adding RNA in NMR chemical shift are shown in magenta; right, the potential RNA-binding surface. (D) RNA-binding surface of yeast Chp1 (PDB code: 3G7L) is shown for comparison. Left, residues of Chp1 perturbed by adding RNA are shown in magenta; right, the potential RNA-binding surface.

chromatin-associated CBX proteins do not colocalize with H3 K27 trimethylation outside the inactive X [27]. Thus the mammalian Pc chromodomains may not be involved in specific recruitment of PRC1 but instead in stabilization of the complex.

CROSS TALK BETWEEN HISTONE METHYLATION AND OTHER POSTTRANSLATIONAL MODIFICATIONS

Histones can undergo different PTMs that regulate a variety of physiological processes. These covalent modifications show substantial cross talk, providing a wealth of regulatory potential. The ARKS motifs of H3K9me3 and H3K27me3 can undergo other modifications in addition to methylation on the lysine. First, the serine residue could be phosphorylated. Global phosphorylation of H3S10 is a concomitant of the entry into S phase in eukaryotic cells. This phosphorylation is carried out by the Aurora B kinase. HP1 is released from chromatin during the M phase of the cell cycle, even though trimethylation levels of histone H3 lysine 9 remain unchanged, because the transient serine 10 phosphorylation of histone H3 next to the more stable methyl-lysine 9 mark is sufficient to eject HP1 proteins from their binding sites [28]. Juxtaposition of this mark with the H3K9 methylation led to the hypothesis that H3S10 phosphorylation could function as a switch to regulate the access of the H3K9 methyl mark by the HP1 family members. Structurally, S10 is involved in a hydrogen bond network with a glutamic acid of the HP1 chromodomains. A phosphoryl group could introduce both steric hindrance and electrostatic exclusion to disrupt this interaction. S28 of histone H3 can also undergo phosphorylation. Although no direct report is available, it is reasonable to predict that phosphorylation on S28 may have similar effect on the binding of Pc to H3K27me3, due to their highly conserved interactions. However, phosphorylation on serine does not always exclude histone readers from binding, as the UHRF1 association with H3K9 methylation is insensitive to the adjacent H3 S10 phosphorylation [29], suggesting that the role of serine phosphorylation as a molecular switch is binding-partner dependent.

Secondly, arginine can be mono-, asymmetrically or symmetrically dimethylated in vivo, and these modifications could have different effects on protein binding and biological functions. Arginine may also be converted to citrulline by the calcium-dependent peptidyl-arginine deiminase PADI4. Notably, citrullination would eliminate the positive charge of arginine, whereas methylation would not. Quantitative binding studies on HP1 revealed that the H3R8 citrullination results in a more than 200-fold decrease in the affinity of HP1 for H3cit8K9me3 (K_d:313 $\pm$ 28 μM) compared to H3R8K9me3 (K_d:1.39 $\pm$ 0.06 μM), and a more than 20-fold decrease compared to H3R2meK9me3 (K_d 29.6 $\pm$ 1.3 μM) [30]. Considering the H3R8-interacting negatively charged "polar clasp" of HP1 (Fig. 3.2F), it is easy to understand that the positive charge of this arginine plays a critical role in the interaction. H3R26 is also subject to methylation and citrullination, however, it is hard to predict how these modifications cross talk with H3K27 methylation for the Pc chromodomains simply based on the study on HP1, as they use distinct "clasps" to interact with the arginine. Based on a peptide—array assay, citrullination on R26 reduced the binding of the yeast Pc protein

Ccc1 to H3K27me2/3, whereas methylation on R26 had a limited effect on the binding [31].

In addition to the modifications within the ARKS motif of histone H3, other sites may also provide potential cross talk mechanisms to the methyl-lysine recognition. H3K4ac and H3K23ac, two marks associated with gene activation, have the potential to antagonize the repressive marks H3K9me or H3K27me, respectively. In fission yeast, H3K4ac, mediated by Mst1, is enriched at pericentromeres concomitantly with heterochromatin reassembly. H3K4ac has the ability to weaken the interaction between the chromodomain of Chp1 (a component of the RNA-induced transcriptional silencing complex) or Clr4 (an H3K9 methyltransferase) and the tail of H3, but has no effect on the binding of the HP1 homologs Swi6 and Chp2. These results suggest that H3K4ac deposition facilitates the binding of the HP1 proteins to H3K9me by decreasing the affinity of Chp1 for the H3K9me marker [32]. However, how H3K23ac affects Pc proteins remains unclear.

The HP1/Pc chromodomains themselves may also be subject to PTMs, and phosphorylation of HP1 had been described in 1994 [33]. The mouse Cbx2 has been reported to be highly phosphorylated in some cell lines [34]. Mass spectrometric analysis revealed serine-42, a conserved residue in the chromodomain, as a phosphorylation site. Phosphorylation of this residue in vitro resulted in a reduced binding ability of Cbx2 to H3K9me3 but an increased binding ability to H3K27me3, suggesting that such phosphorylation changes the binding specificity of Cbx2. Phosphorylation of the Cbx2 chromodomain may therefore serve as a molecular switch that affects the reading of the histone modification code and thereby controls epigenetic cellular memory [34]. Such a regulatory role of phosphorylation has also been observed for other chromodomains. The protein acetyltransferase KAT5 (also known as TIP60) harbors a chromodomain in its N-terminus, and its binding ability to H3K9me3 depends on phosphorylation of the highly conserved Tyr44 in its chromodomain. Notably, in response to ionizing radiation, only KAT5 with phosphorylated Tyr44 could stimulate the activation of ATM, a key enzyme in the DNA damage response network, as well as the ATM downstream targets, following binding to H3K9me3 [35]. Such examples suggest that PTMs of the chromodomain themselves also provide potential regulatory mechanisms for their functions. However, this field is largely unexplored.

PUTATIVE NONHISTONE TARGETS OF POLYCOMB GROUP CHROMODOMAINS

Although much attention has been directed toward elucidating the principles underlying histone modifications, it has been known for many years that nonhistone proteins are also targets for many of the same categories of modifications, including methylation [36]. It has been suggested that the principles governing the histone code hypothesis might just represent a special case of a "protein code" [37]. Indeed, mounting evidence suggests the existence of direct interaction relationships between chromodomains and nonhistone proteins via a methylated ARK(S/T) consensus motif [20]. Moreover, the ARK(S/T) motif is present in numerous lysine methyltransferases [20] (Fig. 3.1A). These observations suggest a more general role for the canonical chromodomain in cellular signaling networks.

The linker histone H1.4 can be methylated on K26, and K26 resides in an ARKS motif, making it a potential target of chromodomains, in addition to the H3K9me3 and H3K27me3 sites. HP1 was first identified as a reader of H1.4K26me3 [38]. The histone H3K9 methyltransferase G9a contains an ARKT sequence motif in its N-terminus. As with methylation of H3 lysine 9, autocatalytic G9a methylation is necessary and sufficient to mediate in vivo interaction with the epigenetic regulator HP1 [39]. Structural analysis indicates that the HP1 chromodomain recognizes methyl-G9a in a binding mode similar to that used in the recognition of methyl-H3K9, demonstrating that the chromodomain functions as a generalized methyl-lysine–binding module [40]. In the case of Pc homologs, in vitro screening has identified lysine 1170 of SETDB1 (a histone H3K9 methyltransferase) to be a target of CBX7, and the CBX7/8 chromodomains bind to trimethylated SETDB1 peptides with even greater affinities than to histone peptides [26,41]. Nevertheless, this kind of interaction is not established in vivo yet. Of note, two hydrophobic residues Phe1167 and Leu1169 of SETDB1 interact with the "hydrophobic clasp" of CBX7 (Fig. 3.3B) [41], confirming the loss of selectivity for Pc homologs. In addition to the HP1 and Pc chromodomains, other canonical chromodomains have also been implicated in binding to methylated ARKS motifs in nonhistone targets. Examples of this are the chromodomain Y chromosome family of chromodomains binding to H1.4K26me3 and the binding of G9a-K185me3 [42] and MPP8 to DNMT3A-K47me2 [43]. There are more than 100 polypeptides containing the ARK(S/T) motif encoded in the human genome. Thus, the possibilities of these motifs becoming methylated imply the recognition of their chromodomains potentially as an integral component of various nuclear control mechanisms. Additionally, since it is known that many of these proteins have activities outside of the nucleus, this methylation-dependent chromodomain recognition may likely also play roles in various molecular activities in the cytoplasm, including signaling, transport, and cytoskeleton organization [20].

NONCODING RNA: NONCANONICAL PARTNERS OF POLYCOMB GROUP CHROMODOMAINS

When the first 3D structure of the chromodomain of mouse HP1β was determined in 1997, it revealed an unexpected homology to two archaebacterial DNA-binding proteins, Sac7d and Sso7d, which are also involved in regulating chromatin structure [44]. Other studies on chromodomains further confirmed their nucleic acid—binding abilities. For example, in *Drosophila*, the association of the histone acetyltransferase MOF with the male X chromosome depends on its interaction with RNA, and MOF specifically binds to roX2 RNA through its chromodomain in vivo [45]. In another case, the chromatin remodeling protein Mi-2 could bind to DNA through its chromodomains, which do not bind to methylated histones [46]. As mentioned above, there is a significant difference between HP1 chromodomains and Pc chromodomains. That is, HP1 chromodomains are negatively charged and Pc chromodomains are positively charged under physiological conditions. These clues suggest that Pc chromodomains are potential nucleic acid—binding modules, whereas the HP1 protein uses a hinge region rather than the chromodomain to bind RNA [47].

In mammals, PcG proteins have also been implicated in X inactivation, whereby one of the two female X chromosomes is inactivated to balance gene dosage between the sexes. The noncoding *Xist* transcript, which coats the X chromosome in *cis* and triggers X inactivation during early development, may have a role in recruiting both PRC2 and PRC1 proteins to chromatin, as inducible *Xist* transgenes result in the rapid recruitment of PRC2 and PRC1 proteins on the chromosome [48,49]. Indeed, all mouse Pc chromodomains, with the exception of Cbx2, could bind a 500-nt single-stranded RNA. On the other hand, the chromodomains of mouse Cbx1 and Cbx5 (HP1 homologs) were unable to bind this RNA [25]. The Pc chromodomains bind to RNA in a sequence nonspecific manner, as they bind to various RNAs tested, including *Xist* sequences. Cbx7 appears to prefer binding to single-stranded RNA with a higher affinity (~100 μM) over double-stranded RNA and double-stranded DNA. Furthermore, point mutations of the aromatic cage residues of the Cbx7 chromodomain do not completely abrogate their RNA-binding activity, suggesting that the surface of the chromodomain responsible for RNA interaction may be distinct from that which binds the methyl-lysine peptide [25].

The human CBX7 was also reported to specifically bind to *ANRIL* [50], a long noncoding antisense RNA transcript overlapping the *INK4b/ARF/INK4a* locus [51]. In vitro, purified full-length CBX7 binds to a 26 nt RNA, as does the chromodomain alone. Markedly, the CBX7 chromodomain binds just to RNA, not DNA. The affinity for the CBX7 chromodomain—RNA interaction is estimated to be ~51 μM. CBX7 employs overlapping yet distinct regions within its chromodomain for binding to H3K27me and RNA (Fig. 3.3C). From comparative binding studies, an R17A mutation on the RNA-binding site of CBX7 led to a nearly complete loss of its RNA-binding capability, but had no significant impact on its H3K27me3-binding ability compared to the wild type. Conversely, a W35A mutation of the aromatic cage adversely affected the binding of CBX7 to H3K27me, but not RNA. The three components (CBX7 chromodomain, H3K27me3 peptide, and *ANRIL* loop RNA) are able to form a ternary complex, but the presence of one ligand weakens the affinity of CBX7 for the other [50]. Functionally, mutations affecting CBX7 binding to

H3K27me have a more dramatic effect on cellular survival than those affecting their ability to bind RNA. Nevertheless, ablation of either H3K27me- or RNA-binding activity of CBX7 compromises its capacity to repress the *INK4b/ARF/INK4a* locus and control senescence.

Examination of the yeast Chp1 chromodomain revealed that it possesses unique nucleic acid–binding activities that are essential for heterochromatic gene silencing. Interestingly, the RNA- and DNA-binding activities of Chp1 chromodomain were strongly enhanced when they were bound to the H3K9me peptide [52]. Furthermore, the RNA-binding surface on Chp1 is distinct from that of CBX7 (Fig. 3.3D), suggesting that the intrinsic RNA-binding ability of chromodomains may vary and should be examined case by case.

CHEMICAL PROBES FOR CBX7 CHROMODOMAIN

Epigenetic proteins have been intently pursued as targets in ligand discovery. Research efforts have led to significant success on chromatin-modifying enzymes, the so-called epigenetic 'writers' and 'erasers'. Recently, potent inhibitors of histone-binding modules, the epigenetic 'readers', have also been described. The first inhibitors of a methyl-lysine reader were reported to target a few similar members of a single family (the MBT proteins) and have moved quickly through subsequent rounds of optimization [53,54]. As with bromodomain inhibitors, the first compound against an MBT domain has been demonstrated to have a potent influence on the biology of cancer cells [54]. Considering the important roles of CBX proteins in gene transcription in a wide array of cellular processes, a small molecule–mediated disruption of the binding of CBX chromodomains and H3K27me3 would inhibit the transcriptional activities of CBXs and result in derepression of its target genes. Such small molecules could be used to fine-tune a balance between stem cell self-renewal and differentiation, and could also be developed into potential new therapeutics for cancer treatment. Of the five Pc homologs, CBX7 is the one best studied; it is a master regulator that extends cellular life span, delays senescence, drives proliferation, and bestows pluripotency on adult and embryonic stem cells [50,55]. Importantly, multiple reports show that the proliferative/prosurvival functions of CBX7 depend only on its H3K27me3-binding ability; a single mutation in the methyl-lysine–binding site of CBX7's chromodomain nullifies its positive effects on cellular proliferation [56].

Previous work utilizing peptide arrays revealed that the overall CBX7 consensus–binding sequence is A(R/I/L/F/Y/V)Kme3(S/T), and the binding affinity to a 25 amino acid long peptide sequence derived from the SETDB1 protein is higher than that to the H3K27me3 sequence itself [26]. A group from the University of Victoria conducted a search for peptidic antagonists of the CBX7 chromodomain started from this report [57]. They first truncated the methylated SETDB1 sequence to arrive at a five amino acid long peptide Ac-

FALKme3S-NH2. Then, they made a family of analogs of this peptide that varied at each position except methyl-lysine and tested them for their ability to disrupt the H3K27me3–CBX7 complex. Finally, a selected set of compounds bearing multiple well-tolerated modifications to the FALKme3S scaffold were tested and some compounds exhibited potent binding to CBX7 with a K_d of ~200 nM. The most selective compound is ~10-fold selective over CBX8, the homolog most similar in sequence (~82%) to CBX7 [57]. A group from the University of North Carolina at Chapel Hill further optimized this peptidic antagonist to a high-quality, cellularly active chemical probe, UNC3866 [58].

UNC3866 binds the chromodomains of CBX4 and CBX7 most potently, with a K_d of ~100 nM for each of these domains and is 6- to 18-fold selective versus seven other CBX and CDY chromodomains while being highly selective versus >250 other protein targets. X-ray crystallography revealed that UNC3866 closely mimics the interactions of the methylated H3 tail with these chromodomains and binds via an extended surface groove interaction (Fig. 3.4A). Functionally, UNC3866 engages intact PRC1 complex and inhibits PC3 cell proliferation, a known CBX7 phenotype [58]. These peptidic lead compounds greatly promote the elucidation of the biological effects of CBX protein antagonism, and thus can potentially lead to the advent of improved therapeutic strategies targeted against chromodomain-containing proteins.

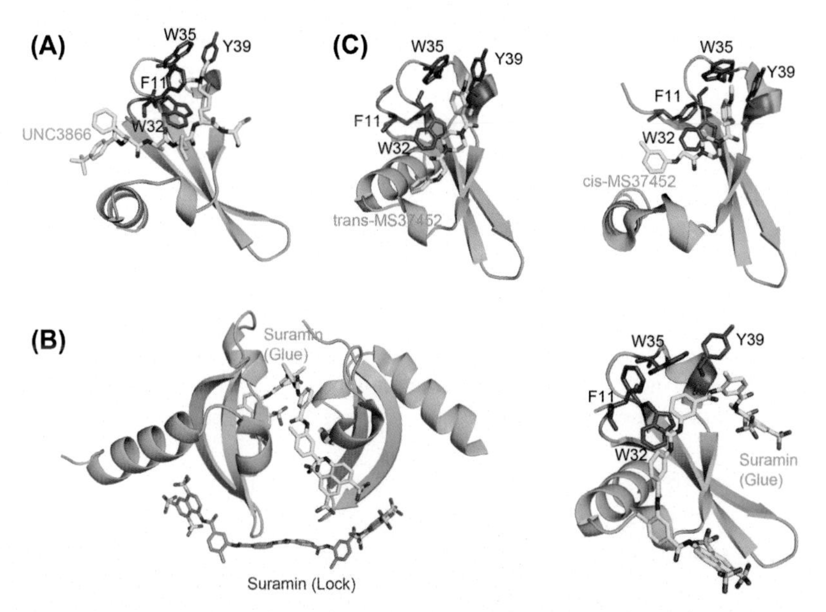

FIGURE 3.4 Chemical probes for CBX7 chromodomain. (A) The peptidic compound UNC3866 (PDB code: 5EPJ). (B) Suramin (PDB code: 4X3U). (C) MS37452 (PDB code: 4X3T).

A high-throughput screening of a library of 2560 Food and Drug Administration (FDA)–approved drug molecules and an L1 library of 100,160 compounds selected from commercial sources was conducted in an attempt to identify inhibitors for the CBX7 chromodomain. This investigation used a fluorescence anisotropy–binding assay with a fluorescein isothiocyanate–labeled SETDB1-K1170me3 peptide as an assay probe [41]. Five hits (sennoside A, suramin, aurin tricarboxylic acid, trypan blue, and Evans blue) from the FDA-approved drug library and one hit (MS37452) from the L1 library were confirmed as binders of the CBX7 chromodomain. Of the five FDA-approved drug hits, suramin was investigated in detail. Suramin binds to CBX7 with an affinity (IC_{50}) of ~8 μM. In the crystal structure of their complex, two suramin molecules bound to two CBX7 molecules in a unit cell (Fig. 3.4B). The two β-sheets of CBX7 create a cleft, to which one suramin molecule binds. This binding interaction sticks the two domains of CBX7 together, and therefore is named as "suramin glue." Meanwhile, the other suramin molecule is called "suramin lock," as it locks the two CBX7 proteins together by binding to the plane formed by the two β-sheets of CBX7 orthogonal to the "suramin glue." Additionally, the CBX7 aromatic cage is occupied and disrupted by the "suramin glue" (Fig. 3.4B).

MS37452 has a binding affinity (K_d) of ~28.9 μM to the CBX7 chromodomain and disrupts CBX7–H3K27me3 or CBX7–H3K9me3 interaction with K_i of 43.0 μM and 55.3 μM, respectively. Strikingly, in the crystal structure of the CBX7–MS37452 complex, MS37452 adapts two rotamer conformations (Fig. 3.4C). The dimethoxylbenzene and piperazine moieties of MS37452 interact with the aromatic cavity of CBX7 in the same orientation. On the other hand, the methylbenzene moiety in a *cis* or *trans* configuration to the dimethoxylbenzene pivots on the carbonyl group that bridges the methylbenzene and the piperazine ring. Specifically, the methylbenzene in the *trans* conformation interacts with the chromodomain similar to the H3K27me3 peptide. However, when it is in the *cis* conformation, the methylbenzene interrupts the N-terminal β-strand structure by an upward swing into CBX7, leading to the destabilization of the complex. In vivo, MS37452 blocks CBX7 binding to its target *INK4/ARF* gene locus and induces derepression of *p14/ARF* and *p16/INK4a* in human PC3 cells. Given the essential roles of CBX7 in PRC-directed transcriptional repression of the *Ink4a/Arf* locus in biology and diseases, such new inhibitors can potentially be used to provide senescence control, cancer prevention, and stem cell lineage specification.

POLYCOMB HOMOLOGS FROM YEAST AND PLANT: EVOLUTIONARILY CONSERVED BIOLOGICAL SIGNIFICANCE OF CHROMODOMAIN

Given that the PRC2 complex represses developmental genes of higher eukaryotes but is not present in unicellular fungi, it was thought that the PRC2 complex appeared in higher eukaryotic organisms together with multicellularity in the course of evolution. The core subunits of the PRC2 complex, nevertheless, were recently identified in unicellular eukaryotes including the *Opisthokonta*, *Chromalveolata*, and *Archaeplastida* supergroups [59]. Moreover, a chromodomain-containing protein Ccc1 was identified as a subunit of the PRC2 complex in the yeast *Cryptococcus neoformans* [31]. The chromodomain of Ccc1 recognizes H3K27me2/3 specifically. Of note, the *C. neoformans* histone H3 sequence differs from the human sequence at residues downstream of K27 (K_{27}QTTT for *C. neoformans* vs. K_{27}SAPA for human). Importantly, the Ccc1 component−mediated binding of PRC2 complex to its product weakens its responsiveness to activation signals from H3K9me2 domains. As a result, chromodomain-mediated recognition of H3K27me confers the ability to differentiate these two different types of repressive chromatin marks, suggesting that Ccc1 may likely have a similar function as Pc in *Drosophila* [31].

In plants, homologs of all four core subunits of animal PRC2 exist. In contrast, plants lack Pc homologs. However, a chromodomain-containing protein Like Heterochromatin Protein 1 (LHP1) is considered to fulfill the role of Pc in plants based on its ability to bind to H3K27me3 in vitro and its genome-wide colocalization with H3K27me3 in vivo [60,61]. LHP1 binding to H3K27me3 is required for its function, and LHP1 is required for repression of several PcG protein targets. Furthermore, LHP1 is linked to PRC2 complex by MSI1, a homolog of *Drosophila* p55. The LHP1−MSI1 interaction forms a positive feedback loop to recruit PRC2 to chromatin that carries H3K27me3 [62]. Plants have little H3K9me3; there seems to have been an evolutionary switch in the specificity of plant HP1 family chromodomain, making it more like the Pc chromodomain.

CONCLUSION

Decades after the first identification of the chromodomain, we have gained some understanding of its specific properties that explain, in some cases, the ability of chromodomains to target specific chromatin compartments. Now it is well known that the chromodomains of Pc and its homologs are specific reader modules of two repressive histone markers, H3K27me3 and H3K9me3, both of which share a consensus sequence motif ARKS with lysine being methylated. In addition to their canonical biological significance in the orchestration of key developmental events as well as cellular lineage specifications, it is increasingly realized that PcG proteins have pleiotropic cellular activities in eukaryotes. Given the complexity of cellular activities associated with PcG proteins, it is not so surprising that a significant number of new components of PcG complexes have been recently identified. As a result, a large repertoire of variant PcG complexes exist in eukaryotes probably to faithfully accomplish all the complex molecular tasks, and a subset of these PRC1 complexes may not even harbor Pc proteins [4]. Studies have suggested that Pc chromodomains have the potential to bind methyl-ARKS motif-containing nonhistone targets, but no direct in vivo evidence for this has been reported so far. In addition to the methylated ARKS motif recognition, the Pc chromodomains also have the potential to bind noncoding RNA, which provides an alternative mechanism to regulate their targeting and function. However, our understanding on this mechanism is very limited, especially considering the diversity among Pc homologs. The nucleic acid–binding activities of Pc proteins may be further complicated due to the presence of an AT-hook-like domain immediately following the N-terminal chromodomain, as an AT-hook may also represent a nucleic acid–binding module.

Despite advances in chemical probe development for CBX7, these efforts are mostly focused on its methyl-lysine–binding activity. Probes against its RNA-binding ability have not been achieved yet. It would also be challenging to enhance the probe selectivity due to the high sequence similarity of CBXs. More structural and functional analyses of chromodomains will be undoubtedly forthcoming, and these continued research efforts should further enlighten our understanding of the functions of Pc proteins in chromatin biology.

LIST OF ACRONYMS AND ABBREVIATIONS

CBX Chromobox
HP1 Heterochromatin-associated protein 1
Pc Polycomb
PcG Polycomb group
PRC Polycomb repressive complexes

GLOSSARY

Chromodomain A chromodomain is a protein structural domain of about 40–50 amino acid residues commonly found in proteins associated with the remodeling and manipulation of chromatin.

Dissociation constant (K_d) A dissociation constant (K_d) is a specific type of equilibrium constant that measures the propensity of a larger object to separate (dissociate) reversibly into smaller components, as when a complex falls apart into its component molecules.

Noncoding RNA A noncoding RNA is an RNA molecule that is transcribed from a DNA template but not translated into a protein.

Chemical probe Chemical probes are potent, selective, and cell-permeable inhibitors of protein function that allow the user to probe into mechanistic and phenotypic questions about their molecular targets.

Posttranslational modification Posttranslational modification refers to the covalent and generally enzymatic modification of proteins during or after protein biosynthesis, which can occur on the amino acid side chains or at the protein's C- or N-termini and extend the chemical repertoire of the 20 standard amino acids by introducing new functional groups such as phosphate, acetate, amide groups, or methyl groups.

Hydrophobic clasp A "hydrophobic clasp" is a structural element formed by two hydrophobic residues (e.g. Val and Leu) to interact with the histone ligand of the Pc-like chromodomains. These two residues can interact with each other through hydrophobic interactions to "clasp" the histone ligand, thus losing selectivity to the n-3 position of the ligand.

Polar clasp A "polar clasp" is a structural element formed by two conserved negatively charged residues to interact with the histone ligand of the HP1-like chromodomains. The two residues with identical charge require a Thr/Val at the n-3 position of the ligand to sandwich their side chains and hold a favorable conformation for methyl-lysine binding.

REFERENCES

[1] Lewis EB. Polycomb. *Drosophila* Inform Serv 1947;21:69.

[2] Lewis EB. A gene complex controlling segmentation in *Drosophila*. Nature December 7, 1978;276(5688):565–70.

[3] Schwartz YB, Pirrotta V. Polycomb silencing mechanisms and the management of genomic programmes. Nat Rev Genet January 2007;8(1):9–22.

[4] Schwartz YB, Pirrotta V. A new world of Polycombs: unexpected partnerships and emerging functions. Nat Rev Genet December 2013;14(12):853–64.

[5] Del Prete S, Mikulski P, Schubert D, Gaudin V. One, two, three: polycomb proteins hit all dimensions of gene regulation. Genes (Basel) 2015;6(3):520–42.

[6] Cao R, Wang L, Wang H, Xia L, Erdjument-Bromage H, Tempst P, et al. Role of histone H3 lysine 27 methylation in Polycomb-group silencing. Science November 1, 2002;298(5595):1039–43.

[7] Czermin B, Melfi R, McCabe D, Seitz V, Imhof A, Pirrotta V. *Drosophila* enhancer of Zeste/ESC complexes have a histone H3 methyltransferase activity that marks chromosomal Polycomb sites. Cell October 18, 2002;111(2):185–96.

[8] Kuzmichev A, Nishioka K, Erdjument-Bromage H, Tempst P, Reinberg D. Histone methyltransferase activity associated with a human multiprotein complex containing the Enhancer of Zeste protein. Genes Dev November 15, 2002;16(22):2893–905.

[9] Paro R, Hogness DS. The Polycomb protein shares a homologous domain with a heterochromatin-associated protein of *Drosophila*. Proc Natl Acad Sci USA January 1, 1991;88(1):263–7.

[10] Bannister AJ, Zegerman P, Partridge JF, Miska EA, Thomas JO, Allshire RC, et al. Selective recognition of methylated lysine 9 on histone H3 by the HP1 chromo domain. Nature March 1, 2001;410(6824):120–4.

[11] Lachner M, O'Carroll D, Rea S, Mechtler K, Jenuwein T. Methylation of histone H3 lysine 9 creates a binding site for HP1 proteins. Nature March 1, 2001;410(6824):116–20.

[12] Messmer S, Franke A, Paro R. Analysis of the functional role of the Polycomb chromo domain in *Drosophila melanogaster*. Genes Dev July 1992;6(7):1241–54.

[13] Platero JS, Hartnett T, Eissenberg JC. Functional analysis of the chromo domain of HP1. EMBO J August 15, 1995;14(16):3977–86.

[14] Ekwall K, Nimmo ER, Javerzat JP, Borgstrom B, Egel R, Cranston G, et al. Mutations in the fission yeast silencing factors clr4+ and rik1+ disrupt the localisation of the chromo domain protein Swi6p and impair centromere function. J Cell Sci November 1996;109(Pt 11):2637–48.

[15] Rea S, Eisenhaber F, O'Carroll D, Strahl BD, Sun ZW, Schmid M, et al. Regulation of chromatin structure by site-specific histone H3 methyltransferases. Nature August 10, 2000;406(6796):593–9.

[16] Nakayama J, Rice JC, Strahl BD, Allis CD, Grewal SI. Role of histone H3 lysine 9 methylation in epigenetic control of heterochromatin assembly. Science April 6, 2001;292(5514):110–3.

[17] Fischle W, Wang Y, Jacobs SA, Kim Y, Allis CD, Khorasanizadeh S. Molecular basis for the discrimination of repressive methyl-lysine marks in histone H3 by Polycomb and HP1 chromodomains. Genes Dev August 1, 2003;17(15):1870–81.

[18] Maurer-Stroh S, Dickens NJ, Hughes-Davies L, Kouzarides T, Eisenhaber F, Ponting CP. The Tudor domain 'Royal Family': Tudor, plant Agenet, Chromo, PWWP and MBT domains. Trends Biochem Sci February 2003;28(2):69–74.

[19] Taverna SD, Li H, Ruthenburg AJ, Allis CD, Patel DJ. How chromatin-binding modules interpret histone modifications: lessons from professional pocket pickers. Nat Struct Mol Biol November 2007;14(11):1025–40.

[20] Blus BJ, Wiggins K, Khorasanizadeh S. Epigenetic virtues of chromodomains. Crit Rev Biochem Mol Biol December 2011;46(6):507–26.

[21] Min J, Zhang Y, Xu RM. Structural basis for specific binding of Polycomb chromodomain to histone H3 methylated at Lys 27. Genes Dev August 1, 2003;17(15):1823–8.

[22] Hughes RM, Wiggins KR, Khorasanizadeh S, Waters ML. Recognition of trimethyllysine by a chromodomain is not driven by the hydrophobic effect. Proc Natl Acad Sci USA July 3, 2007;104(27):11184–8.

[23] Cowell IG, Austin CA. Self-association of chromo domain peptides. Biochim Biophys Acta February 8, 1997;1337(2):198–206.

[24] Yu B, Cassani M, Wang M, Liu M, Ma J, Li G, et al. Structural insights into Rhino-mediated germline piRNA cluster formation. Cell Res April 2015;25(4):525–8.

[25] Bernstein E, Duncan EM, Masui O, Gil J, Heard E, Allis CD. Mouse polycomb proteins bind differentially to methylated histone H3 and RNA and are enriched in facultative heterochromatin. Mol Cell Biol April 2006;26(7):2560–9.

[26] Kaustov L, Ouyang H, Amaya M, Lemak A, Nady N, Duan S, et al. Recognition and specificity determinants of the human cbx chromodomains. J Biol Chem January 7, 2011;286(1):521–9.

[27] Vincenz C, Kerppola TK. Different polycomb group CBX family proteins associate with distinct regions of chromatin using nonhomologous protein sequences. Proc Natl Acad Sci USA October 28, 2008;105(43):16572–7.

[28] Fischle W, Tseng BS, Dormann HL, Ueberheide BM, Garcia BA, Shabanowitz J, et al. Regulation of HP1-chromatin binding by histone H3 methylation and phosphorylation. Nature December 22, 2005;438(7071):1116–22.

[29] Rothbart SB, Krajewski K, Nady N, Tempel W, Xue S, Badeaux AI, et al. Association of UHRF1 with methylated H3K9 directs the maintenance of DNA methylation. Nat Struct Mol Biol November 2012;19(11):1155–60.

[30] Sharma P, Azebi S, England P, Christensen T, Moller-Larsen A, Petersen T, et al. Citrullination of histone H3 interferes with HP1-mediated transcriptional repression. PLoS Genet September 2012;8(9):e1002934.

[31] Dumesic PA, Homer CM, Moresco JJ, Pack LR, Shanle EK, Coyle SM, et al. Product binding enforces the genomic specificity of a yeast polycomb repressive complex. Cell January 15, 2015;160(1–2):204–18.

[32] Xhemalce B, Kouzarides T. A chromodomain switch mediated by histone H3 Lys 4 acetylation regulates heterochromatin assembly. Genes Dev April 1, 2010;24(7):647–52.

[33] Eissenberg JC, Ge YW, Hartnett T. Increased phosphorylation of HP1, a heterochromatin-associated protein of *Drosophila*, is correlated with heterochromatin assembly. J Biol Chem August 19, 1994;269(33):21315–21.

[34] Hatano A, Matsumoto M, Higashinakagawa T, Nakayama KI. Phosphorylation of the chromodomain changes the binding specificity of Cbx2 for methylated histone H3. Biochem Biophys Res Commun June 18, 2010;397(1):93–9.

[35] Kaidi A, Jackson SP. KAT5 tyrosine phosphorylation couples chromatin sensing to ATM signalling. Nature June 6, 2013;498(7452):70–4.

[36] Sims 3rd RJ, Reinberg D. Is there a code embedded in proteins that is based on post-translational modifications? Nat Rev Mol Cell Biol October 2008;9(10):815–20.

[37] Margueron R, Trojer P, Reinberg D. The key to development: interpreting the histone code? Curr Opin Genet Dev April 2005;15(2):163–76.

[38] Daujat S, Zeissler U, Waldmann T, Happel N, Schneider R. HP1 binds specifically to Lys26-methylated histone H1.4, whereas simultaneous Ser27 phosphorylation blocks HP1 binding. J Biol Chem November 11, 2005;280(45):38090–5.

[39] Sampath SC, Marazzi I, Yap KL, Krutchinsky AN, Mecklenbrauker I, Viale A, et al. Methylation of a histone mimic within the histone methyltransferase G9a regulates protein complex assembly. Mol Cell August 17, 2007;27(4):596–608.

[40] Ruan J, Ouyang H, Amaya MF, Ravichandran M, Loppnau P, Min J, et al. Structural basis of the chromodomain of Cbx3 bound to methylated peptides from histone h1 and G9a. PLoS One 2012;7(4):e35376.

[41] Ren C, Morohashi K, Plotnikov AN, Jakoncic J, Smith SG, Li J, et al. Small-molecule modulators of methyl-lysine binding for the CBX7 chromodomain. Chem Biol February 19, 2015;22(2):161–8.

[42] Fischle W, Franz H, Jacobs SA, Allis CD, Khorasanizadeh S. Specificity of the chromodomain Y chromosome family of chromodomains for lysine-methylated ARK(S/T) motifs. J Biol Chem July 11, 2008;283(28):19626–35.

[43] Chang Y, Sun L, Kokura K, Horton JR, Fukuda M, Espejo A, et al. MPP8 mediates the interactions between DNA methyltransferase Dnmt3a and H3K9 methyltransferase GLP/G9a. Nat Commun 2011;2:533.

[44] Ball LJ, Murzina NV, Broadhurst RW, Raine AR, Archer SJ, Stott FJ, et al. Structure of the chromatin binding (chromo) domain from mouse modifier protein 1. EMBO J May 1, 1997;16(9):2473–81.

[45] Akhtar A, Zink D, Becker PB. Chromodomains are protein-RNA interaction modules. Nature September 21, 2000;407(6802):405–9.

[46] Bouazoune K, Mitterweger A, Langst G, Imhof A, Akhtar A, Becker PB, et al. The dMi-2 chromodomains are DNA binding modules important for ATP-dependent nucleosome mobilization. EMBO J May 15, 2002;21(10):2430–40.

[47] Muchardt C, Guilleme M, Seeler JS, Trouche D, Dejean A, Yaniv M. Coordinated methyl and RNA binding is required for heterochromatin localization of mammalian HP1alpha. EMBO Rep October 2002;3(10):975–81.

[48] Plath K, Fang J, Mlynarczyk-Evans SK, Cao R, Worringer KA, Wang H, et al. Role of histone H3 lysine 27 methylation in X inactivation. Science April 4, 2003;300(5616):131–5.

[49] Plath K, Talbot D, Hamer KM, Otte AP, Yang TP, Jaenisch R, et al. Developmentally regulated alterations in Polycomb repressive complex 1 proteins on the inactive X chromosome. J Cell Biol December 20, 2004;167(6):1025–35.

[50] Yap KL, Li S, Munoz-Cabello AM, Raguz S, Zeng L, Mujtaba S, et al. Molecular interplay of the noncoding RNA ANRIL and methylated histone H3 lysine 27 by polycomb CBX7 in transcriptional silencing of INK4a. Mol Cell June 11, 2010;38(5):662–74.

[51] Pasmant E, Laurendeau I, Heron D, Vidaud M, Vidaud D, Bieche I. Characterization of a germ-line deletion, including the entire INK4/ARF locus, in a melanoma-neural system tumor family: identification of ANRIL, an antisense noncoding RNA whose expression coclusters with ARF. Cancer Res April 15, 2007;67(8):3963–9.

[52] Ishida M, Shimojo H, Hayashi A, Kawaguchi R, Ohtani Y, Uegaki K, et al. Intrinsic nucleic acid-binding activity of Chp1 chromodomain is required for heterochromatic gene silencing. Mol Cell July 27, 2012;47(2):228–41.

[53] Herold JM, Wigle TJ, Norris JL, Lam R, Korboukh VK, Gao C, et al. Small-molecule ligands of methyl-lysine binding proteins. J Med Chem April 14, 2011;54(7):2504–11.

[54] James LI, Barsyte-Lovejoy D, Zhong N, Krichevsky L, Korboukh VK, Herold JM, et al. Discovery of a chemical probe for the L3MBTL3 methyllysine reader domain. Nat Chem Biol March 2013;9(3):184–91.

[55] Gil J, Bernard D, Martinez D, Beach D. Polycomb CBX7 has a unifying role in cellular lifespan. Nat Cell Biol January 2004;6(1):67–72.

[56] Bernard D, Martinez-Leal JF, Rizzo S, Martinez D, Hudson D, Visakorpi T, et al. CBX7 controls the growth of normal and tumor-derived prostate cells by repressing the Ink4a/Arf locus. Oncogene August 25, 2005;24(36):5543–51.

[57] Simhadri C, Daze KD, Douglas SF, Quon TT, Dev A, Gignac MC, et al. Chromodomain antagonists that target the polycomb-group methyllysine reader protein chromobox homolog 7 (CBX7). J Med Chem April 10, 2014;57(7):2874–83.

[58] Stuckey JI, Dickson BM, Cheng N, Liu Y, Norris JL, Cholensky SH, et al. A cellular chemical probe targeting the chromodomains of Polycomb repressive complex 1. Nat Chem Biol January 25, 2016;12(3).

[59] Shaver S, Casas-Mollano JA, Cerny RL, Cerutti H. Origin of the polycomb repressive complex 2 and gene silencing by an E(z) homolog in the unicellular alga Chlamydomonas. Epigenetics May 16, 2010;5(4):301–12.

[60] Zhang X, Germann S, Blus BJ, Khorasanizadeh S, Gaudin V, Jacobsen SE. The Arabidopsis LHP1 protein colocalizes with histone H3 Lys27 trimethylation. Nat Struct Mol Biol September 2007;14(9):869–71.

[61] Turck F, Roudier F, Farrona S, Martin-Magniette ML, Guillaume E, Buisine N, et al. Arabidopsis TFL2/LHP1 specifically associates with genes marked by trimethylation of histone H3 lysine 27. PLoS Genet June 2007;3(6):e86.

[62] Derkacheva M, Steinbach Y, Wildhaber T, Mozgova I, Mahrez W, Nanni P, et al. Arabidopsis MSI1 connects LHP1 to PRC2 complexes. EMBO J July 17, 2013;32(14):2073–85.

Chapter 4

Unraveling the Roles of Canonical and Noncanonical PRC1 Complexes

E.M. Conway[1], A.P. Bracken[1,a]
[1]*Trinity College Dublin, Dublin 2, Ireland*

Chapter Outline

Background 57
The Mechanism of Action of Canonical Polycomb Repressive Complexes 1 59
The Mechanism of Action of Noncanonical Polycomb Repressive Complexes 1 63
The Biological Importance of Canonical Polycomb Repressive Complex 1, Noncanonical Polycomb Repressive Complex 1, and H2A Monoubiquitination 65
Perspectives 72
List of Acronyms and Abbreviations 74
References 75

BACKGROUND

One of the key questions in the field of chromatin biology is how cell type–specific genes are differentially regulated in the wide variety of tissue types found in multicellular organisms. This differential gene expression is a central aspect of the ability of multiple cell types to perform diverse functions, yet utilize the same genetic code. Polycomb group proteins are chromatin regulators, which are centrally involved in this transcriptional maintenance of cellular identity [1].

The PRC1 and PRC2 are the two best characterized types of multisubunit Polycomb complexes [1]. The PRC2 complex is composed of the core proteins, EED, SUZ12, and either the EZH1 or EZH2 histone methyltransferase [1–5]. PRC2 is responsible for all di- and trimethylation of H3K27 (H3K27me2/3), which are both histone posttranslational modifications (PTMs)

a. Senior author

Polycomb Group Proteins. http://dx.doi.org/10.1016/B978-0-12-809737-3.00004-0

associated with transcriptional repression [3–5]. While H3K27me3 is localized to the promoters of genes encoding regulators of developmental and cell fate decisions [6,7], H3K27me2 is found both intergenically and on the gene bodies of transcriptionally silent genes [4,5,8]. In addition to the core PRC2 complex members, several substoichiometric components have been identified, such as JARID2, AEBP2, and PCL1-3, which modulate the recruitment and enzymatic activity of the PRC2 complex [9–13].

The PRC1 complex has a more variable biochemical composition than the PRC2 complex [14,15] (Fig. 4.1). The core of all PRC1 complexes consists of a RING-PCGF heterodimer, which can function as an E3 ubiquitin ligase to monoubiquitinate Histone H2A at lysine 119 (or lysine 118 in *Drosophila*) [16,17]. There are two variants of the RING subunit in mammals, RING1A and RING1B, and each can form a heterodimer with one of six variants of the PCGF subunit, PCGF1–6 [14]. The first identified cPRC1 complexes were those that contain PCGF2 (MEL18) or PCGF4 (BMI1), in addition to Chromobox (CBX2, 4, 6, 7, 8), Sex Combs Midleg (SCMH1, SCML1, SCML2), and Polyhomeotic (PHC1–3) subunits [14,18,19].

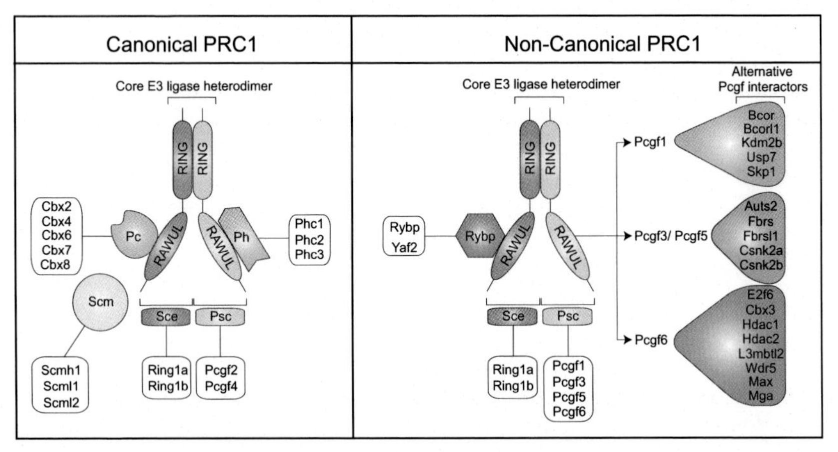

FIGURE 4.1 The biochemical compositions of canonical PRC1 (cPRC1) and noncanonical PRC1 (ncPRC1) complexes. The *Drosophila* PRC1 protein names Pc, Sce, Psc, Ph, and Scm are indicated in the *colored shapes*. The names of their many mammalian homologs are listed in the white boxes, e.g., *Drosophila* Pc is homologous to mammalian Cbx2, Cbx4, Cbx6, Cbx7, and Cbx8. The left panel depicts the protein subunit composition of cPRC1 complexes. The core RING-PCGF heterodimer associates with the Pc and Ph subunits, which are essential for the recruitment of cPRC1 complexes to target genes and subsequent chromatin compaction, respectively. The right panel depicts the protein subunit composition of ncPRC1 complexes. In ncPRC1 complexes, the core RING-PCGF heterodimer associates with either a Rybp or Yaf2 subunit. The different ncPRC1 complexes are defined by their particular PCGF subunit, which interacts with divergent subsets of interacting proteins via its RAWUL domain. For example, Pcgf1 brings Bcor and Kdm2b to its ncPRC1 complex, while Pcgf6 brings E2f6 and L3mbtl2.

THE MECHANISM OF ACTION OF CANONICAL POLYCOMB REPRESSIVE COMPLEXES 1

The precise composition of the cPRC1 complex in each individual cell type appears to depend primarily on the paralogs of each subunit that are expressed in that particular lineage. For example, in embryonic stem (ES) cells, Cbx7 and Pcgf2 are predominantly expressed [20]. However, a switch in the composition of the cPRC1 complex occurs upon differentiation of ES cells such that the Cbx8 and Pcgf4 subunits become the predominantly expressed paralogs, while Cbx7 and Pcgf2 become downregulated [20,21]. Despite the large number of possible cPRC1 complex compositions, it is thought that due to their equivalent biochemistry, they confer the same repressive function in different cell types. Indeed, different cPRC1 complexes, with varying compositions of PHC and CBX paralogs, are known to cooccupy the same target genes in the same cell type [22].

The "hierarchical" model of Polycomb recruitment to target genes explains how the distribution of PRC2-mediated H3K27me3 dictates the genome-wide binding patterns of all cPRC1 complexes. The H3K27me3 modification is specifically recognized or "read" by the chromodomain of the CBX subunit within all cPRC1 complexes, and this leads to the recruitment and consequent colocalization of cPRC1 complexes together with PRC2 and H3K27me3 [18,23–25]. Supporting this model, the majority of H3K27me3- and PRC2-associated genes are also bound by the cPRC1 complex in multiple cell types [7,26–28] (Fig. 4.2). The specificity of the CBX chromodomain was demonstrated in vivo through elegant overexpression studies by the Eissenberg and Khorasanizadeh groups, in which they replaced the chromodomain of dHp1, a chromodomain containing protein known to read heterochromatic H3K9me3, with the chromodomain of the *Drosophila* homolog of the Chromobox subunit, dPc [18]. This domain swap led to the complete relocalization of mutant dHp1 to heterochromatin marked by H3K27me3, while the reciprocal domain swap led to relocalization of mutant dPc to chromatin marked by H3K9me3. Taken together, these experiments clearly established the specificity of the chromodomain of dPc within the cPRC1 complex for H3K27me3 [18,24].

There is some evidence that DNA-binding transcription factors may also contribute to the recruitment of the cPRC1 complex to its target genes independently of H3K27me3. For example, several studies have suggested that the physical association of the cPRC1 complex with transcription factors, such as REST, ZFP277, RUNX1, and CBFβ, could contribute to its recruitment to chromatin [29–32]. All these papers report both physical and functional links between the cPRC1 complex and the transcription factors and present models in which they could directly recruit the cPRC1 complex through their DNA-binding capabilities. However, it is unclear whether the recruitment of the cPRC1 complex by these transcription factors is independent of the CBX

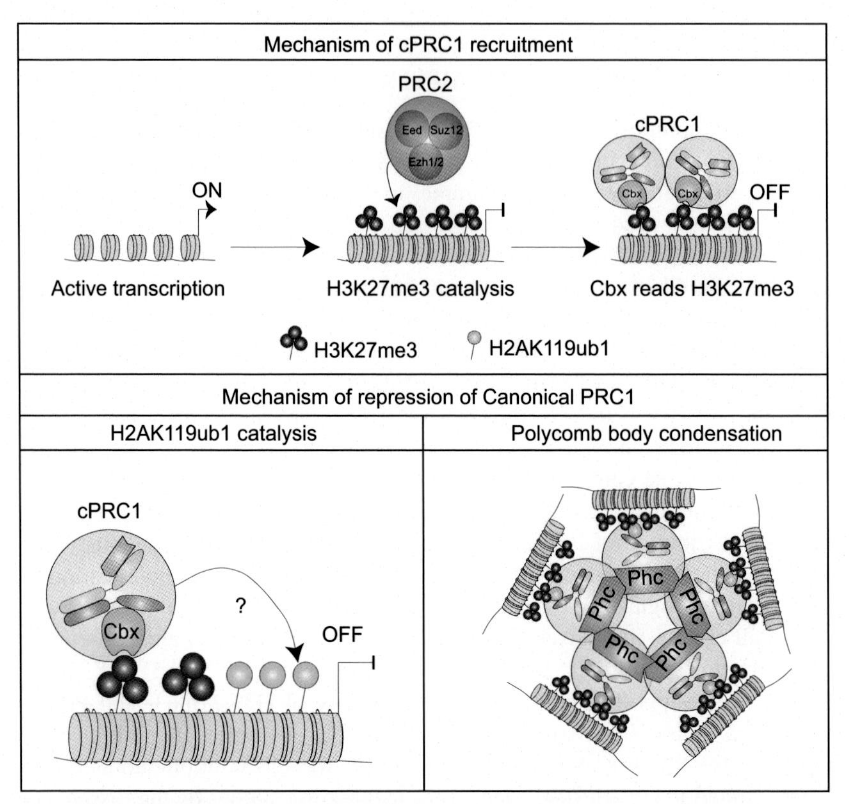

FIGURE 4.2 Recruitment of canonical PRC1 (cPRC1) complex to chromatin and mechanism of gene repression. The top panel depicts how the PRC2 complex catalyzes trimethylation of H3 at lysine 27 (dark blue) and leads to the recruitment of the cPRC1 complex via the ability of its chromodomain containing CBX subunits (light blue) to bind specifically to "read" this mark. The bottom left panel depicts how the cPRC1 complex, containing a Ring-Pcgf (pink and green) heterodimer is capable of monoubiquitinating H2A in vitro (light green). However, it is unclear if this reaction occurs in vivo. The bottom right panel depicts how the cPRC1 complex is understood to mediate gene repression. It confers chromatin condensation and consequent gene repression by the head to tail oligomerization of the Polyhomeotic (orange) subunit of the complex.

subunits and H3K27me3. An alternative explanation is that the transcription factors switch off transcription at their respective target genes and that this could then indirectly lead to an increase of PRC2-mediated H3K27me3 and the consequent recruitment of the cPRC1 complex. Supporting this, the absence of RNA polymerase II elongation activity has been reported to be sufficient to recruit the PRC2 complex to unmethylated CpG islands near the transcription start sites of silent genes [33]. Therefore, the REST, ZFP277, RUNX1, and CBFβ repressors could block transcription, leading to the accumulation of PRC2-mediated H3K27me3 and subsequent cPRC1 recruitment. Furthermore, the reported physical associations between cPRC1 and the transcription factors could be a consequence of their shared occupancy on nucleosomes and therefore be coimmunoprecipitated due to the presence of "bridging" chromatin. In summary, while it is possible the DNA-binding repressors can contribute to the stabilization of cPRC1 at target sites, H3K27me3 is believed to be the dominant driver of cPRC1 recruitment to chromatin.

The cPRC1 complex has a well-established role in the maintenance of transcriptional repression; however, the precise mechanisms by which it mediates this function are not completely understood. It is clear that regions of chromatin that are occupied by cPRC1 form compact regions, referred to as Polycomb bodies [19,34,35]. These Polycomb bodies are higher order chromatin structures in which distal chromatin regions form long-range interactions in a cluster of repressed, compact nucleosomes encompassing multiple Polycomb-binding sites [19,34,35]. Until recently, the RING-PCGF-mediated H2AK119ub1 was thought to be how cPRC1 would somehow mark chromatin for compaction [17,36]. However, we now know that this is unlikely the case as the ncPRC1 complexes confer the majority of genome-wide H2AK119ub1 [15,37–40]. We now know that the PHC1–3 subunits within cPRC1 complexes facilitate this chromatin compaction via oligomerization with other PHC subunits at other repressed target genes drawing distal Polycomb repressed genes into condensed nuclear clusters called Polycomb bodies [19,34,35]. The PHC1–3 subunits are capable of this chromatin compaction as they contain a sterile alpha motif (SAM) domain with two interaction regions, an end helix (EH) and a midloop/helix (ML). The EH of one Polyhomeotic protein hetero/homooligomerizes with another Polyhomeotic protein at the ML helix to enable the formation of a "chain" of helical Polyhomeotic subunits, thereby forming the compact structure of a Polycomb body [19,41,42] (Fig. 4.2).

The formation of Polycomb bodies is understood to be the primary mechanism by which cPRC1 complexes mediate transcriptional repression. For example, they are particularly evident at repressed *HOX* loci where DNA-FISH has established that distal *HOX* genes can associate strongly with one another [19,35]. The Koseki laboratory used two color DNA-FISH in the presence of wild-type Phc2 and the SAM domain mutant Phc2 to show that loss of SAM domain function causes cPRC1 to lose the ability to compact distal *HOX* genes, *HoxB1* and *HoxB13* [19]. This loss of Polycomb body formation was accompanied by derepression of *HoxB1* and *HoxB13* transcription in mouse embryonic fibroblasts [19,42]. Transient expression of the SAM domain mutant Phc2 led to a reduction in Ring1b occupancy, as would be expected due to the loss of the ability of cPRC1 to oligomerize with other cPRC1 complexes within Polycomb bodies. However, surprisingly, it also led to a decrease in the levels of the H3K27me3 PTM at the *HoxB* loci [19]. This suggests that although cPRC1 is recruited to chromatin after PRC2 has catalyzed H3K27me3, it appears to be required for the continued maintenance of

H3K27me3 and subsequent gene repression, perhaps through steric hindrance in Polycomb bodies of transcriptional activators or Histone H3K27 demethylases. An intriguing recent study by the Elderkin group discovered that, within Polycomb bodies, promoter–promoter contacts are maintained between repressed genes, not only *in cis* but also *in trans* [35]. For example, in ES cells, multiple genes within the same *HoxA* locus on chromosome 6 cluster strongly with each other but also with genes from the *HoxD* locus on chromosome 2 [35].

It is becoming clear that the potential of cPRC1 complexes to mediate H2AK119ub1 via their RING-PCGF heterodimer is not the primary mechanism by which they confer gene repression [15]. Supporting this, in *Drosophila*, the depletion of the cPRC1-specific components, dPc and dPh, does not affect the global levels of H2A ubiquitination [43]. Therefore, despite the fact that several studies have shown that a reconstituted cPRC1 complex is capable of ubiquitinating H2A at lysine 119 in vitro [17,36], the cPRC1 complex may not rely on this catalytic activity to confer gene repression (Fig. 4.2). However, in mammals, RNAi-mediated depletion of PCGF4 has been reported to lead to a global reduction in the levels of H2AK119ub1 in HeLa cells [44]. Despite this, no studies have directly linked the catalytic activity of PCGF2 or PCGF4 to the repression of Polycomb target genes, and the effects observed following PCGF4 depletion could be indirect effects due to downstream disruption of ncPRC1 or PRC2 function. Further supporting the idea that the cPRC1 complex primarily functions via chromatin compaction, the Klose group reported that Pcgf2 and Pcgf4 are less capable of mediating H2AK119ub1 activity at an artificial endogenous chromatin locus compared to Pcgf1, Pcgf3, and Pcgf5 [15]. To do this, they fused a TET tag to each PCGF protein, which allowed recruitment to a specific genomic region containing a TET operator. The artificial targeting of Pcgf2 and Pcgf4 led to the recruitment of Ring1b, but did not lead to any increase in the level of H2AK119ub1 around the TET operator locus. They also generated a fusion protein composed of the minimal catalytic region of the Pcgf4 and Ring1b proteins, incapable of interacting with Chromobox and Polyhomeotic subunits, but still capable of ubiquitinating H2A [15]. In contrast to the full-length Pcgf4 protein, the recruitment of the Pcgf4-Ring1b minimal catalytic domain to the TET operator led to a large increase in H2AK119ub1 [15]. Therefore, it seems that Pcgf4 is capable of catalytic activity in vivo, but only in the absence of the CBX and PHC subunits of the cPRC1 complex. These results suggest that there is a subunit within the cPRC1 complex, which inhibits the catalytic activity of the PCGF-RING heterodimer, potentially via steric hindrance due to the chromatin condensation mediated by the PHC subunits. It is possible that the RING and PCGF heterodimers within the cPRC1 complex are primarily required for maintaining structural contacts with the CBX and PHC subunits and with the nucleosome [16], rather than for its ability to mediate H2AK119 ubiquitination. Supporting this, the Bickmore group reported that a catalytic mutant of

Ring1b can rescue the loss of Polycomb body formation observed in *Ring1b* null cells and *HOX* gene repression to the same extent as wild-type *Ring1b* [34,45]. Similarly, in *Drosophila*, the generation of a total knockout of the RING homolog (*dSce*) leads to anteriorization of *HOX* gene expression, characteristic of loss of Polycomb phenotypes; however, a catalytic mutant of *dSce* does not lead to abnormalities in the position of *HOX* gene expression [46]. Taken together, these results suggest that the repressive function of the cPRC1 complex is not dependent on H2Aub catalytic activity.

THE MECHANISM OF ACTION OF NONCANONICAL POLYCOMB REPRESSIVE COMPLEXES 1

More recently, several variant or ncPRC1 complexes have been identified [14,47–49]. These complexes are functionally and biochemically distinct because they lack the CBX and PHC subunits, which are essential for targeting cPRC1 complexes and for mediating Polycomb body formation, respectively (Fig. 4.1). Instead, ncPRC1 complexes contain either an RYBP or YAF2 subunit, close paralogs of each other that interact directly with either RING1A or RING1B. Both RYBP and YAF2 compete for the same interaction pocket as CBX subunits on the surface of RING1A/B. Therefore the interaction between RING1A/B with RYBP/YAF2 is mutually exclusive of the interaction between RING1A/B and CBX proteins [50,51]. The obvious implication of lacking the CBX and PHC subunits is that ncPRC1 complexes cannot be recruited to chromatin or confer chromatin compaction through the same mechanisms of cPRC1 complexes.

Another distinctive feature of ncPRC1 complexes is that they can contain any of PCGF1, PCGF2, PCGF3, PCGF4, PCGF5, or PCGF6 (Fig. 4.1), each of which associate with specific additional subunits. For example, the PCGF6-containing ncPRC1 complexes are composed of several transcription factors (e.g., E2F6, MAX, and MGA), chromatin readers (e.g., CBX3), chromatin modifiers (e.g., HDAC1 and HDAC2) as well as other subunits [14,47,52–54]. The PCGF6-containing ncPRC1 complex has been shown to repress transcription by an unknown mechanism upon artificial recruitment to a Luciferase promoter [47]. The PCGF3- and PCGF5-containing ncPRC1 complexes share many of their associated factors, including FBRS, FBRSL1, AUTS2, CSNK2A, and CSNK2B, and are therefore of particular interest because the *AUTS2* gene is frequently mutated in autism spectrum disorder [14,48]. However, the role of PCGF3 and PCGF5 and their functional interplay with the AUTS2 protein is unclear. Surprisingly, the PCGF5-containing ncPRC1 complex has been reported to have a unique ability among ncPRC1 complexes to promote active transcription, albeit following artificial recruitment to a reporter promoter [48]. It is also likely that ncPRC1 complexes defined by PCGF2 and PCGF4 exist because both RYBP and YAF2 can be immunoprecipitated by PCGF2 and PCGF4 [14,40]. However, the specific complex compositions of such PCGF2/4 ncPRC1 complexes remain to be elucidated. The PCGF1-ncPRC1 complex is the best characterized ncPRC1 complex [14,37,38,49] and is composed of a specific set of subunits including the lysine demethylase KDM2B together with BCOR, BCORL1, SKP1, and USP7 (Fig. 4.1). The KDM2B subunit is essential for the recruitment of the PCGF1-containing ncPRC1 complexes to chromatin [37,38]. It functions by binding to unmethylated CpG islands throughout the genome via its CxxC motif and then recruiting the PCGF1-ncPRC1 complex to these sites where it can then monoubiquitinate H2AK119 [37,38].

In addition to the fact that all ncPRC1 complexes lack a CBX subunit and that Pcgf1-ncPRC1 is recruited to chromatin by KDM2B, several additional lines of evidence further support the idea that ncPRC1 complexes are recruited independently of PRC2. For example, both Rybp and H2AK119ub1 levels are unaffected at Polycomb target genes in the absence of Eed and H3K27me3 [40]. This suggests that H2AK119 ubiquitination and Rybp recruitment occurs independently of H3K27 trimethylation, possibly even prior to PRC2 association [40]. Supporting this latter idea, the artificial targeting of the ncPRC1 complex to chromatin is sufficient to recruit the core components of the PRC2 complex, which then leads to accumulation of H3K27me3 [15]. Furthermore, the artificial recruitment of Kdm2b alone is sufficient to lead to the deposition of H3K27me3 to de novo sites of H2AK119ub1, suggesting that the H2AK119ub1 mark might contribute to PRC2 recruitment [15,55]. Supporting this, the Klose group reported that ES cells with a mutation in the CxxC DNA-binding motif of Kdm2b have reduced Suz12 and H3K27me3 on Polycomb target genes [38]. In contrast, the Helin group found that following RNAi depletion of Kdm2b in ES cells, Suz12 and H3K27me3 are not reduced at a cohort of Polycomb target genes [37]. This apparent discrepancy might be due to the fact that the two groups examined different genes [15,37]. What is clear is that the ability of Kdm2b to promote the PRC2 complex association with chromatin is dependent on the presence of Pcgf1, suggesting that its H2AK119ub1 activity is required [15,38]. Intriguingly, the Müller group discovered that the PRC2 complex can be coimmunoprecipitated with nucleosomes marked by H2A monoubiquitination [13]. This result suggests that a PRC2-associated protein might confer the ability to "read" the H2AK119ub1 PTM to either recruit or stabilize the PRC2 complex chromatin or perhaps promote its activity. Supporting the latter, they showed that mononucleosomes containing ubiquitinated H2A increased the efficiency of PRC2-mediated H3K27me3 catalysis, in the presence of the PRC2 substoichiometric subunits JARID2 and AEBP2.

Importantly, these recent advances in our understanding of the function of ncPRC1 complexes have been proposed to suggest an alternative model of PRC2 targeting to chromatin [56]. In this model, unmethylated CpG islands bound by KDM2B, in the absence of elongating transcription, facilitate the recruitment of the PCGF1-containing ncPRC1 complex (Fig. 4.3). This

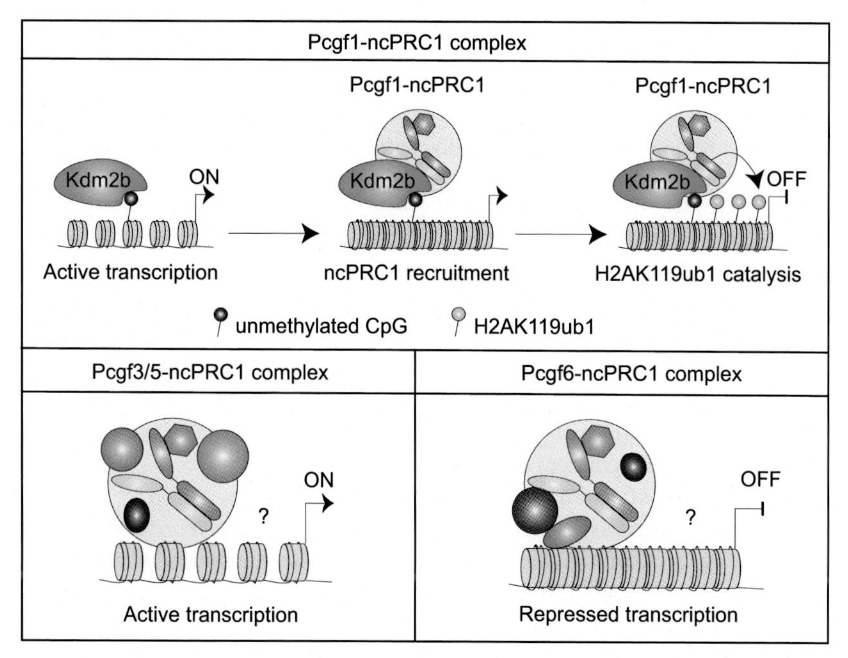

FIGURE 4.3 Mechanisms of action of noncanonical PRC1 (ncPRC1) complexes. The top panel depicts the model of recruitment and H2A monoubiquitination mediated by the Pcgf1-ncPRC1 complex. Kdm2b (blue) binds to chromatin at regions containing unmethylated CpG islands (red) and thereby recruits the Pcgf1-containing ncPRC1 complex leading to the catalysis of H2A monoubiquitination (light green). The bottom left panel depicts the Pcgf3- and Pcgf5-ncPRC1 complexes which have been postulated to be associated with gene activation. The bottom right panel represents the Pcgf6-containing ncPRC1 complex which has been reported to act as a transcriptional repressor by an unknown mechanism.

ncPRC1 complex then catalyzes monoubiquitination of H2A, which is then somehow "read" by the PRC2 complex leading to the accumulation of H3K27me3. Then the H3K27me3 PTM is recognized by the chromodomain of the CBX subunits of cPRC1 resulting in chromatin compaction due to the oligomerization of PHC subunits within cPRC1, leading to the formation of Polycomb bodies and consequent gene repression. However, while this model might be true at some genes, as we will discuss below, it is unlikely to be true at all Polycomb target genes.

THE BIOLOGICAL IMPORTANCE OF CANONICAL POLYCOMB REPRESSIVE COMPLEX 1, NONCANONICAL POLYCOMB REPRESSIVE COMPLEX 1, AND H2A MONOUBIQUITINATION

The cPRC1 and ncPRC1 complexes are believed to have overlapping and distinct roles in the regulation of transcription in metazoans, and this is likely due to the fact that they share the Ring and Pcgf subunits [39]. The first Polycomb protein identified, *dPc*, a subunit unique to the cPRC1 complex, was identified as being required for the correct maintenance of homeotic gene repression in *Drosophila* [57,58]. The loss of *dPc* function led to an embryonic lethal phenotype and homeotic transformations and altered cellular identity in specific segments along the anterior posterior axis of the *Drosophila* larvae due to failure to repress Homeotic (*HOX*) genes. The loss of function of another cPRC1-specific subunit, the Polyhomeotic subunit homolog (dPh), displayed a similar phenotype to the *dPc* mutants [59]. We now know that the aberrant expression of *HOX* genes in the *dPc* and *dPh* mutant larvae outside of the normal anterior expression boundaries is due to a loss of the ability of the cPRC1 complex to maintain their repression in these regions. The embryonic lethal phenotypes of *dPc* and *dPh* mutants revealed for the first time that the cPRC1 complex is essential for the maintenance of correct transcriptional programs during *Drosophila* development.

Each subunit of the mammalian cPRC1 complex has multiple paralogs, suggesting that there is a degree of functional redundancy (Fig. 4.1). This is supported by knockout studies in mice. For example, single knockouts of the Polyhomeotic paralogs *Phc1* and *Phc2* lead to homeotic transformations coupled with perinatal lethality, for *Phc1*, and viable offspring, for *Phc2* [60,61]. These are rather mild phenotypes compared to the *dPc* and *dPh* mutants [59]. However, when both *Phc1* and *Phc2* are knocked out simultaneously, there is a stronger phenotype of embryonic lethality [61]. This synergy between *Phc1* and *Phc2*, coupled with their common biochemical function, shows that they share some functions. Another example is the multiple mammalian paralogs of the *dPC* protein (CBX2, CBX4, CBX6, CBX7, and CBX8). The single knockout of *Cbx2* leads to either perinatal or postnatal lethality, in addition to homeotic transformations [62]. This phenotype is relatively mild most likely due to redundancy with the other CBX proteins. However, the eventual lethal phenotype of *Cbx2* null mice suggests it is essential in at least some tissue types later in development. *Cbx4* knockout mice also exhibit perinatal lethality, but do not appear to have homeotic transformations or derepression of *HOX* genes, again suggesting functional redundancy with other CBX proteins [63]. The *Cbx7* gene is highly expressed in ES cells and appears to be required for maintaining the repression of *HOX* and other Polycomb target genes in these cells [20]. Surprisingly, while *Cbx7* knockout mice are viable and survive into adulthood, they have an increased susceptibility to tumor formation of the liver and lung [64]. The survival of these mice can likely be attributed to functional redundancy among the Chromobox proteins, particularly as it has been demonstrated that depletion of Cbx7 leads to increased expression of *Cbx4* and *Cbx8* genes [20], suggesting they may compensate in certain tissues of the developing *Cbx7* knockout mouse.

The biology of PCGF2 and PCGF4 is more complicated than that of the CBX and PHC subunits of cPRC1 because they may also form part of ncPRC1 complexes. The knockout of *Pcgf2* results in postnatal lethality [65], while the *Pcgf4* knockout results in both perinatal and postnatal lethality [66]. However, the double deletion of *Pcgf2* and *Pcgf4* leads to an embryonic lethal phenotype with more severe homeotic transformations [67], suggesting a functional redundancy between the two proteins in mammals. Taken together with the phenotype of the double knockout of *Phc1* and *Phc2,* it is clear that the function of the cPRC1 complex is essential during embryonic development in mammals for maintaining the repression of development genes such as *HOX* genes [61,67]. The conditional knockout of *Pcgf4* in tissue specific stem cells such as neural (NSC), mammary (MSC), and hematopoietic (HSC) leads to their failure to self-renew [68–70]. This defect in turn leads to pronounced biological consequences such as growth retardation in the neural lineage and failure to repopulate the hematopoietic system when knocked out in HSCs [68,69]. Seminal work by the Van Lohuizen lab identified the *Cdkn2a* gene locus as being dependent on Pcgf4 for its sustained repression [71], and it was subsequently shown to be a direct target gene [6]. The *Cdkn2a* gene locus encodes two distinct proteins generated via two alternatively spliced transcripts that are translated in different reading frames [72]. The *Ink4a* transcript encodes the p16 protein, a key upstream regulator of cellular senescence [72], while the *Arf* transcript encodes the p14 (or p19 in mouse) protein that stabilizes p53 and thereby promotes cell cycle arrest and apoptosis [72]. In the absence of Pcgf4, the *Cdkn2a* gene locus is derepressed, and both the p16 and Arf proteins accumulate resulting in defects in stem cell renewal due to the premature onset of apoptosis and senescence [69,70,73]. A double knockout of *Ink4a-Arf* and *Pcgf4* is sufficient to partially rescue this self-renewal defect in both neural and mammary stem cells [70,73]. The lack of a complete rescue is most likely due to the requirement of the Pcgf4 subunit in the cPRC1 complex to maintain the repression of many hundreds of other Polycomb target genes.

The interpretation of the phenotypes of loss of RING function is complicated due to the fact that they are central components of both the cPRC1 and ncPRC1 complexes. Mammals have two paralogs of the *Drosophila* dSce protein, both capable of forming the RING-PCGF heterodimer. Like Pcgf2 and Pcgf4, Ring1a and Ring1b are required for proper embryonic development [74–76]. The knockout of the *Ring1a* gene alone in mice results in viable offspring with homeotic transformations of the axial skeleton [74]. In contrast, the knockout of *Ring1b* leads to an embryonic lethal phenotype combined with a global reduction of ubiquitinated H2A [75,76]. Interestingly, in an effort to elucidate the importance of Ring1b in maintaining the structure of the cPRC1 complex compared to its catalytic function as an E3 ligase enzyme, the Bickmore group generated a mouse "knock in" of a catalytic mutant of Ring1b incapable of mediating its E3 ubiquitin ligase activity [77]. These catalytic mutant *Ring1b* mice also exhibited an embryonic lethal phenotype demonstrating that the ability of Ring1b to ubiquitinate its substrates is crucial for survival. However, these mice survived to a later embryonic stage than the complete *Ring1b* knockout, and they did not exhibit homeotic transformations. This suggests that while H2A ubiquitination is essential for survival, it may not be essential for the repression of the *HOX* genes. This possibility is supported by the phenotype of the double knockout of *Ring1a* and *Ring1b* in intestinal stem cells, which results in defective villi and dysregulated crypt homeostasis [78]. Unlike the *Pcgf4* knockout mice, this *Ring1a/Ring1b* null phenotype cannot be partially rescued by the deletion of the *Ink4A/Arf* locus, suggesting that essential target genes in addition to the cPRC1 target gene *Cdkn2a* are deregulated in the absence of Ring1a / b.

In *Drosophila*, the knockout of the dSce, the sole homolog of the mammalian Ring1a/b proteins, leads to a derepression of the *HOX* genes, similar to the knockout phenotypes of *dPh* and *dPc* [46]. However, recent work from the Müller group used a catalytic mutant of dSce to delineate the roles of cPRC1 compared to ncPRC1 [46] They observed that despite the loss of dSce catalytic activity, and subsequent global reduction of H2A ubiquitination, there was no derepression of the *HOX* genes [46]. However, consistent with the similar experiment in mice, the *dSce* catalytic mutant flies displayed late-stage embryonic lethality [46], demonstrating again that the catalytic activity of ncPRC1 complexes is essential for organism viability. In yet another effort to evaluate the importance of H2A ubiquitination for development and Polycomb gene repression, Pengelly and colleagues generated flies with mutations of the four lysine residues present in *Drosophila* H2A, thereby preventing all H2A ubiquitination [46]. This resulted in an embryonic lethal phenotype, but did not lead to any detectable derepression of *HOX* genes. This is again in line with the apparent *HOX*-independent role of the ncPRC1 complex in *Drosophila*. Taken together, these results reveal that the global loss of H2A ubiquitination results in a *HOX*-independent embryonic lethal phenotype. Therefore, despite the fact that *HOX* gene repression was maintained in the absence of H2A ubiquitination, it is evident that the H2A ubiquitination mark is absolutely essential for embryogenesis in both *Drosophila* and mice.

Several studies have knocked out specific subunits of the ncPRC1 complex (Table 4.1). For example, knockout of *Rybp* revealed that it is essential for correct embryonic development and the generation of viable offspring in both *Drosophila* and mice [79,80]. However, further supporting the idea that ncPRC1 is not required for *HOX* gene repression, the *Drosophila* embryos lacking a functional *dRybp* gene do not display the classic cPRC1 complex loss of function phenotype of homeotic transformations. Furthermore, the knockout of *dKdm2* in *Drosophila*, similar to the *dRybp* knockout phenotype, does not correlate with homeotic transformations or *HOX* gene derepression [43]. These results are consistent with the fact that there is an absence of H2A ubiquitination on the *Hox* loci in *Drosophila*, despite the strong chromatin association of cPRC1 and PRC2 complexes [8]. However, mice with a

TABLE 4.1 List of Knockout and Mutagenesis Studies of PRC1 Complex Subunits and H2A Genes With Their Associated Phenotypes

Gene (Reference)	In Vivo Phenotype	Effect on *Hox* Genes	Effect on H2A Ubiquitination
Shared cPRC1 and ncPRC1 Complex			
dSce knockout [46]	Unknown	Derepression	Global reduction
dSce catalytic mutant [46]	Embryonic lethal	Normal expression	Global reduction
dPsc knockout [89]	Embryonic lethal	Homeotic transformation	Unknown
mRing1a knockout [74]	Viable	Homeotic transformation	Unknown
mRing1b knockout [75,76]	Embryonic lethal	No homeotic transformation	Global reduction
mRing1b catalytic mutant [77]	Embryonic lethal	No homeotic transformation	Global reduction
cPRC1 Complex Only			
dPc knockout [57,58]	Embryonic lethal	Derepression	Unknown
dPh knockout [59]	Embryonic lethal	Homeotic transformation	Unknown
mPcgf2 knockout [65]	Postnatal lethal	Homeotic transformation	Unknown
mPcgf4 knockout [66]	Perinatal and postnatal lethal	Homeotic transformation	Unknown
mPcgf2 and *mPcgf4* double knockout [67]	Embryonic lethal	Homeotic transformation	Unknown

Continued

TABLE 4.1 List of Knockout and Mutagenesis Studies of PRC1 Complex Subunits and H2A Genes With Their Associated Phenotypes—cont'd

Gene (Reference)	In Vivo Phenotype	Effect on *Hox* Genes	Effect on H2A Ubiquitination
mCbx2 knockout [62]	Perinatal and postnatal lethal	Homeotic transformation	Unknown
mCbx4 knockout [63]	Perinatal lethal	No homeotic transformation	Unknown
mCbx7 knockout [64]	Susceptible to tumor formation	Unknown	Unknown
mPhc1 knockout [60]	Perinatal lethal	Homeotic transformation	Unknown
mPhc2 knockout [61]	Viable	Homeotic transformation	Unknown
mPhc1 and *mPhc2* double knockout [61]	Embryonic lethal	Homeotic transformation	Unknown
mScmh1 knockout [90]	Viable and fertile	No homeotic transformation	Unknown
mScml2 knockout [91]	Viable with fertility defects	Unknown	Unknown
ncPRC1 Complex Only			
dRybp knockout [80]	Embryonic lethal	No homeotic transformation	Unknown
mRybp knockout [79]	Embryonic lethal	Unknown	Unknown

mKdm2b CxxC mutant [15]	Embryonic lethal	Homeotic transformation	Global reduction
mUsp7 knockout [92]	Embryonic lethal	Unknown	Unknown
E2f6 knockout [93]	Viable and fertile	Homeotic transformation	Unknown
Cbx3 knockout [94]	Embryonic lethal	Unknown	Unknown
Hdac1 knockout [95]	Embryonic lethal	Unknown	Unknown
Hdac2 knockout [96]	Perinatal lethal	Unknown	Unknown
L3mbtl2 knockout [54]	Embryonic lethal	Unknown	No change
Max knockout [97]	Embryonic lethal	Unknown	Unknown
Histone 2A lysine Mutants			
dH2A quadruple lysine mutant [46]	Embryonic lethal	Normal repression of *Hox* genes	Unknown
$dH2A_V$ double lysine mutant [46]	Viable and fertile	Unknown	Unknown
dH2A and $dH2A_V$ multiple lysine mutants [46]	Unknown	Normal repression of *Hox* genes	Global reduction

homozygous deletion of the Kdm2b CxxC region are embryonic lethal and display homeotic transformations [15]. This result suggests that the PCGF1-ncPRC1 complex may contribute to PRC2 function and thereby potentially also contribute to the repression of *HOX* genes in mammals. Taken together, these data demonstrate that the recruitment of PCGF1-containing ncPRC1 complexes is essential for H2AK119 ubiquitination and may also contribute to PRC2 recruitment and/or stability at specific target genes, including *HOX* genes. These data also suggest that the cPRC1 and ncPRC1 complex can function at both shared and independent target genes. This latter hypothesis is supported by chromatin immunoprecipitation sequencing (ChIP-seq) studies which identified both independent and shared target genes for the ncPRC1 component Rybp and the cPRC1 complex component Cbx7 in ES cells [39] and similarly, for Pcgf1 and Cbx2 in Leukemic cells [81].

The genetic studies of the cPRC1 and ncPRC1 complex components and histone 2A paint a picture in which the cPRC1 and ncPRC1 complexes share the capacity to repress *HOX* genes, but in the majority of cases, perturbations of cPRC1 lead to stronger homeotic phenotypes than perturbations of ncPRC1. Nonetheless, both complexes are essential for both mouse and *Drosophila* development and viability.

PERSPECTIVES

It is clear that both the cPRC1 and ncPRC1 complexes are essential for the control of correct lineage specification during embryonic development and differentiation (Table 4.1). The loss of cPRC1 complex subunits leads to homeotic gene derepression and developmental defects in both mice and *Drosophila* embryos [57–59,61,67]. Emerging studies suggest that the ncPRC1 complexes and H2Aub are not essential for *HOX* gene repression [43,46,77,80], but they are nonetheless essential for global transcriptional homeostasis and organism viability [15,46,77]. It appears that the oligomerization of the PHC subunits and subsequent Polycomb body formation is the primary mechanism by which cPRC1 contributes to gene repression rather than through H2A monoubiquitination. However, an important open question is how does ncPRC1-mediated H2A monoubiquitination contribute to transcriptional control and development?

A number of mechanistic questions remain unanswered regarding cPRC1- and ncPRC1-mediated transcriptional regulation. For example, it is unclear if the cPRC1 complex has a catalytic activity in vivo and, if so, if this activity is necessary for its repressive function. A good way to address this question would be to mutagenize both the *PCGF2* and *PCGF4* genes such that their respective proteins are incapable of acting as E3 ligases, but still capable of forming the RING-PCGF heterodimer. This question is particularly relevant as selective inhibitors have been developed that inhibit PCGF4 stability as a potential treatment of cancer [82]. Another outstanding question is what role do the PCGF3/5/6-ncPRC1 complexes play in regulating transcription at their target genes? For instance, Pcgf6 is the most abundant Pcgf in ES cells, making it intriguing to explore its role in these cells and its potential interplay with other Polycomb complexes [21,83,84]. It has recently been shown that, unlike any other PRC1 complex component, Pcgf6 is required for the maintenance of ES cell pluripotency. This hints at a unique mechanism of target gene regulation that has yet to be defined [85], and we speculate that it will likely require one or more of its subunits that have DNA-binding activity, including E2F6, MAX, and MGA. Furthermore, genome-wide chromatin occupancy studies are required to identify the target genes of Pcgf3, Pcgf5, and Pcgf6. This would allow comparison with other ChIP-seq data for other components of the cPRC1, ncPRC1, and PRC2 complexes as well as their associated histone PTMs, H2AK119ub1 and H3K27me3. Furthermore, ChIP-seq data of these noncanonical PCGFs, when coupled with expression data from PCGF knockout cells could help to elucidate which PCGFs play a role in transcription activation and which are involved in transcriptional repression, as well as whether or not these complexes mediate H2AK119ub1 in vivo at their respective target genes.

The final, and perhaps most debatable, question is whether H2AK119ub1 contributes to the recruitment, stabilization, and/or increased activity of the PRC2 complex on target genes in vivo. While addressing this will require further studies, it is already clear that ectopically expressed Pcgf1, Pcgf3, Pcgf5 are all capable of recruiting PRC2 to chromatin [15]. Furthermore, the ectopic expression of a RING-PCGF dimer capable of mediating H2Aub, but incapable of associating with cPRC1-specific proteins, is also capable of recruiting the PRC2 complex to chromatin [15]. The discovery that monoubiquitinated H2A in nucleosomes is sufficient to copurify the entire PRC2 complex reinforces the potential link between ncPRC1 and PRC2 complexes [13]. On the other hand, there is no reduction in the global levels of H3K27me3 in *Ring1a/b* or *Sce* knockouts or in catalytic mutants of *Ring1b* and *Sce* [46,78]. This suggests that the majority of PRC2 catalytic activity is not dependent on H2Aub [46,77,78]. However, a potential explanation for this is that the ncPRC1 complex could be responsible for PRC2 recruitment at specific loci at a minority of their many target genes. This hypothesis is supported by the existence of separate as well as shared target genes for the cPRC1 complex and the ncPRC1 complex [39] together with observations that both the catalytic activity of Ring1b and the CxxC domain of Kdm2b are essential for PRC2 occupancy at specific target genes in ChIP experiments [15,77].

The function of the PRC2 complex is deregulated in multiple cancer types [4]. Therefore, the potential functional interplay between the H2AK119ub1 PTM, mediated by ncPRC1, and PRC2 may be another way to deregulate H3K27me3 in cancer [4]. Therefore, it appears that a complete understanding of the molecular functions of ncPRC1 complexes will be vital to fully understand the biology of PRC2 in both development and cancer. This knowledge will be important in designing and optimizing potential therapies that target PRC2 activity in cancer, as well as for progressing our understanding of the function of the Polycombs in cell fate decisions and development [86–88].

LIST OF ACRONYMS AND ABBREVIATIONS

AUTS2 Autism susceptibility candidate 2
BCOR BCL6 corepressor
BCORL1 BCL6 corepressor like 1
CBFβ Core-binding factor β
CBX Chromobox
Ccne1 Cyclin E1
Cdkn2A Cyclin-dependent kinase inhibitor 2A
ChIP Chromatin immunoprecipitation
CpG Cytosine-phosphate-guanine
cPRC1 Canonical Polycomb repressive complex 1
CSNK2A Casein kinase 2A
CSNK2B Casein kinase 2B
E2F6 E2f transcription factor 6
E3 ligase Ubiquitin ligase
EED Embryonic ectoderm development
EH End helix
ES Embryonic stem
EZH1/2 Enhancer of zeste 1/2
FBRS Fibrosin
FBRSL1 Fibrosin like 1
FISH Fluorescent in situ hybridization
H2A Histone 2A
H2AK119ub1 Histone 2A lysine 119 monoubiquitination
HDAC Histone deacetylase
HEK293T Human embryonic kidney cells (with SV40 T antigen)
***Hox* genes** Homeobox genes
Hp1 Heterochromatin protein 1
HSC Hematopoietic stem cell
Ink4a Inhibitor of cyclin-dependent kinase 4
KDM2 Lysine demethylase 2
MEFs Mouse embryonic fibroblasts
MGA Max gene–associated protein
ML Midloop/helix
MSC Mammary stem cell
ncPRC1 Noncanonical Polycomb repressive complex 1
NSC Neural stem cell
Pc Polycomb (*Drosophila* homolog of Chromoboxes)
PCGF/Psc Polycomb Group Ring Finger/Posterior Sex Combs
Ph/Phc Polyhomeotic
pRb Retinoblastoma protein
PRC1 Polycomb repressive complex 1
PRC2 Polycomb repressive complex 2

RAWUL Ring finger- and WD40-associated ubiquitin-like domain
REST RE1 silencing transcription factor
RING Really interesting new gene
RNA PolII Ribonucleic acid polymerase II
RNAi Ribonucleic acid interference
RUNX1 Runt-related transcription factor 1
RYBP Ring1- and Yy1-binding protein
SAM Sterile alpha motif
Sce Sex Combs Extra
SKP1 S-phase kinase–associated protein 1
SUZ12 Suppressor of zeste 12 homolog
USP7 Ubiquitin-specific peptidase 7
YAF2 Yy1-associated factor 2
ZFP277 Zinc finger protein 277

REFERENCES

[1] Margueron R, Reinberg D. The Polycomb complex PRC2 and its mark in life. Nature 2011;469(7330):343–9.

[2] Cao R, Wang L, Wang H, Xia L, Erdjument-Bromage H, Tempst P, et al. Role of histone H3 lysine 27 methylation in Polycomb-group silencing. Science 2002;298(5595):1039–43.

[3] Rastelli L, Chan CS, Pirrotta V. Related chromosome binding sites for zeste, suppressors of zeste and Polycomb group proteins in Drosophila and their dependence on Enhancer of zeste function. EMBO J 1993;12(4):1513–22.

[4] Conway E, Healy E, Bracken AP. PRC2 mediated H3K27 methylations in cellular identity and cancer. Curr Opin Cell Biol 2015;37:42–8.

[5] Ferrari KJ, Scelfo A, Jammula S, Cuomo A, Barozzi I, Stutzer A, et al. Polycomb-dependent H3K27me1 and H3K27me2 regulate active transcription and Enhancer fidelity. Mol Cell 2014;53(1):49–62.

[6] Bracken AP, Kleine-Kohlbrecher D, Dietrich N, Pasini D, Gargiulo G, Beekman C, et al. The Polycomb group proteins bind throughout the INK4A-ARF locus and are disassociated in senescent cells. Genes Dev 2007;21(5):525–30.

[7] Bracken AP, Dietrich N, Pasini D, Hansen KH, Helin K. Genome-wide mapping of Polycomb target genes unravels their roles in cell fate transitions. Genes Dev 2006;20(9):1123–36.

[8] Lee HG, Kahn TG, Simcox A, Schwartz YB, Pirrotta V. Genome-wide activities of Polycomb complexes control pervasive transcription. Genome Res 2015;25(8):1170–81.

[9] Pasini D, Cloos PA, Walfridsson J, Olsson L, Bukowski JP, Johansen JV, et al. JARID2 regulates binding of the Polycomb repressive complex 2 to target genes in ES cells. Nature 2010;464(7286):306–10.

[10] Brien GL, Gambero G, O'Connell DJ, Jerman E, Turner SA, Egan CM, et al. Polycomb PHF19 binds H3K36me3 and recruits PRC2 and demethylase NO66 to embryonic stem cell genes during differentiation. Nat Struct Mol Biol 2012;19(12):1273–81.

[11] Sarma K, Margueron R, Ivanov A, Pirrotta V, Reinberg D. Ezh2 requires PHF1 to efficiently catalyze H3 lysine 27 trimethylation in vivo. Mol Cell Biol 2008;28(8):2718–31.

[12] Li X, Isono K, Yamada D, Endo TA, Endoh M, Shinga J, et al. Mammalian polycomb-like Pcl2/Mtf2 is a novel regulatory component of PRC2 that can differentially modulate polycomb activity both at the Hox gene cluster and at Cdkn2a genes. Mol Cell Biol 2011;31(2):351–64.

[13] Kalb R, Latwiel S, Baymaz HI, Jansen PW, Muller CW, Vermeulen M, et al. Histone H2A monoubiquitination promotes histone H3 methylation in Polycomb repression. Nat Struct Mol Biol 2014;21(6):569–71.

[14] Gao Z, Zhang J, Bonasio R, Strino F, Sawai A, Parisi F, et al. PCGF homologs, CBX proteins, and RYBP define functionally distinct PRC1 family complexes. Mol Cell 2012;45(3):344–56.

[15] Blackledge NP, Farcas AM, Kondo T, King HW, McGouran JF, Hanssen LL, et al. Variant PRC1 complex-dependent H2A ubiquitylation drives PRC2 recruitment and polycomb domain formation. Cell 2014;157(6):1445–59.

[16] McGinty RK, Henrici RC, Tan S. Crystal structure of the PRC1 ubiquitylation module bound to the nucleosome. Nature 2014;514(7524):591–6.

[17] Wang H, Wang L, Erdjument-Bromage H, Vidal M, Tempst P, Jones RS, et al. Role of histone H2A ubiquitination in Polycomb silencing. Nature 2004;431(7010):873–8.

[18] Fischle W, Wang Y, Jacobs SA, Kim Y, Allis CD, Khorasanizadeh S. Molecular basis for the discrimination of repressive methyl-lysine marks in histone H3 by Polycomb and HP1 chromodomains. Genes Dev 2003;17(15):1870–81.

[19] Isono K, Endo TA, Ku M, Yamada D, Suzuki R, Sharif J, et al. SAM domain polymerization links subnuclear clustering of PRC1 to gene silencing. Dev Cell 2013;26(6):565–77.

[20] Morey L, Pascual G, Cozzuto L, Roma G, Wutz A, Benitah SA, et al. Nonoverlapping functions of the Polycomb group Cbx family of proteins in embryonic stem cells. Cell Stem Cell 2012;10(1):47–62.

[21] O'Loghlen A, Munoz-Cabello AM, Gaspar-Maia A, Wu HA, Banito A, Kunowska N, et al. MicroRNA regulation of Cbx7 mediates a switch of Polycomb orthologs during ESC differentiation. Cell Stem Cell 2012;10(1):33–46.

[22] Pemberton H, Anderton E, Patel H, Brookes S, Chandler H, Palermo R, et al. Genome-wide co-localization of Polycomb orthologs and their effects on gene expression in human fibroblasts. Genome Biol 2014;15(2):R23.

[23] Messmer S, Franke A, Paro R. Analysis of the functional role of the Polycomb chromo domain in *Drosophila melanogaster.* Genes Dev 1992;6(7):1241–54.

[24] Platero JS, Hartnett T, Eissenberg JC. Functional analysis of the chromo domain of HP1. EMBO J 1995;14(16):3977–86.

[25] Hansen KH, Bracken AP, Pasini D, Dietrich N, Gehani SS, Monrad A, et al. A model for transmission of the H3K27me3 epigenetic mark. Nat Cell Biol 2008;10(11):1291–300.

[26] Schwartz YB, Kahn TG, Nix DA, Li XY, Bourgon R, Biggin M, et al. Genome-wide analysis of Polycomb targets in *Drosophila melanogaster.* Nat Genet 2006;38(6):700–5.

[27] Boyer LA, Plath K, Zeitlinger J, Brambrink T, Medeiros LA, Lee TI, et al. Polycomb complexes repress developmental regulators in murine embryonic stem cells. Nature 2006;441(7091):349–53.

[28] Lee TI, Jenner RG, Boyer LA, Guenther MG, Levine SS, Kumar RM, et al. Control of developmental regulators by Polycomb in human embryonic stem cells. Cell 2006;125(2):301–13.

[29] Ren X, Kerppola TK. REST interacts with Cbx proteins and regulates polycomb repressive complex 1 occupancy at RE1 elements. Mol Cell Biol 2011;31(10):2100–10.

[30] Dietrich N, Lerdrup M, Landt E, Agrawal-Singh S, Bak M, Tommerup N, et al. REST-mediated recruitment of polycomb repressor complexes in mammalian cells. PLoS Genet 2012;8(3):e1002494.

[31] Negishi M, Saraya A, Mochizuki S, Helin K, Koseki H, Iwama A. A novel zinc finger protein Zfp277 mediates transcriptional repression of the Ink4a/arf locus through polycomb repressive complex 1. PLoS One 2010;5(8):e12373.

[32] Yu M, Mazor T, Huang H, Huang HT, Kathrein KL, Woo AJ, et al. Direct recruitment of polycomb repressive complex 1 to chromatin by core binding transcription factors. Molecular cell 2012;45(3):330–43.

[33] Riising EM, Comet I, Leblanc B, Wu X, Johansen JV, Helin K. Gene silencing triggers polycomb repressive complex 2 recruitment to CpG islands genome wide. Mol Cell 2014;55(3):347–60.

[34] Eskeland R, Leeb M, Grimes GR, Kress C, Boyle S, Sproul D, et al. Ring1B compacts chromatin structure and represses gene expression independent of histone ubiquitination. Mol Cell 2010;38(3):452–64.

[35] Schoenfelder S, Sugar R, Dimond A, Javierre BM, Armstrong H, Mifsud B, et al. Polycomb repressive complex PRC1 spatially constrains the mouse embryonic stem cell genome. Nat Genet 2015;47(10):1179–86.

[36] Cao R, Tsukada Y, Zhang Y. Role of Bmi-1 and Ring1A in H2A ubiquitylation and Hox gene silencing. Mol Cell 2005;20(6):845–54.

[37] Wu X, Johansen JV, Helin K. Fbxl10/Kdm2b recruits polycomb repressive complex 1 to CpG islands and regulates H2A ubiquitylation. Mol Cell 2013;49(6):1134–46.

[38] Farcas AM, Blackledge NP, Sudbery I, Long HK, McGouran JF, Rose NR, et al. KDM2B links the polycomb repressive complex 1 (PRC1) to recognition of CpG islands. eLife 2012;1:e00205.

[39] Morey L, Aloia L, Cozzuto L, Benitah SA, Di Croce L. RYBP and Cbx7 define specific biological functions of polycomb complexes in mouse embryonic stem cells. Cell Rep 2013;3(1):60–9.

[40] Tavares L, Dimitrova E, Oxley D, Webster J, Poot R, Demmers J, et al. RYBP-PRC1 complexes mediate H2A ubiquitylation at polycomb target sites independently of PRC2 and H3K27me3. Cell 2012;148(4):664–78.

[41] Kim CA, Gingery M, Pilpa RM, Bowie JU. The SAM domain of polyhomeotic forms a helical polymer. Nat Struct Biol 2002;9(6):453–7.

[42] Robinson AK, Leal BZ, Chadwell LV, Wang R, Ilangovan U, Kaur Y, et al. The growth-suppressive function of the polycomb group protein polyhomeotic is mediated by polymerization of its sterile alpha motif (SAM) domain. J Biol Chem 2012;287(12):8702–13.

[43] Lagarou A, Mohd-Sarip A, Moshkin YM, Chalkley GE, Bezstarosti K, Demmers JA, et al. dKDM2 couples histone H2A ubiquitylation to histone H3 demethylation during Polycomb group silencing. Genes Dev 2008;22(20):2799–810.

[44] Wei J, Zhai L, Xu J, Wang H. Role of Bmi1 in H2A ubiquitylation and Hox gene silencing. J Biol Chem 2006;281(32):22537–44.

[45] Endoh M, Endo TA, Endoh T, Isono K, Sharif J, Ohara O, et al. Histone H2A mono-ubiquitination is a crucial step to mediate PRC1-dependent repression of developmental genes to maintain ES cell identity. PLoS Genet 2012;8(7):e1002774.

[46] Pengelly AR, Kalb R, Finkl K, Muller J. Transcriptional repression by PRC1 in the absence of H2A monoubiquitylation. Genes Dev 2015;29(14):1487–92.

[47] Akasaka T, Takahashi N, Suzuki M, Koseki H, Bodmer R, Koga H. MBLR, a new RING finger protein resembling mammalian Polycomb gene products, is regulated by cell cycle-dependent phosphorylation. Genes Cell 2002;7(8):835–50.

[48] Gao Z, Lee P, Stafford JM, von Schimmelmann M, Schaefer A, Reinberg D. An AUTS2-Polycomb complex activates gene expression in the CNS. Nature 2014;516(7531):349–54.

[49] Gearhart MD, Corcoran CM, Wamstad JA, Bardwell VJ. Polycomb group and SCF ubiquitin ligases are found in a novel BCOR complex that is recruited to BCL6 targets. Mol Cell Biol 2006;26(18):6880–9.

[50] Junco SE, Wang R, Gaipa JC, Taylor AB, Schirf V, Gearhart MD, et al. Structure of the polycomb group protein PCGF1 in complex with BCOR reveals basis for binding selectivity of PCGF homologs. Structure 2013;21(4):665–71.
[51] Wang R, Taylor AB, Leal BZ, Chadwell LV, Ilangovan U, Robinson AK, et al. Polycomb group targeting through different binding partners of RING1B C-terminal domain. Structure 2010;18(8):966–75.
[52] Trimarchi JM, Fairchild B, Wen J, Lees JA. The E2F6 transcription factor is a component of the mammalian Bmi1-containing polycomb complex. Proc Natl Acad Sci USA 2001;98(4):1519–24.
[53] Trojer P, Cao AR, Gao Z, Li Y, Zhang J, Xu X, et al. L3MBTL2 protein acts in concert with PcG protein-mediated monoubiquitination of H2A to establish a repressive chromatin structure. Mol Cell 2011;42(4):438–50.
[54] Qin J, Whyte WA, Anderssen E, Apostolou E, Chen HH, Akbarian S, et al. The polycomb group protein L3mbtl2 assembles an atypical PRC1-family complex that is essential in pluripotent stem cells and early development. Cell Stem Cell 2012;11(3):319–32.
[55] Cooper S, Dienstbier M, Hassan R, Schermelleh L, Sharif J, Blackledge NP, et al. Targeting polycomb to pericentric heterochromatin in embryonic stem cells reveals a role for H2AK119u1 in PRC2 recruitment. Cell Rep 2014;7(5):1456–70.
[56] Comet I, Helin K. Revolution in the polycomb hierarchy. Nat Struct Mol Biol 2014;21(7):573–5.
[57] Moazed D, O'Farrell PH. Maintenance of the engrailed expression pattern by Polycomb group genes in Drosophila. Development 1992;116(3):805–10.
[58] Denell RE. Homoeosis in Drosophila. II. A genetic analysis of polycomb. Genetics 1978;90(2):277–89.
[59] Dura JM, Randsholt NB, Deatrick J, Erk I, Santamaria P, Freeman JD, et al. A complex genetic locus, polyhomeotic, is required for segmental specification and epidermal development in *D. melanogaster.* Cell 1987;51(5):829–39.
[60] Takihara Y, Tomotsune D, Shirai M, Katoh-Fukui Y, Nishii K, Motaleb MA, et al. Targeted disruption of the mouse homologue of the Drosophila polyhomeotic gene leads to altered anteroposterior patterning and neural crest defects. Development 1997;124(19):3673–82.
[61] Isono K, Fujimura Y, Shinga J, Yamaki M, O-Wang J, Takihara Y, et al. Mammalian polyhomeotic homologues Phc2 and Phc1 act in synergy to mediate polycomb repression of Hox genes. Mol Cell Biol 2005;25(15):6694–706.
[62] Core N, Bel S, Gaunt SJ, Aurrand-Lions M, Pearce J, Fisher A, et al. Altered cellular proliferation and mesoderm patterning in Polycomb-M33-deficient mice. Development 1997;124(3):721–9.
[63] Liu B, Liu YF, Du YR, Mardaryev AN, Yang W, Chen H, et al. Cbx4 regulates the proliferation of thymic epithelial cells and thymus function. Development 2013;140(4):780–8.
[64] Forzati F, Federico A, Pallante P, Abbate A, Esposito F, Malapelle U, et al. CBX7 is a tumor suppressor in mice and humans. J Clin Invest 2012;122(2):612–23.
[65] Akasaka T, Kanno M, Balling R, Mieza MA, Taniguchi M, Koseki H. A role for mel-18, a Polycomb group-related vertebrate gene, during the anteroposterior specification of the axial skeleton. Development 1996;122(5):1513–22.
[66] van der Lugt NM, Domen J, Linders K, van Roon M, Robanus-Maandag E, te Riele H, et al. Posterior transformation, neurological abnormalities, and severe hematopoietic defects in mice with a targeted deletion of the bmi-1 proto-oncogene. Genes Dev 1994;8(7):757–69.
[67] Akasaka T, van Lohuizen M, van der Lugt N, Mizutani-Koseki Y, Kanno M, Taniguchi M, et al. Mice doubly deficient for the Polycomb group genes Mel18 and Bmi1 reveal synergy

and requirement for maintenance but not initiation of Hox gene expression. Development 2001;128(9):1587–97.

[68] Molofsky AV, Pardal R, Iwashita T, Park IK, Clarke MF, Morrison SJ. Bmi-1 dependence distinguishes neural stem cell self-renewal from progenitor proliferation. Nature 2003;425(6961):962–7.

[69] Park IK, Qian D, Kiel M, Becker MW, Pihalja M, Weissman IL, et al. Bmi-1 is required for maintenance of adult self-renewing haematopoietic stem cells. Nature 2003;423(6937): 302–5.

[70] Pietersen AM, Evers B, Prasad AA, Tanger E, Cornelissen-Steijger P, Jonkers J, et al. Bmi1 regulates stem cells and proliferation and differentiation of committed cells in mammary epithelium. Curr Biol 2008;18(14):1094–9.

[71] Jacobs JJ, Kieboom K, Marino S, DePinho RA, van Lohuizen M. The oncogene and Polycomb-group gene bmi-1 regulates cell proliferation and senescence through the ink4a locus. Nature 1999;397(6715):164–8.

[72] Lowe SW, Sherr CJ. Tumor suppression by Ink4a-Arf: progress and puzzles. Curr Opin Genet Dev 2003;13(1):77–83.

[73] Molofsky AV, He S, Bydon M, Morrison SJ, Pardal R. Bmi-1 promotes neural stem cell self-renewal and neural development but not mouse growth and survival by repressing the p16Ink4a and p19Arf senescence pathways. Genes Dev 2005;19(12):1432–7.

[74] del Mar Lorente M, Marcos-Gutierrez C, Perez C, Schoorlemmer J, Ramirez A, Magin T, et al. Loss- and gain-of-function mutations show a polycomb group function for Ring1A in mice. Development 2000;127(23):5093–100.

[75] Leeb M, Wutz A. Ring1B is crucial for the regulation of developmental control genes and PRC1 proteins but not X inactivation in embryonic cells. J Cell Biol 2007;178(2):219–29.

[76] Voncken JW, Roelen BA, Roefs M, de Vries S, Verhoeven E, Marino S, et al. Rnf2 (Ring1b) deficiency causes gastrulation arrest and cell cycle inhibition. Proc Natl Acad Sci USA 2003;100(5):2468–73.

[77] Illingworth RS, Moffat M, Mann AR, Read D, Hunter CJ, Pradeepa MM, et al. The E3 ubiquitin ligase activity of RING1B is not essential for early mouse development. Genes Dev 2015;29(18):1897–902.

[78] Chiacchiera F, Rossi A, Jammula S, Piunti A, Scelfo A, Ordonez-Moran P, et al. Polycomb complex PRC1 preserves intestinal stem cell identity by sustaining Wnt/beta-catenin transcriptional activity. Cell Stem Cell 2016;18(1):91–103.

[79] Pirity MK, Locker J, Schreiber-Agus N. Rybp/DEDAF is required for early postimplantation and for central nervous system development. Mol Cell Biol 2005;25(16):7193–202.

[80] Gonzalez I, Aparicio R, Busturia A. Functional characterization of the dRYBP gene in Drosophila. Genetics 2008;179(3):1373–88.

[81] van den Boom V, Maat H, Geugien M, Rodriguez Lopez A, Sotoca AM, Jaques J, et al. Non-canonical PRC1.1 targets active genes independent of H3K27me3 and is essential for Leukemogenesis. Cell Rep 2016;14(2):332–46.

[82] Kreso A, van Galen P, Pedley NM, Lima-Fernandes E, Frelin C, Davis T, et al. Self-renewal as a therapeutic target in human colorectal cancer. Nat Med 2014;20(1):29–36.

[83] Zdzieblo D, Li X, Lin Q, Zenke M, Illich DJ, Becker M, et al. Pcgf6, a polycomb group protein, regulates mesodermal lineage differentiation in murine ESCs and functions in iPS reprogramming. Stem Cell 2014;32(12):3112–25.

[84] Kloet SL, Makowski MM, Baymaz HI, van Voorthuijsen L, Karemaker ID, Santanach A, et al. The dynamic interactome and genomic targets of Polycomb complexes during stem-cell differentiation. Nat Struct Mol Biol 2016;23:682–90.

[85] Yang CS, Chang KY, Dang J, Rana TM. Polycomb group protein Pcgf6 acts as a master regulator to maintain embryonic stem cell identity. Sci Rep 2016;6:26899.

[86] Knutson SK, Wigle TJ, Warholic NM, Sneeringer CJ, Allain CJ, Klaus CR, et al. A selective inhibitor of EZH2 blocks H3K27 methylation and kills mutant lymphoma cells. Nat Chem Biol 2012;8(11):890–6.

[87] McCabe MT, Ott HM, Ganji G, Korenchuk S, Thompson C, Van Aller GS, et al. EZH2 inhibition as a therapeutic strategy for lymphoma with EZH2-activating mutations. Nature 2012;492(7427):108–12.

[88] Qi W, Chan H, Teng L, Li L, Chuai S, Zhang R, et al. Selective inhibition of Ezh2 by a small molecule inhibitor blocks tumor cells proliferation. Proc Natl Acad Sci USA 2012;109(52):21360–5.

[89] Adler PN, Martin EC, Charlton J, Jones K. Phenotypic consequences and genetic interactions of a null mutation in the Drosophila Posterior Sex Combs gene. Dev Genet 1991;12(5):349–61.

[90] Yasunaga S, Ohtsubo M, Ohno Y, Saeki K, Kurogi T, Tanaka-Okamoto M, et al. Scmh1 has E3 ubiquitin ligase activity for geminin and histone H2A and regulates geminin stability directly or indirectly via transcriptional repression of Hoxa9 and Hoxb4. Mol Cell Biol 2013;33(4):644–60.

[91] Luo M, Zhou J, Leu NA, Abreu CM, Wang J, Anguera MC, et al. Polycomb protein SCML2 associates with USP7 and counteracts histone H2A ubiquitination in the XY chromatin during male meiosis. PLoS Genet 2015;11(1):e1004954.

[92] Kon N, Kobayashi Y, Li M, Brooks CL, Ludwig T, Gu W. Inactivation of HAUSP in vivo modulates p53 function. Oncogene 2010;29(9):1270–9.

[93] Storre J, Elsasser HP, Fuchs M, Ullmann D, Livingston DM, Gaubatz S. Homeotic transformations of the axial skeleton that accompany a targeted deletion of E2f6. EMBO Rep 2002;3(7):695–700.

[94] Brown JP, Bullwinkel J, Baron-Luhr B, Billur M, Schneider P, Winking H, et al. HP1gamma function is required for male germ cell survival and spermatogenesis. Epigenet Chromatin 2010;3(1):9.

[95] Lagger G, O'Carroll D, Rembold M, Khier H, Tischler J, Weitzer G, et al. Essential function of histone deacetylase 1 in proliferation control and CDK inhibitor repression. EMBO J 2002;21(11):2672–81.

[96] Montgomery RL, Davis CA, Potthoff MJ, Haberland M, Fielitz J, Qi X, et al. Histone deacetylases 1 and 2 redundantly regulate cardiac morphogenesis, growth, and contractility. Genes Dev 2007;21(14):1790–802.

[97] Shen-Li H, O'Hagan RC, Hou Jr H, Horner 2nd JW, Lee HW, DePinho RA. Essential role for Max in early embryonic growth and development. Genes Dev 2000;14(1):17–22.

Chapter 5

Structure and Biochemistry of the Polycomb Repressive Complex 1 Ubiquitin Ligase Module

A.G. Cochran
Genentech, Inc., South San Francisco, CA, United States

Chapter Outline

Introduction 81
Ubiquitin Conjugation and Deconjugation 82
PRC1 Is a RING E3 Ligase 85
The Active E3 Ligase Is a Heterodimer of Bml1 and Ring1 89
Biochemical and Structural Studies of the RING Domains of the Bml1-Ring1B Heterodimer 89
Recognition of E2 Enzymes by Bml1-Ring1B 91
Interaction of Bml1-Ring1B-UbcH5c With Substrate 93
Structure of the PRC1 Ubiquitin Ligase Module Bound to Nucleosome 94
Variant PRC1 Complexes and the Mechanism of Ubiquitin Transfer 96
Which PRC1 Contributes Most to H2A Ubiquitination? 99
Is H2A Ubiquitination Really Important? 100
What Does uH2A Do? 102
What's Next for PRC1 and uH2A? 104
References 104

INTRODUCTION

Polycomb group (PcG) proteins are critical regulators of gene expression originally discovered through genetic screens in *Drosophila* [1,2]. During development, PcG proteins and the opposing Trithorax group proteins cooperate to establish transcriptional patterns of key regulators, such as *Hox* (homeobox) genes [3–6]. PcG proteins are implicated in epigenetic inheritance, X-chromosome inactivation, stem cell pluripotency, senescence, and tumorigenesis [7–9]. For example, the loss of the PcG protein Bmi1 (B cell-specific Moloney murine leukemia virus integration site 1) leads to defects in

Polycomb Group Proteins. http://dx.doi.org/10.1016/B978-0-12-809737-3.00005-2

stem cell self-renewal [10–12]. PcG proteins segregate into two major Polycomb repressive complexes, PRC1 and PRC2 [13–15]. Each of these PRCs supports an enzymatic function: PRC2 is the histone H3 lysine 27 (H3K27) methyltransferase [16,17], and PRC1 is an E3 ubiquitin (Ub) ligase that transfers the monoubiquitin mark (uH2A) to the C-terminal tail of histone H2A [18,19]. PRC1 is also capable of producing chromatin compaction, a function not requiring histone tails, and this activity results in gene silencing [20]. Recently, great progress has been made in biochemical and structural characterization of both PRC1 and PRC2. PRC1 will be discussed below with a particular focus on its H2A ubiquitin ligase activity, while PRC2 structural studies [21,22] will be reviewed elsewhere in this volume [23].

UBIQUITIN CONJUGATION AND DECONJUGATION

Ubiquitin modification of protein lysine residues is a versatile and widespread mechanism for regulation of protein function [24–26]. Ubiquitin is itself a small protein (76 amino acids), and it is attached to other proteins through an amide linkage between a target lysine and the Ub C-terminus. Ubiquitin can therefore be considered a form of lysine acyl modification related to acetylation. Similar to acetyl activation through coenzyme A conjugation, ubiquitin must undergo a sequence of reactions to yield a high-energy intermediate that is activated for amide bond formation (Fig. 5.1) [24–26]. In the first part of this cascade, the ubiquitin-activating enzyme, termed E1, converts the

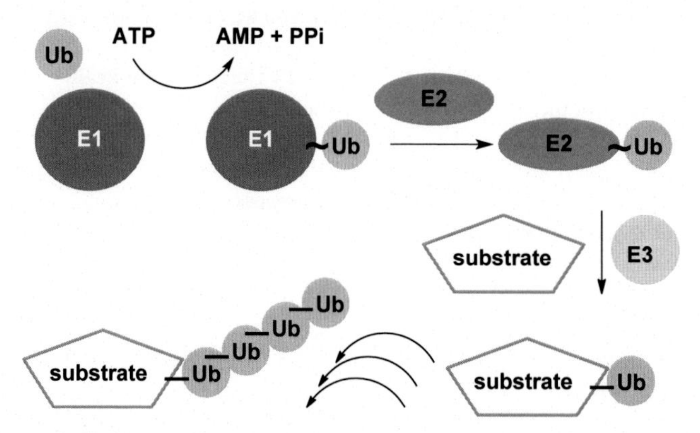

FIGURE 5.1 Ubiquitin modification of proteins. A cascade of three enzymatic reactions is required to transfer ubiquitin. The ubiquitin-activating enzyme (E1) catalyzes a two-step reaction. In the first and ATP-dependent step, the C-terminus of bound ubiquitin is adenylated. In the second step, activated ubiquitin is transferred to a cysteine thiol in the enzyme active site. At this stage, the ubiquitin is transferred to the active-site cysteine of the ubiquitin conjugating enzyme (E2) in a thiol-exchange reaction. The E3 ligase then recruits both E2~Ub and substrate. The E3 catalyzes the transfer of ubiquitin to a substrate lysine to yield a monoubiquitinated product. Depending on the system, additional cycles of activation and transfer may yield a multiply ubiquitinated substrate protein or ubiquitin chains.

ubiquitin C-terminal carboxyl group to a thioester in the E1 active site through a two-step reaction coupled to hydrolysis of ATP. E1-activated ubiquitin then undergoes a thiol-exchange reaction with one of a number of ubiquitin conjugating, or E2, enzymes [27,28]. Finally, one of a very large number of E3 enzymes binds to both ubiquitin-conjugated E2 (E2 ~ Ub) and the appropriate substrate to facilitate transfer [29]. In some cases, the E3 includes a recognizable active site with its own catalytic cysteine residue. In such cases, the ubiquitin transfer to substrate involves an E3-thioester intermediate. However, in the largest class of E3s, Really interesting new gene (RING) E3 ligases [30], there is no catalytic cysteine, and transfer occurs from E2 directly to substrate. Initially, the RING E3s were considered adapter proteins that acted primarily to enforce proximity of E2 and substrate. However, it has been appreciated more recently that the RING domains of RING E3s enforce a "closed" conformation of E2 ~ Ub that enhances reactivity of the thioester bond [31–33]. Thus the E3 may also be considered a catalyst (and a true enzyme) rather than a passive adapter [29,31].

Layered on this basic activation and transfer mechanism are a number of important variations. Ubiquitin itself has multiple potential sites of ubiquitin modification (lysine residues at positions 6, 11, 27, 29, 33, 48, and 63 and the amino terminus), all of which are modified to varying degrees. This modification of ubiquitin by ubiquitin results in the formation of different types of ubiquitin chains with different structural characteristics and functions [34]. For example, chains linked via ubiquitin lysine 48 often confer a degradation signal and direct substrates bearing ubiquitin K48 chains to the proteasome. In contrast, ubiquitin K63 or K11 chains are involved more typically in altering the function of signaling proteins. When one considers the potential effects of chain length, linkage position, multiple sites of ubiquitination on target proteins, mixed-linkage chains, and even branching of ubiquitin chains, an enormous diversity of effect can be imagined, much of which is only now being uncovered [34–37]. A significant challenge is development of the analytical methods for defining and quantifying complex modifications with confidence. However, for the purpose of this review, the modification of greatest relevance is apparently the simplest: monoubiquitination, the conjugation of the C-terminus of ubiquitin directly to a substrate protein. Monoubiquitination has biological functions distinct from those of ubiquitin chain modifications [34,38] and is the common form of ubiquitin modification of histones. In addition to investigating the biological consequences of monoubiquitination, it is interesting to consider why ubiquitin chains do not form on monoubiquitinated proteins despite the fact that chains form so readily on many others [38].

As noted above, RING domain proteins constitute the largest class of E3 ubiquitin ligases [30]. The RING domain is a small zinc-binding module that can be recognized by a characteristic spacing of cysteine and histidine residues (Fig. 5.2A). RING domains often, but not always, function as dimers that can

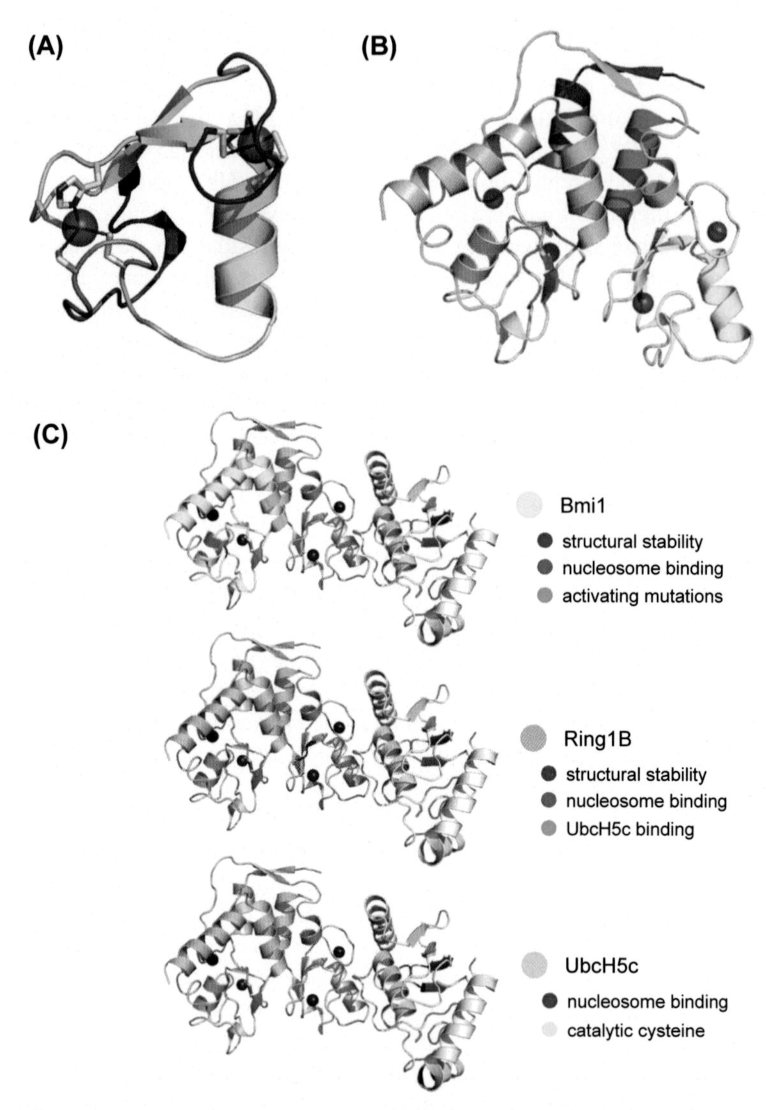

FIGURE 5.2 Structure of the PRC1 ubiquitin ligase module. All images are based on the structure of Bmi1-Ring1B bound to UbcH5c (PDB code: 3RPG) [44] and were prepared using PyMOL (Schrödinger, LLC). (A) RING domain fold. The example shown is that of Ring1B, residues C51–R101. The cartoon color is varied according to primary sequence, with blue at the N-terminus and red at the C-terminus. The two zinc atoms are shown as dark blue spheres, and side chains ligating zinc are shown in stick representation. (B) RING–RING heterodimer. The example shown is that of Bmi1 bound to Ring1B; RINGs are colored white and yellow, respectively. The non-RING elements stabilizing the dimer are shown in shades of pink (Bmi1) or blue (Ring1B). The lighter shaded elements in front are the unusual N-terminal extension of Ring1B (cyan) that wraps around Bmi1 and packs against the long C-terminal extension of Bmi1 (pink). The darker colors (magenta and blue) indicate the more conventional, and largely helical, N-terminal and C-terminal dimerization elements from Bmi1 and Ring1B, respectively. One exception is the short,

be homodimers or heterodimers. The RING domain binds directly to the E2 enzyme and is therefore critical for directing ubiquitin transfer [29]. In the case of heterodimeric RINGs, it can be the case that only one of the two RINGs is competent to bind E2 (see below). Different RINGs apparently act with different preferred E2 enzymes; this preference is usually evaluated by screening E2 proteins biochemically with a RING of interest, which can lead to conclusions that may not apply in vivo [28,30]. The relatively small number of E2s compared to E3s makes a clean determination of "the true E2" in a cellular setting very difficult, as each E2 is likely to be important in a large number of processes. It is also likely that, in some cases, more than one E2 may be capable of supporting a given E3. The difficulty with definitive E2 identification has been noted as well for histone H2A ubiquitination [38]. This high level of uncertainty about relevant participants is unusual when compared to other types of signaling pathways, for example, kinase or proteolytic cascades.

Finally, like other lysine posttranslational modifications, ubiquitination is often reversible. Several distinct families of deubiquitinase (Dub) enzymes exist to carry out a range of chain dismantling and ubiquitin removal functions [39,40]. Newly synthesized, linear oligoubiquitin must be processed to monomers for general use in the pathway. In addition, deubiquitinating activities are important for proper function of the proteasome, transcriptional activation by the regulatory complex Spt-Ada-Gcn5-acetyltransferase (SAGA) [41,42], and regulation of signaling pathways dependent on ubiquitination [34]. Like other forms of ubiquitin modification, ubiquitination of H2A is also reversible and highly regulated: a surprisingly large number of H2A K119 deubiquitinases have been identified, and many of these Dubs appear to be required in important and diverse cellular processes [43].

PRC1 IS A RING E3 LIGASE

A molecular understanding of how PRC1 enforces gene silencing began with the discovery that the complex has ubiquitin ligase activity. Although it had been long appreciated that a substantial fraction of histone H2A was modified

two-stranded β-sheet shown at the top of the image. This interaction involves sequences from the N-terminal extensions of both Bmi1 and Ring1B and is not a typical feature of RING–RING dimers that have been characterized structurally. (C) Structure of the PRC1 ligase module with sites of mutations affecting function indicated (see Table 5.1). In all three images, the proteins from left to right are Bmi1, Ring1B, and UbcH5c, and the catalytic cysteine of the E2 is shown in yellow stick representation. Each image shows sites of mutation in the indicated protein with coloring by type of effect. The activating mutations in Bmi1 (green) fall into two groups; the lighter green indicates the segment swapped in a chimera with PCGF5 that has slightly enhanced H2A ligase activity, while the darker green shows the site of strongly activating mutations that affect the interaction with ubiquitin during transfer from UbcH5c [57].

by a single ubiquitin (monoubiquitination), it was not clear how the modification was introduced. This situation changed in 2004 when reports from two groups clarified fundamental aspects of this process [18,19]. In a biochemically driven approach [18], the group of Yi Zhang fractionated HeLa nuclear extracts by chromatography. These fractions were evaluated in a reconstituted H2A ubiquitin ligase assay supplied with oligonucleosome or histone octamer substrates and subjected to additional purification to isolate an active complex. The reaction product was identified as H2A K119ub (uH2A). (Note that H2A has adjacent lysine residues at positions 119/120 or 118/119, depending on whether one uses numbering based on the gene sequence or the traditional numbering of histones based on mature protein sequence as used by Wang et al. [18]. In addition, recent evidence indicates that the first lysine can also be modified by PRC1 [44]. Together, these have led to some apparent discrepancies in published papers that name the site of modification.) Based on mass spectrometry identification of Ring1 (Ring1A), Ring2 (Ring1B), Bmi1 (PcG RING Finger 4 (PCGF4)), and human polyhomeotic-like 2 (HPH2), the authors concluded that the active complex was related to a human PRC1 complex that had been identified previously and provisionally named it as hPRC1L (human PRC1-like). Additional characterization of the ubiquitin ligase activity revealed some key points. First, the authors determined that, while three of the identified proteins harbor RING domains, the active ligase component appeared to be Ring2 rather than Bmi1 or Ring1. This conclusion was based on assay of individual recombinant GST fusion proteins and, although it was not appreciated at the time that the active unit is a heterodimer of the RING domain of Bmi1 with that of Ring1 or Ring2 (see below), the conclusion turned out to be partially correct. In addition, a strong preference for an intact nucleosomal substrate was revealed. Finally, two different mutations in the Ring2 RING domain proved to abrogate E3 activity (Table 5.1): one of these of a histidine predicted to chelate zinc (and, therefore, to be required for stable structuring of the RING domain) and a second at an adjacent residue analogous to a previously described mutant allele of *Drosophila*. The latter mutation allowed the authors to connect the loss of H2A ubiquitination activity to a loss of silencing of a PRC1 target gene.

Complementary experiments reported by Neil Brockdorff and collaborators [19] solidified the connection between PRC1 and H2A ubiquitination. Studying the inactive X chromosome (Xi) as a model system for silent chromatin, these researchers found Ring1B, as well as ubiquitinated H2A, to be enriched on Xi. *Ring1B* null cell lines showed a strong global reduction (but not elimination) of uH2A, and this loss could be rescued by reexpression of Ring1B in the null cells. A role was defined for the highly similar protein Ring1A in maintaining uH2A in *Ring1B*$^{-/-}$ cells. The combined results (and also those from a related study [45]) indicate that, together, the PRC1 components Ring1A and Ring1B are responsible for uH2A on the inactive X chromosome.

TABLE 5.1 Mutagenesis of the RING Domains of Ring1B, Bmi1, or Both and Effects on H2A E3 Ligase Activity

Mutation	Effect on H2Aub	Likely Mechanism	Source
Ring1B C51F/C54F	Loss of Bmi1 binding[a]	Disrupts RING structure (Zn ligands)	Hemenway et al. [49]
Ring1B R70C	Loss of activity[b]	Disrupts RING structure or dimer interface	Wang et al. [18]
Ring1B H69Y	Loss of activity[b]	Disrupts RING structure (Zn ligand)	Wang et al. [18]
Ring1B R70D	Partial loss of activity	RING–RING interface	Li et al. [52]
Ring1B I53S	Loss of activity	Disrupts E2 binding	Ben-Saadon et al. [48]
Ring1B I53A	Loss of "almost all" activity	Disrupts E2 binding	Buchwald et al. [51]
Ring1B D56K	Loss of activity	Disrupts E2 binding (>8-fold)	Bentley et al. [44]
Ring1B K92A/K93A	Partial loss of activity	Ring1B basic patch (substrate interaction)	Bentley et al. [44]
Ring1B K97A/R98A	Loss of activity	Ring1B basic patch (substrate interaction)	Bentley et al. [44]
Ring1B K59A	Minimal effect	Not in contact with nucleosome	McGinty et al. [62]
Ring1B R81A	Decrease in activity (~50% WT)	Nucleosome contact residue	McGinty et al. [62]
Ring1B K93A	Small decrease in activity (~70% WT)	Nucleosome contact residue	McGinty et al. [62]
Ring1B K97A	Strong decrease in activity (~25% WT)	Nucleosome contact residue	McGinty et al. [62]
Ring1B R98A	Loss of activity	Nucleosome contact residue	McGinty et al. [62]
Ring1B K-to-R at 10 sites	Loss of activity	Substrate binding? Meant to block autoub	Ben-Saadon et al. [48]
Ring1B K112R	Minimal effect	Blocks auto monoubiquitination site	Buchwald et al. [51]
Bmi1 L20A	Minimal effect	E2 binding, if E2 were to bind to Bmi1	Buchwald et al. [51]
Bmi1 C21F	Loss of Ring1B binding[a]	Disrupts RING structure (Zn ligands)	Hemenway et al. [49]

Continued

TABLE 5.1 Mutagenesis of the RING Domains of Ring1B, Bmi1, or Both and Effects on H2A E3 Ligase Activity—cont'd

Mutation	Effect on H2Aub	Likely Mechanism	Source
Bmi1 C53F	Loss of Ring1B binding[a]	Disrupts RING structure (Zn ligands)	Hemenway et al. [49]
Bmi1 T41R	Minimal effect	Interface with Ring1B N-terminal tail	Li et al. [52]
Bmi1 D72R	Minimal effect	RING–RING interface	Li et al. [52]
Bmi1 K62A/R64A	Partial loss of activity	Bmi1 basic patch (substrate recognition)	Bentley et al. [44]
Bmi1 E33A	Strong decrease in activity (~30% WT)	Unclear: Increases nucleosome affinity	McGinty et al. [62]
Bmi1 R45A	Minimal effect	Not in contact with nucleosome	McGinty et al. [62]
Bmi1 H61A	Minimal effect	Nucleosome contact residue	McGinty et al. [62]
Bmi1 K62A	Minimal effect	Nucleosome contact residue	McGinty et al. [62]
Bmi1 R64A	Decrease in activity (~60% WT)	Nucleosome contact residue	McGinty et al. [62]
Bmi1 K73R	Increase in activity[c]	Improved Bmi1–Ub interface	Taherbhoy et al. [57]
Bmi1 K73N/D77E	Increase in activity[c]	Improved Bmi1–Ub interface	Taherbhoy et al. [57]
Bmi1 (PCGF5 aa 1–17)	Small increase in activity[c]	Chimera with more active Bmi1 paralog	Taherbhoy et al. [57]
Bmi1 K73N/D77E (PCGF5 aa 1–17)	Strong increase in activity (~300% WT)	Combined activating mutations	Taherbhoy et al. [57]
Bmi1 K62E/R64N	Strong decrease in activity (~20% WT)	Bmi1 basic patch chimera with PCGF5	Taherbhoy et al. [57]
Bmi1 K62E/R64N/K73N/D77E (PCGF5 aa 1–17)	PCGF5 chimera; ~WT activity	Activating mutations offset by basic patch (substrate recognition) mutations	Taherbhoy et al. [57]
Bmi1 T41R/Ring1B R70D	Loss of activity (complex less stable)	Disrupts two interacting regions	Li et al. [52]
Bmi1 D72R/Ring1B R70D	Activity substantially restored	Restores salt bridge lost in Ring1B R70D	Li et al. [52]

[a] *This study predates the discovery of the E3 ligase activity of PRC1.*
[b] *These assays were conducted with Ring1B alone. The Drosophila equivalent of Ring1B R70C (Sce^{33M2}) is not expressed stably and is therefore functionally null [87].*
[c] *These mutants were assessed by E2-discharge reactions or GST-tag autoubiquitination instead of nucleosome ubiquitination.*

THE ACTIVE E3 LIGASE IS A HETERODIMER OF BMI1 AND RING1

Once it had been established that an important activity of PRC1 was ubiquitination of H2A, more detailed studies of the biochemical and structural properties of the complex followed quickly. Because of the difficulty in purifying large amounts of the native complex from cells, recombinant approaches toward complex reconstitution were initiated with the aim of better defining complex stoichiometry and pairwise interactions between subunits [46,47]. Although some of the resulting models appear not to be fully correct in light of later work (in particular the placement of stoichiometric Ring1A and Ring1B into a single copy of the complex; see below), one important finding was that the biochemical E3 ligase activity of purified Ring1B reported earlier [18] was very low compared to the activity of the reconstituted complex [46]. The most active preparation included copurified Ring1A, Ring1B, Pc3, and Bmi1; in addition, the binary complex of Bmi1 and Ring1B was found to have substantially more activity than that of Ring1B alone. Bmi1 was found to have a central role in stabilizing the complex, forming pairwise interactions with all other members, and, importantly, an additional PCGF protein related to Bmi1, Mel-18 (Melanoma nuclear protein 18, PCGF2), could substitute for Bmi1 to maintain complex integrity. However, the Mel-18 reconstituted complex, while appearing similar in composition to that of Bmi1, was not as active as an E3 ligase, consistent with a lack of functional redundancy of Bmi1 and Mel-18 [46]. An important observation was that deleting individual Ring1 paralogs from the complex reconstitutions yielded complexes with similar levels of activity, suggesting that either Ring1A or Ring1B could support PRC1 function [47]. The finding that a PCGF subunit is needed for robust H2A ubiquitination activity of Ring1A/B is important to recall when interpreting the results of biochemical assays conducted with a Ring1 protein alone (for example, the self-ubiquitination activity of Ring1B [48]).

BIOCHEMICAL AND STRUCTURAL STUDIES OF THE RING DOMAINS OF THE BMI1-RING1B HETERODIMER

As described above, robust ubiquitination activity from PRC1 requires both Bmi1-like and RING1 proteins. However, well before it was appreciated that PRC1 was an E3 ubiquitin ligase, it had been discovered from yeast two-hybrid studies that Bmi1 and Ring1B could heterodimerize and that conserved, N-terminal RING domains present in each were required for the interaction [49,50]. The RING domains alone were not sufficient: stable complex formation required additional regions immediately flanking each of the RING domains [49,50]. Two 2006 structural studies from the groups of Titia Sixma [51] and Rui-Ming Xu [52] clarified how the Bmi1 and Ring1B RING domains associate to form the heterodimeric E3 ubiquitin ligase and revealed important aspects of the reaction mechanism.

Before initiating crystallographic studies, each group identified minimal constructs capable of stable heterodimer formation, either by limited proteolysis [52], or by prediction of domain boundaries through sequence analysis [51]. These two methods yielded similar results (Table 5.2). Importantly, each group concluded that coexpression of the two fragments yielded substantially better quality complex than that obtained on mixing after expressing separately. This observation makes sense in light of extensive interactions in the RING–RING structure (see below), but it is nevertheless possible to prepare an active ligase complex by mixing longer constructs [51]. One cautionary note is that it has not yet been established that simply mixing Bmi1 and Ring1B results in a uniform, maximally active ligase preparation. In our hands, the individual proteins are neither as stable nor as soluble as the complex (unpublished), so we would suggest using coexpressed material whenever possible.

The two crystal structures of the minimal Bmi1-Ring1B heterodimer show conventional RING-domain topology and secondary structural elements. Each RING binds two zinc ions with ligands contributed from CX_2C, CXH, CX_2C, and CX_2C elements (Fig. 5.2A). (Zinc site 1 consists of the first and third elements, while the second zinc ion binds to the other two.) The RINGs themselves are quite short, approximately 50 amino acids, but additional elements are important for stabilization of the heterodimer. Each protein has two short helices that immediately flank the RING in primary sequence and that interact with the heteropartner (Fig. 5.2B). This type of RING–RING dimerization is relatively common but not always sufficient to drive dimerization. Some RING dimers have an additional dimerization element. This can be longer flanking helices (such as those of the BRCA1-BARD1 (Breast cancer type 1 susceptibility protein–BRCA1-associated RING domain protein 1 complex) heterodimer [53]) or a C-terminal β-strand (e.g., those from inhibitor-of-apoptosis proteins [54]). The additional dimerization motifs seen in Bmi1–Ring1B

TABLE 5.2 Minimal Fragments Forming a Bmi1-Ring1B Heterodimer With E3 Activity

Bmi1 Fragment	Ring1B Fragment	Source	PDB Code (Resolution)
1–102	5–115	Li et al. [52]	2H0D (2.5 Å)
1–109	1–114	Buchwald et al. [51]	
1–109	1–159	Buchwald et al. [51]	2CKL[a] (2.0 Å)
1–109	1–116	Bentley et al. [44]	3RPG (2.65 Å; UbcH5c complex)

[a]*Although the longer construct was used for crystallization, Ring1B residues beyond 114 are not visible in the solved structure [51].*

complexes are unique to this family of RING heterodimers. Bmi1 has a long C-terminal helix that juts outward from the rest of the protein, creating a binding cleft for the extended N-terminus of Ring1B. This interaction occurs away from the RING–RING interaction surface on the far side of Bmi1, giving a highly asymmetric appearance to the heterodimer. The total buried surface area in the heterodimer interface is considerable (~2300 Å^2 or more, depending on the report), with a greater contribution from the unique, asymmetric dimerization element [52]. The N-terminal extension of Ring1B is critical for complex formation, as Ring1B lacking the first 29 amino acids fails to bind Bmi1 [51], and a Bmi1 mutant in the interaction surface (T41R) exacerbates the effect of a Ring1B mutation (R70D) in the RING–RING interface (Table 5.1) [52]. The extensive interaction surface between partners is consistent with the observed high stability of the RING–RING complex [51,52] and suggestive of an obligate dimerization rather than an equilibrium association within PRC1.

The biochemical studies accompanying the published crystal structures revealed an important aspect of the histone H2A ubiquitination reaction [51,52]. In each case, the authors tested the minimized RING–RING heterodimers for activity, finding that the RING domains alone were competent to catalyze uH2A formation (in the presence of E1, ATP, E2, and ubiquitin) and apparently fully as active as longer heterodimeric constructs. The finding that Bmi1–Ring1B regions outside the core heterodimer were not required for efficient recognition and specific modification of the nucleosome substrate was unexpected. The RING domain typically recruits E2 to the active complex, and most well-characterized E3s recognize target substrates through domains other than the RING itself [29,30]. In the case of Bmi1-Ring1B, it was concluded that the RING domains must be capable of direct binding to the nucleosome [51].

RECOGNITION OF E2 ENZYMES BY BMI1-RING1B

Early work on the PRC1 ligase complex suggested that Ring1B rather than Bmi1 was the active ligase component. This led to the biochemical hypothesis that Ring1B might be capable of E2 binding, while Bmi1 might act primarily to stabilize the structure of Ring1B, thereby enhancing ligase activity without binding directly to the E2. This hypothesis was tested by mutagenesis (Table 5.1). Based on earlier studies characterizing the interaction of E2 with BRCA1/BARD1 RING–RING heterodimer [55], the Ring1B residue I53 was expected to reside in the interface with E2. Accordingly, the mutations I53A [51] and I53S [48] were found to strongly abrogate formation of uH2A. In contrast, the equivalent mutation on the Bmi1 RING domain, L20A, had no effect in the ubiquitination reaction [51], consistent with the idea that only Ring1B might bind to E2. An interesting observation was that the mutant I53A was able to support a low level of activity [51], suggesting that it might retain residual affinity for E2. The typical assay for uH2A formation is detection by western blot at a fixed time point, and different laboratories use different experimental conditions, making quantitative conclusions about relative activities difficult. In light of in vivo studies with mutants described below, it is worth considering whether "inactivating" mutations result in truly inactive proteins or proteins that are strongly impaired under particular experimental conditions. The latter situation would not counter-indicate the use of mutants to address mechanistic questions but potentially could give misleading results in biological experiments. More detailed characterization of mutants, including biophysical measurements of E2 affinities and quantitative kinetic studies over a broader range of protein concentrations, could provide greater confidence in their suitability for in vivo studies.

An equally important aspect of E2 recognition is to define which is the appropriate E2 for a given E3 ligase. Fortunately, the total number of human E2s (~ 30; reference [56]) is considerably smaller than the number of E3s, and many of these enzymes can be expressed recombinantly or purchased for testing. (However, as noted above, this does not guarantee that the preferred E2 in such a screen will be the relevant E2 in vivo.) The enzyme UbcH5c (ubiquitin conjugating enzyme human 5c) was identified as a suitable E2 for Ring1B in an early study [18]. Evaluation of a small panel of E2s revealed all members of the UbcH5 family, as well as UbcH6, to support uH2A production [51]. Interestingly, the E2 E2-25K bound tightly to the Bmi1−Ring1B RING−RING heterodimer but did not promote uH2A formation [51]. Conceivably, other E2s that were not tested might work in concert with Bmi1-Ring1B.

The mode of E2 binding to Ring1B was confirmed by cocrystallization of the ternary complex of Bmi1−Ring1B−UbcH5c [44]. As had been predicted from earlier structural and mutagenesis studies [51,52], the ternary complex contains a single protomer of UbcH5c bound at the expected site on the RING domain of Ring1B. Indeed, UbcH5c makes no contacts with the Bmi1 fragment in this structure [44]. Comparison of the structures of the Bmi1−Ring1B heterodimer alone to its structure within the ternary complex revealed no major differences, although it did appear from inspection of B-factors that one α helix of Ring1B (that immediately preceding the RING and involved in the dimerization interface with Bmi1) may be more ordered in the ternary complex. The buried surface area in the UbcH5c−Ring1B interface is relatively low (~ 500 $Å^2$), consistent with the modest affinity of the E2−E3 interaction (K_D ~ 5 μM) [44]. Structural details in the interface were quite similar to those reported for other UbcH5−E3 complexes, although the bivalent interaction of Ring1B R91 with UbcH5c Q92 was more extensive than that seen in similar complex structures [44]. Later work demonstrated that recognition of UbcH5c by the PCGF5−Ring1B complex (PCGF5 is closely related to Bmi1; see below) is extremely similar, as might be expected given that all E2 contacts reside within the Ring1B subunit [57].

INTERACTION OF BMI1-RING1B-UbcH5c WITH SUBSTRATE

Understanding how substrate is recognized by the core RING—RING complex of Bmi1-Ring1B is important in understanding how the reaction can target a single site for ubiquitination among the scores of lysine residues in a nucleosome core particle. In addition, as direct substrate recognition by RINGs has not been described broadly, and as it is currently unknown how common it might be, a detailed description in the PRC1 system should provide guidelines for investigating other systems. Recently, it has become clear that other E3s targeting the nucleosome can recognize specific surface features and that this is important for site-specific ubiquitin transfer [58].

The Bmi1-Ring1B E3 ligase shows a strong preference for modification of the intact nucleosome and very little activity toward recombinant H2A or other assemblies of H2A short of the full nucleosome [18,44]. In addition the core RING—RING complex is capable of binding to short duplex DNAs with low μM affinity [44,59]. Taken together, these observations suggested that the RING—RING complex might bind to the DNA portion of the nucleosome, thus explaining why H2A alone is a poor substrate [44]. This model suggested that strongly basic loops in the "bottom" region of the two RING domains (the regions distal to the dimerization helices) might mediate the interactions with both duplex DNA and substrate through electrostatic interactions between the phosphodiester backbone and the E3. Several observations were consistent with this model [44]. First, the interaction between the core RING—RING complex and DNA is highly sensitive to added salt, as is the overall E3 reaction rate. However, the E2—E3 interaction is also sensitive to salt, making it difficult to parse the role of E3-substrate binding. More direct evidence was provided by analysis of mutations in the basic loop regions of the RING domains. Mutation of pairs of basic residues to alanine resulted in either a partial or a complete loss of ligase activity (Table 5.1), and the loss of activity was generally accompanied by lower affinity for DNA [44]. While the evidence that DNA is the feature of the nucleosome recognized by Bmi1-Ring1B was circumstantial (and the model later proven incorrect; see below), it was clear that the basic loops of Bmi1 and, especially, of Ring1B were important in some aspect of the ligase reaction. As all mutants formed stable RING—RING complexes, and as the least active mutant E3 (Ring1B K97A, R98A) bound E2 with affinity very similar to wild-type E3, it was concluded that these loop regions are involved in substrate recognition.

An advance in understanding how the PRC1 ligase module recognizes the nucleosome was made through evaluating recombinant nucleosome substrates made using mutant histones. Ubiquitination of histone H2A or the variant histone H2AX by PCGF4-Ring1B was not observed when the incorporated H2A variant was the E92A mutant [60]. This residue lies in the center of an acidic patch present on histone H2A that is the recognition element for a number of nucleosome-binding proteins, such as the E3 ligase RNF168 (RING finger 168) [59,60] and the 22-amino acid peptide from the Kaposi's sarcoma-associated herpes virus latency-associated nuclear antigen (LANA) [61]. The LANA peptide can compete with RNF168 for nucleosome binding, as shown by inhibition of the ligase reaction (ubiquitination of H2A K13/K15) [59,60]. The Bmi1-Ring1B ligase reaction is also inhibited by LANA peptide [57] but not by control peptide (LANA LRS mutant [61]), supporting the model suggested by the histone mutagenesis study [60], namely that the Bmi1-Ring1B RINGs recognize the nucleosome through the H2A acidic patch. This model has been extended to the E3 ligase BRCA1/BARD1 [62] and variant PRC1 E3 ligases [57] (see below), suggesting that histone ubiquitin ligases may generally recognize nucleosome through the acidic surface patch of H2A.

STRUCTURE OF THE PRC1 UBIQUITIN LIGASE MODULE BOUND TO NUCLEOSOME

Our understanding of nucleosome recognition by Bmi1-Ring1B was advanced substantially by a recent crystal structure Bmi1-Ring1B and the E2 enzyme UbcH5c bound to the nucleosome core particle (NCP) [62]. Crystallization of the E2–E3 complex with nucleosome was facilitated by the engineering of a linker between the C-terminus of the Ring1B minimal RING construct and the N-terminus of UbcH5c. This strategy allowed the authors to coexpress the fusion protein and Bmi1 fragment to yield a stable, homogeneous E2-E3 module [62]. Because of the low affinity of UbcH5c for Bmi1-Ring1B [44], purification of the native E2–E3 complex was not feasible [51]. Crystals of the engineered ligase module bound to the NCP yielded a 3.3 Å structure (Fig. 5.3A, left) that showed two independent views of substrate recognition (these differ slightly, most likely because of crystal packing effects) [62]. This structure is an exciting advance in understanding both the PRC1 ubiquitin ligase module and chromatin-modifying enzymes more broadly.

The structure reveals numerous direct contacts between Bmi1 and, especially, Ring1B with the histone surface in the central area of the NCP (Fig. 5.3B). Indeed, the authors note contacts to each of the four histones, perhaps explaining the requirement for an intact nucleosome substrate for efficient ubiquitination of histone H2A. The structure validates conclusions from each of the mutagenesis studies described in the preceding section. In particular, residues forming an acidic patch centered on histones H2A and H2B make numerous polar contacts with basic side chains of the Bmi1-Ring1B RING domains. Histone H2A E92 was, as expected [60], engaged in a key interaction: this side chain, along with that of H2A D90, forms ionic interactions with the side chain of Ring1B R98. Ring1B R98 and the adjacent K97 had been identified previously as critically important for ligase activity [44], and these basic loop residues interact additionally with H2A residues E61 and E64 [62]. Further ionic interactions were observed between Ring1B K93

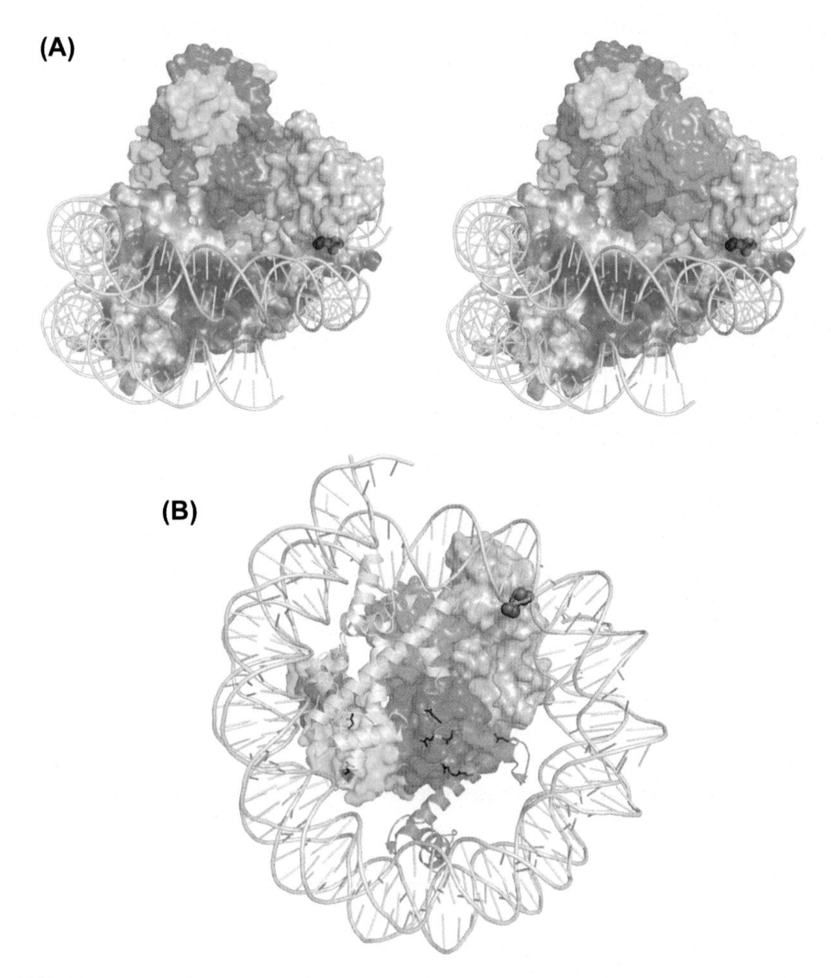

FIGURE 5.3 Structure of the PRC1 ubiquitin ligase module bound to the nucleosome core particle. Images were created in PyMOL (Shrödinger LLC) using the PDB files indicated. (A) Left: The solved structure, including NCP, Bmi1, and the fusion protein of Ring1B and UbcH5c (PDB code: 4R8P) [62]. For clarity, only the "proximal" side PRC1 module is shown, omitting the second copy that is present at the bottom of this view. The histones are shown in surface representation and colored according to electrostatic potential, and DNA is depicted in cartoon format. PRC1 is also in surface representation: Bmi1 is shown in green, Ring1B in purple, and UbcH5c in gray. Right: Same as left, but with ubiquitin (orange) placed in the location expected for the transfer−active complex (see text). The position of ubiquitin was determined by overlaying E2 from the BIRC7−UbcH5b−Ub complex structure (PDB code: 4AUQ) [32] onto E2 of 4R8P and was not energy minimized. (B) Histone-eye view of the interaction surface of PRC1, including the added ubiquitin described in A. Most histones have been omitted to show the underside of the bound PRC1 module. One copy each of histone H2A (brown), histone H3 (tan), and histone H4 (pale cyan) is shown in cartoon depiction. Modified lysines 118/119 of H2A are shown as blue spheres at upper right (note that not all atoms of these side chains are visible in the structure). The C-terminal tail of ubiquitin is enclosed by the UbcH5c surface just to the left of the lysines. Acidic histone residues discussed in the text are shown in stick representation and colored red. Most of these residues are present in H2A and interact with Ring1B, while others present in H3 and H4 interact with Bmi1.

and H2A E91 and between Ring1B R61 and H2A D72. Interactions between Bmi1 and the histone surface were not as extensive, but, as previously suggested [44], the loop region of the Bmi1 RING domain was found in contact. In particular, Bmi1 K62 interacts with histone H4 E74, while Bmi1 R64 interacts with histone H3 D77. In addition to the interactions of charged residues, it was observed that Bmi1 T63 is recognized at the histone surface [62]. Relative contributions from these residues were assessed by evaluating mutants in binding experiments and in H2A ubiquitination assays (see Table 5.1 for a summary of activity data). These studies revealed Ring1B R98 to be of greatest importance; mutation of this residue to alanine reduced both complex affinity and E3 ligase activity by at least 50-fold [62].

In addition to clarifying how the RING domains recognize nucleosome, the crystal structure reveals how the E2 UbcH5c is oriented on the nucleosome, and thus, how it may transfer ubiquitin selectively to the histone H2A C-terminal tail. It is not evident from the structure that the primary sequence immediately surrounding the site of ubiquitination is an important recognition element [62]. Instead it appears that the spatial constraints imposed by the ligase module binding to the nucleosome do not allow ubiquitin-charged E2 sufficient mobility to reach other lysines that could potentially be modified. One surprising aspect of the structure was direct contacts between basic UbcH5c surface patches and nucleosomal DNA. These additional interactions contribute to overall affinity of the E2-E3 module [62], and they may reinforce positioning of the module and, thereby, contribute to specificity. Finally, it should be noted that the same spatial constraints imposed by binding of the ligase module may explain not only selection of the lysine to be modified, but also why the major modification is monoubiquitination rather than ubiquitin chains. Once a single ubiquitin is transferred, it is unlikely that a second module charged with ubiquitin can approach closely enough to form the extensive array of contacts between enzyme and substrate seen in the structure. Thus, one might expect any second event to involve a lower affinity ligase–substrate interaction and, accordingly, to be less efficient. A requirement for the E2-nucleosome interaction might be especially important in guarding the H2A tail against chain formation.

VARIANT PRC1 COMPLEXES AND THE MECHANISM OF UBIQUITIN TRANSFER

To date, no structural studies of the PRC1 ubiquitin ligase module have included ubiquitin, so there is no direct view of how it may be transferred to histone H2A. However, a number of recent studies have characterized UbcH5-family E2s as activated ubiquitin thioesters (or their hydrolytically stable analogues) known as charged E2, or E2~Ub [28,29]; these include crystallographic and NMR studies of E2~Ub–RING E3 complexes [32,33,63–65]. These studies reveal general aspects of the ubiquitin transfer mechanism [29].

The ubiquitin thioester is conformationally mobile in E2~Ub in the absence of E3 binding, and E2~Ub is not especially activated for ubiquitin transfer. A key finding is that the RING E3 interacts directly with ubiquitin in the activated complex [31,66]; in the case of dimeric RINGs, the more distal RING (from E2) is important for the interaction. The distal RING E3–ubiquitin interaction is often hydrophobic and contributes to E3-E2~Ub affinity. Published structures [32,63] show that this hydrophobic interaction positions ubiquitin in a particular orientation relative to the E2 active site, thus activating it for transfer. These structures have been used to model ubiquitin transfer by Bmi1-Ring1B-UbcH5c. One difference between previously characterized RING dimers and Bmi1-Ring1B is that the expected interface with ubiquitin is polar rather than hydrophobic, and there is no enhanced binding of E2~Ub relative to that of E2 [32,57]. Comparison of the properties of variant PRC1 ligase modules has helped to understand the mechanism of ubiquitin recognition and transfer by this family of E3s [57].

In *Drosophila*, PRC1 consists of Polycomb (Pc), Sex Combs Extra (Sce), Posterior Sex Combs (Psc), and Polyhomeotic (Ph) [67,68]. However, duplications of the PcG genes have occurred in higher eukaryotes: in human there are five Pc-like proteins (chromobox 2 (CBX2), CBX4, CBX6, CBX7, CBX8), three Ph-like proteins (Polyhomeotic homolog 1 (PHC1), PHC2, PHC3), two Sce-like proteins (Ring1A, Ring1B), and six Psc proteins (Bmi1/PCGF4, Mel-18/PCGF2, PCGF1, PCGF3, PCGF5, PCGF6) [69]. Because the human PRC1 subunits each have multiple paralogs, many combinations are theoretically possible [69]. Recently, work in many labs has established that these subunits are not indiscriminately combined into complexes, but instead form four distinct types of PRC1 complexes named according to the PCGF family member present [70]. These are grouped into either "canonical" PRC1.2/PRC1.4 complexes, or the "noncanonical" complexes PRC1.1 (related to the *Drosophila* variant complex dRAF (dRING-associated factors) [71], PRC1.3/PRC1.5, and PRC1.6). The noncanonical variant complexes have functions that differ from canonical, or Bmi1/Mel-18 containing, PRC1 [70–82]. Accordingly, the copurifying subunits also differ by complex subtype and include proteins with that are not present in the canonical complexes [70,76,78]. Nevertheless, one common feature of all PRC1 complexes, canonical or noncanonical, is the presence of a Ring1 protein. This is not surprising based on the properties of the Bmi1-Ring1B RING–RING dimer described above: the PCGF RING domain is expected to require a Ring1 partner for stability. All six human PCGF RING domains form biochemically active, H2A-specific E3s in combination with the RING domains from either Ring1B or the closely related Ring1A [57]. Furthermore, several of the endogenous variant complexes, for example PRC1.1 or BCL6 corepressor (BCOR) complex [71,73] and PRC1.6 (E2F6/PRC1L4) [75] have been shown to have H2A K119 ubiquitination activity.

Our group recently compared the human Ring1B-PCGF RING–RING complexes in a range of biochemical assays, both those that report on H2A-independent (or "intrinsic") E3 activities or in assays measuring formation of uH2A on a nucleosome substrate [57]. Readouts of intrinsic activity included autoubiquitination of a GST-tag fused to Ring1B (a pseudosubstrate) and E3-catalyzed transfer of ubiquitin from E2~Ub to excess lysine (E2-discharge assay [83]) Measures of intrinsic activity revealed wide variation in catalytic efficiency among the six PCGF complexes [57]. A kinetic analysis of the E2-discharge rates showed that apparent second-order rate constants for reaction of lysine with PCGF-Ring1B-UbcH5c variants covered a 25-fold range, with rank order (from fastest to slowest ligases) of PCGF1 > PCGF6 ~ PCGF5 > PCGF3 > PCGF4 (Bmi1) > PCGF2 (Mel-18). Intriguingly, activities segregated between canonical and noncanonical PCGF RING–RING complexes, with the canonical variants distinctly less active than the noncanonical group. This difference was especially striking in autoubiquitination assays [57]. Comparing the sequences of the PCGF variants, a difference was observed in residues present at the modeled interface expected for activated ubiquitin in the E2~Ub–E3 complex. Although the residues are generally polar in all six PCGFs, the less active canonical variants Bmi1 and Mel-18 invariably (in human and most other species) have a K73-D77 pair (human Bmi1 numbering) on the surface of a helix ($\alpha 3$) predicted to contact ubiquitin. In most other PCGFs this is an arginine–aspartic acid pair, while in PCGF5 it is an asparagine–glutamic acid pair. The *Drosophila* ortholog Psc has an unusual threonine–alanine pair in these positions, but, like the human canonical PCGF proteins, it forms a poorly efficient ligase [57]. The importance of these putative ubiquitin contacts was tested by mutagenesis, and it was possible to convert slow ligases to more active (or the reverse) simply by substitution of these residues with pairs from the other type of ligase. Mutation of K11 or K33 of ubiquitin to alanine suppressed activity of two of the faster ligases (complexes of PCGF5 or PCGF6). Taken together, the mutagenesis results are consistent with the model of a polar interface between the PCGFs and ubiquitin in the activated complex [57]. Experimental support for a modeled ubiquitin-E3 contact surface allows the confident placement of ubiquitin into the E2-E3-NCP structure (Fig. 5.3A, right).

The evolutionary conservation of "slow" and "fast" PRC1 ligases, and the segregation of the rate differences to canonical and noncanonical PRC1, suggested that the property might be of biological importance. Accordingly, the six PCGF–Ring1B complexes were evaluated in H2A ubiquitination assays. Surprisingly, the rate differences were muted compared to the differences seen in nucleosome-free assays [57]. The explanation was traced to differences in the histone-binding loop of the PCGF subunit. The basic residues present in Bmi1 (described in the previous section) are quite different in the noncanonical PCGF RINGs, in the most extreme examples including acidic residues instead [57,62]. Exchanging these loop residues between PCGF5 and Bmi1 demonstrated that uH2A activity could be increased for the intrinsically fast PCGF5 by providing the more optimal nucleosome binding resides of Bmi1, while Bmi1 was a less efficient H2A ligase when the PCGF5 loop residues were present. Importantly, these loop swaps had no effect in the nucleosome-free assays and had additive effects when combined with mutations in the ubiquitin-binding site. For example, it was possible to engineer a Bmi1 gain-of-function mutant by increasing the "intrinsic" E3 activity [57] (Table 5.1). The PCGF RINGs are therefore highly modular functional units with different surfaces dedicated to binding Ring1B, ubiquitin and nucleosome, each of which is important in the overall rate of H2A ubiquitination. This high density of function may explain why each of the six PCGF RING domains is nearly invariant across species over its entire sequence. However, it remains uncertain why it might be important to regulate overall activity in two different ways and whether there might be any biological consequence to altering the overall rate or changing the balance between nucleosome recognition and intrinsic E3 catalytic competence.

WHICH PRC1 CONTRIBUTES MOST TO H2A UBIQUITINATION?

The discovery of PRC1 variants, all of which include Ring1A/B, prompted investigation of how these different complexes might divide the task of uH2A deposition. The issue is complicated by a subdivision of PRC1.4 and PRC1.2 into further subvariants [70]. A primary differentiator of "canonical" and "noncanonical" PRC1 variants is whether the complex contains RING1B YY1−Binding Protein (RYBP) or the mutually exclusive (and more canonical) combination of a PHC-family and a CBX-family subunit. In the case of PRC1.2/1.4, it appears that each occurs and that the complexes can be physically separated [70]. The isolated PRC1.4 subcomplexes show different levels of activity in a nucleosome ubiquitination assay, with the RYBP-containing PRC1.4 showing substantially more activity at equivalent concentrations of Ring1B [70]. In addition, shRNA knockdown of RYBP in ESCs strongly reduced the global level of uH2A, suggesting that the majority of this mark is placed by the RYBP-containing variants of PRC1 [70]. This finding would appear to be at odds with a later study that reported higher levels of uH2A at CBX7 target genes than at those bound by RYBP [78]. This difference might be explained by mutual reinforcement between PRC1 and PRC2 activities (discovered later; see below) and the ability of CBX7 to bind to the product of PRC2 (H3K27me3); any reinforcing contribution from CBX7 binding would not be present in the assay system involving purified PRC1 variants [70]. It is more difficult to reconcile the strong effect of RYBP knockdown on uH2A with generally lower uH2A at RYBP target loci, although the cells used in the experiments were different (293 cells [70] vs mouse ESCs [78], respectively).

To compare more directly the activities of the PRC1 variant complexes, the Klose group and collaborators used an engineered system that allows targeting to a defined chromatin domain through fusion of query proteins to the sequence-specific DNA-binding protein TetR (tetracycline repressor) [82]. This approach was applied to five of the six PCGF family members (1–5), and Chromatin immunoprecipitation (ChIP) analysis demonstrated that the PCGF fusion proteins bound to the engineered Tet operator array and recruited Ring1B, suggesting that the TetR fusion did not interfere with formation of PRC1 (verified by mass spectrometry for PCGF4) [82]. Strikingly, it was seen in this experimental system that, despite roughly equivalent Ring1B recruitment, PCGF2 (Mel-18) and PCGF4 (Bmi1) complexes did not support significant deposition of uH2A compared to complexes forming around PCGFs 1, 3, and 5. To rule out effects from the fusion partner that might have somehow remained undetected, the PRC2 subunit EED was fused to TetR instead; this approach led to incorporation of H3K27me3 and "natural" recruitment via CBX7 of other "canonical" PRC1 subunits and Ring1B. Nevertheless, the level of uH2A was again very low. These findings were in accord with an earlier study showing the importance of the noncanonical complex dRAF for maintaining uH2A and homeotic gene regulation in *Drosophila* [71], and the authors further show a role for the noncanonical complex PRC1.1 in establishing correct axial skeletal patterning [82], a process traditionally associated with canonical PCGF subunits such as Bmi1 and Mel-18.

IS H2A UBIQUITINATION REALLY IMPORTANT?

Given that ubiquitination of H2A K119 is an enzymatic function shared by all PRC1 variant complexes, the fact that it is evolutionarily conserved, and the observation of developmental defects of varying severity in knockout animals that do not express one or more of the RING-containing subunits [84], one might expect that the ligase activity is an especially important function of PRC1. However, a number of studies have questioned the importance of the ligase activity relative to other functions of PRC1, such as chromatin compaction [20]. Because complete deletion of Ring1B or Bmi1 is expected to affect a number of aspects of PRC1 function, including overall complex stability, recent studies have been aimed at a more targeted disruption of the ligase function and evaluation of the importance of H2A K119ub to PRC1-enforced chromatin compaction and gene repression [85–87] (recently reviewed [88,89]). Using a FISH-based method to relate the distance between the *Hoxb1* and *Hoxb9* loci (and those between other pairs) to Ring1B-dependent chromatin compaction, the Bickmore group and collaborators found that the compaction defect of *Ring1B*$^{-/-}$ embryonic stem cells (ESCs) could be rescued both by wild type and I52A Ring1B [85]. As expected, levels of uH2A in cells expressing the mutant remained low, consistent with the poor ligase activity of this mutant (Table 5.1) [51]. In addition, several *Hox* genes that were activated (relative to parental) in *Ring1B*$^{-/-}$ ESCs were repressed in cells reconstituted with both wild type and I53A Ring1B [85]. Taken together, these results indicate that the ubiquitin ligase function of Ring1B is not required for chromatin compaction or gene repression at *Hox* loci [85] and raise the additional question of whether the ligase function might be completely dispensable.

Working again in murine ESCs, Endoh et al. [86] introduced wild-type Ring1B or the mutants I53A or I53S into *Ring1A*$^{-/-}$ *Ring1B* conditional knockout cells and, after deletion of *Ring1B*, observed that ligase activity and H2A ubiquitination were dispensable for chromatin compaction. In addition, Ring1B mutants associated with other PRC1 proteins, and the complexes bound to PRC1 target genes. These results largely agreed with the findings of Eskeland et al. [85]. However, clear differences were seen between *Ring1B*$^{-/-}$ and *Ring1A/B* double knockout (dKO) cells when each was evaluated for *Hox* gene repression and maintenance of ESCs in a dedifferentiated state. Restoration of *Ring1A* to the dKO reversed these differences, demonstrating that Ring1A could compensate for loss of Ring1B function in ESCs [86]. Evaluation of Ring1B mutants in the dKO cells showed that the catalytically inactive variants could not restore *Hox* gene repression, nor could they support maintenance of the dedifferentiated state. Taken together, these results indicate that uH2A (or Ring1 activity) is required for functions of PRC1 other than chromatin compaction [86] and that the conflicting observations of the earlier study [85] may have been the result of low-level Ring1A activity blunting the effect of loss of Ring1B catalytic activity.

The importance of Ring1 was investigated in parallel in *Drosophila* [89]. Interestingly, PRC1 target genes could be divided into two groups, those whose repression required all PRC1 subunits (Class 1 genes) and those whose repression did not require the sole *Drosophila* Ring1 ortholog Sce or its direct binding partner Pc (Class 2 genes). While there are some clear differences between the *Drosophila* and murine systems, such as stable expression of the Bmi1 ortholog Psc in Sce null animals, it is evident that for regulation of some target genes (Class 2) the Sce E3 ligase activity cannot be required. It remained unclear from this study whether the E3 ligase activity might be important for proper regulation of Class 1 genes [87]. A more recent study from the Müller group [90] addressed the importance of E3 activity directly. Mutant animals were generated that expressed only Sce I48A (analogous to the I53A mutation of human or mouse protein). Expression of all Polycomb proteins tested appeared to be normal in these animals, but the level of uH2A was reduced by >98%; associated with the loss of uH2A was a reduction in global levels of H3K27me3 (1.5-fold) [90]. Phenotypically, the I48A mutants failed to show any of the homeotic transformations typical of Polycomb group mutant embryos and Sce null embryos in particular. However, the I48A embryos showed developmental defects at a later stage. No consistent gross defects were noted, but instead a range of more subtle and sporadic findings was seen [90]. A second mutant that did not eliminate maternally supplied Sce was used to extend observation to postembryonic development. Again defects were relatively subtle, although these mutants failed to develop fully to adulthood. Significantly, *Hox* gene expression appeared to be regulated normally despite the near lack of uH2A. To complete the picture, these investigators generated mutant animals expressing H2A with K-to-R substitutions that block ubiquitination, finding again that *Hox* gene repression was intact [90]. These studies suggest that, while these animals are not completely normal, loss of uH2A alone does not recapitulate the classic Polycomb phenotypes associated with other PRC1 mutants.

The role of Ring1B during mouse development was explored in mice expressing only Ring1B I53A from the endogenous locus, again in a *Ring1A*$^{+/+}$ background [91]. Chromatin immunoprecipitation and sequencing (ChIP-seq) studies in derived ESCs showed that the lack of Ring1B catalytic activity led to reduced levels of both H3K27me3 and Ring1B itself at transcriptional start sites (TSS) normally occupied by wild-type Ring1B. In addition, the ChIP-seq data indicated a redistribution of Ring1B 153A away from TSS and into gene bodies. These differences did not result in major changes in gene repression compared to the large effects seen in *Ring1B*$^{-/-}$ ESCs, although a subset of genes was impacted significantly. *Ring1B*$^{I53A/I53A}$ embryos develop (at sub-Mendelian frequencies) to a later stage (E15.5) than do *Ring1B*$^{-/-}$ embryos (E10.5) but do not yield any live pups. The E9.5 and E12.5 embryos do not show the patterning defects that are characteristic of Polycomb mutations [91], consistent with the finding in *Drosophila* [90]. Overall, these studies point to a critical role for Ring1 ubiquitination activity in later stages of development but to no obvious requirement for this activity in earlier stages (at which the Ring1 protein is nevertheless required).

WHAT DOES uH2A DO?

A number of specific roles for uH2A have been proposed; however, based on the intensity of investigation, it seems the function of uH2A is incompletely defined at present. Several groups report that uH2A can interfere with aspects of RNA polymerase II function [92–94]. More recently "readers" of uH2A have been discovered [82,95–97], a development widely discussed [98–100]. Two collaborative studies from the Brockdorff [97] and Klose [82] groups provided evidence challenging the "hierarchical" model of PRC1 recruitment via CBX proteins to regions previously modified by PRC2 methylation of H3K27. Using the engineered TetR-TetO chromatin targeting system described above, not only did variant PRC1 complexes direct substantial increases in uH2A, they also unexpectedly recruited PRC2 components resulting in increased H3K27me3 [82]. Similar observations were made in a different experimental system (that likewise isolates the observation from effects of host cell genes) involving hypomethylated pericentric heterochromatin (PCH) regions [97]. Tethering KDM2B (a component of noncanonical PRC1.1), but not EZH2, to PCH resulted in an increase in uH2A. An increase in H3K27me3 was also observed in the KDM2B tethering experiment. To test the sufficiency of uH2A for PRC2 recruitment (rather than a protein–protein interaction with bound PRC1), a TetR fusion protein was designed including only the minimal catalytic domains of Bmi1 and Ring1B (linked as a single-chain construct) [97]. Expression of the minimal fusion protein in both the hypomethylated PCH [97] and TetO [82] assay systems supported increased deposition of uH2A and PRC2 components compared to an inactive mutant of the same construct. These studies show that PRC1 ubiquitination can enhance PRC2 recruitment and activity, a counterpoint to the original "hierarchical" model [98,99]. One minor concern is the incorporation of Bmi1 mutation C51G (sic) in addition to the Ring1B I53A catalytic mutation. This type of mutation is expected to destabilize the fold of the RING domain, not just eliminate activity (see Table 5.1, and note that "C53G" is likely what the authors intended: Bmi1 residue 51 is lysine).

A possible mechanism for PRC2 recruitment to sites of PRC1 activity was revealed by proteomics analysis of uH2A-binding partners from *Drosophila* and mESC nuclear extracts [96]. Strong enrichment of PRC2-associated subunits Jarid2 and AEBP2 was observed. The PRC2 core complex was no more active toward chromatin substrates modified by uH2A than toward unmodified substrates, and both Jarid2 and AEBP2 stimulated methylation of the unmodified substrate. However, AEBP2 strongly stimulated methylation of uH2A substrates with an additional contribution seen from added Jarid2. The combined data suggest that AEBP2 might bind to uH2A and thereby recruit PRC2 to uH2A-modified chromatin. The authors note that variant PRC1 components, such as RYBP, also appear in the uH2A pull-downs and suggest that that these interactions may provide a mechanism for feedback regulation of PRC1 activity [96].

Transcription factors may also recognize uH2A [95]. Using a biochemical purification strategy with Flag-H2A mononucleosomes as bait, both H2A and ubiquitin-binding proteins were fractionated and analyzed by mass spectrometry. This approach identified ZRF1 as a possible ubiquitin-dependent interactor. Because the approach might be expected to yield proteins interacting with any ubiquitin-modified nucleosome, specificity for uH2A was established through the dependence of ZRF1 binding on Ring1B and the presence of uH2A at PRC1 target promoters. In addition, affinity purification with tagged, mutant H2A that cannot be ubiquitinated by Ring1B did not enrich ZRF1. Interestingly, it appeared that ZRF1 and Ring1B competed for binding to uH2A target sites (despite an initial requirement for Ring1B modification of H2A), and ZRF1 was found to transcriptionally activate these genes [95], an outcome not typically associated with uH2A. It will be interesting to see whether additional transcriptional regulators emerge as "readers" of uH2A or, perhaps, as locus-specific recruiters of PRC1 [101].

WHAT'S NEXT FOR PRC1 AND uH2A?

The coming years should be an exciting time for the PRC1 ubiquitin ligase. Many key questions need further investigation. For example, we know very little about how H2A ubiquitination might be regulated in the context of intact complexes, and the work cited above indicates that there are major differences among the variant PRC1 complexes identified so far. One possibility, aside from the different subunit composition, is posttranslational modification of PRC1 components, and phosphorylation [102] and ubiquitination [103,104] have each been reported to modulate PRC1 activity. In addition, much work to date has focused on characterizing PRC1 function in whole animals or ESCs, and it is becoming evident that details of function in specific tissues and processes will provide new insights [105]. An exciting direction will be to understand how the many reported H2A deubiquitinases [43] and PRC1 work together, and of especially high interest is BAP1. In *Drosophila*, the BAP1 ortholog Calypso and its activator ASX (which together form the major H2A deubiquitinase complex PR-DUB) are required for *Hox* gene repression [89,106]. Both BAP1 and ASXL (ASX-like) proteins are frequently mutated in human tumors, and the complex may have important regulatory functions in addition to H2A deubiquitination [107]. Beyond BAP1, the finding of context-dependent SCML2-driven regulation of PRC1 by the deubiquitinase USP7 [104,105] suggests that further investigation of the H2A Dubs in different settings will be important. Finally, it seems that Ring1 is not the sole H2A ubiquitin K118/K119 ligase: TRIM37 was recently reported to have this activity, to interact with and recruit PRC2 subunits to target genes, and to be an oncogene frequently overexpressed in breast cancer [108]. This role for TRIM37 raises the very interesting possibility that other E3 ligases may also be found capable of circumventing the many layers of PRC1 regulation to directly impact therapeutically relevant gene expression.

REFERENCES

[1] Jürgens G. A group of genes controlling the spatial expression of the bithorax complex in *Drosophila*. Nature 1985;316:153–5.

[2] Schwartz YB, Pirrotta V. Polycomb silencing mechanisms and the management of genomic programmes. Nat Rev Genet 2007;8:9–22.

[3] Ringrose L, Paro R. Epigenetic regulation of cellular memory by the Polycomb and Trithorax group proteins. Annu Rev Genet 2004;38:413–43.

[4] Schuettengruber B, Chourrout D, Vervoort M, Leblanc B, Cavalli G. Genome regulation by polycomb and trithorax proteins. Cell 2007;128:735–45.

[5] Schuettengruber B, Martinez AM, Iovino N, Cavalli G. Trithorax group proteins: switching genes on and keeping them active. Nat Rev Mol Cell Biol 2011;12:799–814.

[6] Steffen PA, Ringrose L. What are memories made of? How Polycomb and Trithorax proteins mediate epigenetic memory. Nat Rev Mol Cell Biol 2014;15:340–56.

[7] Sparmann A, van Lohuizen M. Polycomb silencers control cell fate, development and cancer. Nat Rev Cancer 2006;6:846–56.

[8] Pietersen AM, van Lohuizen M. Stem cell regulation by polycomb repressors: postponing commitment. Curr Opin Cell Biol 2008;20:201−7.

[9] Bracken AP, Helin K. Polycomb group proteins: navigators of lineage pathways led astray in cancer. Nat Rev Cancer 2009;9:773−84.

[10] Bruggeman SW, Valk-Lingbeek ME, van der Stoop PP, Jacobs JJ, Kieboom K, Tanger E, Hulsman D, Leung C, Arsenijevic Y, Marino S, Van Lohuizen M. Ink4a and Arf differentially affect cell proliferation and neural stem cell self-renewal in Bmi1-deficient mice. Genes Dev 2005;19:1438−43.

[11] Park IK, Qian D, Kiel M, Becker MW, Pihalja M, Weissman IL, Morrison SJ, Clarke MF. Bmi-1 is required for maintenance of adult self-renewing haematopoietic stem cells. Nature 2003;423:302−5.

[12] Liu S, Dontu G, Mantle ID, Patel S, Ahn NS, Jackson KW, Suri P, Wicha MS. Hedgehog signaling and Bmi-1 regulate self-renewal of normal and malignant human mammary stem cells. Cancer Res 2006;66:6063−71.

[13] Simon JA, Kingston RE. Mechanisms of polycomb gene silencing: knowns and unknowns. Nat Rev Mol Cell Biol 2009;10:697−708.

[14] Simon JA, Kingston RE. Occupying chromatin: Polycomb mechanisms for getting to genomic targets, stopping transcriptional traffic, and staying put. Mol Cell 2013;49:808−24.

[15] Di Croce L, Helin K. Transcriptional regulation by Polycomb group proteins. Nat Struct Mol Biol 2013;20:1147−55.

[16] Cao R, Wang L, Wang H, Xia L, Erdjument-Bromage H, Tempst P, Jones RS, Zhang Y. Role of histone H3 lysine 27 methylation in polycomb-group silencing. Science 2002;298:1039−43.

[17] Czermin B, Melfi R, McCabe D, Seitz V, Imhof A, Pirrotta V. *Drosophila* enhancer of Zeste/ESC complexes have a histone H3 methyltransferase activity that marks chromosomal Polycomb sites. Cell 2002;111:185−96.

[18] Wang H, Wang L, Erdjument-Bromage H, Vidal M, Tempst P, Jones RS, Zhang Y. Role of histone H2A ubiquitination in Polycomb silencing. Nature 2004;431:873−8.

[19] de Napoles M, Mermoud JE, Wakao R, Tang YA, Endoh M, Appanah R, Nesterova TB, Silva J, Otte AP, Vidal M, Koseki H, Brockdorff N. Polycomb group proteins Ring1A/B link ubiquitylation of histone H2A to heritable gene silencing and X inactivation. Dev Cell 2004;7:663−76.

[20] Francis NJ, Kingston RE, Woodcock CL. Chromatin compaction by a polycomb group protein complex. Science 2004;306:1574−7.

[21] Ciferri C, Lander GC, Maiolica A, Herzog F, Aebersold R, Nogales E. Molecular architecture of human polycomb repressive complex 2. eLife 2012:1. e00005.

[22] Jiao L, Liu X. Structural basis of histone H3K27 trimethylation by an active polycomb repressive complex 2. Science 2015:350. aac4383.

[23] Huang CS, Nogales E, Ciferri C. Molecular architecture of the polycomb repressive complex 2. 2016 [Chapter 8, this volume].

[24] Pickart CM. Mechanisms underlying ubiquitination. Annu Rev Biochem 2001;70:503−33.

[25] Pickart CM, Eddins MJ. Ubiquitin: structures, functions, mechanisms. Biochim Biophys Acta 2004;1695:55−72.

[26] Kerscher O, Felberbaum R, Hochstrasser M. Modification of proteins by ubiquitin and ubiquitin-like proteins. Annu Rev Cell Dev Biol 2006;22:159−80.

[27] Ye Y, Rape M. Building ubiquitin chains: E2 enzymes at work. Nat Rev Mol Cell Biol 2009;10:755−64.

[28] Wenzel DM, Stoll KE, Klevit RE. E2s: structurally economical and functionally replete. Biochem J 2011;433:31–42.

[29] Metzger MB, Pruneda JN, Klevit RE, Weissman AM. RING-type E3 ligases: master manipulators of E2 ubiquitin-conjugating enzymes and ubiquitination. Biochim Biophys Acta 2014;1843:47–60.

[30] Deshaies RJ, Joazeiro CAP. RING domain E3 ubiquitin ligases. Annu Rev Biochem 2009;78:399–434.

[31] Plechanovova A, Jaffray EG, McMahon SA, Johnson KA, Navratilova I, Naismith JH, Hay RT. Mechanism of ubiquitylation by dimeric RING ligase RNF4. Nat Struct Mol Biol 2011;18:1052–9.

[32] Dou H, Buetow L, Sibbet GJ, Cameron K, Huang DT. BIRC7-E2 ubiquitin conjugate structure reveals the mechanism of ubiquitin transfer by a RING dimer. Nat Struct Mol Biol 2012;19:876–83.

[33] Pruneda JN, Littlefield PJ, Soss SE, Nordquist KA, Chazin WJ, Brzovic PS, Klevit RE. Structure of an E3:E2 ~ Ub complex reveals an allosteric mechanism shared among RING/U-box ligases. Mol Cell 2012;47:933–42.

[34] Komander D, Rape M. The ubiquitin code. Annu Rev Biochem 2012;81:203–29.

[35] Behrends C, Harper JW. Constructing and decoding unconventional ubiquitin chains. Nat Struct Mol Biol 2011;18:520–8.

[36] Kulathu Y, Komander D. Atypical ubiquitylation - the unexplored world of polyubiquitin beyond Lys48 and Lys63 linkages. Nat Rev Mol Cell Biol 2012;13:508–23.

[37] Hospenthal MK, Freund SM, Komander D. Assembly, analysis and architecture of atypical ubiquitin chains. Nat Struct Mol Biol 2013;20:555–65.

[38] Ramanathan HN, Ye Y. Cellular strategies for making monoubiquitin signals. Crit Rev Biochem Mol Biol 2012;47:17–28.

[39] Reyes-Turcu FE, Ventii KH, Wilkinson KD. Regulation and cellular roles of ubiquitin-specific deubiquitinating enzymes. Annu Rev Biochem 2009;78:363–97.

[40] Nijman SM, Luna-Vargas MP, Velds A, Brummelkamp TR, Dirac AM, Sixma TK, Bernards R. A genomic and functional inventory of deubiquitinating enzymes. Cell 2005;123:773–86.

[41] Kohler A, Zimmerman E, Schneider M, Hurt E, Zheng N. Structural basis for assembly and activation of the heterotetrameric SAGA histone H2B deubiquitinase module. Cell 2010;141:606–17.

[42] Samara NL, Datta AB, Berndsen CE, Zhang X, Yao T, Cohen RE, Wolberger C. Structural insights into the assembly and function of the SAGA deubiquitinating module. Science 2010;328:1025–9.

[43] Belle JI, Nijnik A. H2A-DUBbing the mammalian epigenome: expanding frontiers for histone H2A deubiquitinating enzymes in cell biology and physiology. Int J Biochem Cell Biol 2014;50C:161–74.

[44] Bentley ML, Corn JE, Dong KC, Phung Q, Cheung TK, Cochran AG. Recognition of UbcH5c and the nucleosome by the Bmi1/Ring1b ubiquitin ligase complex. EMBO J 2011;30:3285–97.

[45] Fang J, Chen T, Chadwick B, Li E, Zhang Y. Ring1b-mediated H2A ubiquitination associates with inactive X chromosomes and is involved in initiation of X inactivation. J Biol Chem 2004;279:52812–5.

[46] Cao R, Tsukada Y, Zhang Y. Role of Bmi-1 and Ring1A in H2A ubiquitylation and Hox gene silencing. Mol Cell 2005;20:845–54.

[47] Wei J, Zhai L, Xu J, Wang H. Role of Bmi1 in H2A ubiquitylation and Hox gene silencing. J Biol Chem 2006;281:22537–44.

[48] Ben-Saadon R, Zaaroor D, Ziv T, Ciechanover A. The polycomb protein Ring1B generates self atypical mixed ubiquitin chains required for its in vitro histone H2A ligase activity. Mol Cell 2006;24:701–11.

[49] Hemenway CS, Halligan BW, Levy LS. The Bmi-1 oncoprotein interacts with dinG and MPh2: the role of RING finger domains. Oncogene 1998;16:2541–7.

[50] Satijn DP, Otte AP. RING1 interacts with multiple Polycomb-group proteins and displays tumorigenic activity. Mol Cell Biol 1999;19:57–68.

[51] Buchwald G, van der Stoop P, Weichenrieder O, Perrakis A, van Lohuizen M, Sixma TK. Structure and E3-ligase activity of the Ring-Ring complex of polycomb proteins Bmi1 and Ring1b. EMBO J 2006;25:2465–74.

[52] Li Z, Cao R, Wang M, Myers MP, Zhang Y, Xu RM. Structure of a Bmi-1-Ring1B polycomb group ubiquitin ligase complex. J Biol Chem 2006;281:20643–9.

[53] Brzovic PS, Rajagopal P, Hoyt DW, King MC, Klevit RE. Structure of a BRCA1-BARD1 heterodimeric RING-RING complex. Nat Struct Biol 2001;8:833–7.

[54] Mace PD, Linke K, Feltham R, Schumacher FR, Smith CA, Vaux DL, Silke J, Day CL. Structures of the cIAP2 RING domain reveal conformational changes associated with ubiquitin-conjugating enzyme (E2) recruitment. J Biol Chem 2008;283:31633–40.

[55] Brzovic PS, Keeffe JR, Nishikawa H, Miyamoto K, Fox D, Fukuda M, Ohta T, Klevit R. Binding and recognition in the assembly of an active BRCA1/BARD1 ubiquitin-ligase complex. Proc Natl Acad Sci USA 2003;100:5646–51.

[56] van Wijk SJ, Timmers HT. The family of ubiquitin-conjugating enzymes (E2s): deciding between life and death of proteins. FASEB J 2010;24:981–93.

[57] Taherbhoy AM, Huang OW, Cochran AG. BMI1-RING1B is an autoinhibited RING E3 ubiquitin ligase. Nat Commun 2015;6:7621.

[58] Mattiroli F, Sixma TK. Lysine-targeting specificity in ubiquitin and ubiquitin-like modification pathways. Nat Struct Mol Biol 2014;21:308–16.

[59] Mattiroli F, Uckelmann M, Sahtoe DD, van Dijk WJ, Sixma TK. The nucleosome acidic patch plays a critical role in RNF168-dependent ubiquitination of histone H2A. Nat Commun 2014;5:3291.

[60] Leung JW, Agarwal P, Canny MD, Gong F, Robison AD, Finkelstein IJ, Durocher D, Miller KM. Nucleosome acidic patch promotes RNF168- and RING1B/BMI1-dependent H2AX and H2A ubiquitination and DNA damage signaling. PLoS Genet 2014;10:e1004178.

[61] Barbera AJ, Chodaparambil JV, Kelley-Clarke B, Joukov V, Walter JC, Luger K, Kaye KM. The nucleosomal surface as a docking station for Kaposi's sarcoma herpesvirus LANA. Science 2006;311:856–61.

[62] McGinty RK, Henrici RC, Tan S. Crystal structure of the PRC1 ubiquitylation module bound to the nucleosome. Nature 2014;514:591–6.

[63] Plechanovova A, Jaffray EG, Tatham MH, Naismith JH, Hay RT. Structure of a RING E3 ligase and ubiquitin-loaded E2 primed for catalysis. Nature 2012;489:115–20.

[64] Dou H, Buetow L, Sibbet GJ, Cameron K, Huang DT. Essentiality of a non-RING element in priming donor ubiquitin for catalysis by a monomeric E3. Nat Struct Mol Biol 2013;20:982–6.

[65] Buetow L, Gabrielsen M, Anthony NG, Dou H, Patel A, Aitkenhead H, Sibbet GJ, Smith BO, Huang DT. Activation of a primed RING E3-E2-ubiquitin complex by non-covalent ubiquitin. Mol Cell 2015;58:297–310.

[66] Reverter D, Lima CD. Insights into E3 ligase activity revealed by a SUMO-RanGAP1-Ubc9-Nup358 complex. Nature 2005;435:687–92.

[67] Shao Z, Raible F, Mollaaghababa R, Guyon JR, Wu CT, Bender W, Kingston RE. Stabilization of chromatin structure by PRC1, a Polycomb complex. Cell 1999;98:37–46.

[68] Francis NJ, Saurin AJ, Shao Z, Kingston RE. Reconstitution of a functional core polycomb repressive complex. Mol Cell 2001;8:545–56.

[69] Gil J, Peters G. Regulation of the INK4b-ARF-INK4a tumour suppressor locus: all for one or one for all. Nat Rev Mol Cell Biol 2006;7:667–77.

[70] Gao Z, Zhang J, Bonasio R, Strino F, Sawai A, Parisi F, Kluger Y, Reinberg D. PCGF homologs, CBX proteins, and RYBP define functionally distinct PRC1 family complexes. Mol Cell 2012;45:344–56.

[71] Lagarou A, Mohd-Sarip A, Moshkin YM, Chalkley GE, Bezstarosti K, Demmers JA, Verrijzer CP. dKDM2 couples histone H2A ubiquitylation to histone H3 demethylation during Polycomb group silencing. Genes Dev 2008;22:2799–810.

[72] Trimarchi JM, Fairchild B, Wen J, Lees JA. The E2F6 transcription factor is a component of the mammalian Bmi1-containing polycomb complex. Proc Natl Acad Sci USA 2001;98:1519–24.

[73] Gearhart MD, Corcoran CM, Wamstad JA, Bardwell VJ. Polycomb group and SCF ubiquitin ligases are found in a novel BCOR complex that is recruited to BCL6 targets. Mol Cell Biol 2006;26:6880–9.

[74] Sánchez C, Sánchez I, Demmers JAA, Rodriguez P, Strouboulis J, Vidal M. Proteomics analysis of Ring1B/Rnf2 interactors identifies a novel complex with the Fbxl10/Jhdm1B histone demethylase and the Bcl6 interacting corepressor. Mol Cell Proteomics 2007;6:820–34.

[75] Trojer P, Cao AR, Gao Z, Li Y, Zhang J, Xu X, Li G, Losson R, Erdjument-Bromage H, Tempst P, Farnham PJ, Reinberg D. L3MBTL2 protein acts in concert with PcG protein-mediated monoubiquitination of H2A to establish a repressive chromatin structure. Mol Cell 2011;42:438–50.

[76] Vandamme J, Völkel P, Rosnoblet C, Le Faou P, Angrand P-O. Interaction proteomics analysis of Polycomb proteins defines distinct PRC1 complexes in mammalian cells. Mol Cell Proteomics 2011;10:1–23.

[77] Tavares L, Dimitrova E, Oxley D, Webster J, Poot R, Demmers J, Bezstarosti K, Taylor S, Ura H, Koide H, Wutz A, Vidal M, Elderkin S, Brockdorff N. RYBP-PRC1 complexes mediate H2A ubiquitylation at polycomb target sites independently of PRC2 and H3K27me3. Cell 2012;148:664–78.

[78] Morey L, Aloia L, Cozzuto L, Benitah SA, Di Croce L. RYBP and Cbx7 define specific biological functions of polycomb complexes in mouse embryonic stem cells. Cell Rep 2013;3:60–9.

[79] Farcas AM, Blackledge NP, Sudbery I, Long HK, McGouran JF, Rose NR, Lee S, Sims D, Cerase A, Sheahan TW, Koseki H, Brockdorff N, Ponting CP, Kessler BM, Klose RJ. KDM2B links the polycomb repressive complex 1 (PRC1) to recognition of CpG islands. eLife 2012:1. e00205.

[80] Wu X, Johansen JV, Helin K. Fbxl10/Kdm2b recruits polycomb repressive complex 1 to CpG islands and regulates H2A ubiquitylation. Mol Cell 2013;49:1134–46.

[81] He J, Shen L, Wan M, Taranova O, Wu H, Zhang Y. Kdm2b maintains murine embryonic stem cell status by recruiting PRC1 complex to CpG islands of developmental genes. Nat Cell Biol 2013;15:373–84.

[82] Blackledge NP, Farcas AM, Kondo T, King HW, McGouran JF, Hanssen LL, Ito S, Cooper S, Kondo K, Koseki Y, Ishikura T, Long HK, Sheahan TW, Brockdorff N, Kessler BM, Koseki H, Klose RJ. Variant PRC1 complex-dependent H2A ubiquitylation drives PRC2 recruitment and polycomb domain formation. Cell 2014;157:1445–59.

[83] Wenzel DM, Lissounov A, Brzovic PS, Klevit RE. UBCH7 reactivity profile reveals parkin and HHARI to be RING/HECT hybrids. Nature 2011;474:105–8.

[84] van Lohuizen M. Functional analysis of mouse Polycomb group genes. Cell Mol Life Sci 1998;54:71–9.

[85] Eskeland R, Leeb M, Grimes GR, Kress C, Boyle S, Sproul D, Gilbert N, Fan Y, Skoultchi AI, Wutz A, Bickmore WA. Ring1B compacts chromatin structure and represses gene expression independent of histone ubiquitination. Mol Cell 2010;38:452–64.

[86] Endoh M, Endo TA, Endoh T, Isono K, Sharif J, Ohara O, Toyoda T, Ito T, Eskeland R, Bickmore WA, Vidal M, Bernstein BE, Koseki H. Histone H2A mono-ubiquitination is a crucial step to mediate PRC1-dependent repression of developmental genes to maintain ES cell identity. PLoS Genet 2012;8:e1002774.

[87] Gutiérrez L, Oktaba K, Scheuermann JC, Gambetta MC, Ly-Hartig N, Muller J. The role of the histone H2A ubiquitinase Sce in Polycomb repression. Development 2012;139:117–27.

[88] Simon JA. Chromatin compaction at Hox loci: a polycomb tale beyond histone tails. Mol Cell 2010;38:321–2.

[89] Scheuermann JC, Gutiérrez L, Muller J. Histone H2A monoubiquitination and Polycomb repression: the missing pieces of the puzzle. Fly 2012;6:162–8.

[90] Pengelly AR, Kalb R, Finkl K, Muller J. Transcriptional repression by PRC1 in the absence of H2A monoubiquitylation. Genes Dev 2015;29:1487–92.

[91] Illingworth RS, Moffat M, Mann AR, Read D, Hunter CJ, Pradeepa MM, Adams IR, Bickmore WA. The E3 ubiquitin ligase activity of RING1B is not essential for early mouse development. Genes Dev 2015;29:1897–902.

[92] Stock JK, Giadrossi S, Casanova M, Brookes E, Vidal M, Koseki H, Brockdorff N, Fisher AG, Pombo A. Ring1-mediated ubiquitination of H2A restrains poised RNA polymerase II at bivalent genes in mouse ES cells. Nat Cell Biol 2007;9:1428–35.

[93] Zhou W, Zhu P, Wang J, Pascual G, Ohgi KA, Lozach J, Glass CK, Rosenfeld MG. Histone H2A monoubiquitination represses transcription by inhibiting RNA polymerase II transcriptional elongation. Mol Cell 2008;29:69–80.

[94] Lehmann L, Ferrari R, Vashisht AA, Wohlschlegel JA, Kurdistani SK, Carey M. Polycomb repressive complex 1 (PRC1) disassembles RNA polymerase II preinitiation complexes. J Biol Chem 2012;287:35784–94.

[95] Richly H, Rocha-Viegas L, Ribeiro JD, Demajo S, Gundem G, Lopez-Bigas N, Nakagawa T, Rospert S, Ito T, Di Croce L. Transcriptional activation of polycomb-repressed genes by ZRF1. Nature 2010;468:1124–8.

[96] Kalb R, Latwiel S, Baymaz HI, Jansen PW, Muller CW, Vermeulen M, Muller J. Histone H2A monoubiquitination promotes histone H3 methylation in Polycomb repression. Nat Struct Mol Biol 2014;21:569–71.

[97] Cooper S, Dienstbier M, Hassan R, Schermelleh L, Sharif J, Blackledge NP, De Marco V, Elderkin S, Koseki H, Klose R, Heger A, Brockdorff N. Targeting polycomb to pericentric heterochromatin in embryonic stem cells reveals a role for H2AK119u1 in PRC2 recruitment. Cell Rep 2014;7:1456–70.

[98] Schwartz YB, Pirrotta V. Ruled by ubiquitylation: a new order for polycomb recruitment. Cell Rep 2014;8:321–5.

[99] Comet I, Helin K. Revolution in the Polycomb hierarchy. Nat Struct Mol Biol 2014;21:573–5.

[100] Blackledge NP, Rose NR, Klose RJ. Targeting Polycomb systems to regulate gene expression: modifications to a complex story. Nat Rev Mol Cell Biol 2015;16:643–9.

[101] Yu M, Mazor T, Huang H, Huang HT, Kathrein KL, Woo AJ, Chouinard CR, Labadorf A, Akie TE, Moran TB, Xie H, Zacharek S, Taniuchi I, Roeder RG, Kim CF, Zon LI,

Fraenkel E, Cantor AB. Direct recruitment of polycomb repressive complex 1 to chromatin by core binding transcription factors. Mol Cell 2012;45:330–43.

[102] Elderkin S, Maertens GN, Endoh M, Mallery DL, Morrice N, Koseki H, Peters G, Brockdorff N, Hiom K. A phosphorylated form of Mel-18 targets the Ring1B histone H2A ubiquitin ligase to chromatin. Mol Cell 2007;28:107–20.

[103] Zaaroor-Regev D, de Bie P, Scheffner M, Noy T, Shemer R, Heled M, Stein I, Pikarsky E, Ciechanover A. Regulation of the polycomb protein Ring1B by self-ubiquitination or by E6-AP may have implications to the pathogenesis of Angelman syndrome. Proc Natl Acad Sci USA 2010;107:6788–93.

[104] Lecona E, Narendra V, Reinberg D. USP7 cooperates with SCML2 to regulate the activity of PRC1. Mol Cell Biol 2015;35:1157–68.

[105] Hasegawa K, Sin HS, Maezawa S, Broering TJ, Kartashov AV, Alavattam KG, Ichijima Y, Zhang F, Bacon WC, Greis KD, Andreassen PR, Barski A, Namekawa SH. SCML2 establishes the male germline epigenome through regulation of histone H2A ubiquitination. Dev Cell 2015;32:574–88.

[106] Scheuermann JC, de Ayala Alonso AG, Oktaba K, Ly-Hartig N, McGinty RK, Fraterman S, Wilm M, Muir TW, Muller J. Histone H2A deubiquitinase activity of the Polycomb repressive complex PR-DUB. Nature 2010;465:243–7.

[107] Abdel-Wahab O, Dey A. The ASXL-BAP1 axis: new factors in myelopoiesis, cancer and epigenetics. Leukemia 2013;27:10–5.

[108] Bhatnagar S, Gazin C, Chamberlain L, Ou J, Zhu X, Tushir JS, Virbasius CM, Lin L, Zhu LJ, Wajapeyee N, Green MR. TRIM37 is a new histone H2A ubiquitin ligase and breast cancer oncoprotein. Nature 2014;516:116–20.

Chapter 6

Cooperative Recruitment of Polycomb Complexes by Polycomb Response Elements

Y.B. Schwartz
Umeå University, Umeå, Sweden

Chapter Outline

Introduction 111
Polycomb Group Protein Complexes 112
Polycomb Response Elements of *Drosophila* 112
Polycomb Response Elements as Cellular Memory Modules 114
Sequence-Specific DNA-Binding Proteins Implicated in PRE Function 115
PREs as DNA Platforms for Cooperative Recruitment of PcG Complexes 118
Parallels Between Polycomb Targeting in *Drosophila* and Mammals 121
Conclusion 122
List of Acronyms and Abbreviations 123
Glossary 124
Acknowledgments 124
References 124

INTRODUCTION

The key for understanding Polycomb group (PcG) mechanisms is the question of how PcG complexes are targeted to specific genes. The question is not trivial. On one hand, PcG mechanisms are epigenetic. That is they tend to repress a target gene if it has been repressed in the previous cell cycle. On the other hand, PcG proteins are ubiquitous but the sets of genes repressed by PcG mechanisms differ between cell types. Many of the PcG target genes are pivotal regulators of developmental pathways, and the choice of which of the developmental genes to repress underlies the ability of multicellular organisms to shut down alternative gene expression programs as their cells differentiate. This also means that any given cell has to keep active a set of PcG target genes that define its developmental program. Therefore, whatever mechanism is used to recruit PcG proteins to genes it has to accommodate the

Polycomb Group Proteins. http://dx.doi.org/10.1016/B978-0-12-809737-3.00006-4

ability of PcG repression to be epigenetically stable but, at the same time, developmentally plastic.

POLYCOMB GROUP PROTEIN COMPLEXES

PcG proteins are incorporated in multiple protein complexes; some of which share common subunits [1]. A subset of these complexes is essential for PcG repression, but for others the functions are unclear and likely unrelated to PcG mechanisms. Of all these complexes the most studied are Polycomb repressive complexes (PRC) 1 and 2. PRC2 complexes methylate lysine 27 of histone H3 (H3K27) [2–5], and trimethylation of H3K27 (H3K27me3) within the chromatin of target genes is integral for PcG repression [6]. There are two types of PRC2 complexes. Both types contain a core of five proteins, namely, enhancer of zeste [E(z)], extra sexcombs (Esc), Su(z)12, Caf1, and Jing (the *Drosophila melanogaster* ortholog of mammalian AEBP2 protein) but differ in alternative subunits jumonji, AT rich interactive domain 2 (Jarid2) or polycomblike (Pcl). The PRC2–Pcl complexes are recruited to PcG target genes and required for repression [7], but the function of PRC2–Jarid2 complexes is less clear.

The *Drosophila* PRC1 complexes consist of polycomb (Pc), polyhomeotic (Ph) and sex comb on midleg (Scm) proteins together with a heterodimer of RING1 (the product of the sex combs extra (*Sce)* gene) and one of two closely related proteins: posterior sex combs (Psc) or Su(z)2 [8–10]. Flies lacking PRC1 subunits die as embryos and show strong misexpression of PcG target genes, indicating that PRC1 is essential for repression. In addition to PRC1, the RING1-Psc dimers may form a different complex called dRING-associated factors (dRAF) [11]. The dRAF complex lacks Pc, Ph, and Scm subunits and instead contains histone demethylase lysine (K)-specific demethylase 2 (Kdm2), RING-associated factor 2 (RAF2), and Ulp1 proteins. PRC1 and dRAF can both monoubiquitylate lysine 118 of histone H2A (H2AK118ub) in vitro via the RING–Psc catalytic core [11,12]. Finally, *Drosophila* RING1 can potentially dimerize with the *Drosophila* protein called lethal (3) 73Ah (L(3)73Ah). Like Psc and Su(z)2, the L(3)73Ah protein can serve as a cofactor to support catalytic activity of RING1. Although interaction between RING1 and L(3)73Ah has not been demonstrated biochemically, genetic evidence indicates that L(3)73Ah is required to produce approximately 70% of all RING1-mediated H2AK118 ubiquitylation [13]. Interestingly, this H2AK118 ubiquitylation appears to be deposited at genomic sites other than PcG target genes [13].

In addition to PRC2, PRC1, and dRAF, two other *Drosophila* complexes, pho repressive complex (PhoRC) and polycomb repressive deubiquitinase (PR-DUB), are important for PcG repression at least at some target genes [14,15]. PR-DUB is the deubiquitylase specific to H2AK118 while PhoRC has no known enzymatic activity but can bind a broad spectrum of mono- and dimethylated histones [16].

POLYCOMB RESPONSE ELEMENTS OF *DROSOPHILA*

Although we still do not know all details, our best understanding of how PcG proteins are targeted to specific genes comes from studies in *Drosophila*, where PcG system was first discovered. In flies, each PcG target gene is equipped with one or more discrete DNA elements, called Polycomb response elements (or PREs in short), which are required for its repression. When incorporated in a transgenic construct and integrated elsewhere in *Drosophila* genome, PREs autonomously recruit obligatory PcG repressive complexes PRC1 and PRC2 and repress associated reporter genes, indicating that PREs are sufficient to target PcG repression. Based on the genome-wide mapping of PcG proteins by chromatin immunoprecipitation (ChIP) the *Drosophila* genome is estimated to contain about 200 PREs [17]. Roughly 10% of these have been experimentally tested and shown to autonomously recruit PcG complexes in transgenic assays [18]. Although many PREs are known, much of the information regarding PRE structure and function comes from detailed studies of few specific cases.

The *bxd*-PRE from the upstream regulatory region of the homeotic gene *Ultrabithorax* (*Ubx*) was the first to be independently discovered by several research groups in the early 90s [19,20]. *Ubx*, together with two other genes *Abdominal* (*abd*)*-A* and *Abd-B*, forms a gene cluster called the bithorax complex that controls developmental specification of anterior—posterior identity of the last thoracic and all abdominal segments of the fly [21]. Transgenic and ChIP analyses indicate that the *bxd* PRE is approximately a 1000-base pair (bp) long compound element. It consists of a 200-bp core, corresponding to a high-affinity binding site for PRC complexes, and flanking sequences that contribute to recruitment [22—24]. Subsequent studies of PREs from the *engrailed* (*en*), *even skipped* (*eve*), *Abd-B* genes indicate that the organization of *bxd*-PRE is typical for elements of this kind [25—30].

A striking feature of PRE-mediated repression of reporter genes, termed pairing-sensitive silencing, is that repression becomes much stronger in flies homozygous for transgenic insertion [31,32]. The phenomenon was first discovered in studies of *en*-PRE repression of the *white* gene but was later demonstrated for other PREs. The *white* gene encodes a protein that participates in the transport of the precursors of eye color pigment into the eye of *Drosophila*. Flies lacking the endogenous *white* function have completely white eyes but the eye color is partially restored by supplying a transgene copy. Normally, homozygotes, which have two copies of the *white* transgene, have darker eye color compared to heterozygous flies with only one copy of *white*. When combined with a PRE the single copy of the *white* gene is expressed in a variegated fashion. Surprisingly, the eyes become much less colored if the transgene is made homozygous indicating that PRE-mediated repression is enhanced (Fig. 6.1).

For the pairing effect, the two transgenes have to be in close proximity, which is normally the case in homozygous flies due to the strong somatic pairing of homologous *Drosophila* chromosomes [33]. However, pairing-sensitive silencing is not limited to transgenic insertions in the same genomic site and can happen between two transgenes inserted in different

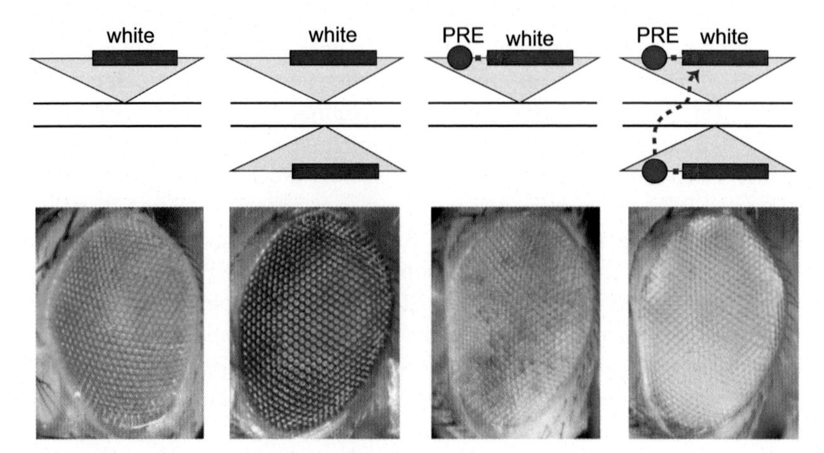

FIGURE 6.1 Pairing-sensitive silencing. Several features of Polycomb group repression are vividly illustrated by the effect of Polycomb response elements (PREs) on expression of the *white* gene. One copy of *white* (*red rectangle*) supplied as a transgene (depicted as *gray triangle*) converts the eye color of flies lacking endogenous *white* function from white to orange. Flies homozygous for the transgenic insertion, hence with two copies of *white*, typically have much darker eye color. In contrast when the transgenic construct contains a PRE (*blue circle*), the *white* gene gets stochastically inactivated. Inactivation happens in precursor cells of the eye specified in the embryo. As precursor cells proliferate to form the adult eye, the repressed state is propagated in the clonal descendants of the cells in which *white* was originally switched off. This leads to variegated pigmentation of the adult eye. Strikingly, when PRE-containing transgenes are made homozygous this results in much greater repression of *white* and, often, in flies with completely white eyes. This suggests that paired PREs reinforce each other's repression.

chromosomes if in these transgenes PREs are coupled to chromatin insulator elements that promote *trans*-interactions [34]. Why the two PREs from paired transgenes are more potent in repression is not entirely understood. However, it suggests that the recruitment of PcG complexes to PREs or the mechanisms of transcriptional repression by PcG proteins are cooperative.

Perhaps because multiple PREs can cooperate and make repression more robust, *Drosophila* PcG target genes are typically equipped with more than one PRE [35]. Often at least one of these PREs resides close to the transcription start site (TSS) of the gene while others may be several thousands or even tens of thousands base pairs away [17,36]. However, in case of the *Ubx* gene, both PREs are ~40 kb from its TSS suggesting that, like transcriptional enhancers, PREs may act over long distances.

POLYCOMB RESPONSE ELEMENTS AS CELLULAR MEMORY MODULES

As we discussed above, PREs are necessary and sufficient to target PcG repression to genes. Thus it came as a surprise that in flies with a transgenic construct that contained the *iab-7* PRE and *β-galactosidase* (*lacZ*) reporter gene, a transient strong activation of the *lacZ* gene released PcG repression and resulted in mitotically inherited derepressed state of the reporter sustained long after withdrawal of the original activator [37]. Additional experiments have shown that this effect is also seen with other PREs [38−40]. Therefore, PREs should be viewed as switchable cellular memory modules (CMMs) that, depending on the initial transcriptional status of a target gene, will provide epigenetic memory of repressed or active state.

The ability of PREs to switch into a derepressed mode depends on the Trithorax (Trx) protein. Genetic evidence indicates that in vivo the role of Trx, together with another protein, Ash1, is not simply to induce the expression of PcG target genes but rather to prevent their inappropriate repression by PcG mechanisms in cells where target genes have to remain active [40,41]. Whether appropriate regulation of all PcG target genes requires the ability of their PREs to act as CMMs is not known. Nevertheless, all PREs bind Trx regardless of the transcriptional status of their target gene [17,42]. Therefore all PREs should also be considered as Trithorax response elements (TREs) and the PRE/TRE functions may be inseparable.

SEQUENCE-SPECIFIC DNA-BINDING PROTEINS IMPLICATED IN PRE FUNCTION

PREs are incredibly busy places. These small elements bind PRC1, PRC2, PhoRC, and PR-DUB complexes responsible for PcG repression. They also bind the large Trx protein involved in counteracting PcG repression and the *Drosophila* ortholog of cAMP response element-binding protein (CREB)-binding protein (CBP, the product of *nejire* gene) [1,43]. The latter is bound to multiple regulatory elements throughout the genome [44,45]. The function of CBP at PREs is not entirely clear. It may counteract PcG repression by catalyzing H3K27 acetylation [46] but may also be linked to dynamic exchange of nucleosomes at PRE cores [47]. Of all PcG complexes, only PhoRC contains a subunit with sequence-specific DNA-binding activity. This raises the question of how all other complexes are recruited to PREs.

It was originally proposed that the PcG complexes are recruited to PREs in a hierarchical order. According to this hierarchical recruitment model, the PhoRC complex would bind first and recruit PRC2, which, via its histone methyltransferase activity would trimethylate lysine 27 of histone H3. The trimethylated H3K27 would serve as a binding platform for PRC1, which can specifically recognize this histone modification via the chromodomain of its Polycomb subunit [2]. This model, however, was later proved wrong. Thus experiments failed to detect direct interactions between PhoRC and PRC2 [14], and it turned out that PhoRC itself requires the presence of PRC1 for efficient binding to PREs [48]. More importantly, PREs are generally depleted of histones and trimethylated H3K27, which makes it impossible to depend on histone methylation for recruitment of PRC1 [23,35,49,50].

Multiple lines of evidence, which we will discuss below, suggest that sequence-specific DNA-binding adaptor proteins play key roles in the recruitment of the PcG and Trx complexes to PREs (Table 6.1).

When compared with each other, PREs show no extended DNA sequence similarity but are distinguished from bulk genomic sequence by enrichment of GAGAG and GCCAT motifs [42,48,51,52]. Transgenic and in vitro binding analyses of *bxd*, *iab-7*, and *Mcp* PREs indicate that GAGA factor (GAF, the protein encoded by the *Trithorax-like* gene) is one of the proteins that binds GAGAG sequences within PREs [22,27,53,54]. Consistently, genome-wide mapping indicates that GAF is bound at many, although not all, PREs [42]. Mutations in the *Trithorax-like* gene cause mild disruption of PRE-mediated repression of transgenic reporter genes suggesting that GAF promotes the recruitment of PcG complexes [22,27,53], possibly, via its ability to displace nucleosomes [55]. Transgenic tests may underestimate GAF contribution because of the large supply of maternal wild-type protein in the egg. In addition, another protein, Pipsqueak (Psq) whose DNA-binding domain also recognizes GAGAG sequences [56,57], was shown to bind *bxd*-PRE in vitro [56] and may partially substitute for GAF function.

A systematic screen for sequence-specific DNA-binding proteins required for pairing-sensitive silencing of the *white* gene by *en*-PRE revealed a number of candidates. Some of them were later shown to bind other PREs and have a general role in PRE function. The first discovered and the most obviously

TABLE 6.1 DNA-Binding Proteins Implicated in PRE Function

Protein	Protein Domains	Sequence Specificity[a]
Pho	Four Zn-fingers, C2H2 type	KBCGCCATWTTK
Phol	Four Zn-fingers, C2H2 type	KBCGCCATWTTK
GAF	BTB/POZ; Zn-finger, C2H2 type	GAGAG
Psq	BTB/POZ; four HTH domains	GAGAG
Spps	Three Zn-fingers, C2H2 type	RRGGYG[b]
Grh	CP2	WCHGGTT
Dsp1	Two HMG domains	no clear sequence preference
Cg	Eleven Zn-fingers, C2H2 type	TGTGHGTG

BTB, Broad complex, Tramtrack, Bric-a-brac; *C2H2*, Zink-finger of C2H2 type; *Cg*, Combgap; *Dsp1*, Dorsal switch protein 1; *GAF*, GAGA factor; *Grh*, Grainy head; *HMG*, high mobility group; *HTH* ,Helix-turn-helix; *Pho*, Pleiohomeotic; *Phol*, Pleiohomeotic-like; *POZ*, Poxvirus and Zink finger Zn-finger; *Psq*, Pipsqueak; *Spps*, pairing-sensitive silencing; *Zn*, Zink.
[a]*K* = *G or T; B* = *C, G or T; W* = *A or T; R* = *G or A; Y* = *C or T; H* = *A, C or T.*
[b]*Predicted from studies of mammalian Sp1/KLF proteins.*

involved in PcG repression are Pleiohomeotic (Pho) and the closely related Pleiohomeotic-like (Phol) proteins [58,59]. Pho is the preferred DNA-binding subunit of PhoRC complex due to its apparent higher affinity to the scm-related gene containing four mbt domains (Sfmbt) subunit [14,48]. When the amount of Pho is reduced by mutation, Phol takes its place and can partially substitute for its function. The loss of both Pho and Phol leads to strong derepression of homeotic genes indicating that they are critical for PcG repression at least at some PREs [58]. Consistently, mutation of the GCCAT sequences, the Pho/Phol recognition sites, within the *en*, *Mcp*, *eve*, and *bxd*-PREs, impairs PcG repression in transgenic constructs [27,29,59,60]. In agreement with functional analyses, genome-wide mapping indicates that Pho and Phol bind many, but not all, PREs [42,48]. Other proteins required for pairing-sensitive silencing by *en*-PRE include Sp1-like factor for pairing-sensitive silencing (Spps), and, potentially, other members of the Sp1/KLF protein family [61,62]. Low-resolution mapping suggests that Spps is bound at many PREs; however, its loss does not affect the binding of PcG proteins and does not lead to the derepression of homeotic genes, possibly, because its function is redundant with other Sp1/KLF proteins [62]. Combgap (Cg) is the most recent addition to the list of sequence-specific proteins known to bind *en*-PRE [63]. Genome-wide ChIP analysis suggests that Cg binds to the majority of PREs and may be critical for the PRC1 recruitment to some of them. Whether pairing sensitive silencing by Spps*en*-PRE requires Cg remains to be investigated.

Expression of the bithorax complex gene *Abd-B* is controlled by a set of segment-specific regulatory modules that consist of transcriptional enhancers, a PRE, and an adjacent chromatin insulator element. Deletion of an insulator element leads to erroneous *Abd-B* hyper-repression in the segment anterior to that specified by the affected regulatory module. A clever genetic screen for mutations that suppress *Abd-B* hyper-repression by *iab-7* PRE in flies lacking the *Fab-7* insulator element identified the DNA-binding protein Grainy head (Grh) as a factor required for repression by the *iab-7* PRE [64]. Subsequent ChIP analysis showed that Grh also binds the *en*-PRE [26] suggesting that Grh may be generally involved in targeting PcG complexes to PREs.

Finally, transgenic studies of *iab*-7 PRE identified Dorsal switch protein 1 (Dsp1) as another DNA-binding protein involved in repression by this element [65]. Dsp1 contains two high mobility group (HMG) box domains. Proteins of this class seem to bind DNA with little or no sequence specificity but may recognize noncanonical DNA structures like Z-DNA. Related mammalian proteins HMGB1 and HMGB2 were shown to bend DNA upon binding and facilitate chromatin remodeling [66]. Genome-wide mapping of Dsp1 confirms that it binds to many PREs suggesting that it is generally implicated in some aspect of PRE function [42]. Interestingly, mutants lacking Dsp1 display homeotic transformations associated with reduced expression if homeotic genes. Therefore, Dsp1 may be important for targeting of Trithorax rather than PcG complexes.

PREs AS DNA PLATFORMS FOR COOPERATIVE RECRUITMENT OF PcG COMPLEXES

What is obvious from genome-wide mapping of the above DNA-binding proteins is that none of them binds to all PREs. This, by itself, suggests that several alternative combinations of DNA-binding proteins can be used to target PcG and Trx complexes to PREs. Also, all of these proteins bind multiple sites other than PREs and some of them, like GAF, have documented functions unrelated to PcG repression [67]. In line with these observations, tethering GAF or Pho to a construct via the LexA DNA-binding domain is not sufficient to recruit PcG repression [68]. The emerging model suggests that PREs are made up of multiple binding sites for different DNA-binding proteins that combine following rules that we do not fully understand. As none of the PRE-associated DNA-binding proteins is strongly associated with PRC1 or PRC2, PREs most likely integrate multiple individually weak interactions, which together result in robust recruitment (Fig. 6.2).

Such integration may involve summing up multiple weak contacts between individual DNA-binding proteins and PRC1 or PRC2 but also weak interactions between PRC1, PRC2, and PhoRC. This principle is vividly exemplified by the recent finding that PREs tend to lack optimal binding sites for the Pho subunit of PhoRC [48,69]. Instead, the robust recruitment of PhoRC to PREs requires interactions with PRC1 [48]. This interaction is mediated by the Scm protein whose SAM domain can simultaneously bind the SAM domains of the PhoRC subunit Sfmbt and the PRC1 subunit Ph [70]. The interaction between Scm and Sfmbt is further strengthened by the additional ability to bind each other via N-terminal Zink (Zn)-finger domains [16]. Interestingly, in mammalian cells, where Scm and Sfmbt have lost their Zn-fingers and cannot interact via multiple domains, the Yin Yang 1 (YY1) protein orthologous to Pho no longer binds PcG target genes [48,71,72]. The role of Scm may extend beyond the crosstalk between the PRC1 and PhoRC complexes. A recent study suggests that it also mediates weak interactions between PRC1 and PRC2 [73].

The view of PREs as flexible collections of recognition sequences is also supported by mapping the genomic distributions of PcG proteins in a diverse set of *Drosophila* species [69]. Although the sets of genes and PREs that regulate them in different *Drosophila* species are strikingly similar, the positions of the motifs recognized by the known sequence-specific DNA-binding proteins are not strictly maintained [69,74]. It appears that within PREs the recognition motifs can be lost and recreated without strict positional constraints.

PRC1 can specifically recognize trimethylated H3K27 produced by PRC2 via the chromodomain of its Pc subunit [75], and PRC2 can specifically recognize monoubiquitylated H2AK118, produced by PRC1 [76]. Therefore both histone modifications may additionally stabilize the binding of PRC complexes to PREs. The extent of this contribution is not entirely clear.

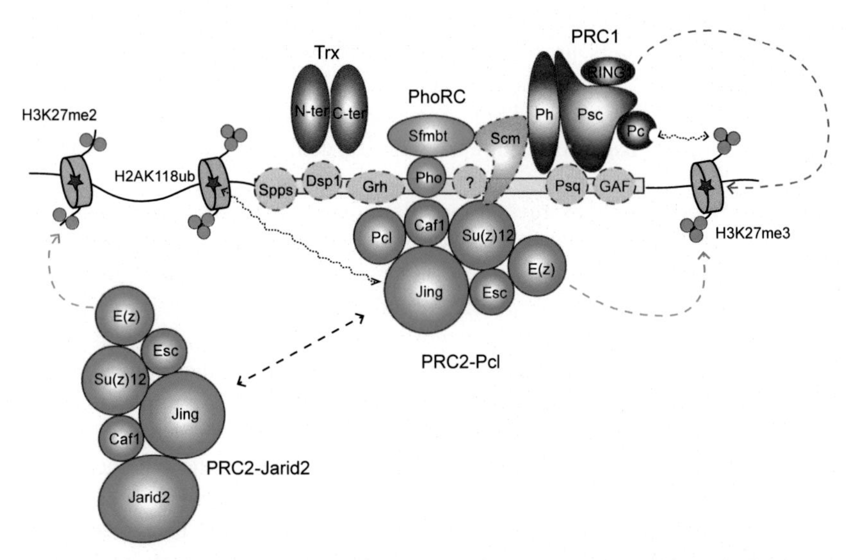

FIGURE 6.2 Cooperative recruitment of Polycomb group (PcG) complexes to Polycomb response elements (PREs). A PRE (*yellow rectangle*) consists of recognition sequences for multiple DNA-binding proteins (*gray circles*). Some of these proteins are still unknown. With the exception of Pleiohomeotic (Pho) or Pleiohomeotic-like (Phol) that are part of the PhoRC complex, none of the other DNA-binding proteins was recovered as a stoichiometric component of any of the core PcG complexes. A current model suggests that DNA-binding proteins combine their weak interactions to provide robust recruitment of Polycomb repressive complexes (PRCs). Which DNA-binding proteins interact with which PRC is not clear but Pipsqueak (Psq) and GAF were proposed to interact with PRC1. An additional layer of cooperative recruitment comes from weak interactions between PRC1, PRC2, and PhoRC. Scm, which may to some extent interact with PREs independently of PRC1 and PRC2 via yet unknown DNA-binding adaptor protein, is a potential candidate to mediate such interactions. The Trithorax (Trx) protein is proteolytically cleaved into two halves (*purple ovals*) both of which are recruited to PREs. Since Dorsal switch protein 1 (Dsp1)—deficient flies show homeotic phenotypes resembling those of Trx mutants, Dsp1 may be involved in Trx recruitment. Finally, chromodomain of Pc subunit of PRC1 can interact with trimethylated H3K27 produced by PRC2 and Jing subunit of PRC2 (the fly ortholog of mammalian AEBP2 protein) can interact with monoubiquitylated H2AK118 produced by PRC1. Although clearly not sufficient for recruitment they can contribute to stable binding of PRC1 and PRC2. Note that there are two flavors of PRC2. One flavor of PRC2 acts by a hit-and-run mechanism and dimethylates H3K27 throughout entire transcriptionally inactive genome, while another is anchored at PREs and is responsible for trimethylation of H3K27 in their vicinity. It is thus possible that the stable binding of PRC2 at PREs results from cooperative retention of a complex that is already in loose contact with chromatin. Although we know that PRC2-Pcl is bound at PRE, the free-roaming nature of PRC2-Jarid2 remains an unproven hypothesis.

As discussed above, PRE cores often lack nucleosomes so only the flanking sequences can potentially be involved. Also, whereas H3K27me3 is always present at genes repressed by PcG mechanisms and is required for repression [6,17], the presence and the requirement for H2AK118ub appears to vary from gene to gene. Thus the majority of the repressed PcG target genes are enriched in H2AK118ub, but the Antennapedia complex and the bithorax complex homeotic gene clusters are completely devoid of H2AK118 ubiquitylation [13]. Consistently, in flies lacking H2A ubiquitylation, PcG repression of homeotic genes is not impaired [77]. Moreover, proper repression of homeotic genes by PcG mechanisms requires H2AK118 deubiquitylation by PR-DUB complex [15].

Noncoding RNAs may further modulate the recruitment of PcG and Trx complexes to PREs. However, the evidence for this has been controversial. One widely publicized early study suggested that specialized noncoding RNAs transcribed from the *bxd*-PRE directly recruited Trx group proteins to this PRE thereby preventing it from repressing the *Ubx* gene [78]. This work, however, was later retracted because of numerous data falsifications [79]. On the other hand, both *Drosophila* and mammalian PRC2 complexes were shown to interact with RNA in vitro and in vivo [80,81]. However, it is unclear whether PRC2 preferentially binds certain RNAs and whether these PRC2–RNA interactions promote or inhibit PRC2 targeting to chromatin [80]. Most of the work on the role of RNAs in repression has focused on mammalian PcG complexes. One recent study suggests that switching between forward and reverse noncoding transcription of the *vestigial* PRE changes its status from repressor (PRE) to antirepressor (TRE) [81]. This property, however, is unlikely to be a general feature of *Drosophila* PREs as most of them are not transcribed [36].

What is the rationale to use such a complex cooperative recruitment strategy when a dedicated high-affinity DNA-binding protein with strong interactions with PRC1 or PRC2 would, arguably, be a much more straightforward way to target PcG complexes to PREs? The answer might be twofold. First, the straightforward unconditional recruitment may not be what the PcG system needs. In *Drosophila*, all PcG components are present in all cells; however, not all PcG target genes are simultaneously repressed. The decision of whether a potential PcG target gene is going to be repressed in a given cell type depends on whether it has been repressed in the precursor cells during the previous cell cycle. This gives the PcG system the critical ability to epigenetically enforce transcriptional decisions made via developmental signaling during early stages of cell differentiation. The implication of this is that the recruitment mechanism cannot be so powerful that it becomes independent of the existing chromatin state. Second, the selection of cooperative recruitment for targeting PcG complexes to PREs may reflect the evolutionary origins of the system. The origin of PRC1 is not well studied but it is clear that PRC2 is an ancient complex already present in the last common ancestor of plants and animals. Recent work has shown that, in addition to PRE-associated action of PRC2 at specific genes, PRC2 has quantitatively a much more important genome-wide role of dimethylating approximately 70% of all H3K27 [13,82]. This is done by an untargeted hit-and-run mechanism, which is likely more evolutionarily ancient than "classical" PcG repression. In this view, PREs do not really recruit PRC2 to chromatin from the nucleoplasm but rather retain the complex that is already transiently engaged with the entire genome.

PARALLELS BETWEEN POLYCOMB TARGETING IN *DROSOPHILA* AND MAMMALS

The core subunits of PRC1 and PRC2 complexes are evolutionarily conserved. Even more remarkably, the PcG system regulates many of the same genes in flies and mammals [83]. It is therefore hard to imagine that targeting of mammalian and *Drosophila* PRC1 and PRC2 complexes to the same genes has evolved completely independently and follows an entirely different logic. Yet, attempts to uncover commonalities in targeting of mammalian and *Drosophila* PcG complexes, so far, have had limited success. In the last seven years a handful of mouse or human DNA fragments were reported to autonomously recruit either PRC1 or PRC2 complexes when integrated elsewhere in the genome and have been hailed as mammalian PREs [71,84–89]. Collectively these studies suggest that DNA elements analogous to *Drosophila* PREs may play a role in targeting mammalian PcG complexes. Nevertheless, more work is needed before we can draw clear parallels. Unlike paradigmatic *Drosophila* PREs that were shown to autonomously recruit both PRC1 and PRC2 in multiple assays by multiple research groups, the recruiting ability of reported mammalian elements has yet to be confirmed by independent investigations. Making the story more complex, some of these elements seem to recruit both PRC1 and PRC2 while others appear to autonomously recruit only one of the PRC complexes. Perhaps, compared to *Drosophila*, the recruitment of mammalian PcG complexes is even more cooperative and involves DNA elements that are more modular or to a greater extent aided by histone modifications and noncoding RNAs.

Little is known regarding sequence motifs and DNA-binding proteins implicated in the targeting of mammalian PcG complexes. Several independent investigations concluded that DNA sequences with high density of unmethylated CpG dinucleotides, lacking bound transcriptional activators, are sufficient to recruit PRC2 although such sequences cannot recruit PRC1 [71,86,87]. The molecular link between the high CpG content and the recruitment of PRC2 is not fully understood. It was first suggested that the JARID2 subunit of PRC2 allows preferential recognition of the CG-rich DNA [90]. However, recent findings shifted the focus toward the recognition of the unmethylated CpG dinucleotides by a Zn-finger CXXC DNA-binding domain of the lysine (K)-specific demethylase 2B (KDM2B) protein. KDM2B forms a stable complex with mammalian RING1/RING2 proteins, which is somewhat similar to *Drosophila* dRAF but contains the PcG RING Finger 1 (PCGF1) protein instead of B lymphoma Mo-MLV insertion region 1 (BMI1)/MEL18 (mammalian orthologs of the Psc/Su(z)2 proteins) [91,92]. According to one popular model, the KDM2B−PCGF1−RING complex can bind to unmethylated CpG-rich DNA, catalyze extensive H2AK119 ubiquitylation (analog of the *Drosophila* H2AK118ub) within neighboring nucleosomes, and thereby trigger the recruitment of PRC2 [93], which can bind H2AK119ub via its AEBP2 subunit [76]. The model is clearly incomplete. For example, it does not explain why, although KDM2B binds all unmethylated CG-rich DNA, only a small subset of these sites, always next to transcriptionally inactive genes, also cobind the RING1/RING2 and PCGF1 proteins [91,92]. It is possible that transcriptional activity, in some way, constrains the formation of the KDM2B−PCGF1−RING complex, providing a degree of specificity. Consistent with this hypothesis, small-molecule inhibitors of transcription were shown to trigger the recruitment of PRC2 and H3K27me3 to the CpG-rich regions upstream of multiple previously active genes [94].

Modeling the dynamic changes of H3K27me3 enrichment throughout the mouse genome during cell differentiation suggests that transcription factors RE1-silencing transcription factor (REST) and SNAIL are involved in PRC2 targeting [84]. The REST protein was also reported to directly interact with PRC1 [95]. *Drosophila* does not have a clear ortholog of mammalian REST protein, and fly members of Snail family have not been implicated in PcG repression. Conversely some of the *Drosophila* proteins involved in PRE function, such as GAF and Spps, do not have clear orthologs in mammals, and YY1, the mammalian ortholog of Pho, appears to have no role in mammalian PcG repression [48,71,72]. Future studies will tell whether the repertoire of mammalian and fly DNA-binding proteins involved in the targeting of PRC complexes is dramatically different or we simply have not yet discovered evolutionary conserved proteins acting at *Drosophila* PREs.

CONCLUSION

Thirty years of research have brought us much closer to an understanding of how PcG regulation is targeted to specific genes. Studies of the PcG system in the classic model organism *D. melanogaster* continue to pave the way for understanding more complex targeting in mammalian cells. Using biochemical, genetic, and genomic approaches we have learned that fly PcG complexes are targeted to genes via specialized PREs. Mounting evidence suggests that the recruitment of PcG complexes by PREs is highly cooperative, and this logic is likely also used to target mammalian PcG complexes. Several challenges lie ahead. First, we still do not know which combination(s) of recognition sequences is sufficient to make a PRE. This is, in part, because we still do not know all DNA-binding proteins that are involved in PRE function. The second challenge is to understand which of the sequence-specific DNA-binding proteins interact with which subunits of the core PcG complexes. Solving this problem is especially challenging because we expect individual interactions to be weak and therefore difficult to be captured by current experimental technologies. Third, the binding of sequence-specific DNA-binding proteins to PREs is likely to be regulated by chromatin remodeling processes. We know that many PREs are nucleosome free, which suggests that nucleosomes are actively removed to allow association of sequence-specific DNA-binding proteins. However, little is known about the molecular machinery involved in this process. Finally, as PREs emerge as flexible collections of binding sites we expect them to differ in their ability to install and maintain PcG repression. They are also likely to differ in the ease with which their CMM state can be switched from PcG-dependent repression to Trx-dependent support of transcriptional activity.

PcG regulation has an undisputable impact on multiple aspects of development and cell differentiation. Not surprisingly, altered dosage of PcG proteins was linked to progression and poor clinical prognosis of certain cancers. Attempts to use small-molecule inhibitors to block general activity of PcG complexes have shown therapeutic promise. Nevertheless, such treatments are likely to disrupt multiple processes controlled by PcG system and, therefore, are likely to cause undesired side effects. Once we know the rules of cooperative recruitment of PcG complexes, we will be in a position to use rapidly developing genome editing tools [for example, clustered regularly interspaced short palindromic repeats (CRISPR)/Cas9-based techniques] to manipulate the targeting of PcG repression to specific loci.

LIST OF ACRONYMS AND ABBREVIATIONS

BMI1 B lymphoma Mo-MLV insertion region 1
ChIP Chromatin Immunoprecipitation
CMM Cellular Memory Module
CREB cAMP response element-binding protein
CRISPR Clustered Regularly Interspaced Short Palindromic Repeats
dRAF dRING-associated factors
E(z) Enhancer of zeste
Esc Extra sexcombs
GAF GAGA factor
H2AK118 Lysine 118 of histone H2A
H3K27 Lysine 27 of histone H3
Jarid2 Jumonji, AT rich interactive domain 2
Kdm2 Lysine (K)-specific demethylase 2
KDM2B Lysine (K)-specific demethylase 2B
L(3)73Ah Lethal (3) 73Ah
lacZ B-galactosidase
Pc Polycomb
PcG Polycomb Group
Pcl Polycomblike
PhoRC Pho repressive complex
PR-DUB Polycomb repressive deubiquitinase
PRC1 Polycomb Repressive Complex 1
PRC2 Polycomb Repressive Complex 2

PRE Polycomb Response Element
Psc Posterior sex combs
RAF2 RING-associated factor
REST RE1-silencing transcription factor
Sce Sex combs extra
Scm Sex comb on midleg
Sfmbt Scm-related gene containing four mbt domains
Spps Sp1-like factor for pairing-sensitive silencing
TRE Trithorax Response Element
TSS Transcription Start Site
YY1 Yin Yang 1

GLOSSARY

Chromatin immunoprecipitation (ChIP) Experimental technique to isolate DNA fragments bound by a chromatin protein in vivo using soluble extract from crosslinked cells and antibodies that recognize the protein of interest.
Chromatin insulators Specialized regulatory elements that can bring distant genomic sites together and interfere with enhancer-promoter communications.
Homeotic genes Genes that control the identity of body segments.
Orthologs Genes in different species that evolved from a single gene of the last common ancestor.

ACKNOWLEDGMENTS

The research in YBS laboratory is supported by grants from Swedish Research Council, Cancerfonden, Knut, and Alice Wallenberg Foundation, Kempestiftelserna, and Umeå University Insamlingsstiftelsen.

REFERENCES

[1] Schwartz YB, Pirrotta V. A new world of Polycombs: unexpected partnerships and emerging functions. Nat Rev Genet 2013;14(12):853–64.
[2] Cao R, Wang L, Wang H, Xia L, Erdjument-Bromage H, Tempst P, et al. Role of histone H3 lysine 27 methylation in polycomb-group silencing. Science 2002;298(5595):1039–43.
[3] Czermin B, Melfi R, McCabe D, Seitz V, Imhof A, Pirrotta V. *Drosophila* enhancer of Zeste/ESC complexes have a histone H3 methyltransferase activity that marks chromosomal Polycomb sites. Cell 2002;111(2):185–96.
[4] Kuzmichev A, Nishioka K, Erdjument-Bromage H, Tempst P, Reinberg D. Histone methyltransferase activity associated with a human multiprotein complex containing the Enhancer of Zeste protein. Genes Dev 2002;16(22):2893–905.
[5] Muller J, Hart CM, Francis NJ, Vargas ML, Sengupta A, Wild B, et al. Histone methyltransferase activity of a *Drosophila* Polycomb group repressor complex. Cell 2002;111(2):197–208.
[6] Pengelly AR, Copur O, Jackle H, Herzig A, Muller J. A histone mutant reproduces the phenotype caused by loss of histone-modifying factor Polycomb. Science 2013;339(6120):698–9.

[7] Nekrasov M, Klymenko T, Fraterman S, Papp B, Oktaba K, Kocher T, et al. Pcl-PRC2 is needed to generate high levels of H3-K27 trimethylation at Polycomb target genes. EMBO J 2007;26(18):4078–88.

[8] Francis NJ, Saurin AJ, Shao Z, Kingston RE. Reconstitution of a functional core polycomb repressive complex. Mol Cell 2001;8(3):545–56.

[9] Lo SM, Ahuja NK, Francis NJ. Polycomb group protein Suppressor 2 of zeste is a functional homolog of Posterior Sex Combs. Mol Cell Biol 2009;29(2):515–25.

[10] Shao Z, Raible F, Mollaaghababa R, Guyon JR, Wu CT, Bender W, et al. Stabilization of chromatin structure by PRC1, a Polycomb complex. Cell 1999;98(1):37–46.

[11] Lagarou A, Mohd-Sarip A, Moshkin YM, Chalkley GE, Bezstarosti K, Demmers JA, et al. dKDM2 couples histone H2A ubiquitylation to histone H3 demethylation during Polycomb group silencing. Genes Dev 2008;22(20):2799–810.

[12] Wang H, Wang L, Erdjument-Bromage H, Vidal M, Tempst P, Jones RS, et al. Role of histone H2A ubiquitination in Polycomb silencing. Nature 2004;431(7010):873–8.

[13] Lee HG, Kahn TG, Simcox A, Schwartz YB, Pirrotta V. Genome-wide activities of Polycomb complexes control pervasive transcription. Genome Res 2015;25(8):1170–81.

[14] Klymenko T, Papp B, Fischle W, Kocher T, Schelder M, Fritsch C, et al. A Polycomb group protein complex with sequence-specific DNA-binding and selective methyl-lysine-binding activities. Genes Dev 2006;20(9):1110–22.

[15] Scheuermann JC, de Ayala Alonso AG, Oktaba K, Ly-Hartig N, McGinty RK, Fraterman S, et al. Histone H2A deubiquitinase activity of the Polycomb repressive complex PR-DUB. Nature 2010;465(7295):243–7.

[16] Grimm C, Matos R, Ly-Hartig N, Steuerwald U, Lindner D, Rybin V, et al. Molecular recognition of histone lysine methylation by the Polycomb group repressor dSfmbt. EMBO J 2009;28(13):1965–77.

[17] Schwartz YB, Kahn TG, Stenberg P, Ohno K, Bourgon R, Pirrotta V. Alternative epigenetic chromatin states of polycomb target genes. PLoS Genet 2010;6(1):e1000805.

[18] Kassis JA, Brown JL. Polycomb group response elements in *Drosophila* and vertebrates. Adv Genet 2013;81:83–118.

[19] Chan CS, Rastelli L, Pirrotta VA. Polycomb response element in the Ubx gene that determines an epigenetically inherited state of repression. EMBO J 1994;13(11):2553–64.

[20] Simon J, Chiang A, Bender W, Shimell MJ, O'Connor M. Elements of the *Drosophila* bithorax complex that mediate repression by Polycomb group products. Dev Biol 1993;158(1):131–44.

[21] Maeda RK, Karch F. The ABC of the BX-C: the bithorax complex explained. Development 2006;133(8):1413–22.

[22] Horard B, Tatout C, Poux S, Pirrotta V. Structure of a polycomb response element and in vitro binding of polycomb group complexes containing GAGA factor. Mol Cell Biol 2000;20(9):3187–97.

[23] Kahn TG, Schwartz YB, Dellino GI, Pirrotta V. Polycomb complexes and the propagation of the methylation mark at the *Drosophila* ubx gene. J Biol Chem 2006;281(39):29064–75.

[24] Sipos L, Kozma G, Molnar E, Bender W. In situ dissection of a Polycomb response element in *Drosophila melanogaster*. Proc Natl Acad Sci USA 2007;104(30):12416–21.

[25] Americo J, Whiteley M, Brown JL, Fujioka M, Jaynes JB, Kassis JA. A complex array of DNA-binding proteins required for pairing-sensitive silencing by a polycomb group response element from the *Drosophila* engrailed gene. Genetics 2002;160(4):1561–71.

[26] Brown JL, Kassis JA. Architectural and functional diversity of polycomb group response elements in *Drosophila*. Genetics 2013;195(2):407–19.

[27] Busturia A, Lloyd A, Bejarano F, Zavortink M, Xin H, Sakonju S. The MCP silencer of the *Drosophila* Abd-B gene requires both Pleiohomeotic and GAGA factor for the maintenance of repression. Development 2001;128(11):2163–73.

[28] Dejardin J, Cavalli G. Chromatin inheritance upon Zeste-mediated Brahma recruitment at a minimal cellular memory module. EMBO J 2004;23(4):857–68.

[29] Fujioka M, Yusibova GL, Zhou J, Jaynes JB. The DNA-binding Polycomb-group protein Pleiohomeotic maintains both active and repressed transcriptional states through a single site. Development 2008;135(24):4131–9.

[30] Mishra RK, Mihaly J, Barges S, Spierer A, Karch F, Hagstrom K, et al. The iab-7 polycomb response element maps to a nucleosome-free region of chromatin and requires both GAGA and pleiohomeotic for silencing activity. Mol Cell Biol 2001;21(4):1311–8.

[31] Kassis JA. Unusual properties of regulatory DNA from the *Drosophila* engrailed gene: three "pairing-sensitive" sites within a 1.6-kb region. Genetics 1994;136(3):1025–38.

[32] Kassis JA, VanSickle EP, Sensabaugh SM. A fragment of engrailed regulatory DNA can mediate transvection of the white gene in *Drosophila*. Genetics 1991;128(4):751–61.

[33] McKee BD. Homologous pairing and chromosome dynamics in meiosis and mitosis. Biochim Biophys Acta 2004;1677(1–3):165–80.

[34] Sigrist CJ, Pirrotta V. Chromatin insulator elements block the silencing of a target gene by the *Drosophila* polycomb response element (PRE) but allow trans interactions between PREs on different chromosomes. Genetics 1997;147(1):209–21.

[35] Schwartz YB, Kahn TG, Nix DA, Li XY, Bourgon R, Biggin M, et al. Genome-wide analysis of Polycomb targets in *Drosophila melanogaster*. Nat Genet 2006;38(6):700–5.

[36] Kharchenko PV, Alekseyenko AA, Schwartz YB, Minoda A, Riddle NC, Ernst J, et al. Comprehensive analysis of the chromatin landscape in *Drosophila melanogaster*. Nature 2011;471(7339):480–5.

[37] Cavalli G, Paro R. The *Drosophila* Fab-7 chromosomal element conveys epigenetic inheritance during mitosis and meiosis. Cell 1998;93(4):505–18.

[38] Cavalli G, Paro R. Epigenetic inheritance of active chromatin after removal of the main transactivator. Science 1999;286(5441):955–8.

[39] Maurange C, Paro R. A cellular memory module conveys epigenetic inheritance of hedgehog expression during *Drosophila* wing imaginal disc development. Genes Dev 2002;16(20):2672–83.

[40] Poux S, Horard B, Sigrist CJ, Pirrotta V. The *Drosophila* trithorax protein is a coactivator required to prevent re-establishment of polycomb silencing. Development 2002;129(10):2483–93.

[41] Klymenko T, Muller J. The histone methyltransferases Trithorax and Ash1 prevent transcriptional silencing by Polycomb group proteins. EMBO Rep 2004;5(4):373–7.

[42] Schuettengruber B, Ganapathi M, Leblanc B, Portoso M, Jaschek R, Tolhuis B, et al. Functional anatomy of polycomb and trithorax chromatin landscapes in *Drosophila* embryos. PLoS Biol 2009;7(1):e13.

[43] Tie F, Banerjee R, Conrad PA, Scacheri PC, Harte PJ. Histone demethylase UTX and chromatin remodeler BRM bind directly to CBP and modulate acetylation of histone H3 lysine 27. Mol Cell Biol 2012;32(12):2323–34.

[44] Heintzman ND, Stuart RK, Hon G, Fu Y, Ching CW, Hawkins RD, et al. Distinct and predictive chromatin signatures of transcriptional promoters and enhancers in the human genome. Nat Genet 2007;39(3):311–8.

[45] Visel A, Blow MJ, Li Z, Zhang T, Akiyama JA, Holt A, et al. ChIP-seq accurately predicts tissue-specific activity of enhancers. Nature 2009;457(7231):854–8.

[46] Tie F, Banerjee R, Stratton CA, Prasad-Sinha J, Stepanik V, Zlobin A, et al. CBP-mediated acetylation of histone H3 lysine 27 antagonizes *Drosophila* Polycomb silencing. Development 2009;136(18):3131–41.

[47] Mito Y, Henikoff JG, Henikoff S. Histone replacement marks the boundaries of cis-regulatory domains. Science 2007;315(5817):1408–11.

[48] Kahn TG, Stenberg P, Pirrotta V, Schwartz YB. Combinatorial interactions are required for the efficient recruitment of pho repressive complex (PhoRC) to polycomb response elements. PLoS Genet 2014;10(7):e1004495.

[49] Mohd-Sarip A, van der Knaap JA, Wyman C, Kanaar R, Schedl P, Verrijzer CP. Architecture of a polycomb nucleoprotein complex. Mol Cell 2006;24(1):91–100.

[50] Papp B, Muller J. Histone trimethylation and the maintenance of transcriptional ON and OFF states by trxG and PcG proteins. Genes Dev 2006;20(15):2041–54.

[51] Mihaly J, Mishra RK, Karch F. A conserved sequence motif in Polycomb-response elements. Mol Cell 1998;1(7):1065–6.

[52] Ringrose L, Rehmsmeier M, Dura JM, Paro R. Genome-wide prediction of Polycomb/Trithorax response elements in *Drosophila melanogaster*. Dev Cell 2003;5(5):759–71.

[53] Hagstrom K, Muller M, Schedl P. A Polycomb and GAGA dependent silencer adjoins the Fab-7 boundary in the *Drosophila* bithorax complex. Genetics 1997;146(4):1365–80.

[54] Hodgson JW, Argiropoulos B, Brock HW. Site-specific recognition of a 70-base-pair element containing d(GA)(n) repeats mediates bithoraxoid polycomb group response element-dependent silencing. Mol Cell Biol 2001;21(14):4528–43.

[55] Tsukiyama T, Becker PB, Wu C. ATP-dependent nucleosome disruption at a heat-shock promoter mediated by binding of GAGA transcription factor. Nature 1994;367(6463):525–32.

[56] Huang DH, Chang YL, Yang CC, Pan IC, King B. Pipsqueak encodes a factor essential for sequence-specific targeting of a polycomb group protein complex. Mol Cell Biol 2002;22(17):6261–71.

[57] Lehmann M, Siegmund T, Lintermann KG, Korge G. The pipsqueak protein of *Drosophila melanogaster* binds to GAGA sequences through a novel DNA-binding domain. J Biol Chem 1998;273(43):28504–9.

[58] Brown JL, Fritsch C, Mueller J, Kassis JA. The *Drosophila* pho-like gene encodes a YY1-related DNA binding protein that is redundant with pleiohomeotic in homeotic gene silencing. Development 2003;130(2):285–94.

[59] Brown JL, Mucci D, Whiteley M, Dirksen ML, Kassis JA. The *Drosophila* Polycomb group gene pleiohomeotic encodes a DNA binding protein with homology to the transcription factor YY1. Mol Cell 1998;1(7):1057–64.

[60] Fritsch C, Brown JL, Kassis JA, Muller J. The DNA-binding polycomb group protein pleiohomeotic mediates silencing of a *Drosophila* homeotic gene. Development 1999;126(17):3905–13.

[61] Brown JL, Grau DJ, DeVido SK, Kassis JA. An Sp1/KLF binding site is important for the activity of a Polycomb group response element from the *Drosophila* engrailed gene. Nucleic Acids Res 2005;33(16):5181–9.

[62] Brown JL, Kassis JA. Spps, a *Drosophila* Sp1/KLF family member, binds to PREs and is required for PRE activity late in development. Development 2010;137(15):2597–602.

[63] Ray P, De S, Mitra A, Bezstarosti K, Demmers JA, Pfeifer K, et al. Combgap contributes to recruitment of Polycomb group proteins in *Drosophila*. Proc Natl Acad Sci USA 2016;113(14):3826–31.

[64] Blastyak A, Mishra RK, Karch F, Gyurkovics H. Efficient and specific targeting of Polycomb group proteins requires cooperative interaction between Grainyhead and Pleiohomeotic. Mol Cell Biol 2006;26(4):1434–44.

[65] Dejardin J, Rappailles A, Cuvier O, Grimaud C, Decoville M, Locker D, et al. Recruitment of *Drosophila* Polycomb group proteins to chromatin by DSP1. Nature 2005;434(7032):533–8.

[66] Stros M. HMGB proteins: interactions with DNA and chromatin. Biochim Biophys Acta 2010;1799(1–2):101–13.

[67] Lehmann M. Anything else but GAGA: a nonhistone protein complex reshapes chromatin structure. Trends Genet 2004;20(1):15–22.

[68] Poux S, McCabe D, Pirrotta V. Recruitment of components of Polycomb Group chromatin complexes in *Drosophila*. Development 2001;128(1):75–85.

[69] Schuettengruber B, Oded Elkayam N, Sexton T, Entrevan M, Stern S, Thomas A, et al. Cooperativity, specificity, and evolutionary stability of Polycomb targeting in *Drosophila*. Cell Rep 2014;9(1):219–33.

[70] Frey F, Sheahan T, Finkl K, Stoehr G, Mann M, Benda C, et al. Molecular basis of PRC1 targeting to Polycomb response elements by PhoRC. Genes Dev 2016;30(9):1116–27.

[71] Mendenhall EM, Koche RP, Truong T, Zhou VW, Issac B, Chi AS, et al. GC-rich sequence elements recruit PRC2 in mammalian ES cells. PLoS Genet 2010;6(12):e1001244.

[72] Vella P, Barozzi I, Cuomo A, Bonaldi T, Pasini D. Yin Yang 1 extends the Myc-related transcription factors network in embryonic stem cells. Nucleic Acids Res 2012;40(8):3403–18.

[73] Kang H, McElroy KA, Jung YL, Alekseyenko AA, Zee BM, Park PJ, et al. Sex comb on midleg (Scm) is a functional link between PcG-repressive complexes in *Drosophila*. Genes Dev 2015;29(11):1136–50.

[74] Schwartz YB, Kahn TG, Dellino GI, Pirrotta V. Polycomb silencing mechanisms in *Drosophila*. Cold Spring Harb Symp Quant Biol 2004;69:301–8.

[75] Fischle W, Wang Y, Jacobs SA, Kim Y, Allis CD, Khorasanizadeh S. Molecular basis for the discrimination of repressive methyl-lysine marks in histone H3 by Polycomb and HP1 chromodomains. Genes Dev 2003;17(15):1870–81.

[76] Kalb R, Latwiel S, Baymaz HI, Jansen PW, Muller CW, Vermeulen M, et al. Histone H2A monoubiquitination promotes histone H3 methylation in Polycomb repression. Nat Struct Mol Biol 2014;21(6):569–71.

[77] Pengelly AR, Kalb R, Finkl K, Muller J. Transcriptional repression by PRC1 in the absence of H2A monoubiquitylation. Genes Dev 2015;29(14):1487–92.

[78] Sanchez-Elsner T, Gou D, Kremmer E, Sauer F. Noncoding RNAs of trithorax response elements recruit *Drosophila* Ash1 to Ultrabithorax. Science 2006;311(5764):1118–23.

[79] McNutt M. Retraction. Science 2014;344(6187):981.

[80] Davidovich C, Cech TR. The recruitment of chromatin modifiers by long noncoding RNAs: lessons from PRC2. RNA 2015;21(12):2007–22.

[81] Herzog VA, Lempradl A, Trupke J, Okulski H, Altmutter C, Ruge F, et al. A strand-specific switch in noncoding transcription switches the function of a Polycomb/Trithorax response element. Nat Genet 2014;46(9):973–81.

[82] Ferrari KJ, Scelfo A, Jammula S, Cuomo A, Barozzi I, Stutzer A, et al. Polycomb-dependent H3K27me1 and H3K27me2 regulate active transcription and enhancer fidelity. Mol Cell January 9, 2014;53(1):49–62.

[83] Schwartz YB, Pirrotta V. Polycomb silencing mechanisms and the management of genomic programmes. Nat Rev Genet 2007;8(1):9–22.

[84] Arnold P, Scholer A, Pachkov M, Balwierz PJ, Jorgensen H, Stadler MB, et al. Modeling of epigenome dynamics identifies transcription factors that mediate Polycomb targeting. Genome Res 2013;23(1):60–73.

[85] Cabianca DS, Casa V, Bodega B, Xynos A, Ginelli E, Tanaka Y, et al. A long ncRNA links copy number variation to a polycomb/trithorax epigenetic switch in FSHD muscular dystrophy. Cell 2012;149(4):819–31.

[86] Jermann P, Hoerner L, Burger L, Schubeler D. Short sequences can efficiently recruit histone H3 lysine 27 trimethylation in the absence of enhancer activity and DNA methylation. Proc Natl Acad Sci USA 2014;111(33):E3415–21.

[87] Lynch MD, Smith AJ, De Gobbi M, Flenley M, Hughes JR, Vernimmen D, et al. An interspecies analysis reveals a key role for unmethylated CpG dinucleotides in vertebrate Polycomb complex recruitment. EMBO J 2012;31(2):317–29.

[88] Sing A, Pannell D, Karaiskakis A, Sturgeon K, Djabali M, Ellis J, et al. A vertebrate Polycomb response element governs segmentation of the posterior hindbrain. Cell 2009;138(5):885–97.

[89] Woo CJ, Kharchenko PV, Daheron L, Park PJ, Kingston RE. A region of the human HOXD cluster that confers polycomb-group responsiveness. Cell 2010;140(1):99–110.

[90] Li G, Margueron R, Ku M, Chambon P, Bernstein BE, Reinberg D. Jarid2 and PRC2, partners in regulating gene expression. Genes Dev 2010;24(4):368–80.

[91] Farcas AM, Blackledge NP, Sudbery I, Long HK, McGouran JF, Rose NR, et al. KDM2B links the polycomb repressive complex 1 (PRC1) to recognition of CpG islands. Elife 2012;1:e00205.

[92] Wu X, Johansen JV, Helin K. Fbxl10/Kdm2b recruits polycomb repressive complex 1 to CpG islands and regulates H2A ubiquitylation. Mol Cell 2013;49(6):1134–46.

[93] Blackledge NP, Farcas AM, Kondo T, King HW, McGouran JF, Hanssen LL, et al. Variant PRC1 complex-dependent H2A ubiquitylation drives PRC2 recruitment and polycomb domain formation. Cell 2014;157(6):1445–59.

[94] Riising EM, Comet I, Leblanc B, Wu X, Johansen JV, Helin K. Gene silencing triggers polycomb repressive complex 2 recruitment to CpG islands genome wide. Mol Cell 2014;55(3):347–60.

[95] Dietrich N, Lerdrup M, Landt E, Agrawal-Singh S, Bak M, Tommerup N, et al. REST-mediated recruitment of polycomb repressor complexes in mammalian cells. PLoS Genet 2012;8(3):e1002494.

Chapter 7

Polycomb Function and Nuclear Organization

F. Bantignies[1,2], G. Cavalli[1,2]

[1]*Institute of Human Genetics, CNRS UPR 1142, Montpellier, France;* [2]*University of Montpellier, Montpellier, France*

Chapter Outline

Introduction 131
Polycomb Complexes and Their Action on Chromatin 132
Polycomb Domains 135
Polycomb and Chromatin Compaction 137
Polycomb Group Target Loci Form Dynamic Multilooped Three-Dimensional Structures 138
Polycomb-Repressed Domains Form a Subset of Topologically Associating Domains 141
Long-Range Chromosomal Interactions and Three-Dimensional Gene Networks 144
Potential Role for Noncoding RNA in Polycomb Group–Dependent Three-Dimensional Organization 148
Polycomb and Three-Dimensional Genomics in Cancer and Other Diseases 151
Concluding Remarks 151
List of Acronyms and Abbreviations 152
Acknowledgments 153
References 153

INTRODUCTION

Initially discovered as repressors of homeotic gene expression in *Drosophila*, Polycomb group (PcG) proteins have later been shown to be involved in a plethora of biological processes, and the interest in their function has literally exploded in the past decade. Indeed, PcG factors negatively regulate a large number of target genes, mainly involved in differentiation and developmental processes, including specific lineage genes. These chromatin factors therefore play major roles in a multitude of cellular functions, such as maintenance of pluripotency, proliferation, differentiation, and reprogramming. Moreover, deregulation of these factors is often associated with cancer [1,2]. Therefore it

Polycomb Group Proteins. http://dx.doi.org/10.1016/B978-0-12-809737-3.00007-6

is crucial to understand PcG mechanisms to gain insights into the epigenetic processes associated with normal development, physiology, and cancer.

In both *Drosophila* and vertebrates, PcG proteins form conserved large multimeric complexes, which act directly on specific chromatin regions via posttranslational histone modifications, chromatin compaction, and higher-order chromatin folding. PcG proteins lead to the formation of large repressive chromosome domains, which are often composed of multiple gene units and covered by their associated histone marks, H3K27me3 and H2AK119ub (H2AK118ub in *Drosophila*). The spatiotemporal regulation of these domains appears dynamic, and target genes can escape their repressive environment to be transcribed in the appropriate cell type or tissue. In the nucleus, PcG components are found aggregated in microscopically visible foci or bodies. PcG foci represent specific nuclear environments for target gene silencing and clustering [3,4], suggesting a specific organization of their target chromatin in the 3D nuclear space. Recent high-throughput 3C data together with microscopy analysis has shed light both on how these chromatin domains are organized along the genome and on their 3D organization. In this chapter, we briefly describe the functional role of PcG proteins, how they form Polycomb-repressive domains, and their role in regulating 3D genome organization and function.

POLYCOMB COMPLEXES AND THEIR ACTION ON CHROMATIN

PcG proteins are conserved in higher eukaryotes and exert their action on chromatin as multimeric complexes. In *Drosophila melanogaster*, PcG proteins assemble into at least five different complexes. These are termed Polycomb repressive complexes 1 (PRC1), Polycomb repressive complex 2 (PRC2), Pho-repressive complex (PHO-RC), dRing-associated factors (dRAF), and Polycomb-repressive deubiquitinase (PR-DUB) (Fig. 7.1A). PRC1 and PRC2 were the first to be identified in *Drosophila* and are the best characterized (for reviews, Refs. [5,6]). PRC1 contains a core quartet of PcG proteins: Polycomb (PC), Polyhomeotic (PH), Posterior sex combs (PSCs), and Sex combs extra (SCE, also termed dRING), whereas PRC2 contains another core quartet: enhancer of zeste (E(Z)), extra sex combs (ESC), suppressor of zeste 12 (SU(Z)12), and Nucleosome-Remodeling Factor 55 (NURF-55). Polycomb-like (PCL) and JARID2 can also be associated with PRC2 and a further protein, Sex comb on midleg (SCM) can associate with both PRC1 and PRC2 complexes [7]. In mammals, PRC1 and PRC2 are well conserved, but the duplication of many PcG genes or the interactions of PcG proteins with various accessory proteins has led to a multiplication of complexes and to more complexity in their composition, which depends on cell type and developmental stage [8]. Several forms of the canonical PRC1 and PRC2 complexes therefore exist in vertebrates, as well as variant or non

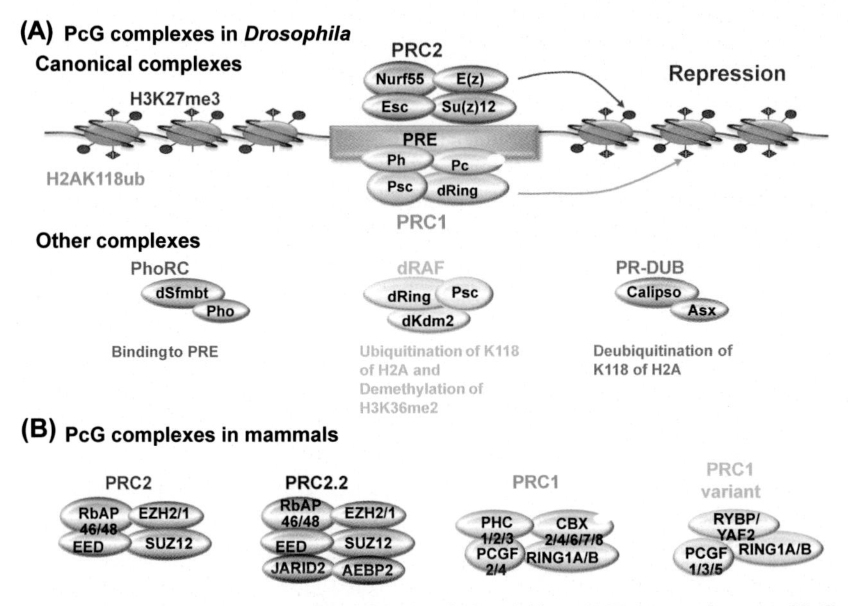

FIGURE 7.1 PcG complexes and their biochemical activities. (A) The two main types of PcG complexes in *Drosophila*, PRC2 and PRC1, bind to specific DNA element called Polycomb-responsive element (PRE) and to their target gene promoters. Three other PcG complexes have been identified in *Drosophila*. Note that H2AK118ub in *Drosophila* corresponds to H2AK119ub in vertebrates. (B) Similarly, PRC2 and PRC1, as well as variant complexes, have been characterized in mammals. For comparison, the position of the different homologous subunits has been kept in PRC2 and PRC1. The presence of different PCGF subunits is thought to define the class of PRC1 complex. Variant complexes include either the RYBP or YAF2 protein, the presence of which is mutually exclusive with the CBX component in the canonical PRC1. Slashes denote "or".

canonical PRC1 complexes, such as the PRC1−RYBP complex [9,10] (Fig. 7.1B). Canonical and variant PRC1 complexes are defined by mutually exclusive subunits (Ref. [9], for review, Ref. [11]). This renders the *Drosophila* system simpler and appropriate to study the basic molecular function of the various PcG subunits and complexes.

PcG recruitment to their target loci was indeed first studied in *Drosophila*. In this organism, PcG components are recruited to Polycomb response elements (PREs), which represent *cis*-regulatory DNA sequences of up to over 1 kilobase (kb) in size. PREs are not defined by consensus sequences but contain a combination of several binding motifs for specific DNA-binding proteins, such as the PcG PHO-RC and GAGA factor (GAF) (for review, Ref. [12]). Given that neither PRC1 nor PRC2 complexes contain DNA-binding proteins, the presence of these DNA-binding factors at PRE may serve as a tethering platform for these PcG complexes. Genome-wide mapping studies in flies show that approximately 50% of putative PREs are located at promoters [13,14]. Other PREs can be located tens of kb upstream or downstream from the promoters of target genes, suggesting that they can exert their repressive action at distance. Once recruited to a target site, the PRC2 complex, via its catalytic subunit E(Z), deposits its characteristic repressive chromatin mark, histone H3 trimethylation at lysine 27 (H3K27me3). This mark can be read by the PRC1 complex, through the chromodomain of its PC subunit [15]. Both PRC1 and dRAF contain SCE/dRING, an active E3 ubiquitin ligase subunit, responsible for the mono-ubiquitination of lysine 118 of H2A (H2AK118ub), but the role for this histone mark in Polycomb mediated silencing, at least in *Drosophila*, is not clear (reviewed in Ref. [3]). The functional importance of these epigenetic marks was well illustrated by the generation of mutant flies with point mutations in the corresponding modified residues. H3K27 mutants (H3K27R or H3K27M) cause derepression of PcG target genes and homeotic transformations, whereas H2Aub-deficient animals fully maintain PcG silencing with no PcG-dependent developmental defects [16–18]. This suggests that, in *Drosophila*, H3K27me3 represents the main epigenetic signature for PcG silencing.

Although the molecular activities of PRC complexes appear broadly similar in metazoans, the mode of PcG recruitment to chromatin seems to have notable differences. In mammals, some PRE-like elements containing motifs for YY1, a PHO homolog, have been described [19,20], but the regions that recruit PcG are in general poorly defined. In fact, genome-wide mapping studies in mammalian cells reveal that PcG-binding sites are mainly associated with transcriptional start sites corresponding to regions of dense CpG dinucleotides or CpG islands, provided they are not DNA methylated [21–23]. Moreover, the classical hierarchical model described in *Drosophila* seems inadequate, at least in some loci, to recapitulate recruitment of PcG proteins. First, PRC1 and PRC1-like complexes can bind a subset of their targets in a PRC2- and H3K27me3-independent manner [10,24,25]. Second, a reverse mechanism has been proposed, whereby PRC1 is recruited first, followed by PRC2. KDM2B, a component of the variant PCGF1/PRC1 complex, which possesses a CxxC domain able to recognize unmethylated CpGs, represents an attractive candidate for the recruitment of mammalian PcG complexes [26–28]. This variant PRC1 complex, also called PRC1-RYBP, deposits H2AK119ub (the equivalent of the H2AK118 in *Drosophila*), leading to the subsequent recruitment of the PRC2 complex and H3K27me3 [29–31]. At present, it is still unclear how PRC2 reads the H2AK119ub mark, possibly associated factors such as JARID2 and AEBP2 might assist this recognition [31]. The absence of transcription could also be sensed by PcG complexes for their recruitment to CpG islands [32].

Finally, other modes of recruitment might involve RNA species. In mammalian cells, the PRC2 complex can physically associate with a large number long noncoding RNAs (lncRNAs), and some of them may participate in PcG recruitment in vertebrates (reviewed in Ref. [8]). Recent reports indicate that RNA entities may also participate to the 3D organization of PcG-bound regions, expanding the role of lncRNAs in PcG regulation (see below).

POLYCOMB DOMAINS

In both *Drosophila* and mammals, genome-wide ChIP analysis has shown that Polycomb members bind to discrete sites on chromatin. However, their associated histone marks, H3K27me3 and H2AK119ub, form large chromatin domains, which extend far beyond PcG-recruitment sites. The positive feedback loops observed among PRC1 and PRC2 complexes are certainly important for the establishment and spreading of PcG-repressive domains. For example, RYBP found in variant PRC1 complexes is a ubiquitin-binding protein able to bind H2AK119ub [33] and embryonic ectoderm development protein (EED) of the PRC2 contains an aromatic cage that is able to specifically bind H3K27me3 [34]. These feedback mechanisms may contribute to maintain the PcG repressive marks through DNA replication. It was also proposed that H2A monoubiquitination could stimulate H3K27 trimethylation, which in turn could facilitate the recruitment of canonical PRC1 complexes [31], and the activity of the PRC2 complex can be stimulated by dense chromatin via an interaction of the SUZ12 subunit with the histone H3 tail [35]. Positive reinforcement mechanisms and spreading of these chromatin marks may provide robustness for the maintenance of repressed states through the cell cycle and during development.

In *Drosophila*, PRC1 and PRC2 core components colocalize at PREs that cluster to form large H3K27me3-coated chromatin domains. H3K27me3 domains cover 10–20% of the genome [36], with the two *Hox* gene clusters, the Antennapedia complex (ANT-C) and the Bithorax complex (BX-C) representing by far the largest PcG domains [14]. The PRC1 complex is also present at transcription start sites (TSSs) with a stalled RNA Polymerase II (PolII) and is linked to the production of short RNAs [13,36]. These sites are also enriched in H3K4me1/me2, and these specific signatures at TSSs might signal transcriptional pausing of key regulatory and developmental genes. In mammalian cells, the H3K27me3 domains appear highly dynamic during cell differentiation. In embryonic stem cells (ESCs), PcG proteins predominantly bind to small regions surrounding promoters, and H3K27me3 domains are rather small [37,38]. In fact, the classical PcG state is depleted in ESCs, with bivalent domains combining active H3K4me3 and repressive H3K27me3 marks being the most abundant Polycomb state at the promoter of target genes [39,40]. These bivalent domains represent poised states for lineage-specific activation of key regulatory and developmental genes, and therefore reflect the pluripotent potential of ESCs. Bivalent domains are then resolved during cell lineage commitment, where genes that have a function in the specific lineage become monovalent for H3K4me3, whereas genes with a function in other lineages tend to lose their active mark. In differentiated cells, although some bivalent domains remain at some target loci [38,41,42], H3K27me3 domains become larger and more widespread over silent loci [43]. Therefore, restricted promoter-proximal regions in pluripotent cells could then spread to larger domains to maintain stable silencing in lineage-committed cells [44].

The dynamic properties of large H3K27me3 are well illustrated during the spatiotemporal control of *Hox* gene transcriptional activation. In *Drosophila* embryos, the H3K27me3 profiles across the BX-C in successive parasegments show a "stairstep" pattern characterized by sharp boundaries and enrichment of acetylated H3K27, in a pattern complementary to the repressive mark, to demarcate the active and repressed regions along the antero—posterior axis [45]. In mouse embryos, activation is associated with a directional transition in chromatin composition, with activation marks progressively replacing the repressive H3K27me3 mark along *Hox* gene clusters [46]. In contrast to this dynamic regulation, H3K27me3-repressive domains can be highly stable throughout development; for example, *Hox* gene clusters remain completely silent in specific parts of the embryos where all genes need to be switched off at all times. In vertebrates, another example of permanent Polycomb silencing involves H3K27me3 enrichment on the inactive X chromosome, which represents a clear example of facultative heterochromatin formation and maintenance [47].

Inside the nucleus, PcG foci represent the physical sites of PcG target gene silencing [3]. PcG foci are highly colocalized with the H3K27me3 mark in the nuclear environment. Moreover, the amount of PcG proteins within PC foci correlates with the size of the genomic domains forming them. Large genomic domains such as the Hox complexes are located in the most intense PcG foci, whereas small genomic domains are located in structures of weaker intensity [48]. With conventional microscopy, the number of PcG foci is, however, lower than the number of PcG domains, suggesting the possibility of a higher-order organization of PcG chromatin domains (Ref. [49], see Long-Range Chromosomal Interactions section). Recent superresolution microscopy of PcG foci in *Drosophila* cells corroborates this idea, suggesting that these foci correspond to the aggregation of several nanoscale protein clusters [50].

Beyond the PcG/H3K27me3-repressive domains, other epigenetic marks and chromatin factors form additional linear domains along the chromatin fiber. In both *Drosophila* and mammals, epigenetic marks and proteins bound to specific regions were used to define different chromatin types, which correspond to various transcriptional states, ranging from highly expressed to constitutively repressed [36,40,51]. Accordingly, the chromatin fiber can now be subdivided into four main types or colors: a first color corresponds to the active "Red chromatin," which contains numerous transcribed genes and is enriched in active marks H3K4me3/H3K36me3 and hyperacetylation of histone tails. Three other chromatin colors correspond to the different types of gene repression: The "Green chromatin" represents heterochromatin, characterized by di/trimethylation on lysine 9 of histone H3 (H3K9m2/H3K9me3) and the heterochromatin protein HP-1; the "Blue chromatin" represents PcG/H3K27me3 domains; the "Black chromatin," also called "void" or "null" chromatin is enriched in the H1 histone linker but lacks specific epigenetic marks. Actually, recent genomic data from *Drosophila* cells provide evidence that H3K27me2-enriched regions largely correspond to Black chromatin [52] and the same might be true for mammalian cells [53]. Although PRC2 is not found in these regions, it is proposed that H3K27me2 is produced by opposing roaming activities of PRC2 and H3K27 demethylase UTX-containing complexes [52]. This linear chromatin organization into functional subdomains truly represents an epigenetic landscape of the genome, and this conserved chromatin organization on the linear genomic scale could impinge on 3D chromosomal organization in the nucleus (see below).

POLYCOMB AND CHROMATIN COMPACTION

Chromatin compaction was proposed to be a major mode of action of PcG complexes, by reducing or interfering with DNA accessibility to the transcriptional machinery. Several lines of evidence indicate that PRC1 represents a major player for chromatin compaction. In vitro studies suggested that the *Drosophila* PRC1 core components can cause compaction of a nucleosome array [54], correlating well with in vivo analysis [55]. This activity is dependent on direct binding to nucleosomes but seems to be independent of the presence of histone tails, at least in vitro, and consistent with in vivo studies in both flies and mouse ESCs [18,56]. *In Drosophila*, the C-terminal region of the PRC1 subunit PSC plays a central role in this process. The role of PRC1 in the compaction of nucleosomal arrays is conserved across metazoans and plants but is carried out by different PRC1 subunits, such as CBX2 in mouse and EMF in *Arabidopsis thaliana* [57,58]. PSC and CBX2 share an unstructured highly basic domain, which allows the binding to single nucleosomes and then promotes oligomerization and the formation of higher-order structures. PH, another PRC1 component, is also able to promote oligomerization via its sterile alpha motif (SAM) domain, facilitating PRC1 and PRC2 binding to chromatin and compaction [59]. The question of compaction was also addressed with fluorescent in situ hybridization (FISH) analysis. Using the distance between two FISH probes spaced along a chromosome locus as readout, these analyses support Polycomb-dependent contraction at the silent *Drosophila* and murine *Hox* gene clusters, as well as in a silent imprinted locus in mouse [56,60,61]. At the murine *Hoxb* and *Hoxd* loci, contraction was shown to be dependent on RING1B, the catalytic subunit of PRC1. However, locus compaction and gene silencing were both shown to be independent on the ubiquitination activity of RING1B [56]. PRC2 components have also been involved in chromatin compaction. For instance, a variant of the mammalian PRC2 complex containing EZH1 (an EZH2 paralog) was reported to mediate repression through its ability to compact chromatin independently of its enzymatic activity [34]. These observations illustrate that PcG complexes can affect chromatin organization beyond histone tail modifications. On the other hand, reconstituted *Drosophila* PRC2 exhibits improved histone methylation activity on dense rather than dispersed oligonucleosomes, suggesting that dense chromatin may favor PRC2 function [35]. Finally, compaction of PcG chromatin domains is compatible with recent FISH observation using superresolution microscopy in *Drosophila* cells. This study revealed that Polycomb chromatin is highly compact [62], together with a strong degree of intermixing within the domain. This highly compact structure may serve to create nuclear subcompartments to prevent unscheduled activation that could emanate by transcription activators present in the surrounding nuclear space. Therefore architecture and the biochemical activities of Polycomb complexes might collaborate to stably maintain gene silencing.

However, in vivo repression of Polycomb target genes cannot simply reflect chromatin condensation into an impenetrable architecture. Indeed, low but significant levels of RNA PolII are found at many PcG-repressed promoters [13,36,63], including the silenced *Hox* genes in *Drosophila* [64]. Moreover, *Drosophila* PREs are nuclease hypersensitive, depleted in nucleosomes and represent peaks of histone replacement [65–67], although global histone turnover at Polycomb domains is slower than that in active regions of the genome [68]. Finally, PcG proteins themselves have short residence times on chromatin [69]. Therefore, the architecture of PcG domains may not be totally inaccessible but may instead adopt more specific higher-order chromatin conformations to block transcription globally while providing access to critical regulatory regions. This feature may be critical to allow reprogramming gene regulation in response to the appropriate regulatory cues.

POLYCOMB GROUP TARGET LOCI FORM DYNAMIC MULTILOOPED THREE-DIMENSIONAL STRUCTURES

In *Drosophila*, PRE-promoter looping to drive silencing was first suggested in studies that used sophisticated transgenic systems [70,71]. FISH and chromosome conformation capture (3C), a method that allows DNA–DNA proximity ligation of genomic fragments that are in close spatial proximity [72], confirmed and extended some of these original observations. Contacts between distal PREs and promoters by looping were first demonstrated by tethering Dam methyltransferase to the *Drosophila Fab-7* element, which contains a core PRE of the BX-C [73]. A direct proof of these loops between PcG-bound sites was given by high-resolution 3C [74].

Many of the PcG target loci, such as the Hox complexes in *Drosophila* and mammals, are made of multiple gene units and contain several PcG protein-binding sites, some of which are located at promoters whereas others are located at large distances upstream or downstream of the coding unit. Using FISH combined with Polycomb immunostaining and 3C approaches, the BX-C was shown to adopt PcG-dependent multilooped structures [60] (Fig. 7.2A). This "repressive chromatin hub" is composed of chromatin loops involving the major PcG-bound regulatory elements in the BX-C, including PREs and core promoters, and is found within large PcG foci. Specific higher-order chromatin

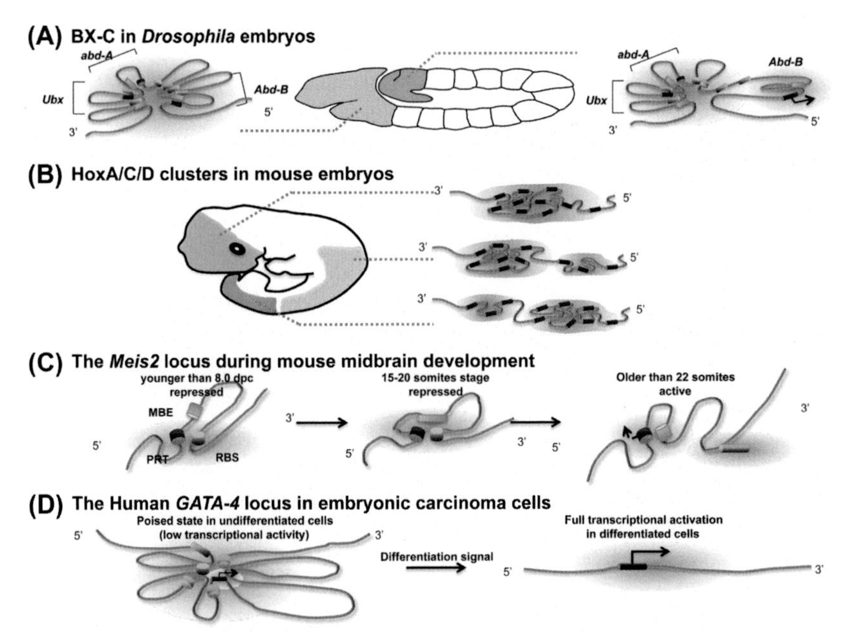

FIGURE 7.2 PcG-repressed loci adopt different three-dimensional conformations, which are dynamics during development and differentiation. (A) The *Drosophila* BX–C complex is folded in dynamic multilooped structures. In the head of embryos, the three *Hox* genes of the complex, *Ubx*, *abd-A*, and *Abd-B* are repressed and form a PcG-dependent repressed chromatin hub, depicted by the shaded blue circle. In the most posterior parasegment of embryos, *Abd-B* is active and is displaced from the RCH. *Red*, *Blue* and *Green cylinders* indicate promoters, 3′ ends and PREs of the *Hox* genes, respectively. (B) The dynamics of looped structures is well illustrated by the Hox clusters (*HoxA*, *HoxC*, *HoxD*) in mouse embryos. Each cluster, located on different chromosomes, contains 9–11 genes (*red cylinders*) that are differentially expressed along the antero–posterior axis of the embryo. In the head, where all the genes are repressed, Hox clusters are covered by H3K27me3 and form an RCH, depicted in blue. In the trunk, the most distal genes, 5′ of the clusters, are activated and switch to an active chromatin hub marked with the H3K4me3 mark, depicted in red. In the tail, the majority of the genes are active, forming a larger ACH. This bimodal 3D organization may be essential to process the collinear activation of Hox clusters. Note that the HoxB cluster shows a slight variation in this organization with a large intergenic region looping out from the two hubs (not shown). (C) Higher-order chromatin transitions regulate the *Meis2* gene during mouse midbrain development. The topological switch involves the promoter (PRT, *red cylinder*) and the 3′ end of the gene (RBS, *green cylinder*), both sequences bound by the PRC1 subunit RING1B, and a midbrain-specific enhancer (MBE, *blue cylinder*). The *shaded blue circle* represents the accumulation of PRC1 and the *shaded red circle* represents the activation of the locus. (D) A repressive hub is formed at the human *GATA-4* locus in undifferentiated embryonic carcinoma cells. This hub is associated with a poised transcriptional state, with RNA PolII bound at the promoter, depicted by a *shaded red circle*. The *shaded blue circle* represents the accumulation of PcG complexes. Upon differentiation, the repressive hub is dissolved leading to an almost linear conformation and full transcriptional activation. *Green cylinders* represent regions with strong enrichment for PcG proteins and the red cylinder represents the promoter of the *GATA-4* locus. *(A) Adapted from Lanzuolo C, Roure V, Dekker J, Bantignies F, Orlando V. Polycomb response elements mediate the formation of chromosome higher-order structures in the bithorax complex. Nat Cell Biol 2007;9(10):1167–74; (B) Adapted from Noordermeer D, Leleu M, Splinter E, Rougemont J, De Laat W, Duboule D. The dynamic architecture of Hox gene clusters. Science 2011;334(6053):222–5; (C) Adapted from Kondo T, Isono K, Kondo K, Endo TA, Itohara S, Vidal M, et al. Polycomb potentiates meis2 activation in midbrain by mediating interaction of the promoter with a tissue-specific enhancer. Dev Cell 2014;28(1):94–101; (D) Adapted from Tiwari VK, McGarvey KM, Licchesi JD, Ohm JE, Herman JG, Schubeler D, et al. PcG proteins, DNA methylation, and gene repression by chromatin looping. PLoS Biol 2008;6(12):2911–27.*

structures also exist among mammalian silent Hox clusters [75,76]. Importantly, these structures appear to be dynamic and strictly depend on the transcriptional status of the genes they contain, with active genes being displaced from the repressive hub. In *Drosophila* embryos, in tissues where they are active, *Hox* genes are absent from the repressive hub and are found outside PcG foci [60,77]. The dynamic features of the mouse *Hox* gene clusters were nicely illustrated using a high-resolution 4C approach [76] (Fig. 7.2B). This study confirmed that repressed *Hox* clusters adopt a single 3D structure well delimited from the flanking genomic regions. Once transcription starts, newly activated genes escape from the repressive hub and progressively cluster into a transcriptionally active compartment, leading to a bimodal 3D organization. This structural switch coincides with the transition in chromatin marking, with H3K27me3 covering the repressive structure and the active one progressively gaining H3K4me3. Moreover, the dynamics of this Hox bimodal structure leads to a switch in interactions with the regulatory landscape flanking the gene cluster, which contributes to the patterning of proximal and distal limb structures [78]. Therefore, specific looping patterns of Hox clusters play an important role in orchestrating the spatially and temporally collinear induction of *Hox* genes during development.

Other repressive multilooped conformations involving PcG-bound regions were also described for several loci, including the human *GATA-4* and *INK4-ARF* loci [79,80], and for the mouse *Meis2* and imprinted *Igf2* loci [81,82]. These repressive higher-order chromatin structures are plastic in nature, in the sense that they can acquire new topologies or be dissolved when transcription of the genes they contain needs to be activated. For example, a topological transition upon dissociation of RING1B is observed at the *Meis2* locus when it is activated during midbrain development (Fig. 7.2C), whereas high-level transcription of the *GATA-4* gene in differentiated cells leads to a strong decrease of PcG occupancy and H3K27me3 enrichment, with the locus acquiring an almost linear conformation (Fig. 7.2D). In the *GATA-4* locus, chromatin loops may serve to maintain the gene in a silent but poised state, with paused RNA PolII bound at the promoter, in preparation for its activation during differentiation. This may actually represent a common theme for looping structures. In *Drosophila*, stable enhancer–promoter chromatin loops, which are frequently associated with paused RNA PolII, have been described for several developmental regulated loci [83], leading to the idea that transcription initiates from preformed enhancer–promoter loops through the release of paused polymerase. Many PcG target genes, including *Hox* genes, contain a paused PolII [13,36,64], and it was proposed that paused promoters confer intrinsic insulator activity to some *Hox* promoters, which may help promote higher-order chromatin organization [64]. On one hand, these multilooped structures maintain stable silencing by clustering multiple PcG-binding sites in the same nuclear compartment and reinforcing PcG function. On the other, they can prepare for gene activation that will be induced later during differentiation and development, with PcG chromatin structures preventing these genes from premature transcriptional activation.

The fact that these chromatin hubs involve PcG-bound sites and are sensitive to PcG protein dosage [60,75,79,80] suggests that PcG components contribute to their establishment and/or maintenance. Other factors, such as the architectural CCCTC binding factor (CTCF), may also be involved in the formation of these chromatin structures. For instance, CTCF binds to the BX-C and the *HOXA* loci at multiple sites [75,84,85] adjacent to or overlapping PcG-binding sites, and a role of CTCF in repressive chromatin hubs has been suggested for these loci [75,86]. In agreement with this idea, CTCF is able to interact with the PRC2 subunit SUZ12 and regulates the mouse imprinted *Igf2* locus by orchestrating a PcG-dependent chromatin loop [82]. CTCF also plays a role in the formation of active chromatin hubs [87], placing this factor as a general mediator of chromatin loops (see below). In mammals, cohesin proteins, which share many binding sites with CTCF [88–90], could help stabilize chromatin loops [91]. Cohesin rings might seal and stabilize chromatin loops in the same manner in which they maintain the pairing of sister chromatids. In *Drosophila*, the cohesin complex does not colocalize with CTCF [92] but biochemically interacts with PRC1 [93]. However, on chromatin, cohesin complex subunits show only limited overlap with PcG proteins [94,95], suggesting that other factors are involved in the regulation of PcG-dependent chromatin structures.

POLYCOMB-REPRESSED DOMAINS FORM A SUBSET OF TOPOLOGICALLY ASSOCIATING DOMAINS

The formation of higher-order chromatin structures appears to be a general feature of 3D genome organization. Indeed, the application of high-throughput 3C (Hi-C) techniques [96] in *Drosophila* and mammalian cells has identified numerous physical or topological domains, also called "topologically associating domains" or TADs [97–100]. TADs represent structural domains within which the chromatin fiber has a higher probability to interact when compared to any surrounding regions of the genome (Fig. 7.3A). Importantly, such physical entities were confirmed at the single cell level using structured illumination microscopy, as FISH probes were found to intermingle more frequently within TADs than between two adjacent TADs [99,101]. More than 1000 TADs partition the *Drosophila* genome [98,100] and many more partition the mammalian genome [97,102]. TADs represent conserved structures, but their size varies among metazoans chromosomes [103]. TADs in *Drosophila* genome range from tens to hundreds of kb with an average size of around 100 kb, with the two *Hox* gene clusters being among the largest TADs. In mammalian genomes, these domains are larger than those in *Drosophila*. Depending on the sequencing depth and method used to call TAD borders, mammalian TAD numbers vary greatly, ranging from 100 to 200 kb to around

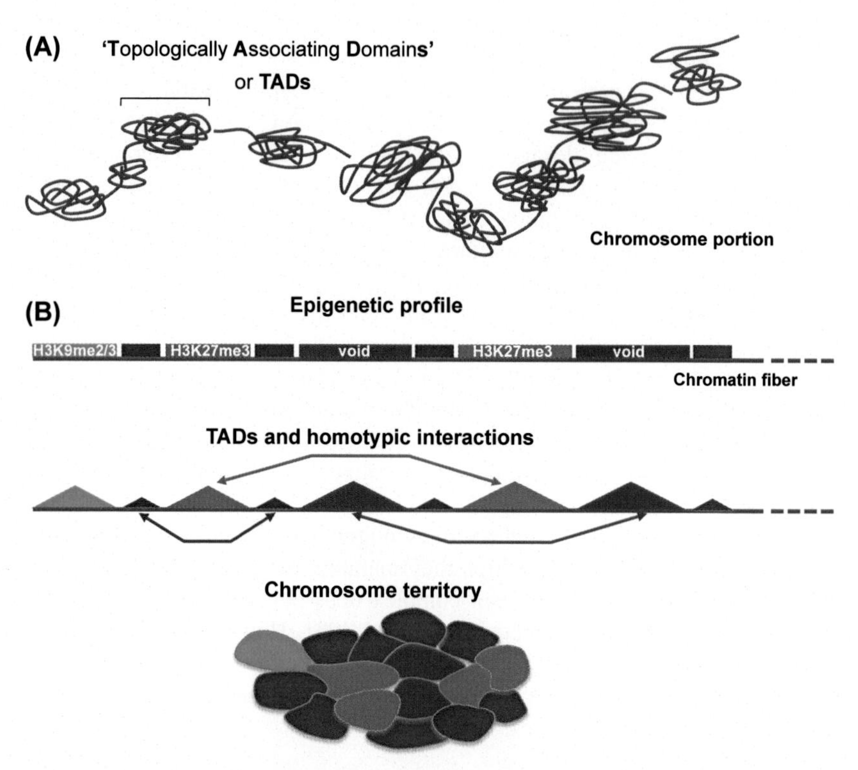

FIGURE 7.3 "Topologically associating domains" in multiple colors. (A) TADs represent physical 3D structures along the chromatin fiber, within which chromatin interactions are more frequent when compared to neighboring regions. (B) The chromatin fiber can be subdivided into four main chromatin types or colors: Green for heterochromatin and the associated mark H3K9me2/3; Red for the active chromatin and many associated marks, including H3K4me3, H3K36me3, and histone acetylation; Blue for the Polycomb repressed domains and the canonical mark H3K27me3; Black/Void or Null chromatin, devoid of specific epigenetic marks. TADs, represented as triangles along the chromatin fiber, correlate with epigenetic domains. TADs therefore may partition the genome into functional genomic units. A second layer of organization corresponds to the tendency of TADs of the same type to interact more frequently with one another, leading to the formation of chromosome compartments, territories and the functional 3D folding of the genome.

800 kb. Furthermore, better resolution in Hi-C analysis tends to reveal sub-TADs conformations, called "loop domains," which are reminiscent of the multilooped 3D structures described above.

TADs are well demarcated and separated by sharp domain borders with variable composition among species. In mammals, TAD boundaries are enriched in active transcription, housekeeping genes, tRNA genes, and short interspersed nuclear elements retrotransposons [97]. In addition, most boundaries are enriched in binding sites for the architectural proteins CTCF and cohesin. In *Drosophila*, these border regions are also enriched in

H3K4me3 reflecting active transcription, as well as in high gene density and chromatin accessibility. They are also enriched in architectural proteins, including CTCF, CP190, Mod(mdg4), Su(Hw), BEAF-32, as well as in Chromator, a protein of the mitotic spindle that possesses a chromodomain [98,100]. Common features of these TAD border regions are therefore high level of transcription and the presence of architectural proteins. However, their presence is not exclusive to boundaries. Many binding sites for architectural proteins also exist within TADs, and CTCF and cohesin depletion reduce the intensity of intra-TAD interactions without affecting the overall 3D organization [104–106], consistent with their role in multilooped structures. However, more work needs to be done to understand the role of these factors at domain boundaries.

TADs appear to be rather stable across different cell types and conserved during evolution [97,99,107]. Single-cell Hi-C showed that chromosomes maintain these structures in every cell [108], and that TADs are maintained as intact modules in syntenic regions during evolution [107]. From this universal and cell-type invariant architecture, TADs can be considered as intrinsic architectural units of chromosomes.

Most importantly, TADs are well correlated with the chromatin states that were previously defined by epigenomic profiling and therefore physically partition the genome into functional epigenetic domains [98–100] (Fig. 7.3B). In mammals, this correlation is not so obvious at the megabase (Mb) resolution [97] but becomes clearer when using higher-resolution Hi-C maps and considering the loop domains, which correspond to the smaller TADs or sub-TADs [102]. *Drosophila* or mammalian TADs have therefore specific chromatin signatures. In *Drosophila*, they have been categorized in four different types or colors: the Active/Red TADs, and three types of repressive TADs, including Polycomb/Blue, Void/Black, and Heterochromatin/Green TADs [100]. In mammals, TADs have been subdivided into active A1-A2 and repressed B1-4 TADs, with B1 correlating positively with H3K27me3 and B3 corresponding to the Black chromatin, with enrichment at the nuclear periphery [102]. The relative stability of TADs indicates that they may provide the basic structural organization of chromosomes onto which epigenomic domains are laid down. Another possibility is that epigenomic demarcation could be the driving force for TADs formation or maintenance. In support of the former model, the disruption of large H3K27me3 or H3K9me2 domains at the Xic locus on the inactive X chromosome did not affect topological architecture [99]. In support of the second idea, epigenomic marks are relatively stable during cell cycle, whereas TADs are completely absent on the highly condensed mitotic chromosomes [109]. Further invalidation experiments of architectural proteins as well as enzymes responsible for the deposition of histone marks should inform us on the mechanisms behind TAD establishment and maintenance during the cell cycle and development.

Despite their relative stability, interactions within TADs or sub-TADs can be dynamic, reflecting cell-type specific structural changes dependent on gene expression [97,102]. This is reminiscent of what was described for multi-looped structures, suggesting that TADs are directly linked to gene regulation. In support of this idea, deletion of a border region or disruption of boundary elements can produce ectopic contacts between previously separated TADs, leading to misregulation of genes [99,110]. The fact that TADs appear well conserved during evolution [107] further supports this notion of functional structural units.

The comparison of three different TADs in *Drosophila* cells (Red, Blue, and Black) in single cells using superresolution stochastic optical reconstruction microscopy (STORM) in combination with Oligopaint DNA FISH [111] reveals different degrees of packaging and intermingling, with Polycomb TADs forming the most compact configuration. This specific structure is probably imposed by the specific biochemical properties of PcG complexes bound at these regions as well as their associated chromatin marks [62]. On the other hand, the active chromatin, mostly composed of housekeeping genes (also called Yellow chromatin), builds a much less compact structure, which may lead to a better accessibility and therefore allow transcription. Black chromatin is characterized by an intermediate degree of compaction. Therefore, chromosomes are organized as strings of topological domains, ranging from the highly compact configuration organized by PcG proteins to less compact structures more favorable to transcription. Future analysis of these TADs with superresolution microscopy should help to better understand the structural and dynamic properties of TADs during differentiation and development.

LONG-RANGE CHROMOSOMAL INTERACTIONS AND THREE-DIMENSIONAL GENE NETWORKS

PcG-dependent long-range interactions were first suggested by work in *Drosophila*. FISH analysis in *Drosophila* embryos indicated that homologous pairing of the BX−C complex still occurs in the presence of a heterozygous translocation, in which the two homologous copies are located on different chromosomes [112]. In addition, genetic analysis indicated that the PcG-dependent regulation at transgenes containing PREs could be influenced by the presence of the endogenous copy located on a different chromosome [113]. FISH analysis or live imaging using green fluorescent protein tagged loci indeed reveals close proximity between transgenic PREs, even when inserted in different chromosomes [114], or between a transgenic PRE and its endogenous copy [49,113]. These long-range interactions depend on PcG function [113] as well as architectural protein components such as CTCF that binds to regulatory sequences flanking interacting PREs [115,116]. Notably, it was shown that the insulator portions of well-known regulatory elements containing PREs are necessary and sufficient to establish long-range contacts, suggesting an important role of CTCF and other insulator-binding proteins for long-range chromatin contacts of PcG-binding sites. PcG proteins may therefore contribute to their stabilization rather than to their establishment. Most importantly, these chromosomal contacts strengthen PcG-mediated silencing at transgenic loci, indicating that they may play a functional role.

These long-range contacts also depend on the RNAi pathway, although the molecular role of this machinery at PREs or at insulator elements flanking them is unknown [49,116]. Concerning this link, Ago2, a component of the RNAi machinery, is bound at many PcG target genes, including PREs of the BX-C and, together with Cp190, another insulator factor, Ago2 may contribute to the 3D organization of PcG target genes [117]. Intriguingly, its role does not seem to require its enzymatic activity, raising a question on the putative role of noncoding RNAs in these phenomena.

Long-range interactions also characterize endogenous PcG target genes. First, FISH experiments indicated that the two Hox complexes, ANT-C and BX-C, located 10 Mb apart on the same chromosome arm, were able to cluster in the same PcG foci in a significant portion of the nuclei [77]. Importantly, *Hox* gene interactions were strongest when the two gene clusters were completely repressed, suggesting corepression. Long-range interactions were reinforced during development and were shown to be conserved during evolution in the *Drosophila* genus, suggesting that they may have functional significance. In *D. melanogaster*, *Hox* genes are the most prominent PcG target domains, but chromosomal interactions were not restricted to these regions. 3C on chip (4C) experiments confirmed the association between the Hox clusters and identified additional interactions with other PcG domains on the same chromosome arm [77,118] (Fig. 7.4A). Importantly, these interactions are highly specific among PcG target genes and exclude active chromatin and poorly expressed regions not bound by PcG proteins. Moreover, it was demonstrated that these long-range interactions involving the Hox clusters are the most abundant among genes on the same chromosome arm, although low-frequency interchromosomal interactions among PcG domains do exist [118]. This result indicates that these chromatin interactions are topologically constrained by the overall chromosome architecture. The chromatin interactome is, however, plastic and can be reprogrammed in response to diverse stimuli. Indeed, a recent study identifies a global reshaping of chromosome interactions upon heat shock in *Drosophila*-cultured cells, with a decrease in the strength of TAD boundaries and increase in long-range interactions, concomitant with a rearrangement in the binding locations of architectural and Polycomb proteins [119]. It will be interesting in the future to study what stimuli induce such nuclear architecture changes and how these changes are regulated.

In the mammalian system, initial studies using EZH2 [120] or RNF2 (also known as RING1B, a core component of PRC1) [121] antibodies combined

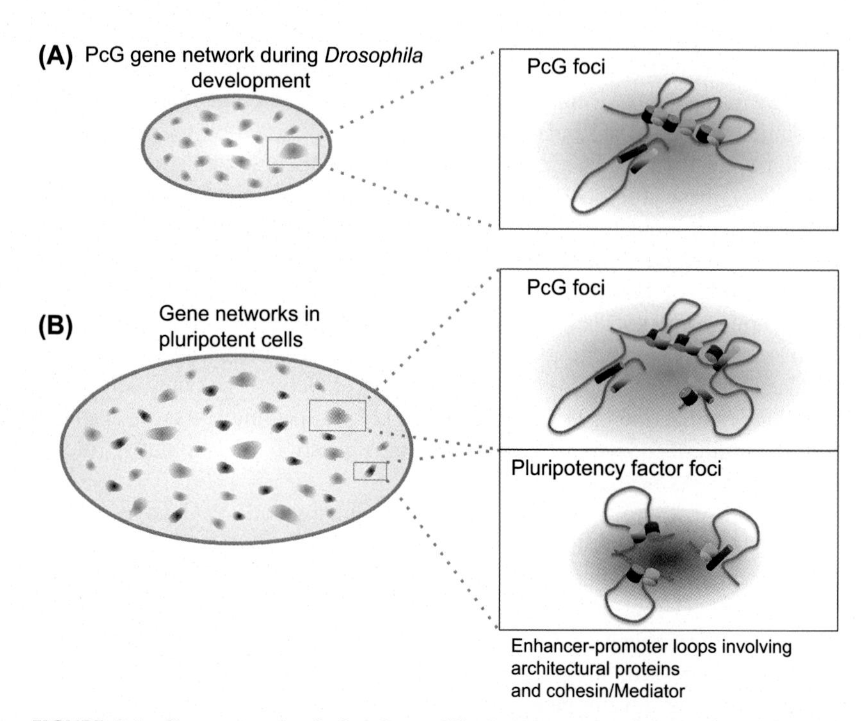

FIGURE 7.4 Gene networks during *Drosophila* development and in mammalian ESCs. (A) In *Drosophila*, Polycomb group (PcG) proteins are organized in aggregates called PcG foci. These foci are the site of gene repression. Inside these foci, Promoter-PRE interactions are forming repressive chromatin Hubs (promoters and PREs are represented by *red and green cylinders*, respectively). PcG foci are also the site of clustering where multiple loci can interact. The *shaded blue circle* represents the accumulation of PcG complexes. (B) In ESCs, two prominent spatial gene networks have been discovered: the "pluripotency" gene network and the PcG gene network. The *shaded red circle* represents the accumulation of pluripotent factors such as Oct4, Sox2, Nanog, or Klf4 (promoters and enhancers are represented by *red and green cylinders*, respectively). This configuration of enhancer–promoter loops also involves architectural proteins and cohesin/Mediator binding and provides a permissive environment for transcriptional activation of the pluripotency genes. Inside PcG foci, depicted by the *shaded blue circle*, enhancer–promoter loops keep the target genes in a poised but repressed state. These two gene networks contribute to the maintenance of ESCs.

with 3C and 4C respectively, also suggested the possibility of a Polycomb target gene network, including both intra- and interchromosomal interactions. More recently, 4C-seq or promoter capture Hi-C (CHi-C) experiments in ESCs expanded the notion of a mammalian PcG interactome [122–125]. The strongest Polycomb network was composed of the four *Hox* gene clusters and early developmental transcription factor genes. Strikingly, the majority of these genes were in contact with poised enhancers, suggesting that this PcG contact network contributes to constrain developmental genes and their enhancer in a silenced but poised state in ESCs [124,126]. Of note, long-range interactions between Polycomb/H3K27me3 chromatin domains were also reported in plants, indicating that this phenomenon goes beyond the metazoan kingdom [127].

3D gene networks may have important functions for stem cell biology and differentiation. In pluripotent stem cells, several groups uncovered a network of long-range contacts, which occur between chromatin regions characterized by high occupancy of pluripotency factors [122,128–132]. This pluripotency gene network involves chromatin loops, which depend on cohesin, CTCF, and/or Mediator. In ESCs, this specific chromatin interactome coexists with the PcG target gene network (Fig. 7.4B). Therefore, chromatin interactomes may enhance the robustness of pluripotent cell properties, as well as orchestrate differentiation pathways.

The application of Hi-C technologies further extends this notion of gene network. Indeed, in *Drosophila* embryos, PcG TADs have a higher propensity to interact with each other than with TADs of different chromatin types, and specific inter-TADs interactions are highly privileged between active TADs or between Black TADs [100]. In *Drosophila* Kc cells, numerous interactions also occur between TAD boundaries [98]. In mammals, recent high-resolution Hi-C indicates that chromosomal regions with similar chromatin signatures preferentially cluster in the nucleus [102]. In further support to this notion, the artificial tethering of EZH2, that deposits H3K27me3, or of SUV39H1, the enzyme responsible for H3K9me3 in constitutive heterochromatin, leads to the repositioning of the targeted locus to nuclear subcompartments of similar epigenetic profiles [133]. Therefore, on top of TAD demarcation, long-range interactions between topological domains of the same type or between boundaries are taking place, a property that could lead to the functional folding of the genome in interphase nuclei (Fig. 7.3B). Noteworthy, single-cell Hi-C analysis indicates that these large-scale chromosome structures are subject to high cell-to-cell variability [108], underscoring the probabilistic nature of chromosome folding.

Although it is often difficult to ascertain the causal relationship between this compartmentalization and gene expression, several lines of evidence indicate that this specific 3D organization contributes to biological function. For instance, in *Drosophila*, it was shown that the deletion of a PRE sequence, which perturbs the interaction between BX-C and ANT-C, has mild but significant effects on gene expression at distant PcG target genes and exacerbates the homeotic phenotypes that are observed in sensitized genetic backgrounds [77]. In mammals, deletion or mutation of genetic elements on one chromosome could occasionally be shown to affect the expression of interacting genes in *trans*, also suggesting the functional significance of these interchromosomal interactions [134–136]. In general, these transcriptional effects were rather subtle, suggesting that long-range interactions might serve as a fine-tuning system to reinforce gene regulation mediated by *cis*-regulatory elements. It is possible that there is redundancy in the interaction among several distal regulatory elements along the chromatin fiber and that the loss of one interaction may be compensated by others. Future work with clustered regularly interspaced short palindromic repeat (CRISPR)/Cas9 genome engineering approaches to delete interacting elements is required to shed light on the functional aspect of these interactions.

A further challenge is to understand the molecular mechanisms guiding these chromosomal interactions. PcG foci can be seen as the physical manifestation of Polycomb TAD interactions, and PcG components are most probably playing a critical role in the organization of these networks. Pioneer in vitro studies showed that the mouse and *Drosophila* PRC1 complexes were able to bridge in *trans* polynucleosome templates [137]. The PRC1 subunit PH contains a SAM domain that induces its oligomerization and seems to play an important role in mediating 3D interactions. Consistent with this view, the use of a PH mutant in the SAM domain in *Drosophila* cells disrupts the clustering of PcG complexes and long-range chromosomal contacts, whereas PH overexpression increases chromatin interactions [50]. In mouse ESCs, the PRC2 subunit EED was shown to contribute to the specific maintenance of the PcG gene network, and loss of EED decreases contact between Polycomb-regulated regions without altering the overall chromosome conformation [122,123]. The double knockout of both RING1A and RING1B in mouse ESCs also disrupts the *Hox* gene network, concomitant with significant upregulation of the genes that were in contact [124]. In both cases, the pluripotency network was largely unaltered, underlining the segregation of the PcG gene network from the pluripotency interactome in ESCs. Finally, a high-resolution microscopy genome-wide RNAi screen identified 129 genes that regulate the nuclear organization of PcG foci. Candidate genes include PcG components and chromatin factors, as well as many protein-modifying enzymes, including components of the SUMOylation pathway [138]. The invalidation of SUMO led to the aggregation of PcG foci, suggesting that SUMOylation may act as fluidifier of PRC1. Future work in this direction should uncover new components and complexes involved in the mechanisms of long-range chromosomal interactions and nuclear compartmentalization.

POTENTIAL ROLE FOR NONCODING RNA IN POLYCOMB GROUP–DEPENDENT THREE-DIMENSIONAL ORGANIZATION

Although FISH and 3C analysis strongly implicate direct interactions between distal chromosomal elements via multimeric protein complexes, this does not exclude possible indirect mechanisms via noncoding RNAs (ncRNAs). In support of this idea, PRC2 was reported to associate with hundreds to thousands of RNAs in various cell types [139–142], and PRC2 was shown to exhibit both specificity and promiscuity in RNA binding in vitro [143].

Among them, *HOTAIR* is a 2.2-kb ncRNA expressed by the *HOXC* locus and associated with PRC2. It was reported that *HOTAIR* can act in *trans* to repress the *HOXD* locus located on a different chromosome, as well as in numerous other regions throughout the genome [144,145]. The physiological significance of this finding is not clear, since deletion of the mouse *HOTAIR* homolog has little effect in vivo [146]. Whether *HOTAIR* exploits the proximity of Hox clusters in the nucleus to exert its function or whether it can directly participate to their clustering in specific cells remains to be tested.

X Chromosome inactivation represents a paradigm of ncRNA-mediated chromatin regulation. *Xist* represents a long ncRNA (lncRNA) expressed from the Xic (X chromosome inactivation center). *Xist* is able to induce X chromosome inactivation by recruiting several repressive activities, including the PRC2/H3K27me3 system. The *Xist* lncRNA first binds to the proximal chromatin region around its site of transcription, probably through the architectural protein hnRNP U/scaffold attachment factor-A (SAF-A), leading to initiation of repression [147]. *Xist* then exploits the 3D conformation of the X chromosome to facilitate its binding on more distal regions and rapidly propagate silencing [148]. The Polycomb silencing system is then recruited on the inactive X chromatin (Fig. 7.5A). Early work suggested that PRC2 recruitment might occur by direct *Xist* interactions, via the A-repeat domain, which forms stem-loop structures, present on the *Xist* lncRNA [149]. More recent work has suggested that *Xist* might instead recruit other repressive factors (such as SHARP/SPEN), some of which may then recruit PRC2 and PRC1 [150−152]. Genomic regions containing lncRNA genes, which contain CTCF-binding tandem repeats that bind CTCF only on the inactive X, are part of a network of long-range superloops, and these lncRNAs might also participate to the 3D conformation of the inactive X Chromosome [102].

On the other side of the coin, ncRNA may also participate in transcriptional activation of PcG target sites. *HOTTIP* is a lncRNA transcribed from the 5′ tip of the *HoxA* locus and also highlights how ncRNA can exploit the 3D architecture of the locus [153]. Chromosomal looping brings the *HOTTIP* lncRNA into close proximity to its target genes, leading to transcriptional activation. The lncRNA *XACT*, which is expressed from and coats the active X chromosome specifically in human pluripotent cells, could be the counterpart of *Xist* for the establishment of X chromosome activation [154]. Whether *XACT* exploits the 3D conformation of the X chromosome to recruit activities for transcriptional activation is awaiting further investigation.

Recently, an ncRNA was implicated in the folding of chromatin regions bound by PcG proteins in *Drosophila* [155] (Fig. 7.5B). At the *vg* PRE (a PRE regulating the *vestigial* gene), ectopic expression of the forward strand causes stabilization of long-range interactions and repression of the endogenous *vg* locus. In contrast, reverse strand transcription antagonizes this activity. Bioinformatic analysis revealed forward and reverse noncoding transcription at several hundred PcG-binding sites in the fly and mouse genomes, suggesting that this ncRNA-

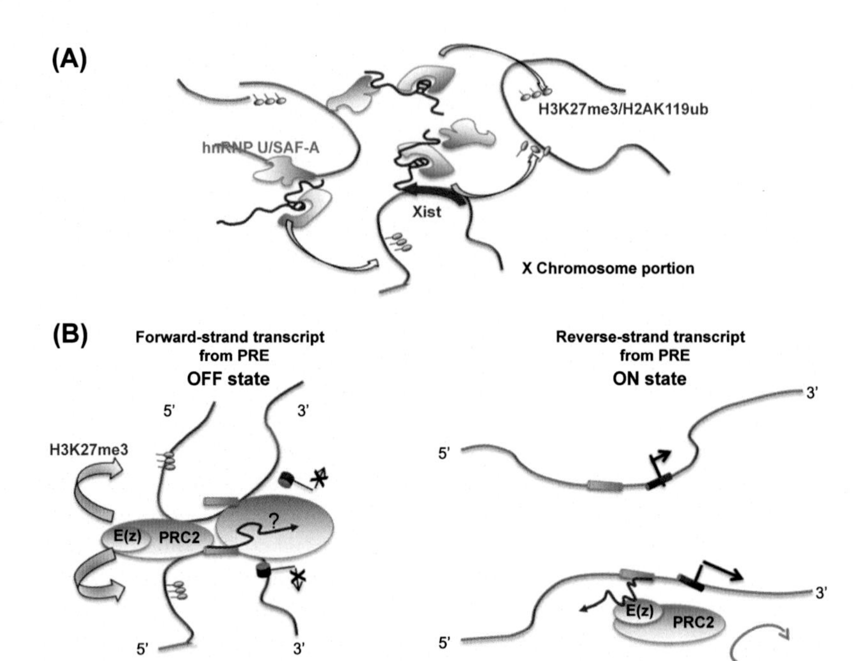

FIGURE 7.5 Role of ncRNA in chromatin organization and PcG-mediated silencing. (A) X chromosome inactivation represents a paradigm for the interplay between ncRNA, PcG regulation and chromosome architecture. The *Xist* lncRNA exploits and alters 3D genome architecture to spread over large distances across the X chromosome. *Xist* interacts with chromatin, possibly through matrix proteins, such as hnRNP U/SAF-A and then recruits multiple components (depicted by a *shaded green rectangle*), including PRC1 and PRC2, via its A-repeat domain, either directly or indirectly through repressor proteins (such as SHARP/SPEN and HDAC activities) to modify and compact chromatin. (B) Role of noncoding-RNA at the *vestigial Drosophila* locus (*vg*). Long-range chromosomal interactions occur between the endogenous locus and a transgene containing the *vg* PRE located approximately 3 Mb away. They could also occur between two *vg* loci located on homologous chromosomes (promoters and PREs are represented by *red and blue cylinders*, respectively). This interaction correlates with PcG-dependent silencing and depends on the transcription of an ncRNA from the PRE in the forward orientation. E(z) does not interact with the forward-strand transcript, but instead this ncRNA might facilitate or stabilize interaction by binding to unknown bridging proteins (depicted by an *orange circle*). This loop or pairing is disrupted upon transcription of an ncRNA in the reverse orientation, leading to the eviction of E(z) from chromatin and transcriptional activation. This mechanism might be widespread in fly and mammalian genomes. *(A) Adapted from Engreitz JM, Pandya-Jones A, McDonel P, Shishkin A, Sirokman K, Surka C, et al. The Xist lncRNA exploits three-dimensional genome architecture to spread across the X chromosome. Science 2013;341(6147):1237973; (B) Adapted from Herzog VA, Lempradl A, Trupke J, Okulski H, Altmutter C, Ruge F, et al. A strand-specific switch in noncoding transcription switches the function of a Polycomb/Trithorax response element. Nat Genet 2014;46(9):973–81.*

dependent switch in chromatin structure and transcriptional states may potentially represent a widespread mechanism for PcG function [155].

Altogether, these data illustrate the interplay between genome folding and lncRNA. Future research should determine how widespread the role of lncRNAs is for the establishment and/or maintenance of various epigenetic processes, including cell differentiation and cancer.

POLYCOMB AND THREE-DIMENSIONAL GENOMICS IN CANCER AND OTHER DISEASES

Spatial proximity has been directly linked to chromosomal translocations and genome instability related to cancers [156–159]. In particular, specific chromosomal translocations correlate with the spatial proximity of transcribed genes at transcription factories [158] or with the degree of intermingling between different chromosomal regions, which depends on transcription-dependent interactions and chromosome decondensation [156]. Since PcG target domains represent compacted repressed structures [62], they are expected to be underrepresented among chromosomal rearrangements and future analysis is needed to analyze this issue.

Another intriguing link between 3D genome contacts and cancer is represented by the ectopic somatic pairing of homologous portions of chromosomes in renal oncocytoma [160], which is linked to misregulation of genes involved in this pairing. This is reminiscent of a *Drosophila* phenomenon called pairing-sensitive silencing, which consists in increased silencing when PREs are present in the homozygous state as opposite to heterozygosis (for review, Ref. [161]). More needs to be done to better understand the relationship between somatic pairing and genome function and, in particular, whether PcG target loci might also be sensitive to ectopic somatic pairing in cancer. *Drosophila*, in which somatic pairing is naturally present, represents an interesting model system to study this phenomenon.

Finally, in addition to loops and networks of chromatin contacts, the "geographical" location of genes in the 3D nuclear space may play a role in gene regulation and disease. It is well established that mammalian genes located at the nuclear periphery tend to be silenced [162]. Furthermore, perturbation of the nuclear lamina induces strong changes in genome organization and gene expression [163]. The recent discovery of a cross talk between lamin A/C and the nuclear organization of PcG proteins may open new perspectives for the understanding of laminopathies [164].

CONCLUDING REMARKS

Because of lower complexities of the PcG complex compositions and of the genome, *Drosophila* has been an invaluable model to pioneer the analysis of PcG function and nuclear organization. Similar functions are now being uncovered in mammals and other organisms, including plants, revealing universal roles of PcG complexes in genome function and nuclear organization.

PcG nuclear organization appears to be a multilayered folding, ranging from kb-size loops inside Polycomb TADs to larger intrachromosomal loops of several megabases between TADs, which can bring genes in proximity with distal regulatory elements. These specific contacts appear to be well segregated from other topological domains and 3D gene networks of different chromatin types. More work is needed to elucidate the mechanism of TAD and 3D gene network formation/maintenance. Beyond PcG complexes and architectural proteins, including cohesin and CTCF, other molecular machines are likely to be involved in regulating this organization, and screens aiming at identifying such factors are necessary.

Today, understanding the 3D folding of the genome remains a big challenge. Indeed, genome organization represents a highly complex system in dynamic equilibrium. Addressing the function of chromatin factors involved in this organization is a challenging task, because it is difficult to uncouple effects on spatial chromosome organization from those on transcriptional regulation. In the last few years, genomic editing has become increasingly accessible with the recent developments of the artificial transcription activator-like effector nucleases system to target specific activities, such as histone-modifying enzymes or the clustered regularly interspaced short palindromic repeats (CRISPRs) technology, to target specific deletions or protein-binding sites in the genome. These technologies will help to better understand the relationships between genome structure and function through differentiation and development.

Finally, a recurrent problem in molecular pathology is the lack of mutations that can be linked to specific diseases. This is because most genome-wide association studies (GWAS) can deliver mutations in gene coding regions, but they fail to identify mutations in gene regulatory regions that can be very large in mammals and for which multiple rare mutations may perturb the regulation of relevant genes. Recent work has shown that noncoding mutations and structural variations disrupting TAD structures can lead to misregulation of genes and play a causative role in malformation syndromes [110]. Moreover, combination of Hi-C technology with GWAS studies can reveal novel classes of disease-associated mutations by the pattern of chromatin contacts established by their underlying genome position [165]. Therefore the recent discovery of TADs as functional structural units of the genome and the identification of 3D gene networks is opening new avenues for the interpretation of GWAS data. This specific 3D genome organization might uncover at least a part of the "missing heritability" and help understanding diseases that cannot be explained by mutations in coding sequences.

LIST OF ACRONYMS AND ABBREVIATIONS

3C Chromosome conformation capture
3D Three-dimensional
4C Circular chromosome conformation capture or chromosome conformation capture on chip

ANT-C Antennapedia complex
BX-C Bithorax complex
CBX2/4/6/7/8 The mammalian homologs of PC
CHi-C Capture Hi-C
CRISPRs Clustered regularly interspaced short palindromic repeats
CTCF CCCTC binding factor
EED Embryonic Ectoderm Development Protein, the mammalian homolog of ESC
ESC Extra sex combs
ESCs Embryonic stem cells
E(Z) Enhancer of zeste
EZH2/1 The mammalian homologs of E(Z)
FISH Fluorescent in situ hybridization
GWAS Genome-wide association studies
H2AK118ub/H2AK119ub Histone H2A monoubiquitination of lysine 118 (in *Drosophila*) or lysine 119 (in mammals)
H3K4me3/H3K9me3/H3K36me3 Histone H3 trimethylation at lysine 4 or 9 or 36
H3K27me3 Histone H3 trimethylation at lysine 27
Hi-C High-throughput 3C
kb Kilobase
lncRNAs Long noncoding RNAs
Mb Megabase
ncRNAs Noncoding RNAs
PC Polycomb
PcG Polycomb group
PCGF1/2/3/4/5 The mammalian homologs of PSC
PH Polyhomeotic
PHC1/2/3 The mammalian homologs of PH
PRC1/PRC2 Polycomb repressor complexes 1 or 2
PSC Posterior sex combs
RING1A/B The mammalian homologs of dRING
RNA PolII RNA Polymerase II
SCE/dRING Sex comb extra, also called dRING
SU(Z)12 Suppressor of zeste 12
SUZ12 The mammalian homolog of SU(Z)12
TADs Topologically associating domains

ACKNOWLEDGMENTS

F.B. and G.C. are supported by the CNRS. The research in G.C. laboratory is supported by grants from the European Horizon 2020 (H2020) MuG project, the CNRS, the European Network of Excellence EpiGeneSys, the Agence Nationale de la Recherche (EpiDevoMath), the Fondation pour la Recherche Médicale (DEI20151234396), the INSERM/Plan Cancer Epigenetics and cancer program (Grant acronym "MM&TT"), the Laboratory of Excellence EpiGenMed, and the Fondation ARC pour la Recherche sur le Cancer.

REFERENCES

[1] Mills AA. Throwing the cancer switch: reciprocal roles of polycomb and trithorax proteins. Nat Rev Cancer 2010;10(10):669–82.

[2] Scelfo A, Piunti A, Pasini D. The controversial role of the Polycomb group proteins in transcription and cancer: how much do we not understand Polycomb proteins? FEBS J 2015;282(9):1703–22.

[3] Bantignies F, Cavalli G. Polycomb group proteins: repression in 3D. Trends Genet 2011;27:454–64.

[4] Pirrotta V, Li HB. A view of nuclear Polycomb bodies. Curr Opin Genet Dev 2012;22(2):101–9.

[5] Margueron R, Reinberg D. The Polycomb complex PRC2 and its mark in life. Nature 2011;469(7330):343–9.

[6] Simon JA, Kingston RE. Occupying chromatin: Polycomb mechanisms for getting to genomic targets, stopping transcriptional traffic, and staying put. Mol Cell 2013;49(5): 808–24.

[7] Kang H, McElroy KA, Jung YL, Alekseyenko AA, Zee BM, Park PJ, et al. Sex comb on midleg (Scm) is a functional link between PcG-repressive complexes in *Drosophila*. Genes Dev 2015;29(11):1136–50.

[8] Beisel C, Paro R. Silencing chromatin: comparing modes and mechanisms. Nature Rev Genet 2011;12(2):123–35.

[9] Gao Z, Zhang J, Bonasio R, Strino F, Sawai A, Parisi F, et al. PCGF homologs, CBX proteins, and RYBP define functionally distinct PRC1 family complexes. Mol Cell 2012;45(3):344–56.

[10] Tavares L, Dimitrova E, Oxley D, Webster J, Poot R, Demmers J, et al. RYBP-PRC1 complexes mediate H2A ubiquitylation at polycomb target sites independently of PRC2 and H3K27me3. Cell 2012;148(4):664–78.

[11] Zhang T, Cooper S, Brockdorff N. The interplay of histone modifications – writers that read. EMBO Rep 2015;16(11):1467–81.

[12] Kassis JA, Brown JL. Polycomb group response elements in *Drosophila* and vertebrates. Adv Genet 2013;81:83–118.

[13] Enderle D, Beisel C, Stadler MB, Gerstung M, Athri P, Paro R. Polycomb preferentially targets stalled promoters of coding and noncoding transcripts. Genome Res 2011;21(2): 216–26.

[14] Schuettengruber B, Ganapathi M, Leblanc B, Portoso M, Jaschek R, Tolhuis B, et al. Functional anatomy of polycomb and trithorax chromatin landscapes in *Drosophila* embryos. PLoS Biol 2009;7(1):e13.

[15] Wang L, Brown JL, Cao R, Zhang Y, Kassis JA, Jones RS. Hierarchical recruitment of polycomb group silencing complexes. Mol Cell 2004;14(5):637–46.

[16] Herz HM, Morgan M, Gao X, Jackson J, Rickels R, Swanson SK, et al. Histone H3 lysine-to-methionine mutants as a paradigm to study chromatin signaling. Science 2014;345(6200):1065–70.

[17] Pengelly AR, Copur O, Jackle H, Herzig A, Muller J. A histone mutant reproduces the phenotype caused by loss of histone-modifying factor Polycomb. Science 2013; 339(6120):698–9.

[18] Pengelly AR, Kalb R, Finkl K, Muller J. Transcriptional repression by PRC1 in the absence of H2A monoubiquitylation. Genes Dev 2015;29(14):1487–92.

[19] Sing A, Pannell D, Karaiskakis A, Sturgeon K, Djabali M, Ellis J, et al. A vertebrate Polycomb response element governs segmentation of the posterior hindbrain. Cell 2009;138(5):885–97.

[20] Woo CJ, Kharchenko PV, Daheron L, Park PJ, Kingston RE. A region of the human HOXD cluster that confers polycomb-group responsiveness. Cell 2010;140(1):99–110.

[21] Ku M, Koche RP, Rheinbay E, Mendenhall EM, Endoh M, Mikkelsen TS, et al. Genomewide analysis of PRC1 and PRC2 occupancy identifies two classes of bivalent domains. PLoS Genet 2008;4(10):e1000242.

[22] Lynch MD, Smith AJ, De Gobbi M, Flenley M, Hughes JR, Vernimmen D, et al. An interspecies analysis reveals a key role for unmethylated CpG dinucleotides in vertebrate Polycomb complex recruitment. EMBO J 2012;31(2):317–29.

[23] Mendenhall EM, Koche RP, Truong T, Zhou VW, Issac B, Chi AS, et al. GC-rich sequence elements recruit PRC2 in mammalian ES cells. PLoS Genet 2010;6(12):e1001244.

[24] Schoeftner S, Sengupta AK, Kubicek S, Mechtler K, Spahn L, Koseki H, et al. Recruitment of PRC1 function at the initiation of X inactivation independent of PRC2 and silencing. EMBO J 2006;25(13):3110–22.

[25] Trojer P, Cao AR, Gao Z, Li Y, Zhang J, Xu X, et al. L3MBTL2 protein acts in concert with PcG protein-mediated monoubiquitination of H2A to establish a repressive chromatin structure. Mol Cell 2011;42(4):438–50.

[26] Farcas AM, Blackledge NP, Sudbery I, Long HK, McGouran JF, Rose NR, et al. KDM2B links the polycomb repressive complex 1 (PRC1) to recognition of CpG islands. eLife 2012;1:e00205.

[27] Wu X, Johansen JV, Helin K. Fbxl10/Kdm2b recruits polycomb repressive complex 1 to CpG islands and regulates H2A ubiquitylation. Mol Cell 2013;49(6):1134–46.

[28] He J, Shen L, Wan M, Taranova O, Wu H, Zhang Y. Kdm2b maintains murine embryonic stem cell status by recruiting PRC1 complex to CpG islands of developmental genes. Nat Cell Biol 2013;15(4):373–84.

[29] Blackledge NP, Farcas AM, Kondo T, King HW, McGouran JF, Hanssen LL, et al. Variant PRC1 complex-dependent H2A ubiquitylation drives PRC2 recruitment and polycomb domain formation. Cell 2014;157(6):1445–59.

[30] Cooper S, Dienstbier M, Hassan R, Schermelleh L, Sharif J, Blackledge NP, et al. Targeting polycomb to pericentric heterochromatin in embryonic stem cells reveals a role for H2AK119u1 in PRC2 recruitment. Cell Rep 2014;7(5):1456–70.

[31] Kalb R, Latwiel S, Baymaz HI, Jansen PW, Muller CW, Vermeulen M, et al. Histone H2A monoubiquitination promotes histone H3 methylation in Polycomb repression. Nat Struct Mol Biol 2014;21(6):569–71.

[32] Riising EM, Comet I, Leblanc B, Wu X, Johansen JV, Helin K. Gene silencing triggers polycomb repressive complex 2 recruitment to CpG islands genome wide. Mol Cell 2014;55(3):347–60.

[33] Arrigoni R, Alam SL, Wamstad JA, Bardwell VJ, Sundquist WI, Schreiber-Agus N. The Polycomb-associated protein Rybp is a ubiquitin binding protein. FEBS Lett 2006;580(26): 6233–41.

[34] Margueron R, Justin N, Ohno K, Sharpe ML, Son J, Drury 3rd WJ, et al. Role of the polycomb protein EED in the propagation of repressive histone marks. Nature 2009;461(7265):762–7.

[35] Yuan W, Wu T, Fu H, Dai C, Wu H, Liu N, et al. Dense chromatin activates Polycomb repressive complex 2 to regulate H3 lysine 27 methylation. Science 2012;337(6097):971–5.

[36] Kharchenko PV, Alekseyenko AA, Schwartz YB, Minoda A, Riddle NC, Ernst J, et al. Comprehensive analysis of the chromatin landscape in *Drosophila melanogaster.* Nature 2011;471(7339):480–5.

[37] Lee TI, Jenner RG, Boyer LA, Guenther MG, Levine SS, Kumar RM, et al. Control of developmental regulators by Polycomb in human embryonic stem cells. Cell 2006;125(2): 301–13.

[38] Mikkelsen TS, Ku M, Jaffe DB, Issac B, Lieberman E, Giannoukos G, et al. Genome-wide maps of chromatin state in pluripotent and lineage-committed cells. Nature 2007;448(7153):553–60.

[39] Ernst J, Kellis M. Discovery and characterization of chromatin states for systematic annotation of the human genome. Nat Biotechnol 2010;28(8):817–25.

[40] Ernst J, Kheradpour P, Mikkelsen TS, Shoresh N, Ward LD, Epstein CB, et al. Mapping and analysis of chromatin state dynamics in nine human cell types. Nature 2011;473(7345):43–9.

[41] Barski A, Cuddapah S, Cui K, Roh TY, Schones DE, Wang Z, et al. High-resolution profiling of histone methylations in the human genome. Cell 2007;129(4):823–37.

[42] Roh TY, Cuddapah S, Cui K, Zhao K. The genomic landscape of histone modifications in human T cells. Proc Natl Acad Sci USA 2006;103(43):15782–7.

[43] Pauler FM, Sloane MA, Huang R, Regha K, Koerner MV, Tamir I, et al. H3K27me3 forms BLOCs over silent genes and intergenic regions and specifies a histone banding pattern on a mouse autosomal chromosome. Genome Res 2009;19(2):221–33.

[44] Hawkins RD, Hon GC, Lee LK, Ngo Q, Lister R, Pelizzola M, et al. Distinct epigenomic landscapes of pluripotent and lineage-committed human cells. Cell Stem Cell 2010;6(5):479–91.

[45] Bowman SK, Deaton AM, Domingues H, Wang PI, Sadreyev RI, Kingston RE, et al. H3K27 modifications define segmental regulatory domains in the *Drosophila* bithorax complex. eLife 2014;3:e02833.

[46] Soshnikova N, Duboule D. Epigenetic temporal control of mouse *Hox* genes in vivo. Science 2009;324(5932):1320–3.

[47] Chow J, Heard E. X inactivation and the complexities of silencing a sex chromosome. Curr Opin Cell Biol 2009;21(3):359–66.

[48] Cheutin T, Cavalli G. Progressive polycomb assembly on H3K27me3 compartments generates polycomb bodies with developmentally regulated motion. PLoS Genet 2012;8(1):e1002465.

[49] Grimaud C, Bantignies F, Pal-Bhadra M, Ghana P, Bhadra U, Cavalli G. RNAi components are required for nuclear clustering of Polycomb group response elements. Cell 2006;124(5):957–71.

[50] Wani AH, Boettiger AN, Schorderet P, Ergun A, Munger C, Sadreyev RI, et al. Chromatin topology is coupled to Polycomb group protein subnuclear organization. Nat Commun 2016;7:10291.

[51] Filion GJ, van Bemmel JG, Braunschweig U, Talhout W, Kind J, Ward LD, et al. Systematic protein location mapping reveals five principal chromatin types in *Drosophila* cells. Cell 2010;143(2):212–24.

[52] Lee HG, Kahn TG, Simcox A, Schwartz YB, Pirrotta V. Genome-wide activities of Polycomb complexes control pervasive transcription. Genome Res 2015;25(8):1170–81.

[53] Ferrari KJ, Scelfo A, Jammula S, Cuomo A, Barozzi I, Stutzer A, et al. Polycomb-dependent H3K27me1 and H3K27me2 regulate active transcription and enhancer fidelity. Mol Cell 2014;53(1):49–62.

[54] Francis NJ, Kingston RE, Woodcock CL. Chromatin compaction by a Polycomb group protein complex. Science 2004;306(5701):1574–7.

[55] King IF, Emmons RB, Francis NJ, Wild B, Muller J, Kingston RE, et al. Analysis of a polycomb group protein defines regions that link repressive activity on nucleosomal templates to in vivo function. Mol Cell Biol 2005;25(15):6578–91.

[56] Eskeland R, Leeb M, Grimes GR, Kress C, Boyle S, Sproul D, et al. Ring1B compacts chromatin structure and represses gene expression independent of histone ubiquitination. Mol Cell 2010;38(3):452–64.

[57] Beh LY, Colwell LJ, Francis NJ. A core subunit of Polycomb repressive complex 1 is broadly conserved in function but not primary sequence. Proc Natl Acad Sci USA 2012;109(18):E1063–71.

[58] Grau DJ, Chapman BA, Garlick JD, Borowsky M, Francis NJ, Kingston RE. Compaction of chromatin by diverse Polycomb group proteins requires localized regions of high charge. Genes Dev 2011;25(20):2210–21.

[59] Isono K, Endo TA, Ku M, Yamada D, Suzuki R, Sharif J, et al. SAM domain polymerization links subnuclear clustering of PRC1 to gene silencing. Dev Cell 2013;26(6):565–77.

[60] Lanzuolo C, Roure V, Dekker J, Bantignies F, Orlando V. Polycomb response elements mediate the formation of chromosome higher-order structures in the bithorax complex. Nat Cell Biol 2007;9(10):1167–74.

[61] Terranova R, Yokobayashi S, Stadler MB, Otte AP, van Lohuizen M, Orkin SH, et al. Polycomb group proteins Ezh2 and Rnf2 direct genomic contraction and imprinted repression in early mouse embryos. Dev Cell 2008;15(5):668–79.

[62] Boettiger AN, Bintu B, Moffitt JR, Wang S, Beliveau BJ, Fudenberg G, et al. Super-resolution imaging reveals distinct chromatin folding for different epigenetic states. Nature 2016;529(7586):418–22.

[63] Stock JK, Giadrossi S, Casanova M, Brookes E, Vidal M, Koseki H, et al. Ring1-mediated ubiquitination of H2A restrains poised RNA polymerase II at bivalent genes in mouse ES cells. Nat Cell Biol 2007;9(12):1428–35.

[64] Chopra VS, Cande J, Hong JW, Levine M. Stalled Hox promoters as chromosomal boundaries. Genes Dev 2009;23(13):1505–9.

[65] Mishra RK, Mihaly J, Barges S, Spierer A, Karch F, Hagstrom K, et al. The *iab*-7 polycomb response element maps to a nucleosome-free region of chromatin and requires both GAGA and pleiohomeotic for silencing activity. Mol Cell Biol 2001;21(4):1311–8.

[66] Mito Y, Henikoff JG, Henikoff S. Histone replacement marks the boundaries of cis-regulatory domains. Science 2007;315(5817):1408–11.

[67] Mohd-Sarip A, van der Knaap JA, Wyman C, Kanaar R, Schedl P, Verrijzer CP. Architecture of a polycomb nucleoprotein complex. Mol Cell 2006;24(1):91–100.

[68] Deal RB, Henikoff JG, Henikoff S. Genome-wide kinetics of nucleosome turnover determined by metabolic labeling of histones. Science 2010;328(5982):1161–4.

[69] Ficz G, Heintzmann R, Arndt-Jovin DJ. Polycomb group protein complexes exchange rapidly in living *Drosophila*. Development 2005;132(17):3963–76.

[70] Comet I, Savitskaya E, Schuettengruber B, Negre N, Lavrov S, Parshikov A, et al. PRE-mediated bypass of two Su(Hw) insulators targets PcG proteins to a downstream promoter. Dev Cell 2006;11(1):117–24.

[71] Kyrchanova O, Toshchakov S, Podstreshnaya Y, Parshikov A, Georgiev P. Functional interaction between the Fab-7 and Fab-8 boundaries and the upstream promoter region in the *Drosophila* Abd-B gene. Mol Cell Biol 2008;28(12):4188–95.

[72] Naumova N, Dekker J. Integrating one-dimensional and three-dimensional maps of genomes. J Cell Sci 2010;123(Pt 12):1979–88.

[73] Cleard F, Moshkin Y, Karch F, Maeda RK. Probing long-distance regulatory interactions in the *Drosophila melanogaster* bithorax complex using Dam identification. Nat Genet 2006;38(8):931–5.

[74] Comet I, Schuettengruber B, Sexton T, Cavalli G. A chromatin insulator driving three-dimensional Polycomb response element (PRE) contacts and Polycomb association with the chromatin fiber. Proc Natl Acad Sci USA 2011;108(6):2294–9.

[75] Ferraiuolo MA, Rousseau M, Miyamoto C, Shenker S, Wang XQ, Nadler M, et al. The three-dimensional architecture of Hox cluster silencing. Nucleic Acids Res 2010;38(21):7472–84.

[76] Noordermeer D, Leleu M, Splinter E, Rougemont J, De Laat W, Duboule D. The dynamic architecture of Hox gene clusters. Science 2011;334(6053):222–5.

[77] Bantignies F, Roure V, Comet I, Leblanc B, Schuettengruber B, Bonnet J, et al. Polycomb-dependent regulatory contacts between distant Hox loci in *Drosophila*. Cell 2011;144(2):214–26.

[78] Andrey G, Montavon T, Mascrez B, Gonzalez F, Noordermeer D, Leleu M, et al. A switch between topological domains underlies HoxD genes collinearity in mouse limbs. Science 2013;340(6137):1234167.

[79] Kheradmand Kia S, Solaimani Kartalaei P, Farahbakhshian E, Pourfarzad F, von Lindern M, Verrijzer CP. EZH2-dependent chromatin looping controls INK4a and INK4b, but not ARF, during human progenitor cell differentiation and cellular senescence. Epigenet Chromatin 2009;2(1):16.

[80] Tiwari VK, McGarvey KM, Licchesi JD, Ohm JE, Herman JG, Schubeler D, et al. PcG proteins, DNA methylation, and gene repression by chromatin looping. PLoS Biol 2008;6(12):2911–27.

[81] Kondo T, Isono K, Kondo K, Endo TA, Itohara S, Vidal M, et al. Polycomb potentiates meis2 activation in midbrain by mediating interaction of the promoter with a tissue-specific enhancer. Dev Cell 2014;28(1):94–101.

[82] Li T, Hu JF, Qiu X, Ling J, Chen H, Wang S, et al. CTCF regulates allelic expression of Igf2 by orchestrating a promoter-polycomb repressive complex 2 intrachromosomal loop. Mol Cell Biol 2008;28(20):6473–82.

[83] Ghavi-Helm Y, Klein FA, Pakozdi T, Ciglar L, Noordermeer D, Huber W, et al. Enhancer loops appear stable during development and are associated with paused polymerase. Nature 2014;512(7512):96–100.

[84] Holohan EE, Kwong C, Adryan B, Bartkuhn M, Herold M, Renkawitz R, et al. CTCF genomic binding sites in *Drosophila* and the organisation of the bithorax complex. PLoS Genet 2007;3(7):e112.

[85] Negre N, Brown CD, Shah PK, Kheradpour P, Morrison CA, Henikoff JG, et al. A comprehensive map of insulator elements for the *Drosophila* genome. PLoS Genet 2010;6(1):e1000814.

[86] Kyrchanova O, Ivlieva T, Toshchakov S, Parshikov A, Maksimenko O, Georgiev P. Selective interactions of boundaries with upstream region of Abd-B promoter in *Drosophila* bithorax complex and role of dCTCF in this process. Nucleic Acids Res 2011;39(8):3042–52.

[87] Handoko L, Xu H, Li G, Ngan CY, Chew E, Schnapp M, et al. CTCF-mediated functional chromatin interactome in pluripotent cells. Nat Genet 2011;43(7):630–8.

[88] Parelho V, Hadjur S, Spivakov M, Leleu M, Sauer S, Gregson HC, et al. Cohesins functionally associate with CTCF on mammalian chromosome arms. Cell 2008;132(3):422–33.

[89] Stedman W, Kang H, Lin S, Kissil JL, Bartolomei MS, Lieberman PM. Cohesins localize with CTCF at the KSHV latency control region and at cellular c-myc and H19/Igf2 insulators. EMBO J 2008;27(4):654–66.

[90] Wendt KS, Yoshida K, Itoh T, Bando M, Koch B, Schirghuber E, et al. Cohesin mediates transcriptional insulation by CCCTC-binding factor. Nature 2008;451(7180):796–801.

[91] Hadjur S, Williams LM, Ryan NK, Cobb BS, Sexton T, Fraser P, et al. Cohesins form chromosomal cis-interactions at the developmentally regulated IFNG locus. Nature 2009;460(7253):410–3.

[92] Bartkuhn M, Straub T, Herold M, Herrmann M, Rathke C, Saumweber H, et al. Active promoters and insulators are marked by the centrosomal protein 190. EMBO J 2009;28(7):877–88.
[93] Strubbe G, Popp C, Schmidt A, Pauli A, Ringrose L, Beisel C, et al. Polycomb purification by in vivo biotinylation tagging reveals cohesin and Trithorax group proteins as interaction partners. Proc Natl Acad Sci USA 2011;108(14):5572–7.
[94] Misulovin Z, Schwartz YB, Li XY, Kahn TG, Gause M, MacArthur S, et al. Association of cohesin and Nipped-B with transcriptionally active regions of the *Drosophila melanogaster* genome. Chromosoma 2008;117(1):89–102.
[95] Schaaf CA, Misulovin Z, Gause M, Koenig A, Gohara DW, Watson A, et al. Cohesin and polycomb proteins functionally interact to control transcription at silenced and active genes. PLoS Genet 2013;9(6):e1003560.
[96] Lieberman-Aiden E, van Berkum NL, Williams L, Imakaev M, Ragoczy T, Telling A, et al. Comprehensive mapping of long-range interactions reveals folding principles of the human genome. Science 2009;326(5950):289–93.
[97] Dixon JR, Selvaraj S, Yue F, Kim A, Li Y, Shen Y, et al. Topological domains in mammalian genomes identified by analysis of chromatin interactions. Nature 2012;485(7398):376–80.
[98] Hou C, Li L, Qin ZS, Corces VG. Gene density, transcription, and insulators contribute to the partition of the *Drosophila* genome into physical domains. Mol Cell 2012;48(3):471–84.
[99] Nora EP, Lajoie BR, Schulz EG, Giorgetti L, Okamoto I, Servant N, et al. Spatial partitioning of the regulatory landscape of the X-inactivation centre. Nature 2012;485(7398):381–5.
[100] Sexton T, Yaffe E, Kenigsberg E, Bantignies F, Leblanc B, Hoichman M, et al. Three-dimensional folding and functional organization principles of the *Drosophila* genome. Cell 2012;148:458–72.
[101] Giorgetti L, Servant N, Heard E. Changes in the organization of the genome during the mammalian cell cycle. Genome Biol 2013;14(12):142.
[102] Rao SS, Huntley MH, Durand NC, Stamenova EK, Bochkov ID, Robinson JT, et al. A 3D map of the human genome at kilobase resolution reveals principles of chromatin looping. Cell 2014;159(7):1665–80.
[103] Dekker J, Heard E. Structural and functional diversity of topologically associating domains. FEBS Lett 2015;589(20 Pt A):2877–84.
[104] Sofueva S, Yaffe E, Chan WC, Georgopoulou D, Vietri Rudan M, Mira-Bontenbal H, et al. Cohesin-mediated interactions organize chromosomal domain architecture. EMBO J 2013;32(24):3119–29.
[105] Zuin J, Dixon JR, van der Reijden MI, Ye Z, Kolovos P, Brouwer RW, et al. Cohesin and CTCF differentially affect chromatin architecture and gene expression in human cells. Proc Natl Acad Sci USA 2014;111(3):996–1001.
[106] Seitan VC, Faure AJ, Zhan Y, McCord RP, Lajoie BR, Ing-Simmons E, et al. Cohesin-based chromatin interactions enable regulated gene expression within preexisting architectural compartments. Genome Res 2013;23(12):2066–77.
[107] Vietri Rudan M, Barrington C, Henderson S, Ernst C, Odom DT, Tanay A, et al. Comparative Hi-C reveals that CTCF underlies evolution of chromosomal domain architecture. Cell Rep 2015;10(8):1297–309.
[108] Nagano T, Lubling Y, Stevens TJ, Schoenfelder S, Yaffe E, Dean W, et al. Single-cell Hi-C reveals cell-to-cell variability in chromosome structure. Nature 2013;502(7469):59–64.

[109] Naumova N, Imakaev M, Fudenberg G, Zhan Y, Lajoie BR, Mirny LA, et al. Organization of the mitotic chromosome. Science 2013;342(6161):948–53.

[110] Lupianez DG, Kraft K, Heinrich V, Krawitz P, Brancati F, Klopocki E, et al. Disruptions of topological chromatin domains cause pathogenic rewiring of gene-enhancer interactions. Cell 2015;161(5):1012–25.

[111] Beliveau BJ, Boettiger AN, Avendano MS, Jungmann R, McCole RB, Joyce EF, et al. Single-molecule super-resolution imaging of chromosomes and in situ haplotype visualization using Oligopaint FISH probes. Nat Commun 2015;6:7147.

[112] Gemkow MJ, Verveer PJ, Arndt-Jovin DJ. Homologous association of the Bithorax-Complex during embryogenesis: consequences for transvection in *Drosophila melanogaster.* Development 1998;125(22):4541–52.

[113] Bantignies F, Grimaud C, Lavrov S, Gabut M, Cavalli G. Inheritance of Polycomb-dependent chromosomal interactions in *Drosophila.* Genes Dev 2003;17(19):2406–20.

[114] Vazquez J, Muller M, Pirrotta V, Sedat JW. The Mcp element mediates stable long-range chromosome-chromosome interactions in *Drosophila.* Mol Biol Cell 2006;17(5): 2158–65.

[115] Li HB, Muller M, Bahechar IA, Kyrchanova O, Ohno K, Georgiev P, et al. Insulators, not Polycomb response elements, are required for long-range interactions between Polycomb targets in *Drosophila melanogaster.* Mol Cell Biol 2011;31(4):616–25.

[116] Li HB, Ohno K, Gui H, Pirrotta V. Insulators target active genes to transcription factories and polycomb-repressed genes to Polycomb bodies. PLoS Genet 2013;9(4):e1003436.

[117] Moshkovich N, Nisha P, Boyle PJ, Thompson BA, Dale RK, Lei EP. RNAi-independent role for Argonaute2 in CTCF/CP190 chromatin insulator function. Genes Dev 2011;25(16):1686–701.

[118] Tolhuis B, Blom M, Kerkhoven RM, Pagie L, Teunissen H, Nieuwland M, et al. Interactions among Polycomb domains are guided by chromosome architecture. PLoS Genet 2011;7(3):e1001343.

[119] Li L, Lyu X, Hou C, Takenaka N, Nguyen HQ, Ong CT, et al. Widespread rearrangement of 3D chromatin organization underlies Polycomb-mediated stress-induced silencing. Mol Cell 2015;58(2):216–31.

[120] Tiwari VK, Cope L, McGarvey KM, Ohm JE, Baylin SB. A novel 6C assay uncovers Polycomb-mediated higher order chromatin conformations. Genome Res 2008;18(7): 1171–9.

[121] Choi D, Goo HG, Yoo J, Kang S. Identification of RNF2-responding loci in long-range chromatin interactions using the novel 4C-ChIP-Cloning technology. J Biotechnol 2011;151(4):312–8.

[122] Denholtz M, Bonora G, Chronis C, Splinter E, de Laat W, Ernst J, et al. Long-range chromatin contacts in embryonic stem cells reveal a role for pluripotency factors and polycomb proteins in genome organization. Cell Stem Cell 2013;13(5):602–16.

[123] Joshi O, Wang SY, Kuznetsova T, Atlasi Y, Peng T, Fabre PJ, et al. Dynamic reorganization of extremely long-range promoter-promoter interactions between two states of pluripotency. Cell Stem Cell 2015;17(6):748–57.

[124] Schoenfelder S, Sugar R, Dimond A, Javierre BM, Armstrong H, Mifsud B, et al. Polycomb repressive complex PRC1 spatially constrains the mouse embryonic stem cell genome. Nature Genet 2015;47(10):1179–86.

[125] Vieux-Rochas M, Fabre PJ, Leleu M, Duboule D, Noordermeer D. Clustering of mammalian Hox genes with other H3K27me3 targets within an active nuclear domain. Proc Natl Acad Sci USA 2015;112(15):4672–7.

[126] Cavalli G. PRC1 proteins orchestrate three-dimensional genome architecture. Nat Genet 2015;47(10):1105–6.

[127] Rosa S, De Lucia F, Mylne JS, Zhu D, Ohmido N, Pendle A, et al. Physical clustering of FLC alleles during Polycomb-mediated epigenetic silencing in vernalization. Genes Dev 2013;27(17):1845–50.

[128] Apostolou E, Ferrari F, Walsh RM, Bar-Nur O, Stadtfeld M, Cheloufi S, et al. Genome-wide chromatin interactions of the Nanog locus in pluripotency, differentiation, and reprogramming. Cell Stem Cell 2013;12(6):699–712.

[129] de Wit E, Bouwman BA, Zhu Y, Klous P, Splinter E, Verstegen MJ, et al. The pluripotent genome in three dimensions is shaped around pluripotency factors. Nature 2013;501(7466):227–31.

[130] Phillips-Cremins JE, Sauria ME, Sanyal A, Gerasimova TI, Lajoie BR, Bell JS, et al. Architectural protein subclasses shape 3D organization of genomes during lineage commitment. Cell 2013;153(6):1281–95.

[131] Wei Z, Gao F, Kim S, Yang H, Lyu J, An W, et al. Klf4 organizes long-range chromosomal interactions with the oct4 locus in reprogramming and pluripotency. Cell Stem Cell 2013;13(1):36–47.

[132] Zhang H, Jiao W, Sun L, Fan J, Chen M, Wang H, et al. Intrachromosomal looping is required for activation of endogenous pluripotency genes during reprogramming. Cell Stem Cell 2013;13(1):30–5.

[133] Wijchers PJ, Krijger PH, Geeven G, Zhu Y, Denker A, Verstegen MJ, et al. Cause and consequence of tethering a SubTAD to different nuclear compartments. Mol Cell 2016;61(3):461–73.

[134] Ling JQ, Li T, Hu JF, Vu TH, Chen HL, Qiu XW, et al. CTCF mediates interchromosomal colocalization between Igf2/H19 and Wsb1/Nf1. Science 2006;312(5771):269–72.

[135] Spilianakis CG, Lalioti MD, Town T, Lee GR, Flavell RA. Interchromosomal associations between alternatively expressed loci. Nature 2005;435(7042):637–45.

[136] Zhao Z, Tavoosidana G, Sjolinder M, Gondor A, Mariano P, Wang S, et al. Circular chromosome conformation capture (4C) uncovers extensive networks of epigenetically regulated intra- and interchromosomal interactions. Nat Genet 2006;38(11):1341–7.

[137] Lavigne M, Francis NJ, King IF, Kingston RE. Propagation of silencing; recruitment and repression of naive chromatin in trans by Polycomb repressed chromatin. Mol Cell 2004;13(3):415–25.

[138] Gonzalez I, Mateos-Langerak J, Thomas A, Cheutin T, Cavalli G. Identification of regulators of the three-dimensional polycomb organization by a microscopy-based genome-wide RNAi screen. Mol Cell 2014;54(3):485–99.

[139] Kaneko S, Son J, Shen SS, Reinberg D, Bonasio R. PRC2 binds active promoters and contacts nascent RNAs in embryonic stem cells. Nat Struct Mol Biol 2013;20(11):1258–64.

[140] Kanhere A, Viiri K, Araujo CC, Rasaiyaah J, Bouwman RD, Whyte WA, et al. Short RNAs are transcribed from repressed Polycomb target genes and interact with Polycomb repressive complex-2. Mol Cell 2010;38(5):675–88.

[141] Khalil AM, Guttman M, Huarte M, Garber M, Raj A, Rivea Morales D, et al. Many human large intergenic noncoding RNAs associate with chromatin-modifying complexes and affect gene expression. Proc Natl Acad Sci USA 2009;106(28):11667–72.

[142] Zhao J, Ohsumi TK, Kung JT, Ogawa Y, Grau DJ, Sarma K, et al. Genome-wide identification of Polycomb-associated RNAs by RIP-seq. Mol Cell 2010;40(6):939–53.

[143] Davidovich C, Wang X, Cifuentes-Rojas C, Goodrich KJ, Gooding AR, Lee JT, et al. Toward a consensus on the binding specificity and promiscuity of PRC2 for RNA. Mol Cell 2015;57(3):552–8.

[144] Li L, Liu B, Wapinski OL, Tsai MC, Qu K, Zhang J, et al. Targeted disruption of hotair leads to homeotic transformation and gene derepression. Cell Rep 2013;5(1):3–12.

[145] Rinn JL, Kertesz M, Wang JK, Squazzo SL, Xu X, Brugmann SA, et al. Functional demarcation of active and silent chromatin domains in human HOX loci by noncoding RNAs. Cell 2007;129(7):1311–23.

[146] Schorderet P, Duboule D. Structural and functional differences in the long non-coding RNA hotair in mouse and human. PLoS Genet 2011;7(5):e1002071.

[147] Hasegawa Y, Brockdorff N, Kawano S, Tsutui K, Tsutui K, Nakagawa S. The matrix protein hnRNP U is required for chromosomal localization of Xist RNA. Dev Cell 2010;19(3):469–76.

[148] Engreitz JM, Pandya-Jones A, McDonel P, Shishkin A, Sirokman K, Surka C, et al. The Xist lncRNA exploits three-dimensional genome architecture to spread across the X chromosome. Science 2013;341(6147):1237973.

[149] Zhao J, Sun BK, Erwin JA, Song JJ, Lee JT. Polycomb proteins targeted by a short repeat RNA to the mouse X chromosome. Science 2008;322(5902):750–6.

[150] McHugh CA, Chen CK, Chow A, Surka CF, Tran C, McDonel P, et al. The Xist lncRNA interacts directly with SHARP to silence transcription through HDAC3. Nature 2015;521(7551):232–6.

[151] Minajigi A, Froberg JE, Wei C, Sunwoo H, Kesner B, Colognori D, et al. Chromosomes. A comprehensive Xist interactome reveals cohesin repulsion and an RNA-directed chromosome conformation. Science 2015;349(6245).

[152] Chu C, Zhang QC, da Rocha ST, Flynn RA, Bharadwaj M, Calabrese JM, et al. Systematic discovery of Xist RNA binding proteins. Cell 2015;161(2):404–16.

[153] Wang KC, Yang YW, Liu B, Sanyal A, Corces-Zimmerman R, Chen Y, et al. A long noncoding RNA maintains active chromatin to coordinate homeotic gene expression. Nature 2011;472(7341):120–4.

[154] Vallot C, Huret C, Lesecque Y, Resch A, Oudrhiri N, Bennaceur-Griscelli A, et al. XACT, a long noncoding transcript coating the active X chromosome in human pluripotent cells. Nat Genet 2013;45(3):239–41.

[155] Herzog VA, Lempradl A, Trupke J, Okulski H, Altmutter C, Ruge F, et al. A strand-specific switch in noncoding transcription switches the function of a Polycomb/Trithorax response element. Nat Genet 2014;46(9):973–81.

[156] Branco MR, Pombo A. Intermingling of chromosome territories in interphase suggests role in translocations and transcription-dependent associations. PLoS Biol 2006;4(5):e138.

[157] Lin C, Yang L, Tanasa B, Hutt K, Ju BG, Ohgi K, et al. Nuclear receptor-induced chromosomal proximity and DNA breaks underlie specific translocations in cancer. Cell 2009;139(6):1069–83.

[158] Osborne CS, Chakalova L, Mitchell JA, Horton A, Wood AL, Bolland DJ, et al. Myc dynamically and preferentially relocates to a transcription factory occupied by Igh. PLoS Biol 2007;5(8):e192.

[159] Roix JJ, McQueen PG, Munson PJ, Parada LA, Misteli T. Spatial proximity of translocation-prone gene loci in human lymphomas. Nat Genet 2003;34(3):287–91.

[160] Koeman JM, Russell RC, Tan MH, Petillo D, Westphal M, Koelzer K, et al. Somatic pairing of chromosome 19 in renal oncocytoma is associated with deregulated EGLN2-mediated [corrected] oxygen-sensing response. PLoS Genet 2008;4(9):e1000176.

[161] Kassis JA. Pairing-sensitive silencing, Polycomb group response elements, and transposon homing in *Drosophila*. Adv Genet 2002;46:421–38.
[162] Towbin BD, Meister P, Gasser SM. The nuclear envelope – a scaffold for silencing? Curr Opin Genet Dev 2009;19(2):180–6.
[163] Burke B, Stewart CL. Functional architecture of the cell's nucleus in development, aging, and disease. Curr Top Dev Biol 2014;109:1–52.
[164] Cesarini E, Mozzetta C, Marullo F, Gregoretti F, Gargiulo A, Columbaro M, et al. Lamin A/C sustains PcG protein architecture, maintaining transcriptional repression at target genes. J Cell Biol 2015;211(3):533–51.
[165] Martin P, McGovern A, Orozco G, Duffus K, Yarwood A, Schoenfelder S, et al. Capture Hi-C reveals novel candidate genes and complex long-range interactions with related autoimmune risk loci. Nat Commun 2015;6:10069.

Chapter 8

Molecular Architecture of the Polycomb Repressive Complex 2

C.S. Huang[1], E. Nogales[2,3,4], C. Ciferri[1]
[1]*Genentech, Inc., South San Francisco, CA, United States;* [2]*University of California, Berkeley, CA, United States;* [3]*Lawrence Berkeley National Laboratory, Berkeley, CA, United States;* [4]*Howard Hughes Medical Institute, UC Berkeley, Berkeley, CA, United States*

Chapter Outline

Introduction **165**
PRC2 Electron Microscopy Studies **169**
PRC2 X-ray Crystallography Studies **171**
EED Recognizes a Number of Histone Motifs 171
Drosophila Nurf55 (RbAp48) + Histone H3 or Suz12 173
Architecture of a Ternary Ezh2–EED–Suz12 Complex 174
EED 176
Suz12[VEFS] 177
Ezh2 177
Mechanism of H3K27M Inhibition **179**
Mechanism of H3K27me3 Activation **180**
Summary and Outlook **182**
List of Acronyms and Abbreviations **184**
References **185**

INTRODUCTION

Eukaryotic cells have developed several mechanisms to package their DNA into compact structures known as chromatin. The building block of chromatin is the nucleosome, which comprises 145–147 base pairs (bp) of DNA wrapped 1.8 times around an octamer of four evolutionarily conserved core histone proteins: H2A, H2B, H3, and H4 [1,2]. Nucleosomes are separated by DNA spacers of approximately 10–90 bp in length. Histone H1 binds this linker DNA and subsequently promotes further compaction of the chromatin into a higher-order structure [3–5]. Histones possess flexible and positively charged tails at their N-termini that undergo a variety of posttranslational modifications (PTMs). These modifications influence several nuclear processes, including gene transcription, DNA compaction, DNA replication, and DNA repair. The importance of PTMs is twofold. On one hand they may alter the charge of

Polycomb Group Proteins. http://dx.doi.org/10.1016/B978-0-12-809737-3.00008-8

histone tails, which ultimately modify the accessibility rate of chromatin to other cellular factors. On the other, PTMs can also generate docking sites for protein domains that may be contained within larger protein complexes, including chromatin remodeling complexes, transcription factors, or enzymes that catalyze the deposition of additional histone tail modifications [6].

The antagonistic activities of the Trithorax group (TrxG) and the Polycomb group (PcG) proteins contribute to cell fate and embryonic development through the activation and repression of genes such as the homeotic (Hox) genes, respectively [7]. TrxG and PcG complexes are evolutionarily conserved across all eukaryotes and are responsible for histone tail modification and gene regulation. TrxG mediates the deposition of a trimethyl group on lysine 4 {K4} of histone H3, generating the H3K4me3 mark. H3K4me3 is typically associated with actively transcribed chromatin. PcG proteins assemble into two distinct families of transcriptional–repressive complexes, Polycomb repressive complex 1 (PRC1) (Reviewed in Refs. [8,9]) and Polycomb repressive complex 2 (PRC2) [8]. Within the PcG group, PRC2 catalyzes the di- and trimethylation of lysine 27 of Histone H3, generating the H3K27me2/3 marks [8,10–12]. H3K27me3 is associated with chromatin compaction and can be recognized by the PRC2 subunit embryonic ectoderm development (EED), which propagates the repressive H3K27me3 histone marks to adjacent histone tails [13]. H3K27me3 can also recruit the PRC1 complex by interacting with the chromobox domain (CBX) of one of its subunits. PRC1 then catalyzes the monoubiquitination of H2A on K119 (H2AK119ub1) through its E3 ligases RING1a and RING1b [9,14,15]. H2AK119ub1 is important for blocking RNA polymerase II activity [16] and transcription elongation [17]. In addition, the presence of H2AK119ub1 prevents H3K4me3 deposition and therefore further inhibits gene transcription [18].

PRC2 is a ~250 kDa complex that is found in all eukaryotes, including humans. PRC2 is composed of four core components: EED, either one of the Enhancer of Zeste 1 and 2 (EZH1/EZH2) subunits, Suppressor of Zeste (Suz12), and Retinoblastoma-binding proteins 46 and 48 (RbAp46/48) (Fig. 8.1). EED exists as four isoforms, which are generated by alternative translation of a single mRNA [19] and contains a WD40 domain involved in the binding of trimethylated peptides [20]. The Ezh1 and Ezh2 proteins contain the methyltransferase activity in their SET domain and also feature additional domains, described in detail in the following sections, which are involved in interactions with other PRC2 subunits [46] (Fig. 8.1). Ezh1 and Ezh2 possess different methyltransferase activities, with Ezh2 being more active, both in vivo and in vitro [21,22]. These subunits are mutually exclusive in the PRC2 complex and the incorporation of either subunit is dependent on the specific cell type in which PRC2 is expressed. Ezh2 is highly expressed in embryonic tissues and actively dividing cells, while Ezh1 is primarily found in adult tissues and nondividing cells [21,22]. The Suz12 subunit contains two highly conserved domains: a Cys_2-His_2 zinc finger and an ~150-amino-acid

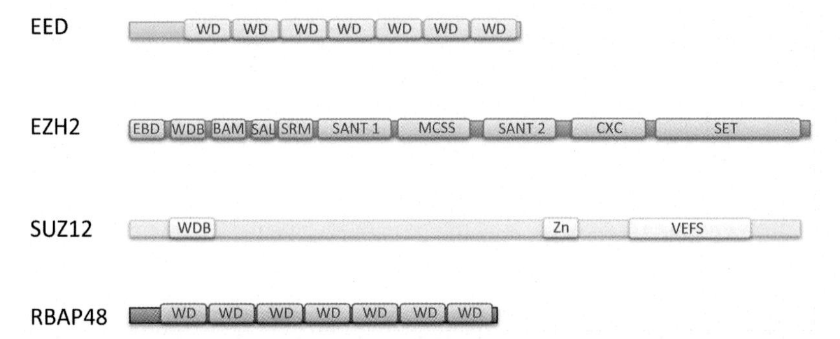

FIGURE 8.1 Domain architecture of the Polycomb Repressive Complex 2 components. Schematic representation of the individual components and their domain organization. *BAM*, β-addition motif; *CXC*, cysteine-rich domain; *MCSS*, motif connecting SANT1L and SANT1L; *SAL*, SET activation loop; *SANT1L*, SANT1-like; *SANT2L*, SANT2-like; *SBD*, SANT1L-binding domain; *SET*, suppressor of variegation 3-9, enhancer of zeste and trithorax domain; *SRM*, stimulation-responsive motif; *VEFS*, VRN2-EMF2-FIS2-Su(z)12 domain; *WBD*, WD40-binding domain; *Zn*, zinc finger domain.

VEFS (VRN2-EMF2-FIS2-Su(z)12) domain, which is highly similar among plants, flies, and humans. Removal of the VEFS domain results in the dissociation of the PRC2 assembly [23], suggesting that it cofolds with other PRC2 components. The Cys_2-His_2 zinc finger does not appear to be essential for DNA binding in vitro and may be responsible for mediating additional protein–protein interactions [24]. RbAp46/48, the mammalian homologs of *Drosophila* Nurf55, is similar to EED in that both contain a prominent WD40 domain. RbAp46/48 binds to Suz12 and enhances PRC2 activity in vitro [10,11,26]. Sequence alignments of PRC2 components reveal a high degree of conservation across species, particularly within key functional domains such as the SANT, SET, WD40, VEFS, and zinc-finger motifs.

Given the number of EED, Ezh, and RbAp isoforms, there are several alternative PRC2 complexes that can be assembled, creating the possibility for fine-tuning specific gene repression programs in a cell- or tissue-dependent manner.

PRC2 modifies nucleosomal H3K27 in a stepwise manner, from the monomethylated to trimethylated forms. Some studies have indicated PRC2 may catalyze the monomethylation process as well, but the exact details are an area of active research. In vivo, PRC2 deposits H3K27me2 very rapidly during DNA synthesis [27] and does not require a stable association with genomic DNA. In contrast to H3K27me2 deposition, H3K27me3 deposition is less efficient and is achieved only on the stable interaction of PRC2 with chromatin [28].

Studies in mouse embryonic stem cells (mESC) have shown that the H3K27 population is composed of approximately ~70% in the dimethylated form and ~7% in the trimethylated form. It was proposed that H3K27me2 deposition acts as a mechanism to prevent unwanted gene activation [27], which is consistent with the finding that acetylated H3K27 (H3K27ac) levels are increased on the loss of PRC2 activity [29].

PRC2 catalysis of H3K27me3 deposition is regulated by several additional proteins [30], which differentially affect PRC2 activity at different developmental stages or in different cell types [31]. The accessory components identified so far include adipocyte enhancer-binding protein 2 (AEBP2), Jumonji AT-rich interactive domain 2 (Jarid2), PCLs (divided into three distinct families), and two mammalian-specific proteins, C17orf96 and C10orf12 [31].

AEBP2 increases PRC2 enzymatic activity [10,11] by recruiting PRC2 to specific DNA loci through its zinc finger domains [32]. Electron microscopy analysis of the PRC2-AEBP2 complex reveals that AEBP2 binding also increases the stability of the PRC2 complex [33].

Jarid2 physically interacts with the PRC2 complex [33–35] and stabilizes PRC2's occupation at chromatin sites through its specific interactions with nucleosomes [36], ncRNAs [37,38], and H2Aub1 [39]. Jarid2 activity is modulated by Ezh2-mediated methylation and is dependent on the chromatin environment [40].

Polycomb-like proteins (PCL) include PHF1, MTF2, and PHF19, which are mammalian homologs of *Drosophila* PCL. All three contain a TUDOR domain, which is involved in binding H3K36me3, a histone mark that is deposited on nucleosomes of actively transcribed genes [41–43]. Some PCL members have been shown to recruit PRC2 to regions of actively transcribing chromatin and, in cooperation with a histone demethylase, were able to remove the H3K36me3 mark, allowing for H3K27me3-induced transcription silencing [41,42].

The final two interacting partners for H3K27me3 deposition are C17orf96 and C10orf12, which are still poorly characterized and require additional studies to identify their roles in PRC2 regulation.

In addition to these components, recent years have witnessed an increase in the number of long noncoding RNAs (lncRNA) reported to interact directly with PRC2 subunits. As a result, the PRC2 complex is recruited to specific regions of DNA for silencing (reviewed in Ref. [44]). An example of this class of lncRNA is the long ncRNA Xist, which functions in *cis* to confer transcriptional silencing onto one entire X chromosome in female mammals in a process known as dosage compensation [45].

Recent efforts employing EM and X-ray crystallography have furthered our understanding of PRC2 regulation. Negative staining electron microscopy (EM) was used to describe the first structure of the entire human PRC2 holoenzyme in complex with AEBP2 [32]. This study represented the first effort to orient and position existing high-resolution X-ray structures in the context of the holoenzyme and understand the resulting interactions between different subunits. A recent high-resolution crystal structure of the EED/

Ezh2/Suz12-VEFS subcomplex from the yeast *Chaetomium thermophilum* reveals important atomic level details of PRC2 methyltransferase activity [46]. In the following sections, we will describe the structural details of the PRC2 complex and its component proteins, obtained by different laboratories over several years, using a combination of EM and X-ray crystallographic techniques.

PRC2 ELECTRON MICROSCOPY STUDIES

In past 20 years, numerous biochemical and molecular studies have targeted the PRC2 complex in attempts to elucidate its structure. Despite these efforts, until recently, little was known regarding the protein's overall architecture and the way its different components interact to coordinate histone binding and methylation.

Negative stain EM and single particle analysis were used to determine the molecular architecture of the human PRC2 complex bound to the cofactor AEBP2 at a resolution of 21 Å [32]. The PRC2 complex measures roughly 160 by 120 by 90 Å in size and is composed of four discrete lobes (Lobes A–D, Fig. 8.2), each approximately 55 Å in the longest dimension. The lobes are interconnected through two structurally distinct arms (Arm1 and Arm2, Fig. 8.2). Arm1 connects Lobe A and Lobe B in the upper portion of the structure, while Arm two connects the top and bottom portions of the PRC2 complex (Fig. 8.2). Lobes A and D feature distinct ring-like shapes and represent the WD40 domains of EED and RbAp46/48 (Fig. 8.2).

Chemical crosslinking coupled to mass spectrometry (Reviewed in Ref. [47]) was used to determine the spatial proximity of different domains within the PRC2 complex [32]. These studies confirmed the reported interactions between the C-terminal WD40 propeller of EED and the N-terminal region of Ezh2 [48], but also showed that EED interacts with the first SANT (SANT1) domain of Ezh2 as well as the Zn fingers of AEBP2. In addition, it

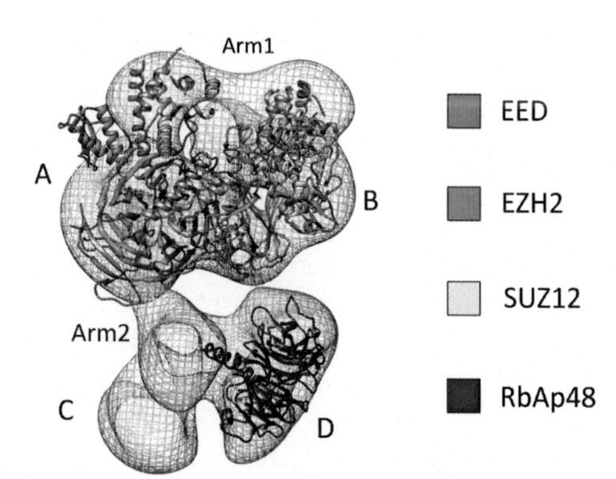

FIGURE 8.2 Architecture of the human PRC2–AEBP2 complex by electron microscopy. Model of the PRC2 complex obtained from a combination of EM and X-ray structural biology data. Docking of available crystal structure for EED, EZH2, Suz12-VEFS, and RbAp48 domains (PDB: 5CH1, 2YB8) are indicated.

was revealed that the Ezh2 SANT domain is situated in close proximity to the second SANT domain (SANT2), the VEFS domain of Suz12, and the Zn-finger region of AEBP2. The latter represents an important hub for multiple interactions and coordinates several regions of PRC2, including the Ezh2 methyltransferase (SET) domain. This finding could explain AEBP2's role in binding and stimulating the activity of the PRC2 complex [10,11,31,32].

Protein tagging and single particle EM analysis were used to localize the individual domains of the PRC2–AEBP2 complex. Specifically, individual domains of each of the subunits were either fused at their N-termini to MBP (Maltose-Binding Protein) or internally and at their C-termini to GFP (Green Fluorescent Protein). The resulting complexes were imaged by single particle EM, and reference-free 2D classification was used to determine the positions of the specific tags, following comparisons to the corresponding class averages of the untagged complex [32,49]. These analyses showed that the EED-WD40 domain and the N-terminal WD40-binding region of Ezh2 were localized within Lobe A (Fig. 8.2). In addition, the RbAp48 WD40 domain was localized to Lobe D, consistent with its donut-shape density. Crystal structures of the mouse EED-Ezh2 (PDB: 2QXV, [48]) and *Drosophila* RbAp48–Suz12 complex (PDB: 2YB8, [25]) could be docked into these EM densities and properly oriented, based on domain mapping data (Fig. 8.2, [32,49]). The WD40 domains of EED and RbAp46/48 bind to H3K27me3 and H3K4me3, respectively, and their binding regulates PRC2 activity. Therefore, the localization of these WD40 domains indicates the positions of the PRC2 complex in which histone binding and regulation occurs. Domain localization of the other components suggests that the Ezh2 domain starts at Lobe A, where it interacts with EED, continues in Arm1 with its SANT domains, reaches Lobe B, where its methyltransferase SET domain begins, and returns to Lobe A, where the Ezh2 C-terminal region is located (Fig. 8.2, [32,49]). Protein labeling of Suz12 showed that this component runs along the length the PRC2 complex. Suz12 originates at Lobe D, moves with its central region at Lobe C, and continues with its Zn finger domain in Arm two before ending with its VEFS domain in Lobe B, near the SET domain of Ezh2. Complexes carrying protein tags in AEBP2 showed that this accessory protein runs in the opposite direction of Suz12, presenting its Zn finger domains at Lobe A and terminating in the proximity of Lobe C [32,49]. Interestingly, EM studies suggest that the PRC2–AEBP2 complex contains only one copy of each subunit because the labeling of individual subunits resulted in only one additional density per PRC2–AEBP2 complex.

While these initial EM studies were limited to relatively low resolution, they were fundamental in addressing important functional questions, based on the domain organization and relative positions of individual PRC2 domains. Given the intrinsic flexibility of the PRC2 complex, high-resolution cryo-EM studies have proven very difficult. The introduction of direct detectors and multimodel refinement to overcome conformational heterogeneity will hopefully provide, in the near future, a high resolution map of the entire complex, possibly bound to other accessory subunits or noncoding RNAs, both of which are important for PRC2 function and regulation. Elucidation of the molecular architecture of the PRC2 complex and the relative interactions between its domains has been instrumental for engineering novel and stable PRC2 subcomplexes, which are amenable to high-resolution X-ray structure determination. In the next section, we will review the atomic details of various human PRC2 domains and the crystal structure of a recently determined ternary complex of EED/Ezh2/Suz12 (VEFS) from *Chaetomium thermophilum*, in both the constitutive and activated states [46]. These data can be used in combination with the negative stain PRC2 electron density, as a framework by which to understand PRC2 nucleosome binding, its methyltransferase activity, and regulation.

PRC2 X-RAY CRYSTALLOGRAPHY STUDIES

EED Recognizes a Number of Histone Motifs

The EED subunit of PRC2 is characterized by a prominent C-terminal WD40 beta-propeller, a closed solenoid of seven blades, each containing a four-stranded antiparallel beta sheet. Most beta-propellers possess remarkably similar folds despite low sequence homology and are typically used as scaffolds on which multisubunit protein complexes are assembled [50]. In addition, EED contains a small 80-residue subdomain at its N-terminus, and studies in *Drosophila* reveal that this subdomain interacts directly with histone H3 to ensure the trimethylation of H3K27 [51].

Developmental studies have shown that missense mutations in one of the WD40 repeats of EED result in an embryonic lethal phenotype, and reconstitution of these mutations in both mammalian and yeast two-hybrid systems disrupted binding between EED and Ezh2 [52]. Interestingly, the structural mediators of EED−Ezh2 interaction are conserved in Ezh1, as well as in the *Drosophila* homolog E(Z) [48].

There are currently 13 EED structures registered in the Protein Data Bank (PDB), many of them bound to ligands such as trimethylated nucleosomal peptides. These co−crystal structures provide valuable information on the contributions of EED to PRC2 biochemistry. For instance, a structure of EED bound to a 30-residue fragment of Ezh2 reveals that the Ezh2 peptide is wedged into a narrow binding groove along the "front" face of EED. This interaction is maintained by a stable network of hydrogen bonding and van der Waals interactions and contributes to the EED-Ezh2 pairing that is necessary for PRC2 methyltransferase activity [48].

In comparison, several trimethylated nucleosomal histone peptides (H1K26, H3K4, H3K9, H3K27, H4K20) are observed to bind across the rear face of the beta-propeller, with the trimethylated lysine side chains honing in on an aromatic cage at the EED surface [13]. The cage consists of three aromatic residues (Phe97, Tyr148, Tyr365) that interact with the terminal trimethyl ammonium group of the lysine through van der Waals and cation-π interactions. An additional fourth aromatic residue (Trp364) stabilizes the aliphatic portion of the lysine side chain. Of note, the conformations of the side chains in the EED residues involved in these different interactions are essentially the same in the various co−crystal structures (Fig. 8.3).

Interestingly, the *Drosophila* homolog of EED (ESC) is not required for PRC2 recruitment to the nucleosome. However, once bound, ESC appears to interact with histone H3 through its N-terminal domain and may be required to present the lysine substrate to the catalytic subunit in EZH2 [51]. A related model was proposed for the human complex, in which EED binding to trimethylated histone tails leads to an allosteric activation of PRC2 activity. Specifically, H3K27me3 was found to stimulate PRC2 methyltransferase activity against purified nucleosomes by approximately threefold [20].

In contrast, the addition of H1K26me3 to the reaction mixture inhibited the reaction, but resulted in the heterogeneous (mono-, di-, and tri-) methylation of four surface lysines on EED [20]. It is possible that H1K26me3 affects the overall PRC2 structure to preferentially accept EED as a substrate, but confirmation of this hypothesis will necessitate the structural determination of the PRC2 holoenzyme in complex with the requisite peptide ligands. The interplay between H3K27me3 and H1K26me3 binding comprises an attractive model for an enzymatic switch, and will likely act alongside the activities of

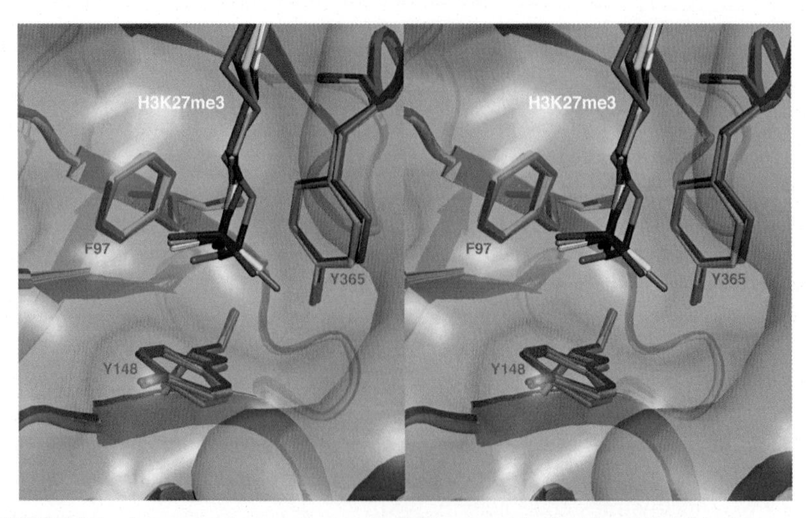

FIGURE 8.3 A conserved aromatic cage in EED surrounds the trimethylated lysine group of histone substrates. Structures of EED bound to various trimethylated lysine histone substrates shown in stereo, including H1K26me3 (magenta), H3K4me3 (blue), H3K9me3 (orange), H3K27me3 (yellow), and H4K20me3 (cyan). EED is shown as a transparent surface.

histone lysine demethylases for the sensitive and specific control of PRC2 methyltransferase activity [53].

Drosophila Nurf55 (RbAp48) + Histone H3 or Suz12

Like EED, the PRC2 subunit RbAp48 features a beta-propeller domain. Structural studies on the *Drosophila* ortholog Nurf55 revealed an N-terminal alpha helix followed by a seven-stranded WD40 beta-propeller [54]. Because this domain was previously reported to bind the N-terminus of unmodified histone H3, Schmitges et al. sampled a variety of peptide lengths and determined that Nurf55 binds histone $H3^{1-14}$ with low micromolar affinity. A subsequent structure showed that the histone peptide is embedded within an acidic pocket across a face of the Nurf55 beta-propeller, with the N-terminal Arg2 inserted into the central pore (Fig. 8.4).

The first 10 residues of histone H3 engage in clearly defined charge- and geometry-specific contacts with their complementary residues in Nurf55. Beyond these, the electron density of the histone peptide becomes less ordered as it extends into the solvent region, before returning to the binding face of Nurf55. This abrupt loss of stabilizing forces likely signals the boundary of substrate-specific recognition between the two components.

To study the interactions between Nurf55 and histone H3 in a physiologically appropriate context, Nurf55 was bound to a fragment of its PRC2-binding partner, Su(z)12 (known as Suz12 in humans). In this structure, the N-terminal helix of Nurf55 undergoes a noticeable swing away from the bound Su(z)12 compared to the structure of Nurf55 bound to histone $H3^{1-14}$ (Fig. 8.4). A subtle shift of the PP-loop is also observed, but the two Nurf55

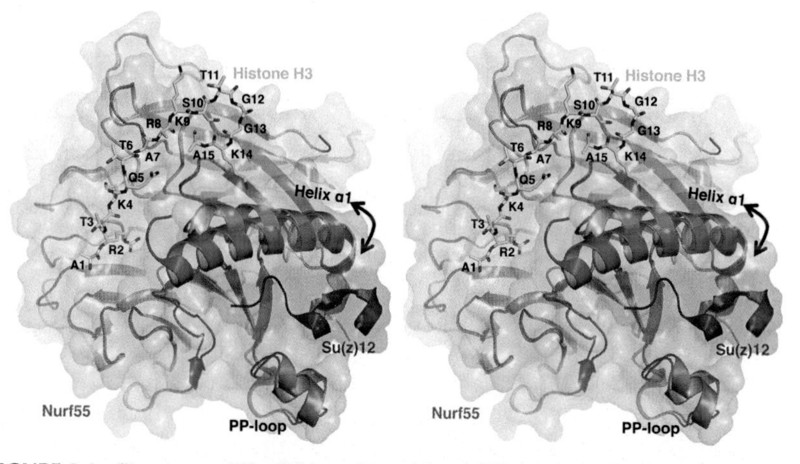

FIGURE 8.4 Structures of Nurf55 bound to a histone H3 peptide and a Su(z)12 fragment. Stereo view of the Nurf55-histone H3 complex (blue and cyan; PDB code: 2YBA) overlaid with the Nurf55–Su(z)12 complex (orange and red; PDB code: 2YB8). The surfaces of Nurf55 and bound Su(z)12 are shown as transparent surfaces.

complexes are otherwise very similar, with an overall rmsd of 0.27 Å. The *apo* Nurf55 and Nurf55-Su(z)12 complexes displayed similar affinities toward a histone $H3^{1-15}$ peptide, indicating that Su(z)12 and histone H3 bind independently of each other [25].

Subsequent studies used isothermal calorimetry to identify interactions between Nurf55 and a trimethylated version of another histone H3 mark, H3K4me3. This mark is associated with active chromatin and appropriately reduces the level of H3K27me3 [55]. The ability of Nurf55 to bind a variety of active (H3K4me3) and repressive (H3K9me3, H3K27me3) histone marks, but still manage to discriminate between them in a stimulus-specific manner, implies varied roles for the PRC2 complex in chromatin modification and regulation. In addition, Nurf55, also known as Caf1, is a component of several other chromatin-related complexes (i.e., CAF1, NuRD, NURF, and DREAM), confirming its role in chromatic regulation.

Interestingly, Su(z)12 and histone H4 share overlapping binding sites on the Nurf55 surface, and competition experiments with increasing concentrations of either component were able to displace binding of the other [55]. Such a phenomenon was not observed with histone H3, indicating distinct binding interfaces for the various portions of the nucleosome. It is not yet known, however, whether Nurf55 is able to accommodate both Su(z)12 and histone H4 as binding partners in the context of the PRC2 holoenzyme, or if conformational rearrangements may open up additional surfaces.

While there exist significant amounts of data chronicling the effects of various histone marks on Nurf55-Su(z)12 mediation of PRC2 activity, the lack of a high-resolution PRC2 holoenzyme structure precludes a full understanding of Nurf55/RbAp48's contribution to PRC2 substrate recognition and catalysis. In the next section, we will detail the structural efforts toward this goal, starting with a discussion of a ternary PRC2 complex.

Architecture of a Ternary Ezh2—EED—Suz12 Complex

Previous crystallographic investigations have focused on individual PRC2 domains or their pairwise interactions, often with peptide mimics substituting for full-length binding partners. In this section, we will review the global architecture of a ternary PRC2 complex containing the EED—Ezh2—$Suz12^{VEFS}$ domains (Fig. 8.5), proceed through a systematic inspection of its core interactions, and, based on current literature, highlight proposed models for allosteric regulation in both PRC2 substrate recognition and catalysis.

In 2015, Jiao and Liu reported two high-resolution crystal structures of a ternary EED—Ezh2—Suz12 complex from the yeast *Chaetomium thermophilum* [46]. The first structure was described in a "stimulated" state and features a captured H3K27me3 peptide, which is a product of the Ezh2 methyltransferase activity and is important for propagating PRC2 activity along chromatin. The second structure lacks this stimulating peptide, and is referred

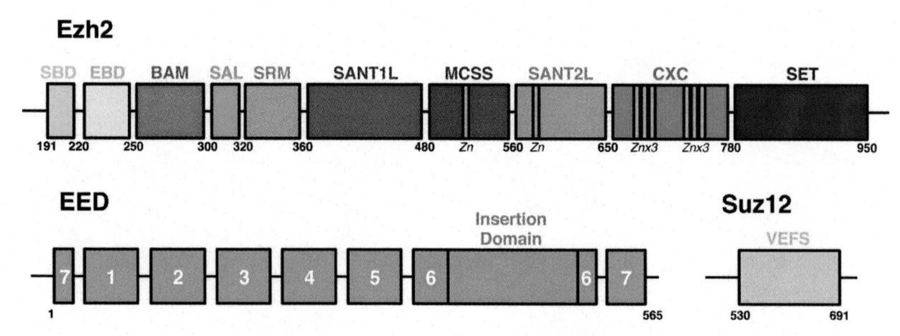

FIGURE 8.5 Subunit organization of the crystallized PRC2 ternary complex from the yeast *Chaetomium thermophilum*. *BAM*, β-addition motif; *CXC*, cysteine-rich domain; *EBD*, EED-binding domain; *MCSS*, motif connecting SANT1L and SANT1L; *SAL*, SET activation loop; *SANT1L*, SANT1-like; *SANT2L*, SANT2-like; *SET*, suppressor of variegation 3-9, enhancer of zeste and trithorax domain; *SRM*, stimulation-responsive motif; *VEFS*, VRN2-EMF2-FIS2-Su(z) 12 domain; *WD SBD*, SANT1L-binding domain. Locations of zinc-coordinating elements are marked with *gray bars*. The numbers indicate domain boundaries of the cloned constructs. *Adapted from Jiao L, Liu X. Structural basis of histone H3K27 trimethylation by an active polycomb repressive complex 2. Science 2015;350(6258):aac4383.*

to as the "basal" state. These structures were determined at 2.3 and 2.7 Å resolution, respectively. Both contain a bound S-adenosyl-L-homocysteine (SAH) cofactor as well as an inhibiting H3K27 M peptide, whose missense mutation has been detected in several pediatric brain tumors [56,57]. Both the basal and stimulated crystal structures, with exception of a yeast-specific EED-WD40 insertion domain, fit very well into the electron density of the human PRC2 complex that was determined by electron microscopy [32] (Fig. 8.2).

The subunit boundaries of the *Chaetomium thermophilum* ternary complex were determined by sequence alignment with their corresponding human orthologs and trimmed to generate a catalytically active module based on previous literature [58]. This reconstituted minimal PRC2 contains EED^{1-565}, $Ezh2^{191-950}$, and $Suz12^{530-691}$ ($Suz12^{VEFS}$), as shown in Fig. 8.5. All subsequent numbering in the text, unless otherwise noted, refers to the amino acid sequences from yeast. Likewise, all structural discussions are targeted toward the stimulated state of the ternary complex (PDB code: 5CH1) unless explicitly noted.

The overall dimensions of the EED–Ezh2–$Suz12^{VEFS}$ ternary complex are approximately 120 by 90 by 50 Å, and correspond to Lobes A and B of the EM map (Fig. 8.2). At the core of the complex is a dimer of $Suz12^{VEFS}$ and EED, and an elongated Ezh2 envelops this dimer pair to form intimate associations using all 10 of its subdomains (Figs. 8.5–8.7). Ezh2 can be further divided into an N-terminal regulatory region and a C-terminal catalytic moiety. The regulatory region assumes a belt-like formation around the WD40 propeller feature of EED and includes the SBD, EBD, beta-addition motif (BAM), SAL, SRM, and SANT1L domains. In contrast, the catalytic moieties form a compact arrangement on the rear surface of $Suz12^{VEFS}$ and include the *m*otif

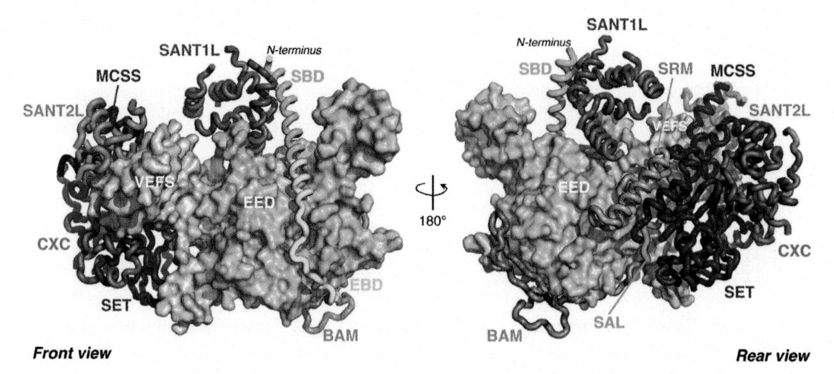

FIGURE 8.6 Molecular architecture of the Suz12^{VEFS}-EED core of the PRC2 ternary complex. Both Suz12^{VEFS} and EED are shown as cartoon representation and transparent surfaces. The coloring and view are identical to those of Fig. 8.7, with Ezh2 omitted for clarity. The yeast-specific insertion domain at the sixth blade of the EED-WD40 propeller is indicated.

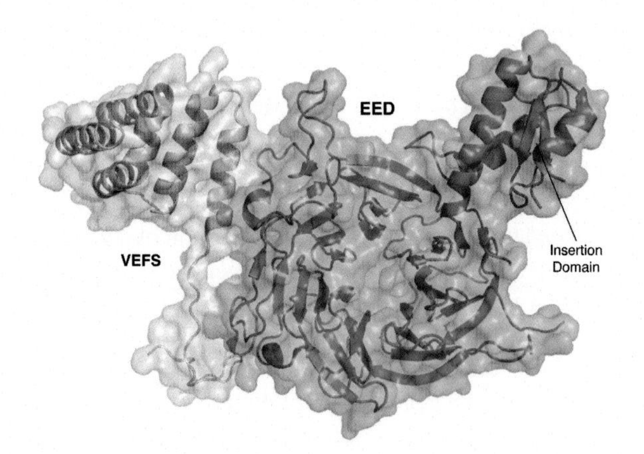

FIGURE 8.7 Ezh2 surrounds the EED-WD40 and Suz12-VEFS domains of PRC2 to form a catalytically competent module. In this activated complex, EED and the VEFS domain of Suz12 are shown in surface representation and are girdled by the N-terminal regulatory portions of Ezh2, which are shown as cartoon loops. The SRM is largely disordered in the basal complex structure (rear view) (PDB code: 5CH1).

*c*onnecting *S*ANT1L and *S*ANT2L (**MCSS**), SANT2L, cysteine-rich (CXC), and SET domains. The definitions of these domains are provided in Fig. 8.5.

In the following paragraphs we will highlight the structural features of each of the minimal PRC2 core components.

EED

The yeast PRC2 EED domain also consists of a seven-bladed beta-propeller but instead contains a ~150 residue subdomain extending from its outer edge, visible in Fig. 8.6 as the lone protrusion from the central disc-shaped propeller. A comparison of *Chaetomium thermophilum* EED against those of human and mouse reveals high levels of structural conservation, with rmsd values of 1.1 and 0.98 Å, respectively.

As predicted from existing structures of the isolated domain, the front face of yeast EED contains a long central groove that accommodates the EED-binding domain (EBD) of Ezh2 (Fig. 8.7). In the crystal structure of the stimulated state, an H3K27me3 product peptide is observed binding the back face of EED and is sandwiched by a helical motif from the regulatory portion of Ezh2 (SRM domain). This is in agreement with the reported role for EED in augmenting the catalytic potential of PRC2, as well as in the subsequent proliferation of H3K27me3 histone marks along repressed chromatin [20].

The insertion within Blade six of the beta-propeller consists of a three-helix bundle wrapped around a beta-strand core and appears to be specific to thermophilic fungi, based on protein BLAST queries. A subsequent search for structural homologs in the DaliLite database [59,60] did not reveal any potential hits, and the relevance of the insertion is currently unknown.

Suz12^{VEFS}

The structure of the minimal PRC2 complex reported by Jiao and Liu also revealed the first high-resolution structure of the Suz12 VEFS domain and shed light on why this domain is required as part of the minimal module for PRC2 activity [46]. Suz12^{VEFS} comprises a 160-residue domain near the C-terminal end of the full-length protein, and in the ternary complex, folds into an extended loop region followed by a tightly packed five-helix bundle. Within the context of the PRC2 holoenzyme, Suz12^{VEFS} forms a scaffold onto which the catalytic moieties of Ezh2 assemble; therefore, it is plausible to conclude that Suz12^{VEFS} is necessary for the proper folding and assembly of Ezh2. In addition, Suz12^{VEFS} also makes local contacts with the SET activation loop (SAL) domain of Ezh2, further supporting its essential role in the organization and positioning of PRC2 catalytic components.

Ezh2

Ezh2 wraps around the EED-Suz12^{VEFS} core dimer to form the ternary PRC2 complex. Here we proceed through a systematic overview of Ezh2's local structure, contributions to overall complex stability, and possible implications for the specificity and regulation of its methyltransferase activity.

The first defined domain in Ezh2 is the SANT1L-binding domain (**SBD**), a helical element that contacts the downstream SANT1L domain to form the junction of Ezh2 N-terminal regulatory components (Fig. 8.7). Sequence alignments between human, fly, and fungal SBD/SANT1L reveal that their junction is rich in hydrophobic residues, indicating an energetically favorable association between the two domains. The SBD continues into the

EED-binding domain (**EBD**), which consists of a short helix followed by a flexible loop region (Fig. 8.7). The EBD occupies a compact surface groove that runs the length of the EED propeller, and existing literature suggests that this interaction is driven by both electrostatic factors and surface geometry complementarity [48].

Following the EBD is a flexible loop that forms a cradle along the bottom edge of the WD40 propeller and continues into a triad of three beta strands known as the **BAM** (Fig. 8.7). This motif forms an extension to Blade five in the EED-WD40 beta-propeller and presumably increases the available surface area for binding EED modulators. A similar model of beta-sheet augmentation is found in the PDZ domains of signaling proteins, in which the association of binding partners contributes additional beta strands to the existing sheet [61]. However, this may be a species-specific phenomenon, as the BAM of human Ezh2 is not predicted to form similar secondary structure elements. The subsequent SET activation loop (**SAL**) extends away from the EED beta-propeller and snakes through the core of the ternary complex, making close contacts with the SET, MCSS, and Suz12^{VEFS} domains (Fig. 8.7). Comparisons of the basal and stimulated structures of the ternary complex reveal that the SAL is required to maintain the active conformation of the catalytic components [46].

The stimulation-responsive motif (**SRM**) is a mobile helix that follows the SAL, and is noteworthy in that its structure was elucidated only in the presence of a stimulating H3K27me3 peptide, a direct product of PRC2 catalysis. Evidence for SRM translocation is found in comparisons of the two ternary complex structures: in the presence of H3K27me3, the SRM domain traps the peptide against the rear face of the EED propeller in a sandwich arrangement, while in the absence of the peptide, the SRM remains flexible and is disordered in the crystal structure (PDB code: 5CH2). Given that the site of the bound H3K27me3 peptide is far from the SET domain, and that this stable histone mark affects the overall methyltransferase activity in a feedforward manner [13,62], it is likely that PRC2 requires some degree of allostery to transmit the regulatory signals from the SRM domain back to the SET catalytic center.

Following the SRM is the SANT1-like (**SANT1L**) domain of Ezh2, which forms a helical bundle and interacts with the SBD to close off the N-terminal regulatory "belt" around EED (Fig. 8.7). Previous literature has implicated the SANT1L domain in DNA binding, where it may have a role in recruiting the PRC2 complex to sites of DNA damage [63,64].

At the start of the catalytic moiety of Ezh2, the **MCSS** links its namesake domains and contains a pair of bent alpha helices as well as a zinc-binding motif (Fig. 8.7). The SANT2-like (**SANT2L**) domain of Ezh2 also contains a zinc-binding motif and occupies a distal portion of the ternary complex, forming an extended helical bundle with elements from MCSS and Suz12^{VEFS}. In contrast to SANT1L, the yeast SANT2L domain is more similar to its human ortholog, based on secondary structure predictions for human Ezh2 [46].

However, the yeast SANT2L lacks a ~80-residue insertion that is present in humans, and to a less degree, in *Drosophila*. The structural and functional significance of this divergence is not known. Both SANT1L and SANT2L, however, pack bulky hydrophobic residues into the cores of their helical bundles, which is a classical feature of SANT domains [65].

The **CXC** domain follows the SANT2L and contains two zinc-sulfur clusters (Zn_3Cys_8His and Zn_3Cys_9) within a compact domain of short helical segments interspersed among long flexible loops (Fig. 8.7). CXC is conserved across multiple species and shows evolutionary relationships to the DNA-binding CXC domains in *Drosophila*, wherein they recruit the dosage compensation complex (DCC) to high-affinity sites on the X chromosome [66]. In the context of PRC2, the CXC domain makes direct contacts with $Suz12^{VEFS}$, MCSS, SANT2L, and the SET domain and may help direct the complex to its nucleosomal substrates.

Lastly, the *s*uppressor of variegation 3-9, *e*nhancer of zeste and trithorax (**SET**) domain house the active site and catalytic hub of the PRC2 complex. This domain is common to nearly all lysine methyltransferases, with relatively few exceptions [67,68]. The SET domain is defined by two signature motifs: (ELxF/YDY and NHS/CxxPN), as well as an internal pseudoknot fold [69].

Studies in *Drosophila* and other systems have demonstrated that the SET domain, even in the context of additional EZH2 components, lacks appreciable activity and must associate with ESC/EED and SU(Z)12/Suz12 for efficient catalysis [23,70]. The crystal structure of the ternary complex fully supports this notion, as the architecture of the catalytic domain is maintained by interactions with both EED (particularly through the fourth blade of the WD40 propeller) and $Suz12^{VEFS}$. Analysis of the SET domain architecture reveals that the SAH cofactor and the lysine substrate are bound in a narrow hydrophobic channel with the SAH thioether positioned immediately opposite the lysine ε-amino group and is suggestive of a catalytically competent state.

MECHANISM OF H3K27M INHIBITION

The H3K27M mutation was originally identified from a survey of pediatric brain tumors harboring missense mutations in histone H3 and was found to be a significant cause of diffuse intrinsic pontine gliomas as well as nonbrain stem gliomas [71,72]. Samples from affected patients were shown to display significantly lower levels of H3K27me3 as a result of inhibited PRC2 activity.

In the structure of the ternary complex, an H3K27M-containing substrate peptide (KQLATATKAARMSAPATGGVKK) was observed to bind the SET domains of Ezh2 in both the basal and stimulated states (Fig. 8.8). The peptide density reveals that H3R26 occupies the deep substrate-binding cavity formed by SET domain residues, while the mutant H3M27 makes nonspecific contacts with the remainder of the SET domain. This finding contradicts the expectation that the methionine side chain would occupy the cleft typically reserved

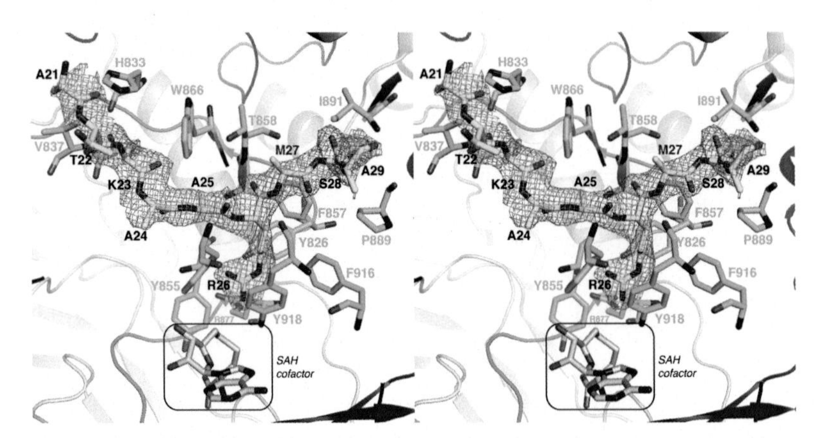

FIGURE 8.8 Binding interactions of the disease-associated, inhibiting H3K27M peptide. Stereo view of the H3K27M mutant peptide bound in the SET domain of Ezh2. The mesh surrounding the peptide is contoured at 1σ. SET domain residues within 4 Å of the bound peptide were extracted and colored in light green to indicate relevant stabilizing interactions. The residues of the SET domain are labeled in light green and those of the mutant peptide are labeled in black, with the exception of Met27, which is labeled in red. An S-adenosyl-L-homocysteine cofactor (box) is also bound near Tyr855 of the SET domain.

for the wild-type lysine substrate. Indeed, structural comparisons with several other SET domains indicate that this lysine access channel is effectively unchanged in location and geometry [73]. The chemical properties of the trimethylated lysine's aromatic cage are also conserved between human and yeast EED (Phe97, Tyr148, Tyr365 in human; Phe41, Cys96, Phe328 in yeast), and the exact reason for the shift in residue binding is unclear.

From a clinical perspective, mutation of H3K27 to Met is much more inhibitory than other residue substitutions at this site [46] but the exact reason for this phenomenon is not known. Others have shown that the mutant H3K27M peptide is able to suppress in vitro methyltransferase activity in a dose-dependent manner and that additional SET-domain-containing methyltransferases are similarly sensitive to methionine substitution at the sites of lysine methylation (e.g. H3K9 and H3K36; [56]).

MECHANISM OF H3K27me3 ACTIVATION

The end product of PRC2 catalysis is H3K27me3, which has been shown to interact with EED and SRM in a modified sandwich structure at the rear face of the EED domain [46] (Fig. 8.7). This interaction effectively traps the SRM from an otherwise flexible state and subsequently stimulates the methyltransferase activity of the PRC2 complex [13,62]. This positive feedback mechanism may explain the proliferation of H3K27me3, silencing of PRC2 target genes, and ultimately chromatin compaction.

Comparisons of the basal and stimulated structures in the SRM region reveal that the entirety of the Ezh2 catalytic moiety (MCSS, SANT2L, CXC, SET) as well as Suz12^{VEFS} undergoes a conformational shift on binding the stimulating H3K27me3 peptide (TKAAR**Kme3**SAPAT) to the ternary complex. The movement is especially pronounced when compared to an inactive SET domain from human Ezh2 [74], and is visible in Fig. 8.9 as an ~20-degree counterclockwise rotation of helix α21 (SET-I) between the inactive and stimulated states.

The H3K27me3 binding site is located approximately 25 Å from the hinge region of helix α21 and approximately 35 Å from the SET substrate–binding cavity. This substantial separation implies some form of long-range communication across domains and indicates that PRC2 uses allostery to transmit the regulatory signals from the SRM domain back to the SET catalytic center. A survey of SET-domain–containing methyltransferases also reveals that the requirement of an adjacent domain, or even a binding platform, for catalysis is frequently found in large protein complexes [75].

Closer inspection of the H3K27me3 peptide reveals that the trimethylated lysine is cradled within an aromatic pocket formed by Phe41, Cys96, Phe327, and Phe328 of EED [46]. Additional hydrogen bonding and backbone interactions stabilize the peptide against the SRM helix. Sequence alignments of this region show near-absolute conservation of the residues comprising the SRM helix in human and fly Ezh2; the fungal sequence is less conserved and may indicate a different mode or extent of response to the trimethylated lysine peptide.

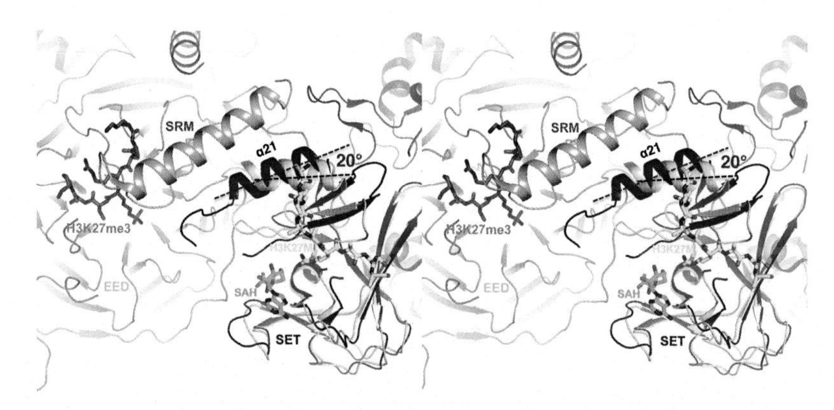

FIGURE 8.9 Local conformational shifts in the SET domain upon H3K27me3 stimulation. The stereo view is similar to that of Fig. 8.8 but expanded to show the binding of the H3K27me peptide. The structure of the stimulated ternary complex (PDB code: 5CH1) is colored as in previous figures, and the structure of the human SET domain (PDB code: 4MI0) is colored gray. On binding, helix α21 of the SET domain undergoes an ~20-degree rotation toward the SRM motif. *Adapted from Jiao L, Liu X. Structural basis of histone H3K27 trimethylation by an active polycomb repressive complex 2. Science 2015;350(6258):aac4383.*

SUMMARY AND OUTLOOK

The Trithorax (TrxG) and Polycomb groups (PcG) of protein complexes orchestrate the transcriptional activation and silencing of the genome through trimethylation of H3K4 (H3K4-me3)/H3K36 (H3K36-me2/3), and H3K27 (H3K27-me3), respectively. PRC2 is a critical enzyme for transcriptional repression and coordinates with additional cellular factors to execute and maintain gene-silencing programs. In humans, PRC2 assembles as an ~250-kDa holoenzyme of four core components (EED, Ezh2, Suz12, RbAp46/48) and is responsible for conferring di- and trimethyl marks onto lysine 27 of histone H3, which is a defining feature of silenced chromatin.

PRC2 methyltransferase activity is enhanced by H3K27me3 and is inhibited by H3K4me3 and H3K36me3 [13,25]. The WD40 domain of EED within PRC2 binds the H3K27me3 [13] via an allosteric mechanism that is likely mediated by the stabilization of the Ezh2 SRM domain [46]. This activation results in the propagation of the repressive state to neighboring nucleosomes [13]. Schmitges and colleagues showed that binding of the chromatin activation marks H3K4me3 and H3K36me3 on RbAp46/8 inhibits PRC2 activity and that the Suz12-VEFS domain is responsible for mediating this inhibition [25].

PRC2 can therefore simultaneously integrate activating and inhibitory chromatin marks and fine-tune its enzymatic activity based on the surrounding chromatin state. In this context, the spreading of H3K27me3 marks is prevented in the presence of transcriptionally active chromatin, which carries the H3K4me3/H3K36me3 marks. If these marks are removed, transcriptionally active regions are converted into silent chromatin domains, where the presence of repressive H3K27me3 marks further activate PRC2 activity.

As a matter of fact, PRC2's role as an essential complex is manifested in the various clinical phenotypes that arise from residue substitutions and premature truncations. The deregulation and resulting transcriptional instability of Ezh2 has been detected in a number of maladies, including breast and prostate cancer and T-acute lymphoblastic leukemia [76,77]. Fig. 8.10 depicts a collection of ~40 missense mutations that have been identified and profiled from clinical studies, and mapped onto the structure of the PRC2 ternary complex. The majority of mutations are clustered in the catalytic SET domain, the zinc-coordinating sites, and the histone peptide recognition motifs of EED. Because these are all sites of residue-specific interactions, it is likely that the effects of distal mutations are shielded by the surrounding environment, and have more subtle consequences on PRC2 stability and activity.

Several PRC2 subunits have now been characterized at the structural level, lending insight into their architecture and contributions to PRC2 activity. EED consists of a beta-propeller domain that binds a stabilizing EBD motif on its front face, and a variety of trimethylated histone peptides on its rear surface. Co–crystal structures revealed that a conserved aromatic cage within EED

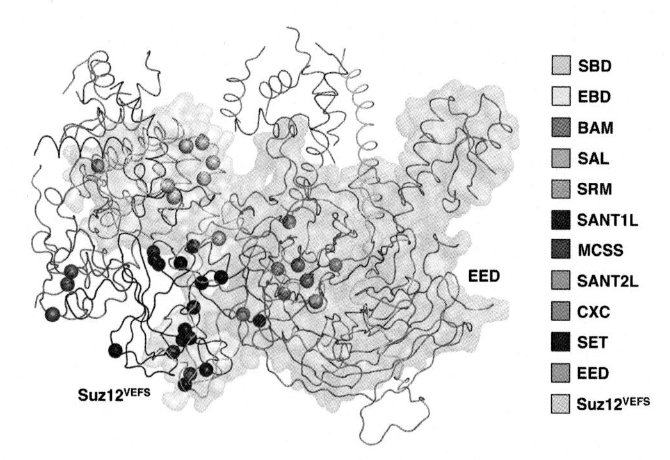

FIGURE 8.10 Locations of disease-associated gene mutations in the PRC2 ternary complex. Cartoon ribbon representation of the domains and view in Fig. 8.7 with EED and Suz12^{VEFS} also shown as transparent surfaces for orientation purposes. *Spheres* indicate the locations of published clinical missense mutations.

recognizes the trimethylated functionality. RbAp48 was studied from the perspective of its *Drosophila* ortholog Nurf55 and also assumes the shape of a beta-propeller. Nurf55 was captured in complexes with a histone H3 peptide or a fragment of Su(z)12, and responds to its bound ligands with movement of a rigid helix between the two binding sites.

The first understanding of PRC2 global architecture was revealed through EM studies, which succeeded in pinpointing the relative arrangements and orientations of the EED, Ezh2, Suz12, and RbAp48 domains, as well as a bound AEBP2 cofactor, in a low-resolution EM envelope (Fig. 8.2). Many of these findings were confirmed and extended in a recent report of a high-resolution crystal structure of a ternary EED−Ezh2−Suz12 complex, which exhibits in vitro PRC2 methyltransferase activity and was observed in both the constitutive and activated states.

The structure of the ternary complex features a dimer of EED and Suz12^{VEFS} at its core, while Ezh2 wraps around the exterior of the dimer. In particular, the N-terminal regulatory moiety of Ezh2 forms an extended girdle around the EED beta-propeller, while the C-terminal catalytic regions assemble on the distal surfaces of Suz12^{VEFS} and EED as discrete domains. The basal state of ternary complex was studied in the presence of a bound SAH cofactor and an inhibiting H3K27M-containing peptide bound within the catalytic SET domain. A separate activated state of the complex also captured a stimulating H3K27me-containing peptide bound to the rear face of EED, making intimate contacts with the SRM and suggesting the possibility of long-range communication in amplifying the H3K27me3 repressive mark during gene silencing.

While the elucidation of the PRC2 ternary complex structure represents significant progress in the understanding of its architecture and interactions, many questions remain to be addressed. The high-resolution structure of a holo-PRC2 complex, and even larger assemblies with cofactors, lncRNA, and nucleosomal substrates, would unveil a treasure trove of structural data and shed considerable insight on PRC2 activity and its regulation. For instance, the exact contributions of full-length Suz12 and RbAp46/48 remain unclear. In addition, the atomic-level details of the interactions between PRC2 and additional nuclear components are still unknown. Efforts toward high-resolution structures of the PRC2 complex bound to the nucleosome and to important regulatory proteins (e.g. Jarid2, AEBP2, and PCLs) and ncRNA are currently ongoing in several laboratories. In addition, the functional consequences of PRC2 clinical mutations are still poorly understood. The basis for substrate turnover and catalysis has only been established at the elementary level, and additional work is required to understand why certain histone marks are processed in place of others. The interplay between various domains of PRC2 represents an attractive model for long-range communication between bound ligands and methyltransferase activity and would extend our knowledge of allosteric regulation in large holoenzyme complexes. Once complete, these findings will increase our understanding of PRC2 activity and regulation and reveal the molecular mechanisms by which PcG genetic alterations may contribute to cancer development.

LIST OF ACRONYMS AND ABBREVIATIONS

AEBP2 Adipocyte enhancer-binding protein 2
BAM β-addition motif
CAF1 Chromatin assembly factor 1 subunit A
CBX Chromobox-domain
CXC Cysteine-rich domain
EED Embryonic Ectoderm Development
EZH1, 2 Enhancer of Zeste 1 and 2
GFP Green Fluorescent Protein
H2A Histone 2A
H2AK119ub1 Monoubiquitination of H2A on K119
H2B Histone 2B
H3 Histone 3
H3K27me2/3 Di and tri methylation at lysine 27 of Histone H4
H3K4me3 Tri methylation at lysine 4 of Histone H3
H4 Histone 4
Hox Homeotic genes
Jarid2 Jumonji AT-rich interactive domain 2
lncRNA Linear non coding RNA
MBP Maltose-binding protein
MCSS Motif connecting SANT1L and SANT1L
mESC Mouse embryonic stem cells

MTF2 Metal response element binding transcription factor 2
ncRNAs Non coding RNAs
NuRD Nucleosome remodeling deacetylase
NURF Nucleosome remodeling factor
PcG Polycomb group
PDB Protein Data Bank
PHF1 PHD finger protein 1
PRC1 Polycomb Repressive Complex 1
PRC2 Polycomb Repressive Complex 2
PTMs Post-Translational Modifications
RbAp46/48 Retinoblastoma binding proteins 46 and 48
SAL SET activation loop
SANT Swi3, Ada2, N-Cor, and TFIIIB
SANT1L SANT1-like
SANT2L SANT2-like
SBD SANT1L-binding domain
SET Suppressor of variegation 3-9, enhancer of zeste and trithorax domain
SRM Stimulation-responsive motif
Suz12 Suppressor of Zeste
TrxG Trithorax group
VEFS VRN2-EMF2-FIS2-Su(z)12 domain
WBD WD40-binding domain
WD40 Trp-Asp 40mer containing peptide
Zn Zinc finger domain

REFERENCES

[1] Luger K, Mader AW, Richmond RK, Sargent DF, Richmond TJ. Crystal structure of the nucleosome core particle at 2.8 Å resolution. Nature 1997;389(6648):251–60.
[2] Kornberg RD. Chromatin structure: a repeating unit of histones and DNA. Science 1974;184(4139):868–71.
[3] Thoma F, Koller T, Klug A. Involvement of histone H1 in the organization of the nucleosome and of the salt-dependent superstructures of chromatin. J Cell Biol 1979;83:403–27.
[4] Widom J, Klug A. Structure of the 300A chromatin filament: X-ray diffraction from oriented samples. Cell 1985;43(1):207–13.
[5] Song F, Chen P, Sun D, Wang M, Dong L, Liang D, Xu RM, Zhu P, Li G. Cryo-EM study of the chromatin fiber reveals a double helix twisted by tetranucleosomal units. Science 2014;344(6182):376–80.
[6] Musselma CA, Avvakumov N, Watanabe R, Abraham CG, Lalonde ME, Hong Z, Allen C, Roy S, Nunez JK, Nickoloff J, Kulesza CA, Yasui A, Cote J, Kutateladze TG. Molecular basis for H3K36me3 recognition by the Tudor domain of PHF1. Nat Struct Mol Biol 2012;19(12):1266–72.
[7] Schuettengruber B, Chourrout D, Vervoort M, Leblanc B, Cavalli G. Genome regulation by polycomb and trithorax proteins. Cell 2007;128(4):735–45.
[8] Margueron R, Reinberg D. The polycomb complex PRC2 and its mark in life. Nature 2011;469(7330):343–9.
[9] Cochran AG. Structure and biochemistry of the polycomb repressive complex 1 ubiquitin ligase module. 2016 [This book].

[10] Cao R, Zhang Y. The functions of E(Z)/EZH2-mediated methylation of lysine 27 in histone H3. Curr Opin Genet Dev 2004;14(2):155–64.
[11] Cao R, Zhang Y. SUZ12 is required for both the histone methyltransferase activity and the silencing function of the EED-EZH2 complex. Mol Cell 2004;15(1):57–67.
[12] Czermin B, Melfi R, McCabe D, Seitz V, Imhof A, Pirrotta V. *Drosophila* enhancer of Zeste/ESC complexes have a histone H3 methyltransferase activity that marks chromosomal Polycomb sites. Cell 2002;111(2):185–96.
[13] Margueron R, Justin N, Ohno K, Sharpe ML, Son J, Drury 3rd WJ, Voigt P, Martin SR, Taylor WR, De Marco V, Pirrotta V, Reinberg D, Gamblin SJ. Role of the polycomb protein EED in the propagation of repressive histone marks. Nature 2009;461(7265):762–7.
[14] Wang H, Wang L, Erdjument-Bromage H, Vidal M, Tempst P, Jones RS, Zhang Y. Role of histone H2A ubiquitination in Polycomb silencing. Nature 2004;431(7010):873–8.
[15] Cao R, Tsukada Y, Zhang Y. Role of Bmi-1 and Ring1A in H2A ubiquitylation and Hox gene silencing. Mol Cell 2005;20(6):845–54.
[16] Stock JK, Giadrossi S, Casanova M, Brookes E, Vidal M, Koseki H, Brockdorff N, Fisher AG, Pombo A. Ring1-mediated ubiquitination of H2A restrains poised RNA polymerase II at bivalent genes in mouse ES cells. Nat Cell Biol 2007;9(12):1428–35.
[17] Zhou W, Zhu P, Wang J, Pascual G, Ohgi KA, Lozach J, Glass CK, Rosenfeld MG. Histone H2A monoubiquitination represses transcription by inhibiting RNA polymerase II transcriptional elongation. Mol Cell 2008;29(1):69–80.
[18] Nakagawa T, Kajitani T, Togo S, Masuko N, Ohdan H, Hishikawa Y, Koji T, Matsuyama T, Ikura T, Muramatsu M, Ito T. Deubiquitylation of histone H2A activates transcriptional initiation via trans-histone cross-talk with H3K4 di- and trimethylation. Genes Dev 2008;22(1):37–49.
[19] Kuzmichev A, Jenuwein T, Tempst P, Reinberg D. Different EZH2-containing complexes target methylation of histone H1 or nucleosomal histone H3. Mol Cell 2004;14(2):183–93.
[20] Xu C, Bian C, Yang W, Galka M, Ouyang H, Chen C, Qiu W, Liu H, Jones AE, MacKenzie F, Pan P, Li SS, Wang H, Min J. Binding of different histone marks differentially regulates the activity and specificity of polycomb repressive complex 2 (PRC2). Proc Natl Acad Sci USA 2010;107(45):19266–71.
[21] Margueron R, Li G, Sarma K, Blais A, Zavadil J, Woodcock CL, Dynlacht BD, Reinberg D. Ezh1 and Ezh2 maintain repressive chromatin through different mechanisms. Mol Cell 2008;32(4):503–18.
[22] Shen X, Liu Y, Hsu HJ, Fujiwara Y, Kim J, Mao X, Yuan GC, Orkin SH. EZH1 mediates methylation on histone H3 lysine 27 and complements EZH2 in maintaining stem cell identity and executing pluripotency. Mol Cell 2008;32(4):491–502.
[23] Ketel CS, Andersen EF, Vargas ML, Suh J, Strome S, Simon JA. Subunit contributions to histone methyltransferase activities of fly and worm polycomb group complexes. Mol Cell Biol 2005;25(16):6857–68.
[24] Birve A, Sengupta AK, Beuchle D, Larsson J, Kennison JA, Rasmuson-Lestander A, Muller J. Su(z)12, a novel *Drosophila* Polycomb group gene that is conserved in vertebrates and plants. Development 2001;128(17):3371–9.
[25] Schmitges FW, Prusty AB, Faty M, Stutzer A, Lingaraju GM, Aiwazian J, Sack R, Hess D, et al. Histone methylation by PRC2 is inhibited by active chromatin marks. Mol Cell 2011;42(3):330–41.

[26] Ferrari KJ, Scelfo A, Jammula S, Cuomo A, Barozzi I, Stutzer A, Fischle W, Bonaldi T, Pasini D. Polycomb-dependent H3K27me1 and H3K27me2 regulate active transcription and enhancer fidelity. Mol Cell 2014;53(1):49–62.

[27] Sneeringer CJ, Scott MP, Kuntz KW, Knutson SK, Pollock RM, Richon VM, Copeland RA. Coordinated activities of wild-type plus mutant EZH2 drive tumor-associated hypertrimethylation of lysine 27 on histone H3 (H3K27) in human B-cell lymphomas. Proc Natl Acad Sci USA 2010;107(49):20980–5.

[28] Tie F, Banerjee R, Stratton CA, Prasad-Sinha J, Stepanik V, Zlobin A, Diaz MO, Scacheri PC, Harte PJ. CBP-mediated acetylation of histone H3 lysine 27 antagonizes *Drosophila* Polycomb silencing. Development 2009;136(18):3131–41.

[29] Smits AH, Jansen PW, Poser I, Hyman AA, Vermeulen M. Stoichiometry of chromatin-associated protein complexes revealed by label-free quantitative mass spectrometry-based proteomics. Nucleic Acids Res 2013;41(1):e28.

[30] Vizan P, Beringer M, Ballare C, Di Croce L. Role of PRC2-associated factors in stem cells and disease. FEBS J 2015;282(9):1723–35.

[31] Kim H, Kang K, Kim J. AEBP2 as a potential targeting protein for Polycomb Repression Complex PRC2. Nucleic Acids Res 2009;37(9):2940–50.

[32] Ciferri C, Lander GC, Maiolica A, Herzog F, Aebersold R, Nogales E. Molecular architecture of human polycomb repressive complex 2. Elife 2012:1. e00005.

[33] Peng JC, Valouev A, Swigut T, Zhang J, Zhao Y, Sidow A, Wysocka J. Jarid2/Jumonji coordinates control of PRC2 enzymatic activity and target gene occupancy in pluripotent cells. Cell 2009;139(7):1290–302.

[34] Li G, Margueron R, Ku M, Chambon P, Bernstein BE, Reinberg D. Jarid2 and PRC2, partners in regulating gene expression. Genes Dev 2010;24(4):368–80.

[35] Pasini D, Cloos PA, Walfridsson J, Olsson L, Bukowski JP, Johansen JV, Bak M, Tommerup N, Rappsilber J, Helin K. JARID2 regulates binding of the Polycomb repressive complex 2 to target genes in ES cells. Nature 2010;464(7286):306–10.

[36] Son J, Shen SS, Margueron R, Reinberg D. Nucleosome-binding activities within JARID2 and EZH1 regulate the function of PRC2 on chromatin. Genes Dev 2013;27(24):2663–77.

[37] Kaneko S, Bonasio R, Saldana-Meyer R, Yoshida T, Son J, Nishino K, Umezawa A, Reinberg D. Interactions between JARID2 and noncoding RNAs regulate PRC2 recruitment to chromatin. Mol Cell 2014;53(2):290–300.

[38] Kaneko S, Son J, Bonasio R, Shen SS, Reinberg D. Nascent RNA interaction keeps PRC2 activity poised and in check. Genes Dev 2014;28(18):1983–8.

[39] Kalb R, Latwiel S, Baymaz HI, Jansen PW, Muller CW, Vermeulen M, Muller J. Histone H2A monoubiquitination promotes histone H3 methylation in Polycomb repression. Nat Struct Mol Biol 2012;21(6):569–71.

[40] Sanulli S, Justin N, Teissandier A, Ancelin K, Portoso M, Caron M, Michaud A, Lombard B, da Rocha ST, Offer J, Loew D, Servant N, Wassef M, Burlina F, Gamblin SJ, Heard E, Margueron R. Jarid2 methylation via the PRC2 complex regulates H3K27me3 deposition during cell differentiation. Mol Cell 2015;57(5):769–83.

[41] Ballare C, Lange M, Lapinaite A, Martin GM, Morey L, Pascual G, Liefke R, Simon B, Shi Y, Gozani O, Carlomagno T, Benitah SA, Di Croce L. Phf19 links methylated Lys36 of histone H3 to regulation of Polycomb activity. Nat Struct Mol Biol 2012;19(12):1257–65.

[42] Brien GL, Gambero G, O'Connell DJ, Jerman E, Turner SA, Egan CM, Dunne EJ, Jurgens MC, Wynne K, Piao L, Lohan AJ, Ferguson N, Shi X, Sinha KN, Loftus BJ,

Cagney B, Bracken AP. Polycomb PHF19 binds H3K36me3 and recruits PRC2 and demethylase NO66 to embryonic stem cell genes during differentiation. Nat Struct Mol Biol 2012;19(12):1273–81.

[43] Musselman CA, Lalonde ME, Cote J, Kutateladze TG. Perceiving the epigenetic landscape through histone readers. Nat Struct Mol Biol 2012;19(12):1218–27.

[44] Davidovich C, Cech TR. The recruitment of chromatin modifiers by long noncoding RNAs: lessons from PRC2. RNA 2015;21(12):2007–22.

[45] Zhao J, Sun BK, Erwin JA, Song JJ, Lee TJ. Polycomb proteins targeted by a short repeat RNA to the mouse X chromosome. Science 2008;322(5902):750–6.

[46] Jiao L, Liu X. Structural basis of histone H3K27 trimethylation by an active polycomb repressive complex 2. Science 2015;350(6258):aac4383.

[47] Leitner A, Faini M, Stengel F, Aebersold R. Crosslinking and mass spectrometry: an integrated technology to understand the structure and function of molecular machines. Trends Biochem Sci 2016;41(1):20–32.

[48] Han Z, Xing X, Hu M, Zhang Y, Liu P, Chai J. Structural basis of EZH2 recognition by EED. Structure 2007;15(10):1306–15.

[49] Ciferri C, Lander GC, Nogales E. Protein domain mapping by internal labeling and single particle electron microscopy. J Struct Biol 2015;192(2):159–62.

[50] Smith TF, Gaitatzes C, Saxena K, Neer EJ. The WD repeat: a common architecture for diverse functions. Trends Biochem Sci 1999;24(5):181–5.

[51] Tie F, Stratton CA, Kurzhals RL, Harte PJ. The N terminus of *Drosophila* ESC binds directly to histone H3 and is required for E(Z)-dependent trimethylation of H3 lysine 27. Mol Cell Biol 2007;27(6):2014–26.

[52] Denisenko O, Shnyreva M, Suzuki H, Bomsztyk K. Point mutations in the WD40 domain of Eed block its interaction with Ezh2. Mol Cell Biol 1998;18(10):5634–42.

[53] Dimitrova E, Turberfield AH, Klose RJ. Histone demethylases in chromatin biology and beyond. EMBO Rep 2015;16(12):1620–39.

[54] Song JJ, Garlick JD, Kingston RE. Structural basis of histone H4 recognition by p55. Genes Dev 2008;22(10):1313–8.

[55] Nowak AJ, Alfieri C, Stirnimann CU, Rybin V, Baudin F, Ly-Hartig N, Lindner D, Muller CW. Chromatin-modifying complex component Nurf55/p55 associates with histones H3 and H4 and polycomb repressive complex 2 subunit Su(z)12 through partially overlapping binding sites. J Biol Chem 2011;286(26):23388–96.

[56] Lewis PW, Muller MM, Koletsky MS, Cordero F, Lin S, Banaszynski LA, Garcia BA, Muir TW, Becher JO, Allis CD. Inhibition of PRC2 activity by a gain-of-function H3 mutation found in pediatric glioblastoma. Science 2013;340(6134):857–61.

[57] Bechet D, Gielen GG, Korshunov A, Pfister SM, Rousso C, Faury D, Fiset PO, Benlimane N, Lewis PW, Lu C, Allis CD, Kieran MW, Ligon KL, Pietsch T, Ellezam B, Albrecht S, Jabado N. Specific detection of methionine 27 mutation in histone 3 variants (H3K27M) in fixed tissue from high-grade astrocytomas. Acta Neuropathol 2014;128(5):733–41.

[58] Yuan W, Wu T, Fu H, Dai C, Wu H, Liu N, Li X, Xu M, Zhang Z, Niu T, Han Z, Chai J, Zhou XJ, Gao S, Zhu B. Dense chromatin activates Polycomb repressive complex 2 to regulate H3 lysine 27 methylation. Science 2012;337(6097):971–5.

[59] Holm L, Sander C. Protein structure comparison by alignment of distance matrices. J Mol Biol 1993;233(1):123–38.

[60] Dietmann S, Park J, Notredame C, Heger A, Lappe M, Holm L. A fully automatic evolutionary classification of protein folds: Dali Domain Dictionary version 3. Nucleic Acids Res 2001;29(1):55–7.

[61] Cowburn D. Peptide recognition by PTB and PDZ domains. Curr Opin Struct Biol 1997;7(6):835–8.

[62] Zee BM, Levin RS, Xu B, LeRoy G, Wingreen NS, Garcia BA. In vivo residue-specific histone methylation dynamics. J Biol Chem 2010;285(5):3341–50.

[63] Ntziachristos P, Tsirigos A, Van Vlierberghe P, Nedjic J, et al. Genetic inactivation of the polycomb repressive complex 2 in T cell acute lymphoblastic leukemia. Nat Med 2012;18(2):298–301.

[64] Campbell S, Ismail IH, Young LC, Poirier GG, Hendzel MJ. Polycomb repressive complex 2 contributes to DNA double-strand break repair. Cell Cycle 2013;12(16):2675–83.

[65] Boyer LA, Latek RR, Peterson CL. The SANT domain: a unique histone-tail-binding module? Nat Rev Mol Cell Biol 2004;5(2):158–63.

[66] Zheng S, Wang J, Feng Y, Wang J, Ye K. Solution structure of MSL2 CXC domain reveals an unusual Zn_3Cys_9 cluster and similarity to pre-SET domains of histone lysine methyltransferases. PLoS One 2012;7(9). e45437.

[67] Feng Q, Wang H, Ng HH, Erdjument-Bromage H, Tempst P, Struhl K, Zhang Y. Methylation of H3-lysine 79 is mediated by a new family of HMTases without a SET domain. Curr Biol 2002;12(12):1052–8.

[68] Patel A, Vought VE, Dharmarajan V, Cosgrove MS. A novel non-SET domain multi-subunit methyltransferase required for sequential nucleosomal histone H3 methylation by the mixed lineage leukemia protein-1 (MLL1) core complex. J Biol Chem 2011;286(5):3359–69.

[69] Xiao B, Wilson JR, Gamblin SJ. SET domains and histone methylation. Curr Opin Struct Biol 2003;13(6):699–705.

[70] Muller J, Hart CM, Francis NJ, Vargas ML, Sengupta A, Wild B, Miller EL, O'Connor MB, Kingston RE, Simon JA. Histone methyltransferase activity of a *Drosophila* Polycomb group repressor complex. Cell 2002;111(2):197–208.

[71] Khuong-Quang DA, Buczkowicz P, Rakopoulos P, Liu XY, et al. K27M mutation in histone H3.3 defines clinically and biologically distinct subgroups of pediatric diffuse intrinsic pontine gliomas. Acta Neuropathol 2012;124(3):439–47.

[72] Wu G, Broniscer A, McEachron TA, Lu C, Paugh BS, Becksfort J, Qu C, Ding L, et al. Somatic histone H3 alterations in pediatric diffuse intrinsic pontine gliomas and non-brainstem glioblastomas. Nat Genet 2012;44(3):251–3.

[73] Qian C, Zhou MM. SET domain protein lysine methyltransferases: structure, specificity and catalysis. Cell Mol Life Sci 2006;63(23):2755–63.

[74] Wu H, Zeng H, Dong A, Li F, He H, Senisterra G, Seitova A, Duan S, Brown PJ, Vedadi M, Arrowsmith CH, Schapira M. Structure of the catalytic domain of EZH2 reveals conformational plasticity in cofactor and substrate binding sites and explains oncogenic mutations. PLoS One 2013;8(12). e83737.

[75] Schapira M. Structural chemistry of human SET domain protein methyltransferases. Curr Chem Genomics 2011;5(Suppl. 1):85–94.

[76] Simon JA, Lange CA. Roles of the EZH2 histone methyltransferase in cancer epigenetics. Mutat Res 2008;647(1–2):21–9.

[77] Hock HA. Complex Polycomb issue: the two faces of EZH2 in cancer. Genes Dev 2012;26(8):751–5.

Chapter 9

Polycomb Repressive Complex 2 Structure and Function

D. Holoch[1,2], R. Margueron[1,2]
[1]*PSL Research University, Paris, France;* [2]*INSERM U934, CNRS UMR3215, Paris, France*

Chapter Outline

Introduction: Discovery of PRC2 **192**
Discovery of the ESC-E(Z) Complex 192
PRC2: A Histone Methyltransferase Activity Required for PcG-Mediated Silencing 193
PRC2 Evolutionary Conservation **195**
The PRC2 Core Complex **197**
Complex Architecture and Requirement of the Subunits for Enzymatic Activity 197
Control of PRC2 Activity by the Chromatin Context 199
Functional Roles of H3K27me3, H3K27me2, and H3K27me1 202
The Alternative PRC2 Catalytic Subunit EZH1 204
PRC2 Cofactors **206**
AEBP2 207
PCL 208
JARID2 209
Recently Identified PRC2 Cofactors 211
PRC2 Within the Polycomb Machinery **212**
Concluding Remarks: On the Deterministic or Responsive Role of PRC2 in Transcriptional Regulation **214**
List of Acronyms and Abbreviations **216**
Acknowledgments **216**
References **216**

In the development of a complex multicellular organism, a single totipotent cell gives rise to a myriad of cell types. The vast majority of them share the same genome, and yet they assume a wide range of structures and functions by implementing distinct transcriptional programs. These diverse cellular identities, once adopted, must be faithfully preserved through cell divisions to ensure the proper formation of specialized organs and tissues. Inherent to the maintenance of cellular identity is a stable memory of gene transcription patterns. Species across the entire eukaryotic lineage achieve this memory using a remarkably similar set of molecular components. Polycomb repressive complex 2 (PRC2), which methylates histone H3 on lysine 27, represents one of the most ancient and developmentally important of these machineries.

Polycomb Group Proteins. http://dx.doi.org/10.1016/B978-0-12-809737-3.00009-X

In this chapter, we review the current state of knowledge on the structure and function of PRC2, with a particular focus on mammalian organisms. We begin by tracing the series of findings that led to the discovery of the complex and mentioning noteworthy examples of its evolutionary conservation. We then discuss the molecular activity and regulation of the core PRC2 complex, the contribution of facultative subunits to PRC2 function, and finally the interactions of PRC2 with other elements of the Polycomb system. We conclude by reflecting on competing views of the role of PRC2 in the regulation of transcription and development.

INTRODUCTION: DISCOVERY OF PRC2

Discovery of the ESC-E(Z) Complex

The multisubunit assembly we now call PRC2 was first discovered in *Drosophila*, a critical model system for studying the genetic basis of body plan development. Although the term "Polycomb" had been coined much earlier to describe a mutant fly line bearing ectopic sex combs, landmark studies in the 1980s and early 1990s revealed that mutations in a large group of additional genes, termed the Polycomb group (PcG), produced similar and mechanistically related phenotypes [1,2]. These experiments pointed to a role for PcG genes in restricting the expression of homeotic proteins to precise regions along the anterior−posterior axis of the developing embryo, thereby ensuring the proper overall configuration of body segments [1,2].

Interestingly, it was noticed that different PcG proteins were expressed at the same time and place in the developing embryo and that they even colocalized at chromosomal sites corresponding to silenced homeotic genes. This suggested that they might act in a concerted manner, as multimeric regulators of chromatin. That idea was cemented by a series of biochemical studies culminating in the isolation from *Drosophila* embryos of Polycomb repressive complex 1 (PRC1), a complex capable of blocking the activity of chromatin remodelers and inducing chromatin compaction in vitro [3−5]. Conspicuously absent from PRC1 were the products of two highly conserved PcG genes with critical roles in embryonic patterning, E(Z) and ESC. But the realization was also rapidly emerging that these two proteins associated to form a complex of their own. Indeed, several groups had uncovered a direct physical link between E(Z) and ESC or between their respective mammalian counterparts Ezh1/Ezh2 and Eed, with binding taking place between the N-terminus of E(Z)/Ezh1/Ezh2 and the WD40 repeats of ESC/Eed [6−10]. Strikingly, point mutations in ESC/Eed previously shown to cause developmental defects in vivo also eliminated its interaction with E(Z)/Ezh1/Ezh2 in yeast two-hybrid or in vitro binding assays [6−9]. This observation strongly implied that the ability of ESC and E(Z) to form a complex was integral to their biological functions.

A more detailed view of the ESC-E(Z) complex was obtained when nuclear extracts from *Drosophila* embryos were analyzed by size-exclusion chromatography. This approach revealed an approximate complex size of 600 kDa [11,12], arguing against a simple heterodimer of ESC (50 kDa) and E(Z) (90 kDa). Consistently, additional factors such as the histone-binding protein Nurf55 were identified as stable interacting partners of ESC-E(Z) [11,12]. But the real breakthrough in understanding the molecular function of the complex came from the characterization of a conserved region that the E(Z) protein shares with a group of other highly studied chromatin regulators, named the SET domain after its most prominent members, SU(VAR)3-9, E(Z), and Trithorax [13,14]. While several lines of evidence suggested that SET-domain proteins influence the structure of chromatin [15], their precise mechanistic role remained a mystery until it was recognized that they possess an enzymatic activity.

PRC2: A Histone Methyltransferase Activity Required for PcG-Mediated Silencing

In a groundbreaking study, Thomas Jenuwein and colleagues demonstrated that the mammalian SET-domain proteins Suv39h1/2 are methyltransferases that specifically modify histone H3 on the lysine residue at position 9 (H3K9) [16]. Methylated H3K9 was subsequently shown to serve as a conserved signal for the recruitment of the core heterochromatin component HP1 [17,18]. These findings suggested that histone methylation might represent a general mechanism for specifying distinct functional chromatin regions, and raised the exciting possibility that E(Z), a SET-domain protein, might also fulfill its role in development through a histone methylation activity. This model was made all the more attractive by the observation that other PcG proteins lose their normal spatial distribution on chromatin when E(Z) is mutated [19], consistent with their recruitment being facilitated by an E(Z)-dependent histone modification. But curiously, under the conditions of the original Suv39h1/2 histone methyltransferase (HMTase) assay, the human E(Z) ortholog EZH2 exhibits no detectable activity [16].

In fact, E(Z) and its homologs are indeed HMTases whose enzymatic activity is essential to their function in gene regulation and development. But unlike Suv39h1/2, E(Z) family proteins are only active within their native, multisubunit complexes [20,21]. As a consequence, the demonstration of their biochemical role coincided with the isolation and reconstitution by several different groups of a core complex consisting of E(Z), ESC/ESCL, Nurf55, and the PcG protein SU(Z)12 (EZH2, EED, RbAp46/48, and SUZ12 in humans). This complex was termed PRC2 [21–24] (Fig. 9.1).

The HMTase activity of PRC2 was determined to be specific for lysine 27 of histone H3 (H3K27) and to require predicted catalytic residues in the SET domain of the E(Z) subunit [21–24]. The amino group of a lysine side chain

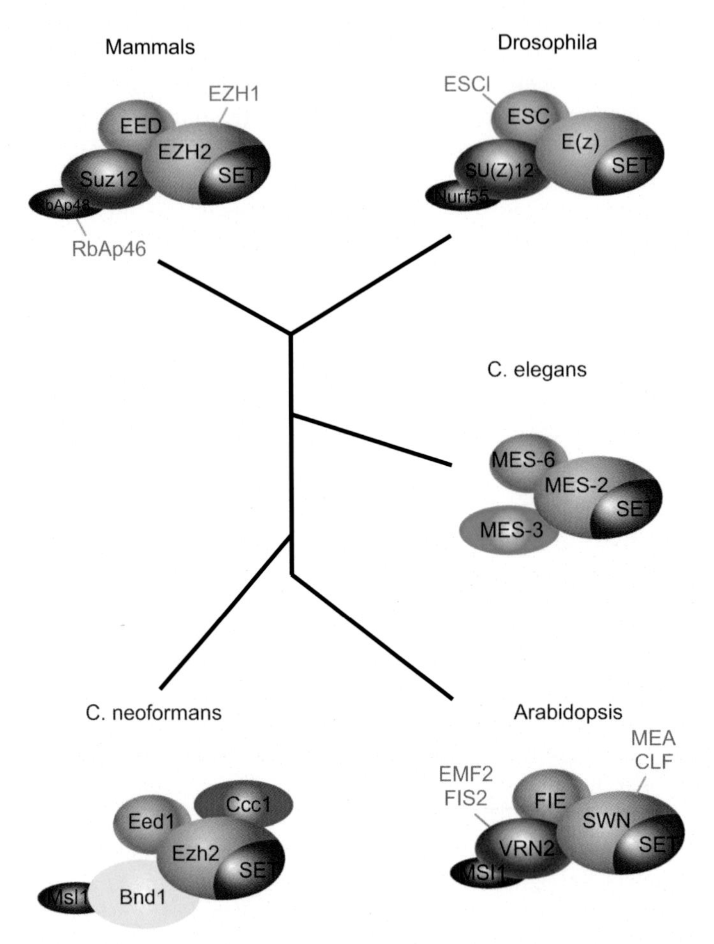

FIGURE 9.1 Evolutionary conservation and divergence of Polycomb repressive complex 2 subunit composition.

can be mono-, di-, or trimethylated in a stepwise manner, and PRC2 can carry out all three reactions on H3K27, with the products denoted as H3K27me1, H3K27me2, and H3K27me3. Both native human PRC2 and recombinant *Drosophila* PRC2 were found to methylate oligonucleosomes more efficiently than simpler substrates such as mononucleosomes, histone octamers, or histone peptides [22,24]. This observation was an early indication that the HMTase activity of PRC2 on a particular nucleosome can be influenced considerably by its physical contacts with features of the surrounding chromatin, a point we will return to in the next section.

Importantly, chromatin immunoprecipitation experiments revealed that the *Drosophila* homeotic gene *Ultrabithorax* (*Ubx*), in cells in which it is normally silenced by the PcG machinery, bears methylated H3K27 in its promoter [22].

Furthermore, a mutation that impairs E(Z) recruitment to *Ubx* also eliminates H3K27 methylation at the locus and leads to a defect in gene silencing [22]. Finally, loss of *Ubx* silencing in the wing discs of *E(z)*-null larvae is rescued by expression wild-type E(Z) but not E(Z) mutants with impaired in vitro HMTase activity [24]. These early results immediately established a strong connection between the H3K27-directed enzymatic activity of PRC2 and its in vivo function in gene silencing and development, which has remained widely accepted ever since. But only much more recently was this link formally demonstrated in *Drosophila* with the generation of mutant flies in which all copies of histone H3 carry a lysine-to-arginine substitution at position 27. Their developmental phenotype, from misexpression of homeotic genes to aberrant body segmentation, faithfully mirrors that of PRC2 mutants, thereby confirming that H3K27 represents a critical substrate for PRC2 in living organisms [25].

The discovery that PRC2 methylates a specific histone residue led to the question of how cells interpret this event as a signal for stable repression of underlying genes. By analogy to H3K9 methylation, which triggers specific recognition and binding of HP1, it was hypothesized that H3K27 methylation could mediate the recruitment of another protein with a specific affinity for the mark. An obvious candidate in *Drosophila* was the PRC1 subunit Pc, because it shares the conserved chromodomain, which in HP1 had been implicated in methyl-lysine binding [17,18]. Moreover, Pc was one of the PcG members whose normal pattern of association with chromosomes was known to be lost in *E(z)* mutant larvae [19]. As predicted, Pc and its mouse ortholog Cbx2 were both found to bind specifically to methylated H3K27 in vitro, observations soon confirmed by an atomic resolution structure [21–23,26]. The importance of the HP1 and Pc chromodomains for specific targeting of the two proteins to distinct chromatin marks was further confirmed by swapping experiments showing that a chimeric fusion of Pc with the HP1 chromodomain exhibits a subnuclear localization pattern mirroring that of HP1 [27].

These results gave rise to a paradigm for the molecular mechanism of PcG-mediated silencing, consisting of two sequential steps, in which PRC2 first marks a locus for repression, and PRC1 is then recruited via its affinity for methylated H3K27 to inhibit transcription of the locus [28,29]. Although this model is compelling, a more complex picture has emerged in recent years, which includes novel modes of interaction between PRC1 and PRC2, as well as evidence that PRC2 can mediate gene silencing through additional mechanisms. Indeed, the fundamental importance of PRC2 for ensuring transcriptional memory is underscored by its near-universal conservation throughout eukaryotic evolution.

PRC2 EVOLUTIONARY CONSERVATION

Orthologs of all four of the core constituents of PRC2 have been identified in all major plant taxa [30]. Their biological functions are best understood in flowering plants, where studies of the classic genetic model organism *Arabidopsis thaliana* have revealed three major developmental roles, each corresponding to a distinct complex defined by its SUZ12-like subunit: VRN2, EMF2, and FIS2 (Fig. 9.1). The VRN2 complex is required for vernalization, the process by which flowering is strictly licensed by an extended period of exposure to cold temperatures through progressive silencing of the master regulator gene *FLC*, accompanied by local H3K27 methylation [31]. EMF2 prevents premature flowering through a separate genetic circuit [32], while FIS2 ensures that the seed development program is repressed until fertilization has taken place [33]. Thus, in keeping with their cousins in metazoans, the plant PRC2 complexes are responsible for maintaining transcriptional states that underpin specific cellular behaviors. Importantly, it was recently demonstrated that silent and active *FLC* alleles can coexist within the same cells during vernalization, with their opposite states remaining stable through cell divisions [34]. These data indicate that the memory of gene silencing is stored in *cis*, which is directly consistent with the idea that this memory could in part reside in the PRC2-dependent H3K27 methylation mark.

The nematode worm *Caenorhabditis elegans* harbors a smaller PRC2 complex, consisting simply of the catalytic H3K27 methyltransferase MES-2, the EED/ESC ortholog MES-6, and the worm-specific subunit MES-3 [35,36] (Fig. 9.1). The best-studied function of PRC2 in *C. elegans* is in the development of a specific tissue, the germline, where its repressive activities principally target the X chromosomes [35]. In other tissues, PRC2 has been reported to play an important role in the regulation of cell plasticity [37]. Recent experiments showed that in embryos engineered to contain methylated H3K27 only on maternally inherited chromosomes, the restriction of the modification to these chromosomes persists throughout embryonic development [38]. Thus, like *FLC* gene silencing in vernalized plants, H3K27 methylation at the whole chromosome level in worms is preserved in *cis* through cell divisions. Together, these examples illustrate that despite some divergence in PRC2 complex composition and physiological roles, both its importance in defining cellular identity and the molecular mechanisms it employs to achieve this task are strongly conserved.

PRC2 and its role in transcriptional memory are classically associated with the needs of multicellular organisms with specialized cell types. But PRC2 is also found in unicellular organisms, suggesting that its inception as a chromatin regulator in fact preceded the appearance of multicellular life [39]. In the pathogenic yeast *Cryptococcus neoformans*, a PRC2 consisting of Ezh2, Eed1, the RbAp48 ortholog Msl1, and two novel subunits (Fig. 9.1) is responsible for H3K27 methylation and gene silencing at all subtelomeric regions of the genome [40]. Interestingly, the lineage-specific subunit Ccc1 contains a chromodomain that confers affinity for methylated H3K27. This suggests a parsimonious mechanism by which *C. neoformans* PRC2 may couple the "reading" and "writing" of its histone mark, thereby promoting its stable persistence at the subtelomeres [40]. In the single-celled green alga *Chlamydomonas reinhardtii*, which also possesses orthologs of PRC2 members, knockdown of the SET-domain protein EZH causes reactivation of transgenes and retrotransposons [39]. This observation led to the proposal that PRC2 first evolved in the last universal eukaryotic ancestor to protect the genome against foreign genetic material and was later repurposed with the emergence of multicellular organisms to mediate cell-type-specific gene silencing [39]. Supporting this view is the case of ciliates such as *Tetrahymena thermophila* and *Paramecium tetraurelia*, which must eliminate repetitive sequences from the macronuclear chromosomes after sexual conjugation to maintain genomic stability. This process requires EZH2 orthologs and their H3K27 methylation activity [41,42], consistent with the notion of an ancestral role for PRC2 in genome defense.

Although we must limit our discussion to these few examples, some common themes in the evolution of PRC2 are worth highlighting. The first is that although certain other PcG proteins are also evolutionarily ancient, PRC2 seems to represent a critical foundation that is universal to organisms endowed with a PcG system. This is illustrated by the existence of species containing PRC2 as their sole apparent PcG machinery, such as *C. neoformans*, with no known converse examples bearing other PcG components but lacking PRC2. Second, although all of the PRC2 core subunits are required for enzymatic activity in *Drosophila* and mammals [20,21], their functions have been bypassed or substituted in certain taxa, with SUZ12 notably missing from *C. elegans* and *C. neoformans*, and EED absent from ciliates [39]. How the corresponding EZH2 homologs are able to carry out H3K27 methylation without these binding partners remains unclear. These exceptions notwithstanding, in cases where EED and SUZ12 are both present the basis for their requirement in the HMTase reaction is now understood at the atomic level [43]. We devote the following section to a detailed discussion of this core PRC2 complex.

THE PRC2 CORE COMPLEX

Complex Architecture and Requirement of the Subunits for Enzymatic Activity

Methylation of H3K27 is the essential biological role of PRC2 [25]. The catalytic center of the complex resides in EZH2, but, as we have noted, its enzymatic activity strongly relies on the presence of the other core subunits, and at a minimum on EED and the conserved EZH2-interacting VEFS domain from the SUZ12 C-terminus [20,21,44]. A series of recent crystal structures has provided long-awaited insights into this dependence. First, structures of the SET domain of human EZH2 have shown that, on its own, this domain adopts an autoinhibitory state and cannot support catalysis. Indeed, in contrast to what is seen in active conformations of other methyltransferases, the substrate-binding channel is obstructed and the target lysine side chain therefore cannot enter the catalytic site [45,46]. Furthermore, the pocket that binds the methyl donor *S*-adenosyl methionine [45] forms only partially [45,46]. Thus, EZH2 lacks intrinsic activity because it simply fails to properly engage its substrate and the methyl source. How EZH2 gains catalytic activity upon assembly of PRC2 was recently revealed by two crystal structures, one active and one inactive, of a ternary Ezh2-Eed-Suz12(VEFS) complex from the thermophilic yeast *Chaetomium thermophilum* [43]. Specifically, the structures identify a SET activation loop within Ezh2, which is required for catalytic function and whose interactions with Eed and Suz12(VEFS) are critical for stabilizing an active conformation of the enzyme [43] (Fig. 9.2). In this active state, the substrate-binding channel becomes accessible and the SAM-binding pocket is fully formed, thereby allowing methyl transfer to lysine 27. These structures reveal precisely why EED and SUZ12 are essential for PRC2 activity, and thus constitute a major advance in the field.

EED and SUZ12 directly enable the catalysis of H3K27 methylation by EZH2, but the PRC2 core subunit RbAp48 is also necessary for full enzymatic activity [20,47,48]. Assembly of RbAp48 into an in vitro reconstituted mouse PRC2 strictly requires Suz12, even though Suz12 is dispensable for association of recombinant Ezh2 with Eed [48]. The same is true of the corresponding *Drosophila* subunits [47]. This suggests that RbAp48 interacts with PRC2 essentially or entirely through Suz12 (Fig. 9.2), a notion that is well supported both by mass spectrometry analysis of intermolecular cross-links in the human complex [49] and mutational and crystallographic analysis of the respective *Drosophila* orthologs Nurf55 and Su(Z)12 [44,50]. These approaches have shown that RbAp48 binds a conserved peptide in the SUZ12 N-terminus and probably makes additional contacts with other N-terminal portions of SUZ12 as well [49,50]. Despite this progress in mapping the association of RbAp48 with PRC2, the nature of its contribution to HMTase activity remains mysterious.

The overall structure of PRC2 has long been depicted as consisting of two spatially distinct entities surrounding EZH2: EED on one side and SUZ12 linked to RbAp48 on the other [47]. But in fact, the atomic-resolution crystal structures of the *C. thermophilum* ternary complex indicate instead that Suz12(VEFS) and Eed interact directly through a small but functionally critical interface [43]. Indeed, Suz12(VEFS) and Eed secure a catalytically

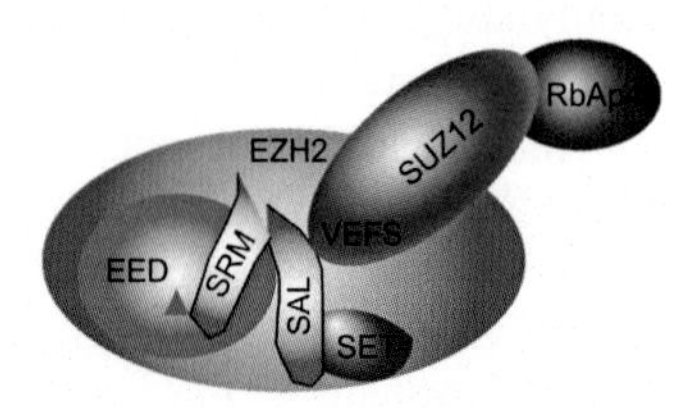

FIGURE 9.2 Architecture of the Polycomb repressive complex 2 core complex.

competent PRC2 conformation not only by each stabilizing the Ezh2 SET activation loop, but also through hydrophobic contacts with each other [43]. One of the amino acid residues involved was previously found to be important for HMTase activity of PRC2 in *Drosophila* [51], further suggesting that this interaction contributes to the proper arrangement of the catalytic site. Therefore, given what is now known, a reasonable provisional model for the architecture of the core PRC2 complex includes all combinations of pairwise interactions between EZH2, EED, and SUZ12, with RbAp48 associating principally and perhaps exclusively via SUZ12 (Fig. 9.2).

Control of PRC2 Activity by the Chromatin Context

Very early on it was noticed that PRC2 catalyzes H3K27 methylation more efficiently on arrays of nucleosomes than on mononucleosomes or histone H3 alone [22,24]. Consistent with this initial observation, a large body of work has since revealed the exquisite capacity of PRC2 to sense the molecular characteristics of the local chromatin environment and integrate this information to modulate its enzymatic output (Fig. 9.3) [52].

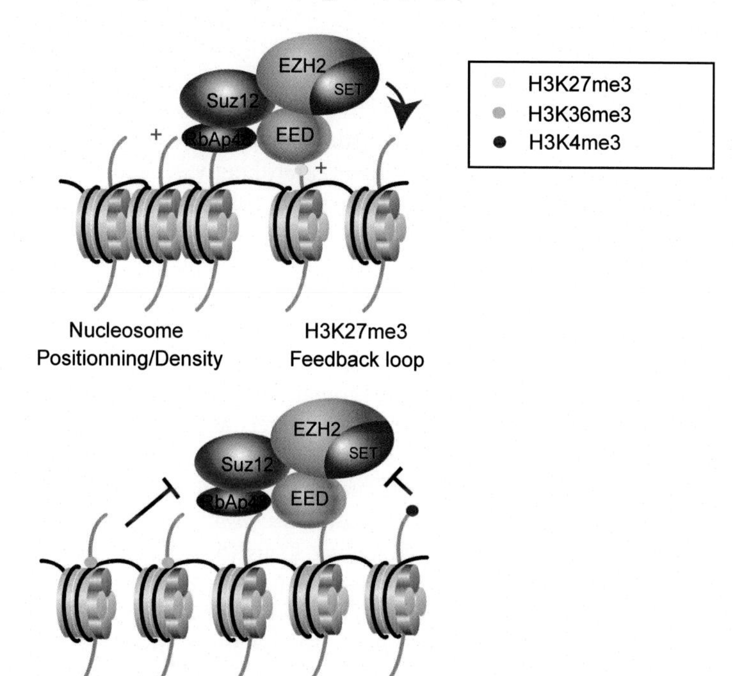

FIGURE 9.3 Polycomb repressive complex 2 enzymatic activity is regulated by the chromatin context.

A key illustration of this principle is the ability of PRC2 to recognize its own catalytic product in a manner that stimulates further catalysis. This feedback hinges on the EED subunit, whose WD40 repeats form an aromatic cage that binds to trimethylated lysines, including in particular H3K27me3 [53,54]. Speaking to the importance of this feature of EED, the affinity of PRC2 for nucleosome arrays is enhanced when the arrays bear H3K27me3, and this sensitivity is lost when the EED aromatic cage is mutated [53]. But in addition to strengthened nucleosome binding, engagement of H3K27me3 by EED triggers a substantial stimulation of PRC2 enzymatic activity (Fig. 9.3), and this also holds true for the *Drosophila* complex [53,54]. Separately, the N-terminal domain of the *Drosophila* EED ortholog ESC is also known to bind to histone H3 in a manner that is important for efficient PRC2 activity, but this interaction does not seem to involve the N-terminal tail of H3 [55]. In contrast, the aromatic cage allows EED to assess the modification status of the H3 tail, and if H3K27me3 is encountered, to signal its presence to the PRC2 catalytic center to significantly increase its activity. The molecular details of this response were illuminated by the recent crystal structures of PRC2 from *C. thermophilum.* In the structure of the active enzyme, an H3K27me3 peptide is nestled between the Eed WD40 repeats and a region of Ezh2 termed the stimulation-responsive motif (SRM). The presence of the EED-anchored histone peptide causes the SRM to undergo significant stabilization, thereby securing an optimal catalytic conformation through an allosteric effect [43].

From a biological perspective, the stimulation of the PRC2 HMTase by recognition of its own product represents a strikingly simple and effective mechanism for maintaining H3K27 methylation patterns through each round of cell division. As existing marks become diluted twofold during chromosome duplication, they can counteract this dilution by directly instructing the deposition of new marks on adjacent, newly deposited nucleosomes. This model is theoretically sufficient to underpin the faithful inheritance of H3K27me3. It is also supported by other experimental observations, notably that the expression of EZH2 is tightly coupled to the rate of cellular proliferation [56,57]. Yet in spite of this positive feedback, the propagation of H3K27me3 is likely not entirely self-driven. Suggesting that it also relies on features of the underlying DNA sequence, excision of genetic elements implicated in PRC2 recruitment in *Drosophila* results in a rapid decay of transcriptional silencing of the surrounding locus [58,59]. Nevertheless, the importance of product recognition for the function of PRC2 in transcriptional memory during development is underscored by the finding that *Drosophila* embryos carrying single-point mutations in the EED aromatic cage fully recapitulate PRC2-null phenotypes [53].

Aside from H3K27me3 itself, another property of chromatin that seems to promote PRC2 enzymatic activity is the spacing and density of nucleosomes. In an in vitro *Drosophila* PRC2 HMTase assay comparing different dinucleosomal substrates, it was found that reducing the length of linker DNA between the two nucleosome-positioning sites resulted in higher activity [44]. This suggested that the nearby presence of additional nucleosomes promotes catalysis, and indeed similar effects could be observed using alternative strategies predicted to increase the spatial proximity of nucleosomes, such as adding magnesium ions or the linker histone H1 [44,60]. Providing a histone H3 peptide consisting of residues 31 through 42 in *trans* could recapitulate the enzymatic stimulation and render PRC2 activity insensitive to nucleosome spacing, suggesting that this portion of the H3 tail mediates the effect of nucleosome concentration [44]. Finally, this activating peptide was found to interact directly with the SU(Z)12 subunit, whose E(Z)-associating VEFS domain appears to be responsible for conferring the response [44]. Altogether, these findings support a model in which densely arranged nucleosomes serve as a signal, interpreted by SUZ12, that promotes increased PRC2 activity (Fig. 9.3). This model is intriguing, because chromatin compaction has often been considered one of the possible results of PcG protein action that presumably contributes to transcriptional silencing [3,57,61,62], whereas the control of PRC2 activity by nucleosome density indicates that chromatin compaction might also act upstream of PRC2. It should be noted, however, that it remains unclear to what extent this mechanism operates in vivo.

The influence of the chromatin context on PRC2 is not limited to stimulatory effects. One example of a PRC2-inhibitory phenomenon involves lysine 4 of histone H3 (H3K4), a residue that is subjected to dynamic, transcription-coupled methylation events, especially at gene promoters. The H3K27-directed HMTase activity of PRC2 is sharply reduced when H3K4me3 is present on the same histone tail [50,63] (Fig. 9.3), suggesting a mechanism whereby actively transcribed areas resist the possible overwriting of their own histone modification patterns by the EED-driven positive feedback [50]. The discovery of this crosstalk came out of a structural analysis of the *Drosophila* RbAp48 ortholog Nurf55 in complex with an N-terminal peptide of histone H3, which revealed extensive contacts between a pocket of Nurf55 and the side chain of H3K4 [50]. Consistently, a complex consisting of Nurf55 and an SU(Z)12 N-terminal peptide showed an affinity for the H3 peptide that was significantly weakened by H3K4 methylation, hinting at a role for this submodule of PRC2 in sensing the H3K4 methylation state [50]. Yet, intriguingly, enzymatic inhibition of PRC2 by H3K4me3 occurs even in the absence of Nurf55; the apparently critical player instead seems to be the SU(Z) 12 C-terminal VEFS domain [50]. Thus, on the one hand the physiological relevance of Nurf55 binding to H3K4 is not yet understood, and on the other hand the molecular details of PRC2 inhibition by H3K4me3 remain obscure.

That said, other features of actively transcribed chromatin besides H3K4me3 also negatively modulate PRC2 activity, further supporting the idea that expressed loci may act as barriers to the propagation of H3K27 methylation. For instance, H3K36me2 and H3K36me3, which occur in the bodies of transcribed genes, both inhibit the HMTase activity of PRC2 in vitro [50,64].

As in the case of H3K4me3, H3K36me2/3 only inhibits methylation of H3K27 on the same histone tail and cannot exert its influence in *trans* [50,63] (Fig. 9.3). Consistent with these observations, a mass spectrometry study of natively isolated nucleosomes found that while the active modifications H3K4me3 and H3K36me2/3 are virtually never found on the same copy of histone H3 as the repressive H3K27me3, these two types of modifications are detected within the same nucleosome in a fraction of cases, indicating that they can be present on opposite copies of H3 [63]. The feedback between these marks may therefore have evolved to keep the self-reinforcing spreading of H3K27me3 in check, while at the same time accommodating the formation of nucleosomes that carry both types of marks. Such nucleosomes and the chromatin domains they inhabit are termed "bivalent" and have been proposed to be important for maintaining the plasticity of gene expression states in pluripotent and multipotent cells [65,66].

Lastly, methylation of H3K27 is blocked in a more straightforward manner through simple chemical incompatibility by H3K27 acetylation, another mark of active genes and gene enhancers. Competition between these modifications actually takes place in vivo, as evidenced by the global increase in H3K27 acetylation observed upon disruption of PRC2 [67,68]. Although H3K27 acetylation is much less abundant than H3K27 methylation at a global level, its prevalence at regulatory elements [69] suggests that this antagonism might have functional consequences. Further pointing to the relevance of this relationship, the H3K27 deacetylase NuRD and PRC2 seem to function in concert during the differentiation of embryonic stem (ES) cells to establish silencing at previously active loci [70,71].

Functional Roles of H3K27me3, H3K27me2, and H3K27me1

Methylated H3K27 in living cells is found in all three possible states (me1, me2, and me3 [63,72]). Whereas some H3K27me1 persists in the absence of PRC2 in mammals, both H3K27me2 and H3K27me3 fully depend on a functional complex [48,72,73]. Nevertheless, most studies have regarded H3K27me3 as the operationally relevant form for transcriptional repression, and H3K27me3 alone is routinely assayed as a surrogate for PRC2-mediated regulation.

The critical importance of H3K27me3, as opposed to H3K27me2, still awaits a formal demonstration, but several lines of evidence suggest that it is strictly necessary for silencing under many circumstances. For example, indirect disruption of the third PRC2 methylation step, leading to the accumulation of H3K27me2 at the expense of H3K27me3, resulted in derepression of classical homeotic gene targets in HeLa cells to an extent that was not further exacerbated by disruption of EZH2 itself [74]. Thus, for these genes, transcriptional repression only takes place when H3K27 is trimethylated. Similar effects were observed for homeotic genes in developing *Drosophila* embryos

[75], indicating a conserved requirement for H3K27me3 in silencing. The notion that H3K27me2 and H3K27me3 produce distinct regulatory outcomes is further suggested by the occurrence of distinct EZH2 point mutants in B-cell lymphomas that dramatically alter the global balance between the two states in favor of H3K27me3, in a manner that is consistent with their in vitro behavior [76,77].

The mechanistic significance of H3K27me3 for transcriptional repression, despite years of investigation, remains unclear. The most obvious reason H3K27me3 might be required is that, compared to lesser H3K27 methylation states, H3K27me3 confers a very high nucleosome-binding affinity to Pc homologs [27]. This can trigger the recruitment of PRC1, which would then ultimately be responsible for maintaining the silent state [28,29]. Although it captures a crucial part of the story, this hierarchical model of silencing by the PcG complexes does not entirely account for the role of H3K27me3. First, a mouse ES cell line lacking PRC1 was found to exhibit wild-type levels of H3K27me3 and Suz12 binding at a number of genes, and these genes became derepressed upon disruption of *Eed* [78]. This suggests that in a subset of cases, PRC2 mediates transcriptional repression that is associated with H3K27me3 but does not involve PRC1-dependent recognition of the mark. Second, as discussed earlier, PRC2 is present in species that apparently lack PRC1, such as *C. neoformans*, and in which the catalytic activity of the complex is necessary for its repressive function. To explain the role of H3K27me3 in gene silencing in these organisms, an alternative mode of action must also be invoked. In general, at least three nonmutually exclusive categories of H3K27me3-dependent, PRC1-independent mechanisms can be envisioned. One is that additional proteins besides Pc homologs are also involved in physically recognizing the H3K27me3 signal and executing or conveying the instruction to inhibit transcription. However, proteomic approaches aimed at identifying proteins specifically interacting with H3K27me3 nucleosomes have not revealed obvious candidates [79]. Conversely, it is also conceivable that the presence of H3K27me3 renders the chromatin structure refractory to binding of factors required for transcription. A third possibility is that PRC2 itself can mediate repression in a nonenzymatic fashion, and that in such PRC1-independent cases H3K27me3 is only essential for ensuring the stable binding of PRC2 to chromatin via the EED subunit.

If H3K27me3 has received the most attention by far, recent data have shed more light on the potential functions of H3K27me2 and H3K27me1. Rather than simply serving as intermediates in the formation of H3K27me3, these modification states seem to regulate the genome in distinct ways [72,80]. One clue regarding the role of H3K27me2 came from a genome-wide analysis of H3K27 acetylation in *Eed*-null mouse ES cells. Loss of PRC2 prompts a widespread redistribution of acetylated H3K27, but interestingly, the regions of ectopic H3K27 acetylation tend to be those normally marked with H3K27me2, rather than H3K27me3 [72]. This enrichment is ostensibly a consequence of the much greater abundance of H3K27me2, found in about 70% of all histone H3 in ES cells compared to only 10% for H3K27me3 [72]. Nevertheless, the majority of these novel H3K27 acetylation sites correspond to enhancer elements, leading the authors to conclude that H3K27me2 ordinarily acts to protect enhancers from inappropriate activation by blocking H3K27 acetylation [72].

In contrast, in cultured *Drosophila* cells, where H3K27me2 also marks 70% of total H3, loss of PRC2 leads to a global upregulation of transcription in both genic and intergenic regions [80]. With the exception of canonical PcG targets marked with abundant H3K27me3, the extent of transcriptional derepression is directly proportional to the wild-type levels of H3K27me2 at a given locus, arguing for a direct mode of action [80]. Although the data point to a link between H3K27me2 and the suppression of pervasive transcription, this interpretation is complicated by the observation that low levels of previously unsuspected H3K27me3 are also present within many of the H3K27me2-marked regions [80]. It is thus formally possible that the global transcriptional upregulation in PRC2-mutant cells is mainly due to the loss of H3K27me3. As discussed earlier, the functional distinction between H3K27me2 and H3K27me3 has not yet been definitively tested, and dissecting the relative roles of these methylation states remains a work in progress.

H3K27me1, on the other hand, is strongly associated with intragenic portions of actively transcribed genes [72,80]. Although global H3K27me1 in mammalian cells is only partially lost upon disruption of PRC2 function [48,72,73], its accumulation at individual loci is sharply reduced, and this is consistently accompanied by lower transcript levels [72]. Although it has not been determined whether these results represent a direct effect, they raise the intriguing possibility that PRC2 could positively regulate gene transcription via H3K27me1. Separately, identifying the enzymatic activity responsible for PRC2-independent H3K27me1 in mammalian cells remains an exciting challenge for the future.

The Alternative PRC2 Catalytic Subunit EZH1

Mammalian genomes encode two E(Z) paralogs: EZH2, which we have discussed exclusively until now, and EZH1, which also assembles into a PRC2 complex together with the other core subunits [57,81]. EZH1 differs from EZH2 in several important ways. First, the catalytic HMTase activity of EZH1, although detectable, is considerably weaker than that of EZH2. This can be observed in vitro in assays using reconstituted complexes [57,82], but it is also evident in cultured cells by comparative analysis of global H3K27me2/3 levels upon knockdown of the two enzymes [57], as well as in rescue experiments in which the loss of H3K27me3 upon Ezh2 deletion is complemented more effectively by overexpression of Ezh2 compared to Ezh1 [83]. Conversely, EZH1 possesses a chromatin compaction ability that EZH2 lacks. This property is detected in vitro by electron microscopy, with one PRC2-Ezh1 complex gathering up to four nucleosomes [57]. Of note, PRC2-Ezh1 displays a tighter affinity for chromatin [82], which may underlie this compaction activity [57]. Finally, in contrast to EZH2, whose expression closely tracks the rate of cellular proliferation and is downregulated in nonproliferating cells [56], EZH1 is present in a wide variety of tissues, both proliferating and nonproliferating [57].

Together, these observations gave rise to a model wherein the functional specialization of the two E(Z) paralogs serves to enable the maintenance of transcriptional repression in the face of particular constraints encountered at different developmental stages. Specifically, the robust HMTase activity of EZH2 is proposed to be required in undifferentiated cells to transmit H3K27me3 efficiently through frequent cell divisions, whereas the weak HMTase activity of EZH1 would be sufficient for maintenance of the mark in nonproliferative tissues. In addition to these complementary but distinct roles, there is also evidence that the two types of PRC2 complexes—containing EZH1 or EZH2—associate with one another [81] and can act in concert to control transcriptional repression. While mouse PRC2-Ezh2 alone interacts only weakly with mononucleosomes in vitro, its binding is vastly enhanced upon addition of PRC2-Ezh1 [82]. This suggests that PRC2-Ezh1, which exhibits strong mononucleosome binding of its own, can directly assist the chromatin association of its more enzymatically active counterpart [82]. Consistent with this idea, knockdown of Ezh1 in myoblasts leads to the displacement of Ezh2 from many target genes, whereas Ezh1 enrichment either remains intact or increases upon Ezh2 knockdown [82]. Therefore, EZH1 may promote H3K27 methylation at two levels: by depositing the mark itself and, under certain conditions, enabling EZH2 to access the nucleosome. Another implication of these observations is that cellular PRC2 might dimerize in some instances.

The relative importance of EZH1 and EZH2 for global gene expression patterns and proper establishment of cellular identity seems to vary according to tissue. Mice with a conditional disruption of Ezh2 in skin, for example, show loss of H3K27me3 in only a subset of cells and only during a short developmental window; in contrast, additional disruption of Ezh1 completely abolishes H3K27me3 and severely impairs the formation of hair follicles [84,85]. These data indicate that Ezh1 and Ezh2 play largely redundant roles in this context. In contrast, ES cells lacking Ezh2 exhibit dramatic defects in classical differentiation assays, arguing against redundancy with Ezh1 [81], while Ezh1 alone appears to be responsible for protecting aging hematopoietic stem cells from premature senescence via repression of Cdkn2a, even though Ezh2 is still present in these cells [86]. The factors that determine the respective influence of EZH1 and EZH2 in a given system are not well understood, but both overlapping and nonoverlapping functions are possible.

A series of recent findings points to a provocative noncanonical role for EZH1 in *activating* gene expression at the transcriptional level. In skeletal muscle formation, differentiation of myoblasts into myotubes involves transcriptional activation of the *MyoG* locus by the MyoD transcription factor, and the recruitment of MyoD has been reported to require EZH1 [87]. A genome-wide study extended these results, revealing that Ezh1 globally occupies actively transcribed loci in differentiating mouse myocytes, and that Ezh1 knockdown blocks the induction of myogenic genes and prevents myotube formation [88]. Curiously, however, proper differentiation in this context depends on EZH1 but not SUZ12 [87], arguing against a role for an EZH1-containing PRC2 complex. Indeed, unlike EZH2, EZH1 is proposed to be stable in the absence of the other members of the PRC2 core complex [89]. In erythroid progenitors, promoters bound by EZH1, either alone or in conjunction with SUZ12 (but not EED), correspond overwhelmingly to active genes, further suggesting that EZH1 may participate in gene activation in a manner that does not involve the full PRC2 [89].

The mechanism by which EZH1 activates gene expression, at least in myocyte differentiation, is proposed to involve transcription elongation [88]. How does one reconcile the clear in vitro and in vivo evidence that EZH1 mediates transcriptional repression with recent data supporting a diametrically opposite role? One can superficially attribute these discrepancies to differences in the cell types examined in different studies. An important clue for solving this riddle lies in the observation that knockdown of EZH2 in erythroid progenitors causes a widespread redistribution of EZH1 from active to silent loci [89]. Thus, one possibility is that while EZH1 has retained its evolutionarily ancient repressive function as part of a full PRC2, in situations where this function is carried out by the more efficient EZH2, EZH1 can perform its alternative, more recently evolved and more recently documented activating role. Teasing apart these contrasting activities, and the circumstances in which each is deployed, will constitute an interesting area of exploration in the coming years.

PRC2 COFACTORS

Remarkably, PRC2 maintains the heritable silencing of distinct sets of genes in different cell types, apparently through the same biochemical mechanism. The basis of this versatility is the subject of intense investigation. Part of the answer may lie in the contributions of facultative subunits or protein cofactors of the complex, which have been found to influence its genomic localization and enzymatic activity (Fig. 9.4). These cofactors are typically substoichiometric [90], so it is possible to think of PRC2 as a family of closely related complexes sharing a common set of core subunits. Although the best-characterized cofactors are AEBP2; the three PCL proteins PHF1, MTF2, and PHF19; and JARID2, additional PRC2 interactors have continued to be uncovered in recent years [91]. We discuss their known and proposed functions below.

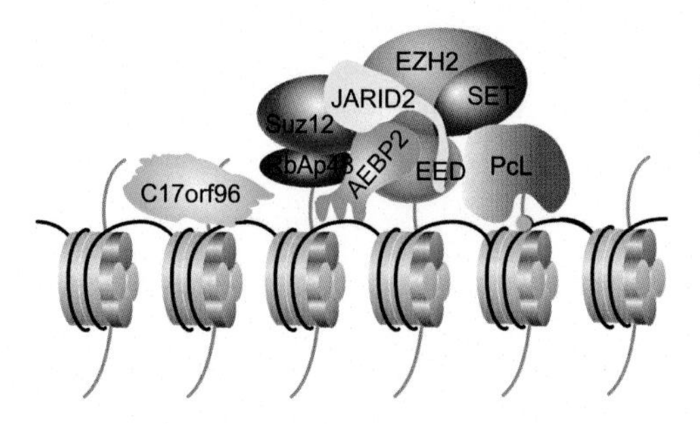

FIGURE 9.4 Polycomb repressive complex 2 cofactors.

AEBP2

AEBP2, a zinc-finger protein, was identified in one of the earliest PRC2 purifications from HeLa cells [22]. Two isoforms resulting from alternative splicing have been reported: an adult-specific 52-kDa species and an embryo-specific 32-kDa form which lacks the N-terminal unstructured acidic domain but retains all three zinc fingers [92]. It was recently suggested that the adult-specific variant might activate transcription, whereas the embryon-specific variant alone would work together with PRC2 [93], but that idea is undermined by the observation that AEBP2 also associates with PRC2 in adult cells [22]. The functional significance of this isoform diversity is thus not yet fully understood.

The C-terminal region of AEBP2 interacts extensively with SUZ12, but direct interactions with all of the other core subunits of PRC2 have also been detected [20,49]. In light of its zinc-finger domain, AEBP2 was hypothesized to contribute to the genomic targeting of PRC2 [20,22] (Fig. 9.4). While a ChIP analysis showed that Aebp2 and Suz12 occupy many of the same genomic loci in the mouse brain, Aebp2 binds to DNA with a sequence preference that is highly degenerate [92], making it difficult to deduce a mechanism by which it might recruit PRC2 to chromatin. Addressing this question directly will require an examination of PRC2 targeting in cells lacking functional AEBP2.

On the other hand, it is clear that AEBP2 enhances the HMTase activity of PRC2. Although this effect is much more pronounced when using oligonucleosomes as substrates rather than mononucleosomes [20,82], the magnitude of Aebp2-mediated stimulation rises if PRC2 and mononucleosomes are preincubated before the start of the reaction [82]. This observation seems to suggest that AEBP2 acts downstream of nucleosome binding to enhance PRC2 enzymatic activity, perhaps by stabilizing an optimal conformation of the complex. Consistent with this notion, AEBP2 was found to substantially stabilize PRC2 and even be required to render it suitable for study by electron microscopy [49]. Newer evidence indicates that AEBP2 does in fact play a part in nucleosome binding and recognition, and particularly in sensing the presence of PRC1-dependent monoubiquitination of histone H2A [94], discussed in more detail in the next part of this chapter.

PCL

PHF1, MTF2, and PHF19 are the three mammalian homologs of a classical PcG member called Polycomblike (Pcl) in *Drosophila*; hence, they are also respectively referred to as PCL1, PCL2, and PCL3. All three contain a series of putative histone-binding regions, namely one Tudor domain and two PHD fingers. Consequently, a great deal of interest has been focused on their potential role in mediating the association of PRC2 with chromatin and its recruitment to target genes. Interestingly, at the major target loci that have been examined in HeLa cells, PHF1 is not required for PRC2 occupancy and conversely its own chromatin localization does not depend on PRC2 [74]. Instead, it is necessary for efficient PRC2-mediated conversion of H3K27me2 into H3K27me3, which in turn is critical for gene silencing [74]. Thus, PHF1 seems to act by positively regulating the enzymatic activity of PRC2, downstream of its recruitment. These findings mirror the situation of *Drosophila* Pcl, whose loss also results in a specific reduction of H3K27me3 without significant changes in PRC2 occupancy [75]. In the case of PHF1, however, a role in PRC2 recruitment is difficult to exclude because of possible redundancy with MTF2 and PHF19.

MTF2 and PHF19 have been studied primarily in ES cells, and their influence on H3K27me3 patterns and transcriptional repression can be traced to a role in helping to direct PRC2 to a subset of its target loci [95–100]. These factors likely interact with chromatin by distinct mechanisms, as PHF19 requires an intact Tudor domain to promote H3K27me3 [95,96,98], whereas MTF2 binds to target genes even in the absence of this moiety, relying instead on the second PHD finger [97]. Mtf2 and Phf19 both appear to be functionally important for successful differentiation of mouse ES cells [95,96,100], yet they seem to regulate ES cell self-renewal in opposite ways, with Phf19 playing a supporting role [98] and Mtf2 an apparently restrictive role as suggested by the enhanced self-renewal phenotype observed upon Mtf2 knockdown [100]. Mtf2 has also been shown to mediate the recruitment of PRC2 to the inactive X chromosome, but Mtf2 disruption does not affect the overall establishment of X-inactivation [97], consistent with the dispensability of PRC2 for this process [101]. Finally, an intriguing report presented evidence that Mtf2 activates the senescence-promoting Cdk2na locus in mouse embryonic fibroblasts [102], but the underlying mechanism is not yet clear.

How are the PCL proteins targeted to particular sites in the genome and what are the implications for PRC2 function? Part of the answer was provided by a series of elegant and surprising papers showing that the Tudor domains of all three paralogs specifically recognize H3K36me3 (Fig. 9.4), a mark of actively transcribed genes [95,96,103,104]. At first glance this finding may seem paradoxical, but it can be rationalized by the hypothesis that the PCL proteins might serve to recruit PRC2 to expressed loci to initiate repression [95,96,103]. Consistent with this interpretation, ectopic overexpression of PHF1 in HeLa cells leads to Tudor domain–dependent spreading of PRC2 and H3K27me3 from silenced homeotic loci into neighboring expressed genes once marked with H3K36me3 [103]. Meanwhile, artificial tethering of PHF19 to a reporter gene in human embryonic kidney (HEK293) cells results in transcriptional repression and recruitment of PRC2 and the H3K36 demethylase NO66 [96], but it is worth noting that a significant association of NO66 with PHF19 was not observed in an independent study [95]. In any event, the H3K36me3-binding pocket seems to be functionally critical, as the widespread losses of H3K27me3 and PRC2 occupancy triggered by Phf19 knockdown in mouse ES cells are rescued only by PHF19 carrying an intact Tudor domain [95,96,103].

Nevertheless, some mysteries regarding PCL proteins and H3K36me3 still persist. One is that H3K36me3 recognition by PHF1 actually inhibits PRC2-mediated H3K27 methylation in human embryonic kidney cells (HEK293) and in vitro using yeast chromatin [104]. This suggests an alternative scenario, in which rather than helping PRC2 overcome existing transcriptional activity, PHF1 senses active loci and inhibits H3K27 methylation, similarly to the inhibition of catalysis by the PRC2 core complex in the presence of H3K36me3 on the same histone tail (discussed earlier). Another discrepancy is that although the Tudor domain of PHF19 binds only to methylated H3K36, not methylated H3K27, on a peptide array [95], it does bind to H3K27me3 in a pull-down assay [96]. Moreover, the PHF1 Tudor domain binds H3K27me3 peptides on histone variant H3t in vitro, and PHF1 colocalizes with H3K27me3, but not H3K36me3, in HEK293 cells [105]. Therefore, the histone-binding activities of the PCL proteins may vary according to the paralog and the cellular context, potentially leading to alternative functional consequences for PRC2. Further studies should provide a fuller picture of the determinants of PCL-dependent regulation.

JARID2

JARID2 (also called JUMONJI) was identified by several groups as a stable PRC2 interactor in ES cells, HEK293 cells, HeLa cells, and mouse thymus [99,106–109]. Although it contains a conserved JmjC histone demethylase domain, JARID2 most likely lacks enzymatic activity because the required catalytic residues are missing. The principal molecular function attributed to JARID2 so far is to mediate the recruitment of PRC2 to target genes, as disruption of Jarid2 in mouse ES cells leads to genome-wide defects in PRC2 occupancy [99,106–109]. A role in PRC2 recruitment is also strongly supported by the observation that artificial tethering of JARID2 to a reporter locus in HEK293 cells is sufficient to trigger de novo PRC2 binding, H3K27 methylation, and transcriptional repression [99,107,110]. But such a sequential mode of recruitment, with JARID2 preceding PRC2, is unlikely to operate in a general manner in vivo, as the occupancy of Jarid2 and PRC2 at natural targets in mouse ES cells is mutually dependent [99,107,109]. A notable exception concerns the targeting of PRC2 to the inactive X chromosome at the onset of X-inactivation in mice. In this case, Jarid2 is recruited first and independently, and its presence on the inactive X chromosome is necessary for subsequent binding of PRC2 [111].

The mechanisms by which JARID2 might contribute to PRC2 target identification are obscure. Direct binding of JARID2 to chromosome-associated noncoding RNAs through a dedicated N-terminal region has been proposed to contribute to the specificity of PRC2 localization [112]. Yet this RNA-binding region of the Jarid2 protein is dispensable for recruitment of Jarid2 to the inactive X chromosome, a process mediated by specific segments of the noncoding RNA Xist [111]. The generality of RNA-dependent targeting of JARID2, and PRC2 in general, is an area of intensive research and debate. Another suggestion is that JARID2 might recognize particular DNA sequence motifs [108]. Indeed, JARID2 possesses both an ARID domain and a C-terminal zinc finger, and a fragment encompassing these features exhibits in vitro DNA-binding activity [99]. However, a SELEX enrichment of JARID2-bound DNAs failed to reveal affinity for a specific sequence beyond a moderate preference for GC-rich species [99]. This is generally consistent with the sequence composition of PRC2-bound sites in the mammalian genome [113–115], but whether the DNA-binding properties of JARID2 are required for PRC2 targeting in vivo has not yet been tested.

Like some of the other PRC2 cofactors we have already mentioned, JARID2 influences the in vitro enzymatic activity of the complex. Initial studies concluded that JARID2 negatively affects HMTase activity [108,109], which was apparently concordant with the observation that mouse ES cells do not show a dramatic loss of H3K27me3 upon Jarid2 knockdown [99,106–110]. But steadily accumulating evidence has shown instead that Jarid2 in fact stimulates the activity of PRC2 under stoichiometric conditions, a notion that now reflects the consensus view in the field [82,99,110,116].

Jarid2 facilitates the in vitro association of PRC2 with nucleosomes, and this contributes to enzymatic stimulation but does not fully account for it [82]. Indeed, remarkably, Jarid2 possesses three distinct activities that impinge on its augmentation of PRC2 catalysis and that are separable in its primary amino acid sequence: nucleosome binding, PRC2 binding, and direct PRC2 stimulation [82]. Given the inherently greater affinity of PRC2-Ezh1 for nucleosomes, as compared to PRC-Ezh2 (see above), one might expect the contribution of Jarid2 to be more critical in the case of PRC2-Ezh2 [82]. In agreement with this, JARID2 associates predominantly with PRC2 complexes containing EZH2, rather than EZH1, in cells [106,107]. In parallel, Jarid2 also potentiates Aebp2-mediated enhancement of in vitro PRC2 activity by enabling nucleosome association [82], consistent with the idea that Aebp2 modulates PRC2 catalysis only downstream of nucleosome binding (see above). As expected in this model, Aebp2 on its own does not significantly increase the activity of PRC2 toward mononucleosomes, but it provides a synergistic enhancement when added together with Jarid2 [82].

Interestingly, Jarid2 also serves as a substrate of the EZH2 methyltransferase, as it is dimethylated and trimethylated in a PRC2-dependent manner both in vitro and in vivo [110]. Mutation of the target lysine (Jarid2-K116A) considerably reduces the ability of Jarid2 to elicit gene repression and H3K27me3 upon artificial tethering to a reporter locus, but not its ability to recruit Ezh2 [110]. These data implicate the methylation status of Jarid2 in its PRC2-stimulatory function. Remarkably, the same aromatic cage of Eed that recognizes H3K27me3 (Ref. [53]; see above) also recognizes Jarid2-K116me3 and is essential for Jarid2-K116me3-dependent stimulation of PRC2 activity [110] (Fig. 9.4). One possible advantage of this mechanism might arise in situations where a target locus does not yet carry H3K27me3 and thus does not autonomously activate PRC2. In these cases, it might be useful for the cell to subordinate the H3K27-directed activity of PRC2 to the specificity of JARID2-mediated recruitment to carefully control de novo establishment of the potentially self-propagating H3K27me3 mark [110]. There is much about JARID2 that remains puzzling, however, including how to make sense of its relatively minor role in governing genome-wide H3K27me3 in light of its substantial role in ensuring normal PRC2 occupancy, and how its recruitment to specific genomic targets is determined.

Recently Identified PRC2 Cofactors

Other recently identified PRC2-associated proteins have also been proposed to regulate the complex at the level of its enzymatic activity or its targeting to chromatin, but our understanding of their functions is even more rudimentary. A recurring interactor is C17orf96, a poorly conserved protein lacking obvious functional domains. The first hint of its connection to PcG-mediated silencing came from a study in which it was noticed among HeLa nuclear proteins exhibiting affinity for H3K27me3-bearing nucleosomes [79], and its physical association with PRC2 in ES cells was discovered not long afterward [116]. Interestingly, exogenous C17orf96 was found to enhance transcription factor—induced reprogramming of fibroblasts into pluripotent stem cells [116], and its expression has been reported to decline sharply upon ES cell differentiation [117,118]. But it is also found in the mouse brain, and it has copurified with PRC2 in nonpluripotent cell lines such as HeLa and HEK293 [79,90,118,119].

Remarkably, quantitative mass spectrometry analysis indicates that the stoichiometry of C17orf96 in PRC2 complexes in HeLa cells is higher than that of AEBP2, the PCL proteins or JARID2, and is surpassed only by the core subunits themselves [90]. This suggests that C17orf96 may play a significant role in controlling PRC2 function, but only a few details have emerged so far. Binding studies using recombinant proteins have mapped at least one interaction with PRC2, between the C-terminus of C17orf96 and the VEFS domain of SUZ12 [118]. *In vitro*, C17orf96 appears to enhance the catalytic activity of PRC2 and, interestingly, this effect is additive with that of JARID2, suggesting that C17orf96 operates by a distinct, nonredundant mechanism [116]. Its genome-wide chromatin binding in HEK293 cells, as measured by ChIP, is identical to that of EZH2 and the H3K27me3 [119], but its contribution to PRC2 recruitment has been difficult to assess because of the lack of a C17orf96 knockout model. Intriguingly, however, C17orf96 knockdown in mouse ES cells leads to *increased* Suz12 occupancy at the genes that were tested and increased levels of H3K27me3 at PRC2 targets genome-wide [118]. This result suggests that despite their matching chromatin association profiles, C17orf96 may act to limit the binding of PRC2 to target genes. Understanding the basis for this unexpected interaction, and how it relates to the cofactor's positive regulation of PRC2 enzymatic activity, will be important questions for future analysis.

Another recently uncovered PRC2-interacting protein is C10orf12. It is present at a much lower stoichiometry in PRC2 purifications than C17orf96 and other known cofactors [90], and it is found at fewer than 50% of genomic EZH2 binding sites in HEK293 cells, compared to 80% for C17orf96 [119]. Its role in PRC2 function remains entirely unknown. It is also tempting to speculate that other PRC2 cofactors, whose expression could be tissue-specific, may have escaped detection and still await discovery.

PRC2 WITHIN THE POLYCOMB MACHINERY

In the extensively studied *Drosophila* and mammalian systems, PRC2 represents only one among several Polycomb complexes whose functions are critical for the establishment and maintenance of cellular identities. As discussed above, a longstanding model has proposed that PRC1 is the effector that directly enforces cell type–specific transcriptional silencing patterns. This hypothesis rests largely on the idea that PRC1 alters chromatin structure in a manner that impedes the process of transcription, either through a chromatin compaction activity [3] or through monoubiquitination of histone H2A on lysine 119 (H2A119Ub) [120,121], which is thought to act as a signal that inhibits RNA polymerase elongation [122]. In this paradigm, the main biological function attributed to H3K27 methylation by PRC2 is to target PRC1 to chromatin, via the high-affinity H3K27me3-binding domain found in the Pc subunit and its homologs [28,29].

A series of recent studies has challenged this hierarchical view of PRC1 and PRC2 function. First, novel variants of PRC1 that lack chromodomain-containing subunits were isolated from mouse ES cells. Importantly, these complexes were shown to ubiquitinate histone H2A independently of H3K27me3, and their defining subunit RYBP was found to be necessary for normal ES cell differentiation [123–125]. These findings demonstrate that PRC1 does not systematically require PRC2 to localize to critical target genes, as previously documented in the specific case of X-inactivation [126]. Conversely, recent evidence has revealed that genome-wide PRC2 recruitment in ES cells is largely dependent on PRC1 [127]. Artificial tethering experiments suggest that the role of PRC1 is direct, since forced recruitment of PRC1 leads to corecruitment of PRC2 and establishment of H3K27me3 [127,128]. Interestingly, PRC2 corecruitment in these cases strictly requires the H2AK119 ubiquitination activity of PRC1 [127,128]. Further pointing to a direct mode of interaction, a biochemical purification of *Drosophila* and mouse nuclear proteins showing affinity for H2AK119Ub-modified oligonucleosomes recovered PRC2 core components and cofactors [94]. Among these, the most highly enriched were AEBP2 and JARID2, and AEBP2 in particular was found to provide H2A119Ub-dependent enhancement of PRC2 HMTase activity in vitro [94].

Collectively, these data support the existence of an alternate hierarchy for the recruitment of PcG proteins to chromatin, in which it is the catalytic product of PRC1 that serves to target PRC2, possibly even serving to stimulate its enzymatic activity. Intriguingly, these two mechanisms do not seem to form a cycle, as artificial tethering of EED to chromatin triggers the recruitment of chromodomain-containing PRC1 complexes, but not H2AK119Ub [127], and thus no H2AK119Ub-mediated reinforcement of PRC2 occupancy. In addition, recent reports in both *Drosophila* and mouse intestine showed that the loss of H2AK119Ub has no substantial effect on H3K27me3 enrichment at a global level, suggesting either the existence of redundant mechanisms for PRC2 recruitment or that the role of PRC1 for this process is context-dependent [129,130]. It is also worth noting that mutation of H2AK118, the ubiquitinated residue in flies, produces much milder developmental defects than mutation of PRC1 [130], while in mouse PRC1 ubiquitination activity is not necessary for early development, whereas PRC1 itself is strictly required [131]. These findings indicate that PRC1 carries out additional tasks, besides histone H2A ubiquitination and consequent PRC2 recruitment, which are essential for its gene silencing function. The molecular activities of PRC1 and PRC2 that underlie the precise execution of global transcriptional silencing programs are complex and intertwined, and sorting out their relative contributions remains a considerable research challenge. Given currently available information, however, we are tempted to infer that PRC2-mediated H3K27me3 constitutes the ancestral underpinning of PcG-dependent repression, while PRC1 evolved both to help generate and help interpret this signal.

In addition to the crosstalk between PRC1 and PRC2 occurring at the level of posttranslational modifications to histones, several reports have also suggested a more direct interaction. In early studies of the PcG machinery in *Drosophila* development, a consensus emerged that the ESC and E(Z) formed a complex separate from the Pc-containing complex that became known as PRC1, but the analyses that led to that conclusion consistently relied on day-old embryos or subsequent developmental stages. A study of 2-h-old embryos revealed an apparently direct association between ESC and Pc [132]. This was proposed to represent an important communication step to allow concerted establishment of silencing by the two complexes [132]. More recently, a novel biochemical technique for purifying chromatin-associated proteins uncovered Scm, previously considered a PRC1 subunit, as a strongly enriched member of PRC2 [133]. Furthermore, Scm was found to interact with reconstituted PRC2 in vitro [133]. Interestingly, it was also observed that an overexpressed Scm fragment consisting of the sterile alpha motif is able to spread autonomously over large portions of chromosomes [133]. Together, these observations raise the possibility that Scm might coordinate PRC1 and PRC2 activities and promote their spreading across the larger PcG-regulated domain characteristic of the *Drosophila* genome [133], a hypothesis to be tested in future studies. It is not clear, however, whether Scm is involved in the early embryonic ESC-Pc interaction.

Finally, there is intriguing recent evidence that the PRC2 subunit EED forms PRC2-independent complexes with PRC1 subunits in prostate carcinoma cells [134]. Remarkably, in vitro binding assays suggest that this interaction is direct and that it can competitively inhibit EED-EZH2 binding and PRC2 HMTase activity [134]. Conversely, the in vitro H2AK119 ubiquitination activity of PRC1 is enhanced upon addition of EED, particularly on H3K27me3 nucleosomes [134], but it is unknown whether this requires the aromatic cage. This last result appears to conflict with the failure of ectopically tethered EED to induce H2AK119Ub in ES cells [127], and it will be important to further probe this unexpected interaction. Altogether, PRC1 and PRC2 communicate through a highly elaborate assortment of direct and indirect channels, and the effort to understand these mutual influences remains ongoing.

CONCLUDING REMARKS: ON THE DETERMINISTIC OR RESPONSIVE ROLE OF PRC2 IN TRANSCRIPTIONAL REGULATION

In this chapter, we have reviewed the discovery and evolutionary conservation of PRC2, its biochemical activity and the mechanisms by which it is controlled, and its physical and functional interactions with a host of protein cofactors and the rest of the PcG machinery. What has emerged, despite the many residual unknowns we have endeavored to highlight, is a picture of a histone-modifying enzyme whose output is highly regulated by the features of its chromatin substrate and its interacting partners.

We close our discussion by reflecting on the function of PRC2 at the level of the whole organism and its molecular implications. PRC2 acts in very disparate cell types to enforce the repression distinct sets of target genes. Its ability to recognize these targets with precision in each cell type is critical for the maintenance of cellular identities and the proper development and fitness of the organism. Yet its targeting is arguably the most enigmatic aspect of its function. As discussed above, the known cofactors of PRC2 may play some role in target determination, but their known properties cannot account for the diversity of cell type—specific outcomes. In general, two kinds of nonmutually exclusive models can be envisioned to explain PRC2 targeting, as proposed by Robert Klose and coauthors [135]. One is an instructive model, in which *trans*-acting factors such as site-specific DNA binding proteins or noncoding RNA molecules dictate target selection in each cell type by interacting with PRC2. The other is a responsive model, in which PRC2 senses particular features of the chromatin at target loci and undergoes stable recruitment in response to these chromatin states, without interacting directly with *trans*-acting factors [135]. In the first model, PRC2 targeting constitutes an initiating event in transcriptional repression; in the second model, it occurs as a consequence of repression and serves to consolidate the silent state.

Both theoretical and empirical arguments suggest that PRC2 plays mainly a responsive role in transcriptional gene silencing and development. First, it is intuitively difficult to suppose that enough dedicated PRC2-interacting factors exist to achieve the observed cell type—specificity of PRC2-mediated repression patterns [135]. Second, early studies of the developmental phenotypes of *Drosophila* PcG mutants showed that PcG targets are initially silenced by distinct repressors before their regulation becomes PcG-dependent, suggesting that PRC2 responds to a preexisting silent configuration [2]. Analogously, Suz12-deficient mouse ES cells undergoing in vitro differentiation are able to globally initiate transcriptional silencing of PRC2 target genes, with abnormal expression arising only after several days [136]. This again suggests that PRC2 acts downstream of the establishment of repression, and that its recruitment depends rather on the transcriptional status than on instructing factors. Further support for this notion comes from the observation that blocking transcription pharmacologically causes ectopic PRC2 recruitment to newly silenced loci genome-wide [136]. Although secondary effects of the transcriptional inhibition are difficult to rule out, this result is nevertheless consistent with the expression state of a gene being sufficient to determine PRC2 targeting.

PRC2 can therefore be thought of as a singular type of transcriptional repressor, whose role is not primarily to execute gene-silencing decisions, but rather to reinforce and preserve them. This activity is crucial for maintaining a memory of cellular identity within the diverse specialized cell types of an organism throughout development and beyond. Fittingly, with an intrinsic ability to propagate its own enzymatic product and ensure its heritability, PRC2 possesses molecular properties ideally suited to its biological function.

LIST OF ACRONYMS AND ABBREVIATIONS

ChIP Chromatin immunoprecipitation
ES cells Embryonic stem cells
H2AK119Ub Histone H2A monoubiquitinated on lysine 119
H3K27 (or 4/9/36) Histone H3 lysine 27 (or 4/9/36)
H3K27me3 (or 1/2) Histone H3 trimethylated (or mono-/dimethylated) on lysine 27
HMTase Histone methyltransferase
PcG Polycomb group
PRC2 (or 1) Polycomb repressive complex 2 (or 1)
SAM *S*-adenosyl methionine

ACKNOWLEDGMENTS

The R.M. laboratory is supported by the European Research Council (ERC-Stg, REPODDID). D.H. is supported by a postdoctoral fellowship from the Fondation pour la Recherche Médicale (SPF20150934266).

REFERENCES

[1] Jürgens G. A group of genes controlling the spatial expression of the bithorax complex in *Drosophila.* Nature July 11, 1985;316:153–5.

[2] Simon J, Chiang A, Bender W. Ten different Polycomb group genes are required for spatial control of the abdA and AbdB homeotic products. Development 1992;114(2):493–505.

[3] Francis NJ, Kingston RE, Woodcock CL. Chromatin compaction by a polycomb group protein complex. Science 2004;306(5701):1574–7.

[4] Franke A, DeCamillis M, Zink D, Cheng N, Brock HW, Paro R. Polycomb and polyhomeotic are constituents of a multimeric protein complex in chromatin of *Drosophila melanogaster.* EMBO J 1992;11(8):2941–50.

[5] Shao Z, Raible F, Mollaaghababa R, Guyon JR, Wu CT, Bender W, et al. Stabilization of chromatin structure by PRC1, a Polycomb complex. Cell 1999;98(1):37–46.

[6] Denisenko O, Shnyreva M, Suzuki H, Bomsztyk K. Point mutations in the WD40 domain of Eed block its interaction with Ezh2. Mol Cell Biol 1998;18(10):5634–42.

[7] Jones CA, Ng J, Peterson AJ, Morgan K, Simon J, Jones RS. The *Drosophila* esc and E(z) proteins are direct partners in polycomb group-mediated repression. Mol Cell Biol 1998;18(5):2825–34.

[8] Sewalt RG, van der Vlag J, Gunster MJ, Hamer KM, den Blaauwen JL, Satijn DP, et al. Characterization of interactions between the mammalian polycomb-group proteins Enx1/EZH2 and EED suggests the existence of different mammalian polycomb-group protein complexes. Mol Cell Biol 1998;18(6):3586–95.

[9] Tie F, Furuyama T, Harte PJ. The *Drosophila* Polycomb Group proteins ESC and E(Z) bind directly to each other and co-localize at multiple chromosomal sites. Development 1998;125(17):3483–96.

[10] van Lohuizen M, Tijms M, Voncken JW, Schumacher A, Magnuson T, Wientjens E. Interaction of mouse polycomb-group (Pc-G) proteins Enx1 and Enx2 with Eed: indication for separate Pc-G complexes. Mol Cell Biol 1998;18(6):3572–9.

[11] Ng J, Hart CM, Morgan K, Simon JA. A *Drosophila* ESC-E(Z) protein complex is distinct from other polycomb group complexes and contains covalently modified ESC. Mol Cell Biol 2000;20(9):3069–78.

[12] Tie F, Furuyama T, Prasad-Sinha J, Jane E, Harte PJ. The *Drosophila* Polycomb Group proteins ESC and E(Z) are present in a complex containing the histone-binding protein p55 and the histone deacetylase RPD3. Development 2001;128(2):275–86.

[13] Jones RS, Gelbart WM. The *Drosophila* Polycomb-group gene Enhancer of zeste contains a region with sequence similarity to trithorax. Mol Cell Biol 1993;13(10):6357–66.

[14] Tschiersch B, Hofmann A, Krauss V, Dorn R, Korge G, Reuter G. The protein encoded by the *Drosophila* position-effect variegation suppressor gene Su(var)3-9 combines domains of antagonistic regulators of homeotic gene complexes. EMBO J 1994;13(16):3822–31.

[15] Jenuwein T, Laible G, Dorn R, Reuter G. SET domain proteins modulate chromatin domains in eu- and heterochromatin. Cell Mol Life Sci 1998;54(1):80–93.

[16] Rea S, Eisenhaber F, O'Carroll D, Strahl BD, Sun ZW, Schmid M, et al. Regulation of chromatin structure by site-specific histone H3 methyltransferases. Nature 2000; 406(6796):593–9.

[17] Bannister AJ, Zegerman P, Partridge JF, Miska EA, Thomas JO, Allshire RC, et al. Selective recognition of methylated lysine 9 on histone H3 by the HP1 chromo domain. Nature 2001;410(6824):120–4.

[18] Lachner M, O'Carroll D, Rea S, Mechtler K, Jenuwein T. Methylation of histone H3 lysine 9 creates a binding site for HP1 proteins. Nature 2001;410(6824):116–20.

[19] Rastelli L, Chan CS, Pirrotta V. Related chromosome binding sites for zeste, suppressors of zeste and Polycomb group proteins in *Drosophila* and their dependence on Enhancer of zeste function. EMBO J 1993;12(4):1513–22.

[20] Cao R, Zhang Y. SUZ12 is required for both the histone methyltransferase activity and the silencing function of the EED-EZH2 complex. Mol Cell 2004;15(1):57–67.

[21] Czermin B, Melfi R, McCabe D, Seitz V, Imhof A, Pirrotta V. *Drosophila* enhancer of Zeste/ESC complexes have a histone H3 methyltransferase activity that marks chromosomal Polycomb sites. Cell 2002;111(2):185–96.

[22] Cao R, Wang L, Wang H, Xia L, Erdjument-Bromage H, Tempst P, et al. Role of histone H3 lysine 27 methylation in Polycomb-group silencing. Science 2002;298(5595): 1039–43.

[23] Kuzmichev A, Nishioka K, Erdjument-Bromage H, Tempst P, Reinberg D. Histone methyltransferase activity associated with a human multiprotein complex containing the Enhancer of Zeste protein. Genes Dev 2002;16(22):2893–905.

[24] Muller J, Hart CM, Francis NJ, Vargas ML, Sengupta A, Wild B, et al. Histone methyltransferase activity of a *Drosophila* Polycomb group repressor complex. Cell 2002;111(2):197–208.

[25] Pengelly AR, Copur O, Jackle H, Herzig A, Muller J. A histone mutant reproduces the phenotype caused by loss of histone-modifying factor Polycomb. Science 2013;339(6120): 698–9.

[26] Min J, Zhang Y, Xu RM. Structural basis for specific binding of Polycomb chromodomain to histone H3 methylated at Lys 27. Genes Dev 2003;17(15):1823–8.

[27] Fischle W, Wang Y, Jacobs SA, Kim Y, Allis CD, Khorasanizadeh S. Molecular basis for the discrimination of repressive methyl-lysine marks in histone H3 by Polycomb and HP1 chromodomains. Genes Dev 2003;17(15):1870–81.

[28] Cao R, Zhang Y. The functions of E(Z)/EZH2-mediated methylation of lysine 27 in histone H3. Curr Opin Genet Dev 2004;14(2):155–64.
[29] Wang L, Brown JL, Cao R, Zhang Y, Kassis JA, Jones RS. Hierarchical recruitment of polycomb group silencing complexes. Mol Cell 2004;14(5):637–46.
[30] Derkacheva M, Hennig L. Variations on a theme: Polycomb group proteins in plants. J Exp Bot 2014;65(10):2769–84.
[31] De Lucia F, Crevillen P, Jones AM, Greb T, Dean C. A PHD-polycomb repressive complex 2 triggers the epigenetic silencing of FLC during vernalization. Proc Natl Acad Sci USA 2008;105(44):16831–6.
[32] Chanvivattana Y, Bishopp A, Schubert D, Stock C, Moon YH, Sung ZR, et al. Interaction of Polycomb-group proteins controlling flowering in Arabidopsis. Development 2004;131(21):5263–76.
[33] Wang D, Tyson MD, Jackson SS, Yadegari R. Partially redundant functions of two SET-domain polycomb-group proteins in controlling initiation of seed development in Arabidopsis. Proc Natl Acad Sci USA 2006;103(35):13244–9.
[34] Berry S, Hartley M, Olsson TS, Dean C, Howard M. Local chromatin environment of a Polycomb target gene instructs its own epigenetic inheritance. eLife 2015:4.
[35] Bender LB, Cao R, Zhang Y, Strome S. The MES-2/MES-3/MES-6 complex and regulation of histone H3 methylation in *C. elegans*. Curr Biol 2004;14(18):1639–43.
[36] Xu L, Fong Y, Strome S. The *Caenorhabditis elegans* maternal-effect sterile proteins, MES-2, MES-3, and MES-6, are associated in a complex in embryos. Proc Natl Acad Sci USA 2001;98(9):5061–6.
[37] Yuzyuk T, Fakhouri TH, Kiefer J, Mango SE. The polycomb complex protein mes-2/E(z) promotes the transition from developmental plasticity to differentiation in *C. elegans* embryos. Dev Cell 2009;16(5):699–710.
[38] Gaydos LJ, Wang W, Strome S. Gene repression. H3K27me and PRC2 transmit a memory of repression across generations and during development. Science 2014;345(6203):1515–8.
[39] Shaver S, Casas-Mollano JA, Cerny RL, Cerutti H. Origin of the polycomb repressive complex 2 and gene silencing by an E(z) homolog in the unicellular alga Chlamydomonas. Epigenetics 2010;5(4).
[40] Dumesic PA, Homer CM, Moresco JJ, Pack LR, Shanle EK, Coyle SM, et al. Product binding enforces the genomic specificity of a yeast polycomb repressive complex. Cell 2015;160(1–2):204–18.
[41] Lhuillier-Akakpo M, Frapporti A, Denby Wilkes C, Matelot M, Vervoort M, Sperling L, et al. Local effect of enhancer of zeste-like reveals cooperation of epigenetic and cis-acting determinants for zygotic genome rearrangements. PLoS Genet 2014;10(9):e1004665.
[42] Liu Y, Taverna SD, Muratore TL, Shabanowitz J, Hunt DF, Allis CD. RNAi-dependent H3K27 methylation is required for heterochromatin formation and DNA elimination in Tetrahymena. Genes Dev 2007;21(12):1530–45.
[43] Jiao L, Liu X. Structural basis of histone H3K27 trimethylation by an active polycomb repressive complex 2. Science 2015;350(6258):aac4383.
[44] Yuan W, Wu T, Fu H, Dai C, Wu H, Liu N, et al. Dense chromatin activates Polycomb repressive complex 2 to regulate H3 lysine 27 methylation. Science 2012;337(6097):971–5.
[45] Antonysamy S, Condon B, Druzina Z, Bonanno JB, Gheyi T, Zhang F, et al. Structural context of disease-associated mutations and putative mechanism of autoinhibition revealed by X-ray crystallographic analysis of the EZH2-SET domain. PLoS One 2013;8(12):e84147.

[46] Wu H, Zeng H, Dong A, Li F, He H, Senisterra G, et al. Structure of the catalytic domain of EZH2 reveals conformational plasticity in cofactor and substrate binding sites and explains oncogenic mutations. PLoS One 2013;8(12):e83737.

[47] Ketel CS, Andersen EF, Vargas ML, Suh J, Strome S, Simon JA. Subunit contributions to histone methyltransferase activities of fly and worm polycomb group complexes. Mol Cell Biol 2005;25(16):6857−68.

[48] Pasini D, Bracken AP, Jensen MR, Lazzerini Denchi E, Helin K. Suz12 is essential for mouse development and for EZH2 histone methyltransferase activity. EMBO J 2004;23(20):4061−71.

[49] Ciferri C, Lander GC, Maiolica A, Herzog F, Aebersold R, Nogales E. Molecular architecture of human polycomb repressive complex 2. eLife 2012;1. e00005.

[50] Schmitges FW, Prusty AB, Faty M, Stutzer A, Lingaraju GM, Aiwazian J, et al. Histone methylation by PRC2 is inhibited by active chromatin marks. Mol Cell 2011;42(3):330−41.

[51] Rai AN, Vargas ML, Wang L, Andersen EF, Miller EL, Simon JA. Elements of the polycomb repressor SU(Z)12 needed for histone H3-K27 methylation, the interface with E(Z), and in vivo function. Mol Cell Biol 2013;33(24):4844−56.

[52] O'Meara MM, Simon JA. Inner workings and regulatory inputs that control Polycomb repressive complex 2. Chromosoma 2012;121(3):221−34.

[53] Margueron R, Justin N, Ohno K, Sharpe ML, Son J, Drury 3rd WJ, et al. Role of the polycomb protein EED in the propagation of repressive histone marks. Nature 2009; 461(7265):762−7.

[54] Xu C, Bian C, Yang W, Galka M, Ouyang H, Chen C, et al. Binding of different histone marks differentially regulates the activity and specificity of polycomb repressive complex 2 (PRC2). Proc Natl Acad Sci USA 2010;107(45):19266−71.

[55] Tie F, Stratton CA, Kurzhals RL, Harte PJ. The N terminus of *Drosophila* ESC binds directly to histone H3 and is required for E(Z)-dependent trimethylation of H3 lysine 27. Mol Cell Biol 2007;27(6):2014−26.

[56] Bracken AP, Pasini D, Capra M, Prosperini E, Colli E, Helin K. EZH2 is downstream of the pRB-E2F pathway, essential for proliferation and amplified in cancer. EMBO J 2003;22(20):5323−35.

[57] Margueron R, Li G, Sarma K, Blais A, Zavadil J, Woodcock CL, et al. Ezh1 and Ezh2 maintain repressive chromatin through different mechanisms. Mol Cell 2008;32(4): 503−18.

[58] Busturia A, Wightman CD, Sakonju S. A silencer is required for maintenance of transcriptional repression throughout *Drosophila* development. Development 1997;124(21): 4343−50.

[59] Sengupta AK, Kuhrs A, Muller J. General transcriptional silencing by a Polycomb response element in *Drosophila*. Development 2004;131(9):1959−65.

[60] Martin C, Cao R, Zhang Y. Substrate preferences of the EZH2 histone methyltransferase complex. J Biol Chem 2006;281(13):8365−70.

[61] Boettiger AN, Bintu B, Moffitt JR, Wang S, Beliveau BJ, Fudenberg G, et al. Super-resolution imaging reveals distinct chromatin folding for different epigenetic states. Nature 2016;529(7586):418−22.

[62] Isono K, Endo TA, Ku M, Yamada D, Suzuki R, Sharif J, et al. SAM domain polymerization links subnuclear clustering of PRC1 to gene silencing. Dev Cell 2013;26(6): 565−77.

[63] Voigt P, LeRoy G, Drury 3rd WJ, Zee BM, Son J, Beck DB, et al. Asymmetrically modified nucleosomes. Cell 2012;151(1):181–93.
[64] Yuan W, Xu M, Huang C, Liu N, Chen S, Zhu B. H3K36 methylation antagonizes PRC2-mediated H3K27 methylation. J Biol Chem 2011;286(10):7983–9.
[65] Bernstein BE, Mikkelsen TS, Xie X, Kamal M, Huebert DJ, Cuff J, et al. A bivalent chromatin structure marks key developmental genes in embryonic stem cells. Cell 2006;125(2):315–26.
[66] Voigt P, Tee WW, Reinberg D. A double take on bivalent promoters. Genes Dev 2013;27(12):1318–38.
[67] Pasini D, Malatesta M, Jung HR, Walfridsson J, Willer A, Olsson L, et al. Characterization of an antagonistic switch between histone H3 lysine 27 methylation and acetylation in the transcriptional regulation of Polycomb group target genes. Nucleic Acids Res 2010;38(15):4958–69.
[68] Tie F, Banerjee R, Stratton CA, Prasad-Sinha J, Stepanik V, Zlobin A, et al. CBP-mediated acetylation of histone H3 lysine 27 antagonizes *Drosophila* Polycomb silencing. Development 2009;136(18):3131–41.
[69] Heintzman ND, Ren B. Finding distal regulatory elements in the human genome. Curr Opin Genet Dev 2009;19(6):541–9.
[70] Kim TW, Kang BH, Jang H, Kwak S, Shin J, Kim H, et al. Ctbp2 modulates NuRD-mediated deacetylation of H3K27 and facilitates PRC2-mediated H3K27me3 in active embryonic stem cell genes during exit from pluripotency. Stem Cells 2015;33(8):2442–55.
[71] Reynolds N, Salmon-Divon M, Dvinge H, Hynes-Allen A, Balasooriya G, Leaford D, et al. NuRD-mediated deacetylation of H3K27 facilitates recruitment of Polycomb Repressive Complex 2 to direct gene repression. EMBO J 2012;31(3):593–605.
[72] Ferrari KJ, Scelfo A, Jammula S, Cuomo A, Barozzi I, Stutzer A, et al. Polycomb-dependent H3K27me1 and H3K27me2 regulate active transcription and enhancer fidelity. Mol Cell 2014;53(1):49–62.
[73] Pasini D, Bracken AP, Hansen JB, Capillo M, Helin K. The polycomb group protein Suz12 is required for embryonic stem cell differentiation. Mol Cell Biol 2007;27(10):3769–79.
[74] Sarma K, Margueron R, Ivanov A, Pirrotta V, Reinberg D. Ezh2 requires PHF1 to efficiently catalyze H3 lysine 27 trimethylation in vivo. Mol Cell Biol 2008;28(8):2718–31.
[75] Nekrasov M, Klymenko T, Fraterman S, Papp B, Oktaba K, Kocher T, et al. Pcl-PRC2 is needed to generate high levels of H3-K27 trimethylation at Polycomb target genes. EMBO J 2007;26(18):4078–88.
[76] McCabe MT, Graves AP, Ganji G, Diaz E, Halsey WS, Jiang Y, et al. Mutation of A677 in histone methyltransferase EZH2 in human B-cell lymphoma promotes hypertrimethylation of histone H3 on lysine 27 (H3K27). Proc Natl Acad Sci USA 2012;109(8):2989–94.
[77] Sneeringer CJ, Scott MP, Kuntz KW, Knutson SK, Pollock RM, Richon VM, et al. Coordinated activities of wild-type plus mutant EZH2 drive tumor-associated hypertrimethylation of lysine 27 on histone H3 (H3K27) in human B-cell lymphomas. Proc Natl Acad Sci USA 2010;107(49):20980–5.
[78] Leeb M, Pasini D, Novatchkova M, Jaritz M, Helin K, Wutz A. Polycomb complexes act redundantly to repress genomic repeats and genes. Genes Dev 2010;24(3):265–76.
[79] Bartke T, Vermeulen M, Xhemalce B, Robson SC, Mann M, Kouzarides T. Nucleosome-interacting proteins regulated by DNA and histone methylation. Cell 2010;143(3):470–84.
[80] Lee HG, Kahn TG, Simcox A, Schwartz YB, Pirrotta V. Genome-wide activities of Polycomb complexes control pervasive transcription. Genome Res 2015;25(8):1170–81.

[81] Shen X, Liu Y, Hsu YJ, Fujiwara Y, Kim J, Mao X, et al. EZH1 mediates methylation on histone H3 lysine 27 and complements EZH2 in maintaining stem cell identity and executing pluripotency. Mol Cell 2008;32(4):491–502.

[82] Son J, Shen SS, Margueron R, Reinberg D. Nucleosome-binding activities within JARID2 and EZH1 regulate the function of PRC2 on chromatin. Genes Dev 2013;27(24):2663–77.

[83] Wassef M, Rodilla V, Teissandier A, Zeitouni B, Gruel N, Sadacca B, et al. Impaired PRC2 activity promotes transcriptional instability and favors breast tumorigenesis. Genes Dev 2015;29(24):2547–62.

[84] Ezhkova E, Pasolli HA, Parker JS, Stokes N, Su IH, Hannon G, et al. Ezh2 orchestrates gene expression for the stepwise differentiation of tissue-specific stem cells. Cell 2009;136(6):1122–35.

[85] Ezhkova E, Lien WH, Stokes N, Pasolli HA, Silva JM, Fuchs E. EZH1 and EZH2 cogovern histone H3K27 trimethylation and are essential for hair follicle homeostasis and wound repair. Genes Dev 2011;25(5):485–98.

[86] Hidalgo I, Herrera-Merchan A, Ligos JM, Carramolino L, Nunez J, Martinez F, et al. Ezh1 is required for hematopoietic stem cell maintenance and prevents senescence-like cell cycle arrest. Cell Stem Cell 2012;11(5):649–62.

[87] Stojic L, Jasencakova Z, Prezioso C, Stutzer A, Bodega B, Pasini D, et al. Chromatin regulated interchange between polycomb repressive complex 2 (PRC2)-Ezh2 and PRC2-Ezh1 complexes controls myogenin activation in skeletal muscle cells. Epigenetics Chromatin 2011;4:16.

[88] Mousavi K, Zare H, Wang AH, Sartorelli V. Polycomb protein Ezh1 promotes RNA polymerase II elongation. Mol Cell 2012;45(2):255–62.

[89] Xu J, Shao Z, Li D, Xie H, Kim W, Huang J, et al. Developmental control of polycomb subunit composition by GATA factors mediates a switch to non-canonical functions. Mol Cell 2015;57(2):304–16.

[90] Smits AH, Jansen PW, Poser I, Hyman AA, Vermeulen M. Stoichiometry of chromatin-associated protein complexes revealed by label-free quantitative mass spectrometry-based proteomics. Nucleic Acids Res 2013;41(1):e28.

[91] Vizan P, Beringer M, Ballare C, Di Croce L. Role of PRC2-associated factors in stem cells and disease. FEBS J 2015;282(9):1723–35.

[92] Kim H, Kang K, Kim J. AEBP2 as a potential targeting protein for Polycomb Repression Complex PRC2. Nucleic Acids Res 2009;37(9):2940–50.

[93] Kim H, Ekram MB, Bakshi A, Kim J. AEBP2 as a transcriptional activator and its role in cell migration. Genomics 2015;105(2):108–15.

[94] Kalb R, Latwiel S, Baymaz HI, Jansen PW, Muller CW, Vermeulen M, et al. Histone H2A monoubiquitination promotes histone H3 methylation in Polycomb repression. Nat Struct Mol Biol 2014;21(6):569–71.

[95] Ballare C, Lange M, Lapinaite A, Martin GM, Morey L, Pascual G, et al. Phf19 links methylated Lys36 of histone H3 to regulation of Polycomb activity. Nat Struct Mol Biol 2012;19(12):1257–65.

[96] Brien GL, Gambero G, O'Connell DJ, Jerman E, Turner SA, Egan CM, et al. Polycomb PHF19 binds H3K36me3 and recruits PRC2 and demethylase NO66 to embryonic stem cell genes during differentiation. Nat Struct Mol Biol 2012;19(12):1273–81.

[97] Casanova M, Preissner T, Cerase A, Poot R, Yamada D, Li X, et al. Polycomb-like 2 facilitates the recruitment of PRC2 Polycomb group complexes to the inactive X chromosome and to target loci in embryonic stem cells. Development 2011;138(8):1471–82.

[98] Hunkapiller J, Shen Y, Diaz A, Cagney G, McCleary D, Ramalho-Santos M, et al. Polycomb-like 3 promotes polycomb repressive complex 2 binding to CpG islands and embryonic stem cell self-renewal. PLoS Genet 2012;8(3):e1002576.

[99] Li G, Margueron R, Ku M, Chambon P, Bernstein BE, Reinberg D. Jarid2 and PRC2, partners in regulating gene expression. Genes Dev 2010;24(4):368–80.

[100] Walker E, Chang WY, Hunkapiller J, Cagney G, Garcha K, Torchia J, et al. Polycomb-like 2 associates with PRC2 and regulates transcriptional networks during mouse embryonic stem cell self-renewal and differentiation. Cell Stem Cell 2010;6(2):153–66.

[101] McHugh CA, Chen CK, Chow A, Surka CF, Tran C, McDonel P, et al. The Xist lncRNA interacts directly with SHARP to silence transcription through HDAC3. Nature 2015;521(7551):232–6.

[102] Li X, Isono K, Yamada D, Endo TA, Endoh M, Shinga J, et al. Mammalian polycomb-like Pcl2/Mtf2 is a novel regulatory component of PRC2 that can differentially modulate polycomb activity both at the Hox gene cluster and at Cdkn2a genes. Mol Cell Biol 2011;31(2):351–64.

[103] Cai L, Rothbart SB, Lu R, Xu B, Chen WY, Tripathy A, et al. An H3K36 methylation-engaging Tudor motif of polycomb-like proteins mediates PRC2 complex targeting. Mol Cell 2013;49(3):571–82.

[104] Musselman CA, Avvakumov N, Watanabe R, Abraham CG, Lalonde ME, Hong Z, et al. Molecular basis for H3K36me3 recognition by the Tudor domain of PHF1. Nat Struct Mol Biol 2012;19(12):1266–72.

[105] Kycia I, Kudithipudi S, Tamas R, Kungulovski G, Dhayalan A, Jeltsch A. The Tudor domain of the PHD finger protein 1 is a dual reader of lysine trimethylation at lysine 36 of histone H3 and lysine 27 of histone variant H3t. J Mol Biol 2014;426(8):1651–60.

[106] Landeira D, Sauer S, Poot R, Dvorkina M, Mazzarella L, Jorgensen HF, et al. Jarid2 is a PRC2 component in embryonic stem cells required for multi-lineage differentiation and recruitment of PRC1 and RNA Polymerase II to developmental regulators. Nat Cell Biol 2010;12(6):618–24.

[107] Pasini D, Cloos PA, Walfridsson J, Olsson L, Bukowski JP, Johansen JV, et al. JARID2 regulates binding of the Polycomb repressive complex 2 to target genes in ES cells. Nature 2010;464(7286):306–10.

[108] Peng J, Valouev A, Swigut T, Zhang J, Zhao Y, Sidow A, et al. Jarid2/Jumonji coordinates control of PRC2 enzymatic activity and target gene occupancy in pluripotent cells. Cell 2009;139:1290–302.

[109] Shen X, Kim W, Fujiwara Y, Simon MD, Liu Y, Mysliwiec MR, et al. Jumonji modulates polycomb activity and self-renewal versus differentiation of stem cells. Cell 2009;139(7):1303–14.

[110] Sanulli S, Justin N, Teissandier A, Ancelin K, Portoso M, Caron M, et al. Jarid2 methylation via the PRC2 complex regulates H3K27me3 deposition during cell differentiation. Mol Cell 2015;57(5):769–83.

[111] da Rocha ST, Boeva V, Escamilla-Del-Arenal M, Ancelin K, Granier C, Matias NR, et al. Jarid2 is implicated in the initial Xist-induced targeting of PRC2 to the inactive X chromosome. Mol Cell 2014;53(2):301–16.

[112] Kaneko S, Bonasio R, Saldana-Meyer R, Yoshida T, Son J, Nishino K, et al. Interactions between JARID2 and noncoding RNAs regulate PRC2 recruitment to chromatin. Mol Cell 2014;53(2):290–300.

[113] Jermann P, Hoerner L, Burger L, Schubeler D. Short sequences can efficiently recruit histone H3 lysine 27 trimethylation in the absence of enhancer activity and DNA methylation. Proc Natl Acad Sci USA 2014;111(33):E3415–21.

[114] Lynch MD, Smith AJ, De Gobbi M, Flenley M, Hughes JR, Vernimmen D, et al. An interspecies analysis reveals a key role for unmethylated CpG dinucleotides in vertebrate Polycomb complex recruitment. EMBO J 2012;31(2):317–29.

[115] Mendenhall EM, Koche RP, Truong T, Zhou VW, Issac B, Chi AS, et al. GC-rich sequence elements recruit PRC2 in mammalian ES cells. PLoS Genet 2010;6(12):e1001244.

[116] Zhang Z, Jones A, Sun CW, Li C, Chang CW, Joo HY, et al. PRC2 complexes with JARID2, MTF2, and esPRC2p48 in ES cells to modulate ES cell pluripotency and somatic cell reprogramming. Stem Cells 2011;29(2):229–40.

[117] De Cegli R, Iacobacci S, Flore G, Gambardella G, Mao L, Cutillo L, et al. Reverse engineering a mouse embryonic stem cell-specific transcriptional network reveals a new modulator of neuronal differentiation. Nucleic Acids Res 2013;41(2):711–26.

[118] Liefke R, Shi R. The PRC2-associated factor C17orf96 is a novel CpG island regulator in mouse ES cells. Cell Discovery 2015:1.

[119] Alekseyenko AA, Gorchakov AA, Kharchenko PV, Kuroda MI. Reciprocal interactions of human C10orf12 and C17orf96 with PRC2 revealed by BioTAP-XL cross-linking and affinity purification. Proc Natl Acad Sci USA 2014;111(7):2488–93.

[120] de Napoles M, Mermoud JE, Wakao R, Tang YA, Endoh M, Appanah R, et al. Polycomb group proteins Ring1A/B link ubiquitylation of histone H2A to heritable gene silencing and X inactivation. Dev Cell 2004;7(5):663–76.

[121] Wang H, Wang L, Erdjument-Bromage H, Vidal M, Tempst P, Jones RS, et al. Role of histone H2A ubiquitination in Polycomb silencing. Nature 2004;431(7010):873–8.

[122] Stock JK, Giadrossi S, Casanova M, Brookes E, Vidal M, Koseki H, et al. Ring1-mediated ubiquitination of H2A restrains poised RNA polymerase II at bivalent genes in mouse ES cells. Nat Cell Biol 2007;9(12):1428–35.

[123] Gao Z, Zhang J, Bonasio R, Strino F, Sawai A, Parisi F, et al. PCGF homologs, CBX proteins, and RYBP define functionally distinct PRC1 family complexes. Mol Cell 2012;45(3):344–56.

[124] Morey L, Aloia L, Cozzuto L, Benitah SA, Di Croce L. RYBP and Cbx7 define specific biological functions of polycomb complexes in mouse embryonic stem cells. Cell Rep 2013;3(1):60–9.

[125] Tavares L, Dimitrova E, Oxley D, Webster J, Poot R, Demmers J, et al. RYBP-PRC1 complexes mediate H2A ubiquitylation at polycomb target sites independently of PRC2 and H3K27me3. Cell 2012;148(4):664–78.

[126] Schoeftner S, Sengupta AK, Kubicek S, Mechtler K, Spahn L, Koseki H, et al. Recruitment of PRC1 function at the initiation of X inactivation independent of PRC2 and silencing. EMBO J 2006;25(13):3110–22.

[127] Blackledge NP, Farcas AM, Kondo T, King HW, McGouran JF, Hanssen LL, et al. Variant PRC1 complex-dependent H2A ubiquitylation drives PRC2 recruitment and polycomb domain formation. Cell 2014;157(6):1445–59.

[128] Cooper S, Dienstbier M, Hassan R, Schermelleh L, Sharif J, Blackledge NP, et al. Targeting polycomb to pericentric heterochromatin in embryonic stem cells reveals a role for H2AK119u1 in PRC2 recruitment. Cell Rep 2014;7(5):1456–70.

[129] Chiacchiera F, Rossi A, Jammula S, Piunti A, Scelfo A, Ordonez-Moran P, et al. Polycomb Complex PRC1 Preserves Intestinal Stem Cell Identity by Sustaining Wnt/beta-Catenin Transcriptional Activity. Cell Stem Cell 2016;18(1):91–103.

[130] Pengelly AR, Kalb R, Finkl K, Muller J. Transcriptional repression by PRC1 in the absence of H2A monoubiquitylation. Genes Dev 2015;29(14):1487–92.

[131] Illingworth RS, Moffat M, Mann AR, Read D, Hunter CJ, Pradeepa MM, et al. The E3 ubiquitin ligase activity of RING1B is not essential for early mouse development. Genes Dev 2015;29(18):1897–902.

[132] Poux S, Melfi R, Pirrotta V. Establishment of Polycomb silencing requires a transient interaction between PC and ESC. Genes Dev 2001;15(19):2509–14.

[133] Kang H, McElroy KA, Jung YL, Alekseyenko AA, Zee BM, Park PJ, et al. Sex comb on midleg (Scm) is a functional link between PcG-repressive complexes in *Drosophila*. Genes Dev 2015;29(11):1136–50.

[134] Cao Q, Wang X, Zhao M, Yang R, Malik R, Qiao Y, et al. The central role of EED in the orchestration of polycomb group complexes. Nat Commun 2014;5:3127.

[135] Klose RJ, Cooper S, Farcas AM, Blackledge NP, Brockdorff N. Chromatin sampling – an emerging perspective on targeting polycomb repressor proteins. PLoS Genet 2013;9(8):e1003717.

[136] Riising EM, Comet I, Leblanc B, Wu X, Johansen JV, Helin K. Gene silencing triggers polycomb repressive complex 2 recruitment to CpG islands genome wide. Mol Cell 2014;55(3):347–60.

Chapter 10

Regulation of PRC2 Activity

N. Liu[1], B. Zhu[1,2]
[1]*Institute of Biophysics, Chinese Academy of Sciences, Beijing, China;* [2]*University of Chinese Academy of Sciences, Beijing, China*

Chapter Outline

Polycomb Repressive Complex 2 and Its Enzymatic Activity **225**
- Activity of Polycomb Repressive Complex 2 226
- Structure of Polycomb Repressive Complex 2 227
- Solo EZH2 229

Role of H3K27 Methylation **230**
Embryonic Ectoderm Development Facilitates the Propagation of H3K27 Methylation **233**
Polycomb Repressive Complex 2 Is Stimulated by Dense Chromatin **234**
Cross Talk Among Histone Modifications **236**
- H2A K119 Ubiquitination Stimulates Polycomb Repressive Complex 2 Activity 237
- H3K4ME3 and H3K36ME2/3 Inhibit Polycomb Repressive Complex 2 Activity 239
- H3S28 Phosphorylation Antagonizes Polycomb Silencing 240

Accessory Components Modulate Polycomb Repressive Complex 2 Activity **241**
- AEBP2 242
- PCL Proteins (PHF1, MTF2, PHF19) 243
- JARID2 244
- EZH1-Containing Polycomb Repressive Complex 2 246

H3K27M Inhibits Polycomb Repressive Complex 2 Activity and Leads to Pediatric Glioblastoma **247**
Conclusion **250**
List of Acronyms and Abbreviations **250**
Glossary **250**
Acknowledgments **251**
References **251**

POLYCOMB REPRESSIVE COMPLEX 2 AND ITS ENZYMATIC ACTIVITY

Polycomb group (PcG) proteins were first discovered as factors that control body segmentation in *Drosophila melanogaster*. They silence HOX gene expression in embryos and can maintain this silencing for the rest of development. Decades after identification by genetic screens, biochemical studies demonstrate that PcG

Polycomb Group Proteins. http://dx.doi.org/10.1016/B978-0-12-809737-3.00010-6

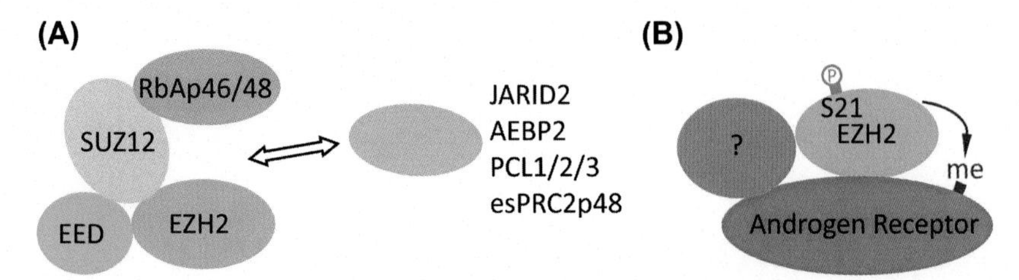

FIGURE 10.1 (A) The composition of the Polycomb repressive complex 2 (PRC2) complex. The core PRC2 is composed of embryonic ectoderm development (EED), EZH2, SUZ12, and RbAp46/48. Accessory components, including JARID2, AEBP2, and PCL1/2/3, associate with a subset of PRC2 in cell type—specific manner. (B) PRC2-independent solo EZH2. In castration-resistant prostate cancer cells, EZH2 exists in a protein complex together with androgen receptor (AR). This protein complex is required for target gene activation. The enzymatic activity of EZH2 is essential for its activating function, likely through methylating AR or other unknown factors, but not H3K27. Akt-mediated S21 phosphorylation of EZH2 is also required for its function in cancer progression.

proteins exist in several multiprotein complexes. One of the complexes, which consists of ESC, E(Z), SU(Z)12, and the non-PcG protein NURF-55, exerts methyltransferase activity toward lysine 27 of histone H3 [1—5]. This protein complex is named Polycomb repressive complex 2 (PRC2). The human orthologs of the PRC2 subunits are embryonic ectoderm development (EED), EZH2 or EZH1, SUZ12, and RbAp46/48. In addition to the four core components, PRC2 also contains several accessory subunits, including AEBP2, JARID2, and PCL proteins (Fig. 10.1A), etc. [6—18]. These subunits exist in a subset of PRC2 complexes or associate with core PRC2 in a cell type—specific manner, modulating PRC2 activity or facilitating PRC2 recruitment.

In this chapter, we will summarize how the core and accessory subunits regulate PRC2 enzymatic activity. Since most of these mechanisms are conserved from fly to human, we will uniformly use the human nomenclature in this chapter unless specified.

Activity of Polycomb Repressive Complex 2

Lysine-specific methyltransferases, with the exception of Dot1L, contain a SET domain. The SET domain is an evolutionarily conserved domain [19] whose name comes from initials of Su(var)3-9, E(Z), and Trithorax, which all contain a SET domain. This domain utilizes *S*-adenosyl-L-methionine (SAM) as a methyl donor to modify the ε-nitrogen of a lysine residue through a bimolecular nucleophilic substitution (S_N2) group transfer reaction, producing *S*-adenosyl-L-homocysteine (SAH) and N-methylated lysine [20]. EZH2 is the PRC2 subunit that contains a SET domain. However, EZH2 itself is catalytically inactive in vitro and in vivo for mediating H3K27 methylation. Two additional factors, SUZ12 and EED, are both essential for PRC2 to methylate H3K27 [6,21—23].

Theoretically, the ε-nitrogen could be methylated to three states. One, two, or three hydrogen atoms could be substituted by methyl groups, generating mono-, di-, and trimethylated products. However, some histone methyltransferases can only modify lysine into mono- or dimethylated states. Structural studies have revealed that the space restrictions or hydrogen-bonding patterns within the substrate-binding pocket determine the different degrees of catalysis [24]. The core PRC2 could catalyze the formation of all three methylation states in vitro. However, the efficiency of producing H3K27me3 is much lower than that of producing H3K27me1/2 [25,26]. The most abundant product of PRC2 in vivo is H3K27me2, which accounts for 45–70% of total histone H3, whereas H3K27me3 accounts for 5–20% [27–30]. This is in accordance with the observed in vitro activity.

Interestingly, two cancer-related mutations of EZH2 (Y641F/N/S/H/C, A677G) have significantly higher enzymatic activities in generating H3K27me3 [25,26,31], suggesting that abnormally high PRC2 activity may be relevant to tumorigenesis. Structural simulations reveal that these mutations lead to the enlargement of the lysine tunnel, permitting the dimethylated lysine to accept the third methyl group [25]. A number of small molecule inhibitors that specifically inhibit EZH2 have been developed, some of which performed well in inhibiting cancer cell growth and are under clinical trials in patients with B-cell lymphoma or advanced solid tumors (http://www.clinicaltrials.gov).

Structure of Polycomb Repressive Complex 2

Recently, the crystal structure of the thermophilic fungus *Chaetomium thermophilum* PRC2 was published [32]. The resolved structure contains three subunits: EED, SUZ12, and EZH2. The overall architecture of the trimeric PRC2 is compact. EZH2 makes extensive contacts with both SUZ12 and EED, with its N-terminal domain stably binding EED and its SET domain residing in the middle of the structure. This arrangement allows the multiple allosteric regulations of enzymatic activity.

The PRC2 structure explains how EZH2 is activated in the complex. EZH2 alone is enzymatically inactive because an EZH2 fragment named SET-I prevents the access of the substrate to the catalytic site. In addition, the post-SET domain is also misplaced, resulting in an incomplete cofactor-binding pocket [33,34]. In contrast, when in complex with EED and SUZ12, the SET-I fragment rotates outward 20 degrees, making the substrate pocket accessible and assembling a complete cofactor-binding pocket. This auto-inhibition and requirement for associating subunits are also observed in another histone methyltransferase, MLL1 [35]. This finding may explain the solo EZH2 activity initially observed in prostate cancer cells [36] (see Solo EZH2 section). Solo EZH2 may associate with other proteins, including androgen receptor [36], which could reposition the SET-I region and make the catalytic site accessible to substrates, thus activating EZH2 (Fig. 10.1B).

EED is a WD repeat domain—containing protein with WD repeat motifs forming a seven-bladed β-propeller structure. This domain is widely implicated in protein—protein interactions. Similar to typical WD repeat proteins, the residues outside the repeats conduct most of the interactions with its partners, including EZH2 and SUZ12 [32]. The central pocket of EED specifically binds peptides with repressive markers [37] and triggers conformational changes in EZH2, resulting in allosteric activation of PRC2 (see EED facilitates the propagation of H3K27 methylation section).

In the reported PRC2 structure [32], SUZ12 is truncated to 140 amino acids and fused to EZH2. The truncated protein contains the conserved VEFS domain that directly binds to the SET domain of EZH2. The VEFS domain, named after VRN2, EMF2, FIS2, and Su(z)12, is also involved in the allosteric regulation of PRC2 activity (see PRC2 is stimulated by dense chromatin and cross talk among histone modifications sections). The N-terminus and the middle portion of SUZ12, which is absent in the structure, are implicated in interacting with RbAp46/48 and PRC2 cofactors such as AEBP2 [6,38].

Another core component of PRC2, RbAp46/48 or *Drosophila* NURF-55, is not present in the current structure. Although the other three subunits are indispensable for enzymatic activity, loss of RbAp46/48 only impaired the activity by two to threefold in vitro [6,23]. In addition, it is also present in other chromatin-modifying complexes, including Chromatin Assembly Factor 1 (CAF1), nucleosome remodeling factor complex (NURF), and nucleosome remodeling deacetylase complex (NuRD). Since RbAp46/48 is also a WD repeat protein, it may serve as a scaffold for these complexes or it may recognize histone modifications in a manner similar to EED. *Drosophila* NURF-55 is reported to bind histone H4 through residues outside the β-propeller [39]. A cocrystal structure of NURF-55 with residues 15-41 of H4 was resolved [40], suggesting that NURF-55 may stabilize the PRC2-nucleosome interaction. However, the binding depends on helix 1 of the histone fold, which is inaccessible in the nucleosome. Later studies found that NURF-55 could bind to the SUZ12 N-terminus via the same region that binds H4 [41,42], suggesting that the H4 binding ability may be utilized by complexes other than PRC2. In addition, the channel within the β-propeller could bind to unmodified or H3K9me3 or H3K27me3 peptides but not H3K4me1/2/3, similar to EED. Despite the ability of NURF-55 to bind unmodified or repressive markers, it cannot translate these signals into enzymatic outputs [42]. One possible explanation is that NURF-55 peptide binding has little, if any, effect on the conformation of the SET domain and thus cannot stimulate enzymatic activity as EED does. The proteins that allosterically activate PRC2 all interact with EZH2 directly. In contrast, binding to repressive markers may be involved in stabilizing PRC2 on chromatin. On the other hand, the ability of NURF-55 to exclude active markers may destabilize PRC2 at active chromatin regions.

Solo EZH2

EZH2 is a core subunit of PRC2. Biochemical and structural studies show that EZH2 tightly associates with EED, SUZ12, and RbAp46/48 to form the PRC2 complex. EZH2 itself is enzymatically inactive, and it requires EED and SUZ12 to exert its enzymatic activity [6,21–23,32]. However, emerging studies shed light on the PRC2-independent roles of EZH2, which are tightly linked to cancer [36,43,44]. EZH2 interacts with proteins such as NF-κB and estrogen receptor in breast cancer cells or androgen receptor in castration-resistant prostate cancer cells to activate gene expression.

EZH2 expression is strongly elevated in castration-resistant prostate cancer (CRPC) [45]. CRPC is a late stage of prostate cancer. It is lethal and cannot be treated by the removal of testicular androgen (castration). A cellular hallmark of CRPC is continued androgen receptor (AR) signaling, which may result from amplification or mutation of the AR gene. This explains the androgen independence of CPRC [46]. Elevated levels of transcriptional coregulators of AR are also involved in AR activation. EZH2 was recently reported as an AR coactivator [36]. The AR coactivator function of EZH2 is PRC2 independent, as EZH2 binds to a small set of genes together with AR in the absence of SUZ12. The genes that are bound by solo EZH2 are depleted of H3K27me3 and enriched for H3K4me3 and Pol II, suggesting active transcription. Indeed, many of these genes are stimulated by EZH2, and their high expression level correlates with poor patient survival [36]. In contrast, the hormone-dependent prostate cancer cell line LNCaP does not have EZH2 solo peaks, and overexpression of active EZH2 in LNCaP cells is sufficient to support its androgen-independent growth, demonstrating the crucial role of EZH2 in CRPC progression. Intriguingly, although H3K27me3 is absent from EZH2 solo peaks, the catalytic activity of EZH2 is required for gene activation [36].

What is the substrate of the solo EZH2? Biochemical experiments show that EZH2 directly binds to AR. They coexist in a large protein complex without detectable PRC2 components besides EZH2. Depletion of EZH2 decreases AR-associated lysine methylation. The EZH2 substrate may be AR and possibly other nonhistone proteins [36]. The mechanisms that regulate the EZH2-AR binding and functional switch of EZH2 are not clear without structural data. One hypothesis is EZH2 phosphorylation. Indeed, EZH2 S21 phosphorylation by Akt kinase increased significantly in CRPC cells compared to that in LNCaP cells [36,47]. Importantly, EZH2-S21P is present at EZH2 solo peaks, associates with AR, and is required for the activation of solo EZH2 target genes. Thus, S21 phosphorylation may be the key of EZH2's functional switch (Fig. 10.1B). Questions regarding the mechanisms remain. How does EZH2 phosphorylation trigger its association with AR? How does EZH2-AR activate gene transcription? Do other proteins associate with EZH2-AR and what are their functions? Why do such events not occur in normal cells? Although the mechanism is unclear, these findings provide exciting new insights into CRPC therapy. Inhibitors of EZH2, especially those that could target the EZH2 activation function, may have good potential in treating CRPC [48].

ROLE OF H3K27 METHYLATION

PRC2 can methylate H3K27 to all three states and is the only enzyme that generates H3K27me2 and H3K27me3 in vivo. Although the three methylation states differ only in the number of methyl groups, they have distinct distributions and functions.

H3K27me1 often marks active gene bodies in association with H3K36me3 [49–51]. The level of H3K27me1 correlates well with gene expression level. However, the role of H3K27me1 in transcription is not clear. It may be a product of the incomplete demethylation of H3K27me2/3 (the H3K27 demethylase UTX is inefficient in demethylating H3K27me1), or it may play functional roles in promoting transcription. In the first scenario, H3K27me1 is only a byproduct of demethylation, reflecting the counteraction of UTX and PRC2. H3K27me1 should be unable to recruit downstream factors and would not have a negative effect on transcription, making it unnecessary to be completely demethylated. In another scenario, H3K27me1 can also arise from either partial demethylation or de novo monomethylation. Importantly, H3K27me1 should be specifically recognized by effectors that could in turn promote transcription. To date, no such factors have been identified. Of note, G9a and GLP produce approximately 30% of the H3K27me1 in mouse embryonic stem (ES) cells [52], with unknown biological consequence.

H3K27me2 is an abundant modification, accounting for up to 45–70% percent of all histone H3 [27–30]. Despite its abundance and widespread distribution, the role of H3K27me2 has long been underappreciated until recently [50,51]. H3K27me2 is present throughout the genome, both on intergenic regions and on non-PcG target genes. It is inversely correlated with transcriptional activity and H3K27me1. Depletion of H3K27me2 by inactivating PRC2 leads to the global transcriptional derepression of both the genic regions of non-PcG targets and intergenic regions [51]. In addition, loss of H3K27me2 causes a global increase of H3K27ac and H3K4me1-that are typical markers of active enhancers [50,51]. Upon loss of H3K27me2 and gain of H3K27ac, nearby genes are preferentially activated [50]. Thus, H3K27me2 not only controls the pervasive transcription of the vast majority of intergenic regions and non-PcG target genes but also protects cell type–specific enhancers.

H3K27me3 is a classic marker of PcG target genes. It accounts for 5–20% of total histone H3 [27–30] and is essential for gene repression [51]. The presence of H3K27me3 is always associated with PRC2 binding. However, PRC2 preferentially binds around the transcription start site (TSS) whereas

H3K27me3 often marks entire gene bodies. Such distribution patterns imply that both recruiting and spreading events are likely involved in establishing H3K27me3 across the gene locus.

The mechanism of H3K27me3-mediated gene silencing has been intensively investigated but yet not fully understood. In one popular model, PRC2 binds to a target gene and deposits H3K27me3, followed by PRC1 binding through its CBX subunits. PRC1 could then ubiquitinate lysine 119 of histone H2A, compact chromatin and inhibit transcription elongation [53]. This simple hierarchy has been challenged by recent studies that emphasize the role of noncanonical PRC1 in the initial recruitment of PcG proteins. The recruitment of noncanonical PRC1 occurs through its KDM2B subunit, which binds unmethylated CpG islands. Noncanonical PRC1 then ubiquitinates histone H2A, followed by PRC2 recruitment by H2A K119 ubiquitination (H2Aub) and H3K27me3 deposition, which in turn recruits canonical PRC1. This generates a positive feedback loop that results in the spreading of H2Aub and H3K27me3 and in the end results in gene silencing [53] (see H2A K119 Ubiquitination Stimulates Polycomb Repressive Complex 2 Activity section). In addition to recruiting PRC1, H3K27me3 may also affect transcription through other mechanisms. For example, H3K27me3 may recruit other silencing factors, antagonize H3K27ac, or impede the binding of transcriptional activators, remodelers, or transcription elongation factors. Therefore, H3K27me3 plays multifaceted roles in transcriptional repression.

The maintenance of H3K27 methylation during cell division is important for maintaining the chromatin landscape of a cell. During S phase, histone modifications are diluted because of the incorporation of new histones. At this stage, PRC2 has been reported to propagate H3K27 methylation through association with replication forks [54]. Although the full recovery of the H3K27me3 level is accomplished during G1 phase [55–57], H3K27me2 increases and exhibits a linear correlation with the amount of DNA [51], in agreement with a faster reestablishment of H3K27me2 during the cell cycle [55–57]. The details of this mechanism will be discussed in the next section. Of note, although H3K27me3 is diluted during S phase, the remaining methylation level is sufficient to maintain a repressive chromatin environment with the downstream effectors bound, as long as the level of H3K27me3 does not drop below a certain threshold. Therefore the limited dilution of H3K27me3 in S phase is not detrimental to gene repression, but it should be restored before the next S phase [55,58].

The final reestablishment of H3K27 methylation occurs after S phase through various mechanisms. An intriguing difference between H3K27me1/2 and H3K27me3 is that regions marked by H3K27me1/2 are often not stably bound by PRC2, but H3K27me3 colocalizes with PRC2. Therefore the reestablishment of H3K27me1/2 can only be attributed to "nonrecruited" PRC2 activity, whereas the reestablishment of H3K27me3 may involve both "nonrecruited" and "recruited" activities. The "nonrecruited" PRC2 activity acts in a "hit-and-run" fashion and deposits H3K27 methylation across the whole genome through transient interactions with nucleosomes. The "nonrecruited" PRC2 activity can be enhanced by the presence of H3K27me2/3 (see Section EED facilitates the propagation of H3K27 methylation) or by condensed chromatin, which may occur after chromatin restores its previous configuration after DNA replication (see Section PRC2 is stimulated by dense chromatin). "Nonrecruited" activity may be sufficient for establishing H3K27me1/2, but "recruited" activity is required for H3K27me3, as PRC2 is efficient in producing H3K27me1/2 but shows low activity in generating H3K27me3 [25,26]. In mammals, the "recruited" activity likely needs assistance from accessory components. For example, PRC2 binding to target genes is stabilized by association with accessory components such as JARID2, AEBP2, and PCL proteins. The persistent binding provides sufficient time for further conversion of H3K27me2 to H3K27me3. In addition, some accessory components could also boost PRC2 activity. JARID2 is more efficient than H3K27me3 in PRC2 stimulation, and PCL1 enhances the ability of PRC2 to generate H3K27me3 [12–14]. Supporting this, depletion of the accessory components in ES cells results in local or global decrease of H3K27me3 [16,17]. In *Drosophila*, the "recruited" activity is predominantly conducted by Polycomb response elements (PREs). PREs are *cis*-regulatory DNA elements that contain consensus motifs for certain DNA-binding proteins that recruit PRC2. In addition, Polycomb repressive complex 1 (PRC1) is also involved in PRC2 targeting [53]. These additional mechanisms ensure the fidelity of H3K27me3 generation at Polycomb target genes and will be discussed in detail in Section H2A K119 ubiquitination stimulates PRC2 activity and accessory components modulate PRC2 activity. Taken together, PRC2 activity could maintain the H3K27 methylation that accounts for up to 80% of total histone H3, mostly without sequence specific targeting. In contrast, the regions that are depleted of H3K27 methylation must actively counteract PRC2 activity through histone demethylation or inhibition of PRC2 activity.

In addition to acute dilution and reestablishment during the cell cycle, the H3K27 methylation level also changes dynamically in other circumstances. During development, the establishment or erasure of H3K27me3 on specific genes occurs in response to external cues such as differentiation signals or stress signals. The need for dynamic control of H3K27 methylation raises the questions of how to efficiently boost PRC2 activity when needed and how to antagonize its activity where necessary.

In the following sections, we will mainly focus on the various mechanisms that regulate PRC2 enzymatic activity. We will discuss how PRC2 enzymatic activities are stimulated or inhibited by various factors and discuss how these mechanisms contribute to PRC2 function.

EMBRYONIC ECTODERM DEVELOPMENT FACILITATES THE PROPAGATION OF H3K27 METHYLATION

A central question in epigenetics is the inheritance of epigenetic information. As mentioned, the mitotic division of a cell results in the dilution of histone modifications. H3K27 methylation accounts for more than 80% of total histone. Therefore, an efficient way to reestablish this modification is essential to maintain the chromatin landscape.

Two models for the mitotic inheritance of histone modifications were proposed [59]. In the semiconservative model, the old H3/H4 tetramer is split into two H3/H4 dimers. During chromatin assembly, these old dimers are assembled with new, unmodified dimers into one nucleosome. The histone modifications could be transmitted from old to new histones with high fidelity. However, studies show that histone modifications are not inherited at near-mononucleosome resolution, and most H3/H4 tetramers essentially remain intact during S phase. This model is therefore incorrect [55,60]. In the other model, which involves the conservative transmission of intact H3/H4 tetramers, the histone modification can be copied from neighboring old nucleosomes to new ones. This copying mechanism requires the cooperation of a reading module and a writing module for histone modification. The reading module reads the marks, and the writing module catalyzes the formation of the same histone modifications. A number of histone-modifying enzymes or complexes possess both reading and writing modules, including Swi6/Clr4 in yeast, G9a/GLP in mammals, as well as PRC2 [37,61–63].

The EED subunit of PRC2 can bind to the H3K27me2/3 peptide [37], leading to the question of whether PRC2 is capable of propagating H3K27 methylation through binding to its products. Indeed, the catalytic activity of PRC2 on oligonucleosome substrates is strongly stimulated in the presence of H3K27me2/3 peptide but not H3K27me0 or H3K27me1 peptides [37,64]. The observed stimulation is dependent on the aromatic cage of EED, which mediates H3K27me2/3 binding. H3K27me3 stimulates PRC2 activity by an allosteric mechanism, because its presence leads to increased maximum reaction rate (V_{max}) of PRC2 without affecting the affinity between enzyme–substrate binding (K_m). PRC2 is therefore capable of propagating H3K27 methylation via H3K27me2/3 binding and allosteric activation. The loss of the product-reading ability in *Drosophila* is detrimental to development. In *esc*$^-$ embryos with only maternal stocks of ESC (ESC and ESCL are the fly homologs of EED), reexpression of wild-type ESC completely rescued the extra sex comb phenotype, while the aromatic cage mutant ESC is ineffectual. In *esc*$^-$ *escl*$^-$ embryos, which are lethal, reexpression of wild-type ESC completely rescued the lethal phenotype, whereas the aromatic cage mutant ESC could not. The levels of H3K27me2/3 and PRC2 binding on target genes are greatly reduced in mutant ESC–expressing embryos, and the total level of

H3K27me2/3 is also reduced in mutants [37]. These findings collectively demonstrate the critical role of PRC2's methylation propagation activity in the maintenance of H3K27 methylation during fly development.

In addition to H3K27me3, EED also binds to other repressive markers, including H1K26me3, H3K9me3, and H4K20me3 but not active markers such as H3K4me3 and H3K36me3 [37,64]. Peptides with these repressive modifications stimulate PRC2 activity to a lesser extent than does H3K27me3 [37], and H1K26me3 is reported to inhibit PRC2 activity [64]. Whether these repressive modifications modulate PRC2 function in vivo is not clear, but it remains an interesting hypothesis that one repressive modification may enhance another repressive mark.

The detailed mechanism of the stimulation is revealed by the comparison of PRC2 structures in basal and stimulated states [32]. In the stimulated state, the H3K27me3 peptide is sandwiched between EED and an exposed EZH2 motif, called the stimulation-responsive motif (SRM), which is flexible and not visible in the basal state structure. H3K27me3 peptide binds to an aromatic cage of EED and forms hydrogen bonds with SRM through its main chain atoms. The formation of this sandwich structure drags the SRM as well as the entire catalytic moiety toward EED, and this conformational change leads to enzymatic activation. Mutations at critical residues in SRM abolish the stimulation. Importantly, some of these residues are mutation hotspots in human diseases [32]. Restoring the enzymatic activity or inhibiting the H3K27 demethylation activity may serve as a choice to treat these diseases.

Product binding generates a positive feedback loop and allosterically stimulates PRC2 activity (Fig. 10.2A). This mechanism provides an ideal solution for the propagation of histone modification during DNA replication. This hypothesis is strengthened by the observation that PRC2 colocalizes with the replication forks in S phase [54]. At the replication forks, PRC2 could be stimulated by H3K27me2/3 on old nucleosomes and modify the nearby newly incorporated histones. Indeed, the H3K27me2 level increases in a linear correlation with DNA content in S phase, supporting the idea that DNA replication and H3K27me2 maintenance cooccur [50]. However, as described previously, this mechanism is insufficient to accomplish the full reestablishment of H3K27me2/3 [55–57], which may require further action outside S phase with other strategies.

POLYCOMB REPRESSIVE COMPLEX 2 IS STIMULATED BY DENSE CHROMATIN

In addition to being stimulated by H3K27me2/3 markers, PRC2 is also able to sense chromatin density. PRC2 prefers oligonucleosome as substrates, and its activity is further enhanced in the presence of linker histone H1 or Mg^{2+} [65], which help to form high-order chromatin structures [2,6,65]. This leads to the question that whether the density of nucleosomes, even in unmodified states,

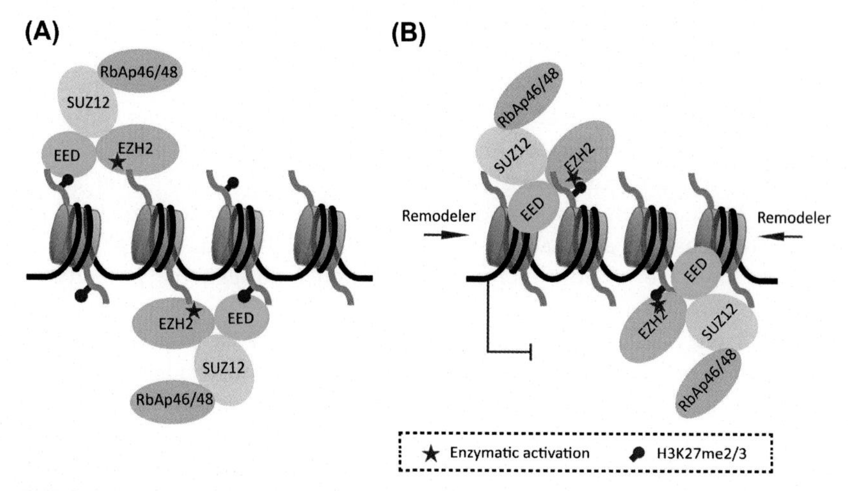

FIGURE 10.2 (A) Propagation of H3K27 methylation. Embryonic ectoderm development (EED) specifically binds to H3K27me2/3, and this binding allosterically activates PRC2, leading to propagation of H3K27me2/3 along the chromatin. (B) Dense chromatin activates PRC2 and facilitates the de novo establishment of H3K27 methylation. Upon cessation of gene transcription, chromatin remodelers begin to compact chromatin. PRC2 could be activated by densely packed nucleosomes, which facilitate the spreading of H3K27 methylation along the gene body. The mechanism of stimulation by dense chromatin involves the sensing of H3 residue 31–42 by SUZ12.

could enhance PRC2 activity. Histone methyltransferase assays show that PRC2 activity is much higher on densely assembled oligonucleosomes [66]. Intriguingly, the feature on a neighboring nucleosome that mediates PRC2 activation independent of prior methylation was discovered to be a peptide fragment of H3 corresponding to amino acids 31 to 42, designated as H3 (31–42). This peptide is necessary and sufficient to stimulate PRC2 activity on dispersed oligonucleosomes and mononucleosomes. It greatly increases the maximum reaction rate (V_{max}) without affecting enzyme-substrate binding (K_m). Thus, the stimulation is also an allosteric activation event. This peptide has no effect on dense substrates, suggesting that a closely placed histone H3 tail is the sole contributing factor for stimulation [66]. Further analyses demonstrated that a small acidic patch within the VEFS domain of SUZ12 is required for stimulation. These acidic residues likely interact with R39 on histone H3. Indeed, mutating either these acidic residues or H3R39 completely abolished the stimulation. The detailed activation mechanism in a structural view remains elusive because of the poor conservation of these acidic residues in *C. thermophilum*. Nevertheless, they are predicted to be located adjacent to a SET-activation loop of EZH2, which supports an allosteric activation mechanism [32].

What is the function of chromatin density–mediated PRC2 activation? An interesting hypothesis is that it can help the de novo deposition of H3K27 methylation during dynamic gene repression (Fig. 10.2B). Gene repression involves the dissociation of transcription machinery, subsequent compaction of local chromatin, and deposition of repressive histone modifications. The above hypothesis predicts that chromatin compaction occurs prior to H3K27me3 deposition. Indeed, this was confirmed by detailed analysis of the dynamics of chromatin accessibility and H3K27me3 establishment at the CYP26a1 gene upon transcription cessation. The enzymatic regulation by chromatin density is an ideal way to recruit PRC2 activity to target genes, not through sequence specific recruitment but rather through the property of substrate chromatin.

PRC2 enzymatic regulation by chromatin density is also involved in maintenance of H3K27me3, as mutating the critical acidic residues of SUZ12 in ES cells leads to dramatically reduced H3K27me3 levels [66]. As discussed in the previous section, EED could partially mediate the maintenance of H3K27me2/3 during S phase. We therefore hypothesize that SUZ12-mediated PRC2 activation could further contribute to H3K27 methylation reestablishment during later stages, when the chromatin condenses and is restored to its previous configuration. Therefore, the stimulation of PRC2 by dense chromatin could serve as one additional mechanism that contributes to epigenetic inheritance. The abovementioned substrate feedback mechanisms are required but may not be sufficient for H3K27me3 inheritance, which also requires PREs in *Drosophila*.

CROSS TALK AMONG HISTONE MODIFICATIONS

Most histone modifications are located at the N-terminal tails of histones. At the histone H3 tail, at least 15 of the 40 residues are reported to be post-translationally modified [67]. Such crowded modifications provide a perfect regulatory platform, allowing another layer of regulation: cross talk.

Cross talk among histone modifications involves several models of action. One modification can directly promote or inhibit the enzymatic activity on another site. For example, PRC2 activity is inhibited by H3K4me3 or H3K36me2/3 on the same histone tail, protecting the active chromatin from silencing. Certain modifications could also create a steric hindrance that interferes with enzyme access or binding to its neighboring residue. For example, H3S28 phosphorylation (H3S28p) in close proximity to the H3K27 residue could displace the PRC2 complex, leading to gene activation. The cross talk could also occur between tails and globular domains. For example, H2A K119 monoubiquitination deposited by PRC1 could stimulate PRC2 activity, creating a positive feedback loop for PcG-mediated gene repression. These cross talk mechanisms allow more flexible regulation of chromatin status in response to external or internal stimuli.

H2A K119 Ubiquitination Stimulates Polycomb Repressive Complex 2 Activity

H2Aub is generated by PRC1 and is required for the repression of target genes. PRC1 and PRC2 colocalize at most of their targets and work in concert. Their collaboration was thought to occur in a linear cascade in which PRC2 is first recruited to target genes and deposits H3K27me3, which is then recognized by the CBX proteins of the PRC1 complex, leading to histone H2A K119 monoubiquitination, chromatin compaction, and repression of gene expression. However, this hierarchical model has been challenged by recent findings. Several studies reported that PRC2 depletion did not affect PRC1 binding at some target genes, suggesting the existence of H3K27 methylation-independent PRC1-targeting mechanisms. Later, a systematic purification study identified noncanonical PRC1 complexes characterized by the presence of RYBP or YAF2 subunits and the lack of CBX proteins, making them incompetent to bind H3K27me3. Thus, the noncanonical PRC1 complexes are unlikely to be recruited through the PRC2-H3K27me3-CBX cascade. Instead, one of the noncanonical PRC1 complexes (PRC1.1) could be directly targeted to CpG islands by KDM2B, a ZF-CxxC domain−containing protein that specifically binds nonmethylated CpG. Artificially targeting KDM2B to a TetO array or pericentric heterochromatin not only recruits PRC1 but also recruits PRC2 and generates an H3K27me3 domain [68,69], suggesting that H2Aub may recruit PRC2. Supporting this, oligonucleosomes containing H2Aub could specifically bind the PRC2 complex, likely through AEBP2 or JARID2, two accessory components of PRC2. Histone methyltransferase assays on oligonucleosome substrates containing H2Aub show that AEBP2-containing PRC2 but not other forms could be strongly stimulated by H2Aub [70].

These findings have revised the previous hierarchical recruiting model and lead to another positive feedback mechanism for PcG protein-mediated gene repression (Fig. 10.3A). In this new model, noncanonical PRC1 is first recruited by KDM2B to CpG islands where it generates H2Aub, which recruits and stimulates PRC2 through AEBP2. H3K27me3 deposited by PRC2 could in turn recruit canonical PRC1, resulting in chromatin compaction. The compact chromatin and H3K27me3 in turn stimulate PRC2 and facilitate the spreading of H3K27me3. Meanwhile, H2Aub may recruit noncanonical PRC1 and lead to the spreading of H2Aub [70,71]. Thus, the formation of H3K27me3 and H2Aub domains involves several layers of positive feedback mechanisms in which PRC1 and PRC2 not only reinforce themselves but also enhance each other's function.

However, the new mechanism does not exclude the earlier one. PRC2 still localizes to a number of genes without the presence of noncanonical PRC1 but

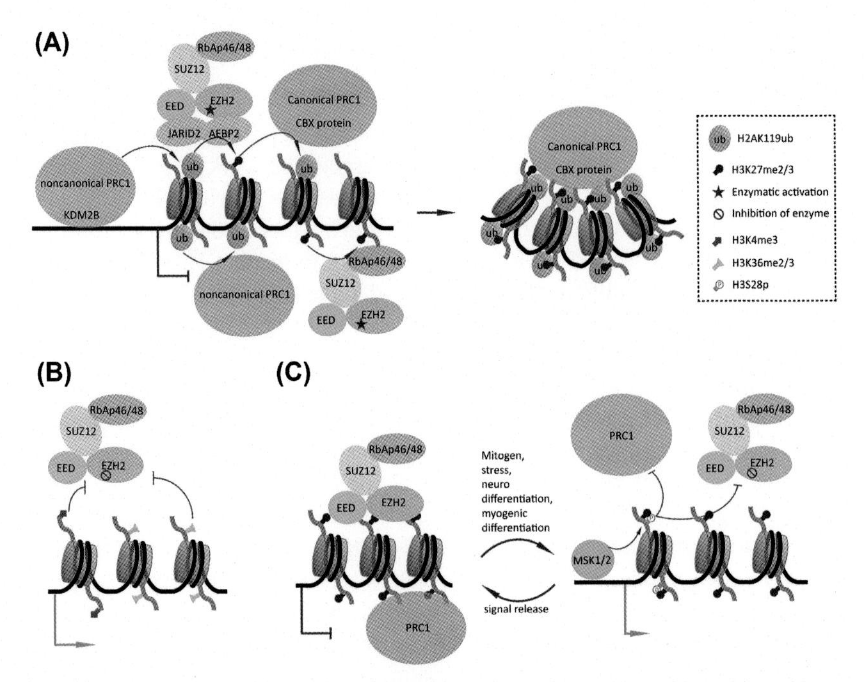

FIGURE 10.3 (A) A multilayered feedback mechanism for Polycomb repression. The initial step of Polycomb silencing depends on KDM2B that binds to CpG islands and recruits noncanonical PRC1 to target genes. Noncanonical PRC1 generates H2A K119 ubiquitination (H2Aub) on nearby nucleosomes. H2Aub in turn promotes the formation of H3K27me3 and H2Aub through recruiting AEBP2-PRC2 and noncanonical PRC1, respectively. H3K27me3 promotes H2Aub through recruiting canonical PRC1 and is also able to spread itself through mechanisms described in Fig. 10.2. Together, multiple positive feedback mechanisms facilitate the formation of H2Aub and H3K27me3 domains along the gene, leading to chromatin compaction and gene repression. (B) Active chromatin inhibits PRC2. Actively transcribed genes harbor H3K4me3 and H3K36me2/3 markers, both of which strongly inhibit PRC2 activity through the VEFS domain of SUZ12. This mechanism prevents PRC2 intrusion to active genes. (C) Signals to the chromatin pathway mediate transient gene activation. When cells are in certain conditions such as mitogen stimulation, stress, and differentiation, the ERK and p38 kinase pathways will activate MSK1/2, which then phosphorylate H3S28 near the promoters of their target genes, generating an H3K27me3-S28p double marker on genes that were previously repressed by Polycomb group proteins (PcG) proteins. This double marker evicts PRC1 and PRC2, leading to open chromatin suitable for transcription. Upon release of the signal, phosphorylation is removed, leaving H3K27me3, which is again bound by PcG proteins. This mechanism ensures rapid activation and resilencing of specific genes in response to external stimuli without the need for demethylation. In some cases, demethylation and acetylation occur, likely when turning on genes that should be activated for the long term.

with the presence of canonical PRC1. These genes are likely silenced through the classical hierarchy. In addition to the different silencing mechanisms, functional analysis shows that genes occupied by noncanonical PRC1 and canonical PRC1 tend to participate in different biological processes. Sole noncanonical PRC1 targets are enriched in germline regulators, whereas sole canonical PRC1 targets are enriched in developmental regulators [72]. These findings lead to the hypothesis that the different PRC1 complexes regulate different sets of genes, thus controlling different biological processes. During cell fate transition, the changes in the abundance of different PRC1 variants (as well as PRC2 variants) facilitate the reshaping of the chromatin landscape. This mechanism greatly enhances the versatility of the function of PcG proteins.

H3K4ME3 and H3K36ME2/3 Inhibit Polycomb Repressive Complex 2 Activity

H3K4me3 and H3K36me2/3 are markers of actively transcribed genes. H3K4me3 marks the TSS, whereas H3K36me2/3 decorates gene bodies. In *Drosophila*, genetic studies and genome-wide profiling studies established the roles of Trx (an H3K4-specific methyltransferase) and Ash1 (an H3K36-specific methyltransferase) in antagonizing PcG silencing [73,74], but the molecular mechanism was only recently revealed.

Biochemical studies show that the presence of H3K4me3 or H3K36me2/3 inhibits the enzymatic activity of PRC2 [29,42]. Unlike the previously described activation of PRC2 that occurs *in trans* [37,66], H3K4me3 or H3K36me2/3 peptides could not inhibit PRC2, implying that the inhibition acts *in cis*. That is to say, only physically proximal H3K4me3 and H3K36me2/3 have inhibitory functions. Indeed, active markers on the sister histone H3 within the same nucleosome are not able to inhibit PRC2 [28,42], and H3K4me3 and H3K27me3 may coexist asymmetrically on sister histones within mononucleosomes, providing one potential explanation for the formation of bivalent domains in ES cells and certain multipotent cells [75]. Intriguingly, enzyme dynamics analysis shows that the H3K4me3-mediated PRC2 inhibition exhibits features of allosteric regulation [42], suggesting that H3K4me3 must be sensed by PRC2. Given that EED and RbAP46/48 both exclude the H3K4me3 modified peptide, it is highly possible that the inhibition is mediated by EZH2 itself or SUZ12.

Detailed analyses reveal that the inhibition of PRC2 activity requires a minimum trimeric complex that contains EED, EZH2, and the VEFS domain of SUZ12, the last of which is also required for the H3 (31−42) mediated activation of PRC2 [42,66]. In plants, changing the SUZ12 homolog EMF2 to its paralog VRN2 abolished the inhibition without affecting basal activity [42], implying an essential role for SUZ12 in sensing the active modifications. Therefore the reason that SUZ12 can sense different fragments of H3 and modulate PRC2 enzymatic activity in opposite directions should be further investigated.

The positive feedback mechanisms, which involve H3K27me3 and nucleosome density−mediated PRC2 activation, are powerful in propagating and spreading H3K27 methylation, resulting in more than 80% of histones methylated by PRC2. On the other hand, the negative regulatory mechanism mediated by H3K4me3 and H3K36me2/3 provides a way to restrain or antagonize PRC2 activity, preventing the intrusion of PRC2-mediated H3K27 methylation into active genes (Fig. 10.3B).

H3S28 Phosphorylation Antagonizes Polycomb Silencing

Histone phosphorylation plays important roles in various biological processes, including chromosome condensation and segregation during mitosis and transcriptional regulation in response to extracellular stimuli [76]. In the latter case, the signal is transmitted to chromatin through a typical signal transduction pathway. It involves ligand sensing by a transmembrane receptor, followed by activation of a kinase cascade that amplifies the signal, then the next element of this cascade translocates into the nucleus. This last element may be a transcription factor, coactivator or corepressor, or a kinase that modifies nuclear proteins. This protein, together with other effectors, changes the expression of a specific set of genes, allowing the cells to respond to the signal.

Gene regulation by external signal-mediated H3S28 phosphorylation is one such example [77–79]. When cells undergo conditions such as stress, mitogen stimulation, or retinoic acid-induced neuronal differentiation, the histone kinases MSK1 and MSK2 are phosphorylated and activated through the ERK and p38 kinases pathways. MSK1 and MSK2 are recruited to the promoters of their target genes by unknown factors (different sets of genes under different conditions) and phosphorylate the S28 residue of histone H3 near the gene promoters. These genes were previously silenced by Polycomb proteins and decorated by H3K27me3. Upon H3S28 phosphorylation, PRC1 and PRC2 binding on gene promoters is compromised, resulting in gene activation. Artificially tethering a constitutively active but not kinase-dead MSK1 to a gene promoter is sufficient to activate target gene transcription [78]. The same pathway also regulates the phosphorylation of H3S10. This transient H3S10 phosphorylation and gene activation upon mitogen treatment is known as the nucleosome response. Interestingly, K9S10 and K27S28 modules share the same ARKS sequence motif [76], suggesting similar regulatory mechanisms.

The cross talk between K27 and S28, or K9 and S10 is based on the stereo counteraction. The phosphate group on S28p might prevent K27 from entering the substrate recognition channel, either through its negative charge or increased size. In addition, S28p might disrupt the stabilization of PRC1 and PRC2 on chromatin, which relies on the recognition of H3K27me3 through CBX proteins [53], EED and RbAP46/48 subunits, respectively. Similarly, phosphorylation of S10 disrupts the interaction between H3K9me3 and HP1 proteins [80,81]. Therefore, K27me3-S28p may act as a molecular switch, functioning to quickly respond to extracellular signals. Supporting this idea, upon mitogen stimulation of quiescent fibroblast cells, the H3K27me3-S28p double mark rapidly appears on the *ATF3* gene promoter within 30 min, accompanied by decreased PcG protein binding and increased transcription. The phosphorylation decreases 3 h after stimulation, concurrent with the rebinding of PcG proteins and repression of transcription. Therefore, this switch allows rapid activation and resilencing of genes through controlling PRC1 and PRC2 binding without the need for histone demethylation (Fig. 10.3C). Nevertheless, demethylation and acetylation of H3K27 may occur in certain conditions. For example, prolonged tethering of constitutively active MSK1 to *α-globin* or *c-fos* gene induces H3K27 acetylation on their promoters. This might also happen to genes that require constitutive activation in natural conditions.

Since Polycomb proteins and the ERK and p38 kinase pathways are all implicated in cancers, this H3K27me3-S28p switch may also be involved. The increased kinase activities might lead to the switch of the H3K27me3-S28p module, resulting in cellular transformation and cancer development [77]. Supporting this idea, in *Ras*-transformed fibroblasts where the ERK pathway is constitutively active, the expression of protooncogenes *JUN* and *FRA-1* is elevated with increased H3S28 phosphorylation at their promoters.

Recognition of H3K27me3 by EZH1-containing PRC2 is not affected by H3S28p [79]. During myoblast differentiation, MSK1-mediated H3S28 phosphorylation displaces EZH2-containing PRC2 from the *MyoG* promoter, which is then bound by EZH1-containing PRC2. The switch from EZH2 to EZH1 is required for proper activation of *MyoG* [79]. Intriguingly, the SET domains of EZH1 and EZH2 share 94% identity. Why the EZH1-containing PRC2 is insensitive to S28 phosphorylation and why the EZH1-containing PRC2 exhibits gene activation functions in this process are interesting questions that remain to be answered.

ACCESSORY COMPONENTS MODULATE POLYCOMB REPRESSIVE COMPLEX 2 ACTIVITY

In addition to the four core components of the PRC2 complex, several other proteins also appear as accessory PRC2 subunits in different purification schemes. These proteins are considered accessory components because some of them display cell type−specific expression patterns, and some exist within a subset of the PRC2 complex. They are not essential for the basal activity of PRC2 but may direct PRC2 recruitment or modulate PRC2 activity under specific conditions. Intensive studies have revealed the critical roles of these proteins in various biological processes. The presence of these accessory components adds another layer of PRC2 regulation, which greatly improves the flexibility and versatility of PRC2 function.

AEBP2

AEBP2 has been purified as a PRC2 component in various cell types [6,10,82]. It is a zinc-finger protein and possesses DNA binding ability, which may help to stabilize PRC2 on chromatin.

Histone methyltransferase assays show that AEBP2 can boost the enzymatic activity of PRC2, including EZH1-containing PRC2 [6,70,83]. The exact mechanism of AEBP2-mediated PRC2 activation is unclear. It is possible that AEBP2 might help stabilize the enzyme–substrate interaction through DNA binding. AEBP2-PRC2 shows little binding with nucleosomal DNA in vitro but does bind naked DNA, which suggests a potential role for AEBP2 in promoter regions where nucleosomes are depleted [84]. In addition, the electron microscopy structure of AEBP2-PRC2 shows that AEBP2 has extensive interactions with SUZ12 and EZH2, suggesting that it may have an allosteric effect on PRC2 [38].

The stimulation by AEBP2 may contribute to the relatively high H3K27me3 levels at PcG target genes. Core PRC2 is less efficient in generating H3K27me3. AEBP2-mediated PRC2 activation may facilitate further conversion of H3K27me1/2 to H3K27me3. Moreover, AEBP2 may stabilize PRC2 at target genes through DNA binding, thereby prolonging the time of catalysis, which may also contribute to H3K27me3 formation (Fig. 10.4). In

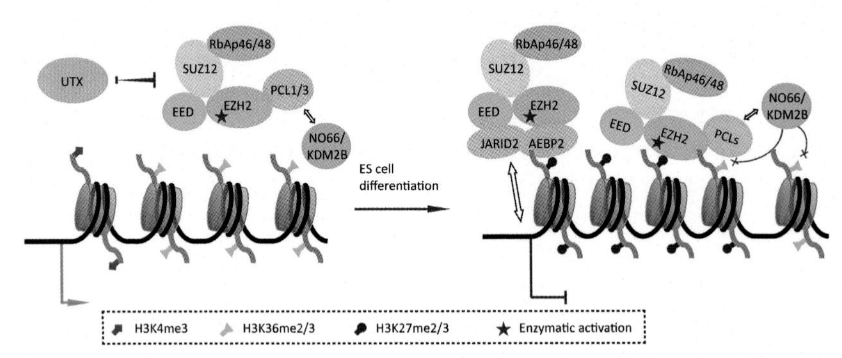

FIGURE 10.4 The roles of PRC2 accessory components in ES cells. (left panel) PRC2 accessory proteins help maintain H3K27 methylation in steady-state ES cells. PCL1 acts through boosting PRC2 to higher trimethylase activity. PCL1/3 could bind to H3K36me2/3 and interact with H3K36me2/3 demethylases NO66 or KDM2B, reflecting a counteraction between "activating" and "repressing" activities. While the "repressing" activity of PCL-PRC2 is dominant on developmental genes, "activating" activities such as Trx and UTX are dominant on pluripotent and housekeeping genes. (right panel) Upon ES cell differentiation, accessory components facilitate PRC2 function through multiple mechanisms. AEBP2 and JARID2K116me3 both stabilize PRC2 on promoters through interacting with DNA or nucleosomes, respectively, and they both stimulate PRC2 enzymatic activity. On gene bodies, PCL proteins are recruited to H3K36me2/3 through their Tudor domains, and the associating H3K36me2/3 demethylases remove the active markers, allowing PRC2 to spread H3K27 methylation. Previously described mechanisms, including H3K27me3, chromatin density and H2Aub-mediated PRC2 activation, may all be involved in the methylation spreading process.

contrast, at intergenic regions and non-PcG target genes the "hit-and-run" activity of PRC2 may not be sufficient to maintain a high level of H3K27me3.

A new role of AEBP2 was revealed by a study reporting that AEBP2 further stimulates PRC2 activity at H2Aub nucleosomes [70]. This finding together with others suggests a positive feedback mechanism for PRC1 and PRC2, in which PRC1 and PRC2 activities are both reinforced by their own products and each other's (see H2A K119 ubiquitination stimulates PRC2 activity section).

PCL Proteins (PHF1, MTF2, PHF19)

PCL1, PCL2, and PCL3, also known as PHF1, MTF2, and PHF19, respectively, are paralogs of *Drosophila* Pcl. They can all associate with PRC2 as accessory components [12–16,85]. The PCL proteins assist PRC2 recruitment in flies and mammalian cells [15,86] and play important roles in various biological processes such as ES cell self-renewal, differentiation, and X chromosome inactivation [17,18].

PCL1 is capable of modulating the enzymatic activity of PRC2 [12–14]. Histone methyltransferase assays demonstrate that PCL1-PRC2 is more efficient in generating H3K27me3 than core PRC2 alone. In PCL1 depleted cells, the PCL1-PRC2 target genes exhibit decreased H3K27me3 and increased H3K27me1/2, supporting the role of PCL1 in converting H3K27me1/2 to H3K27me3. Thus, PCL1 is able to boost PRC2 activity and contribute to H3K27me3 at PcG target genes. PCL2 and PCL3 have not been reported to stimulate PRC2 activity.

PCL proteins possess a Tudor domain and two PHD fingers. It has long been proposed that these domains of PCL protein may be involved in histone modification recognition. Several studies report that the Tudor domains of all three mammalian PCL proteins (but not *Drosophila* Pcl) bind H3K36me2/3 with high affinity [16,87,88]. H3K36me2/3 are active markers found at the bodies of active genes. Importantly, H3K36me2/3 strongly inhibits PRC2 activity to prevent PRC2 intrusion [29]. How to explain this obvious discrepancy and its impact on PRC2 function? One explanation is that PCL-PRC2 is involved in the de novo silencing of active genes during cell differentiation. Supporting this, the PCL3 Tudor domain is essential for proper differentiation of mouse ES cells [87]. Mechanistically, PCL3 could recruit the H3K36me2/3 demethylases NO66 or KDM2B during differentiation [16,87]. H3K36 demethylation eliminates the inhibitory effect on PRC2, allowing PRC2-mediated gene repression. In addition to facilitating gene silencing during differentiation, PCL-PRC2 is also required for H3K27me3 maintenance in steady-state cells [15,17,89]. Depletion or overexpression of PCL1/3 causes global decrease or increase of H3K27me3, respectively, reflecting a counteraction between methyltransferase and demethylase activities inside the cell. Tight regulation of these opposing activities ensures the fine balance between pluripotency and differentiation. Therefore, PCL proteins are not only involved in maintaining the H3K27me3 level at target genes but also required for de novo silencing of genes during differentiation (Fig. 10.4).

JARID2

JARID2 is a founding member of the Jumonji (Jmj) family proteins. It contains a JmjC domain, which is a typical histone demethylase domain, yet the JmjC domain of JARID2 lost its catalytic activity because of mutations at critical residues that are required for cofactor binding. JARID2 also contains an ARID domain and a zinc finger, both of which are DNA-binding domains that are potentially involved in PRC2 recruitment. Although much is known about the role of JARID2 in development [90], no evidence suggested a direct link between JARID2 and PcG proteins until several research teams in parallel reported the copurification of JARID2 with PRC2 in mouse ES cells [7–11]. As JARID2 is preferentially expressed in undifferentiated cells, it is easy to understand why previous studies missed it. The ES cell specificity of JARID2-PRC2 is also consistent with the idea that accessory factors regulate PRC2 in stage-specific or cell type–specific manners. Similar to PCL2 and PCL3, which are also highly expressed in ES cells, JARID2 plays an important role in ES cell differentiation. It is required for shutting down the pluripotent network and for efficiently initiating the expression of lineage-specific genes.

The impact of JARID2 on PRC2 activity started in debate. Several studies reported that JARID2 could inhibit or activate PRC2 enzymatic activities [7,8,10]. The in vivo effect of JARID2 on H3K27me3 was also reported in both directions [8,10,11,91], suggesting that JARID2 may have different functions in different contexts.

Several later studies reported that JARID2 stimulates PRC2 activity in vitro [70,84,91,92]. Detailed biochemistry analyses show that JARID2 could enhance the PRC2-nucleosome association and stimulate PRC2 activity [84]. These two activities are separable, implying that JARID2 may allosterically activate PRC2 as well as assist in substrate binding. Recently, a new finding shed light on the mechanism of the JARID2's stimulation function [92]. It was found that lysine 116 of JARID2, which is conserved from fly to human, is di- or trimethylated by EZH2 in vivo. This methylated form of JARID2 could bind to the aromatic cage of EED, where H3K27me3 could also bind [37]. The dissociation constant of the K116me3 peptide with EED is only 3.4 μM, representing a much higher affinity in comparison to that of the H3K27me3 peptide (36.4 μM). Importantly, JARID2K116me3 allosterically stimulated PRC2 activity approximately 30-fold, whereas nonmethylated JARID2 or JARID2K116A/R only achieved no more than 10-fold. Thus, at pure, unmodified oligonucleosome substrates, JARID2, especially the methylated form, could strongly stimulate PRC2 activity (Fig. 10.4).

However, in the context of native chromatin, the relationship between JARID2 and PRC2 activity becomes more complex, yet can be explained. Unbiased ChIP-seq analysis of H3K27me3 dynamics during ES cell differentiation in wild-type and JARID2 knockout (JARID2 KO) cells reveals that both abnormal gain and loss of H3K27me3 occur upon JARID2 deletion, but they occur at different regions. Abnormal loss of H3K27me3 occurs more at genic regions and especially the TSS, which is the main H3K27me3 target under normal conditions. This indicates a failure in H3K27me3 establishment during differentiation in JARID2 KO cells, supporting the role of JARID2 in PRC2 stimulation or PRC2 recruitment. In contrast, the abnormal H3K27me3 gain of H3K27me3 prefers intergenic regions. This does not contradict the JARID2's stimulatory role, as this abnormal H3K27me3 gain is likely caused by the aberrant recruitment of PRC2 to intergenic regions as a result of loss of JARID2. Thus, the direct impact of JARID2 on PRC2 appears to be mostly positive. During differentiation, JARID2 facilitates the recruitment of PRC2 to pluripotent genes as well as its stable attachment at previously bound developmental genes and boosts PRC2 activity to generate H3K27me3 seeds at the promoter region. Upon JARID2 depletion, the free PRC2 complex is aberrantly recruited to other sites, likely by other accessory factors or by chromatin signatures such as H3K27me3 seeds or high nucleosome density, leading to the formation of abnormal H3K27me3 domains.

The understanding of the function of JARID2 on PRC2 is deepened by the counteraction of JARID2K116me3 and H3K27me3. As mentioned, JARID2K116me3 exhibits an affinity 10-fold higher than that of H3K27me3. Thus, the JARID2K116me3's stimulatory effect is dominant because of the strong occupancy of the methyl lysine binding pocket of EED. Indeed, H3K27me3 titration could not affect JARID2K116me3's stimulation. The insensitivity of JARID2K116me3-PRC2 to H3K27me3 suggests that it may not be involved in propagation of H3K27me3, but is rather responsible for the initial recruitment of PRC2 and robust generation of H3K27me3 seeds at promoters. Subsequent spreading of H3K27me3 may be attributed to other mechanisms as discussed in previous sections or requires the demethylation of JARID2.

Unexpectedly, unknown chromatin signatures severely dampen JARID2K116me3's stimulation when using native chromatin as substrates [92]. Therefore, the JARID2K116me3-PRC2 activity in vivo is restrained by unknown factors. The identity of the chromatin signatures that inhibit JARID2K116me3-PRC2 is an interesting question.

In addition to AEBP2, PCL proteins and JARID2, other proteins are also reported to be accessory proteins of PRC2 [82]. For example, esPRC2p48 is an ES cell-specific PRC2 component that alone or in combination with JARID2 and PCL2 could stimulate PRC2 activity [91]. These accessory proteins could work in combination or in a mutually exclusive fashion, providing various means to regulate PRC2 recruitment and activity.

EZH1-Containing Polycomb Repressive Complex 2

The core catalytic subunit EZH2 has a variant called EZH1 [93]. EZH1 is a paralog of EZH2 and likely arises from an EZH2 gene duplication event. They share 63% overall identity and 94% identity of their SET domain. EZH1-containing PRC2 includes all the other three core components, EED, SUZ12, and RbAp46/48 [83,94,95].

Despite sequence similarities between EZH1 and EZH2, their functions are quite divergent. Analysis of the histone methyltransferase activity of EZH1-containing PRC2 shows that this complex displays weak activity at octamers and very low activity at nucleosomes compared to EZH2-containing PRC2 [84,95]. Interestingly, artificially tethering EZH1, even the catalytically dead version, causes target gene repression, while EZH2 is activity dependent. Because enzymatic activity is not required, it is reasonable to speculate that EZH1-containing PRC2 may affect chromatin structure. Indeed, EZH1-containing PRC2 is able to compact chromatin in vitro and in vivo, as demonstrated by electron microscopy and DNase I sensitivity assay [95]. The chromatin compaction activity depends on the presence of all four subunits but not the catalytic activity. In addition, the presence of histone tails is also required, which is different from PRC1-mediated compaction [96].

EZH1-containing PRC2-mediated chromatin compaction is dependent on all subunits and histone tails [95]. Actually, all four subunits of PRC2 can bind to histone tails. EED and RbAp46/48 bind to H3 tails using their central pockets [37,42]. EZH1 is able to bind nucleosomes [84], whereas EZH2 requires JARID2 to do so [84]. SUZ12 also interacts with H3 (31–42) [66]. If each subunit binds to an H3 tail of a different nucleosome, this quaternary binding will result in the compaction of four nucleosomes, consistent with the electron microscopy observation that one EZH1-containing PRC2 complex brings together three to four nucleosomes at a time [95].

In addition to the molecular activities, the in vivo roles of EZH1 and EZH2 also seem to be different, suggesting that they are not functionally redundant. First, H3K27me2/3 and PRC2 target gene repression are mostly dependent on EZH2 in ES cells. EZH1 only contributes to H3K27me3 at a subset of EZH2 target genes, and loss of EZH1 has little effect on target gene repression. EZH1 could only partially compensate for the loss of EZH2 during ES cell differentiation [83]. Second, EZH1 is widely expressed in virtually all adult cells, whereas EZH2 is preferentially expressed in proliferating cells but barely expressed in terminally differentiated cells [79,83,95,97]. This is also consistent with the fact that EZH1 lacks most of the cell cycle–regulated phosphorylation sites that are targeted by CDKs in EZH2. The different expression profiles suggest that there is a switch between EZH1 and EZH2 in different biological contexts (quiescence versus proliferation). Interestingly, in support of this, when NIH 3T3 cells are induced into quiescence by serum starvation, EZH2 levels are strongly reduced, whereas EZH1 is slightly elevated [95]. All these findings suggest that EZH2 and EZH1 may have distinct roles, rather than being redundant. While EZH2 establishes and maintains H3K27 methylation in the dividing cells of embryos, EZH1 maintains this methylation in terminally differentiated nonproliferating cells [84].

Our understanding of EZH1 is further complicated by the finding that catalytically active EZH1-containing PRC2 is required for gene activation during myogenic differentiation [79,97,98]. Genome-wide profiling shows that EZH1 is directly recruited to myogenic genes, including the master differentiation transcription factor *MyoG*, and is required for the activation of a subset of them. EZH1 occupancy at active genes is also observed in ES cells and developing hippocampal neurons [79,99,100]. Mechanistic studies suggest that EZH1 may function by interacting with elongating Pol II [97] and that an H3K27me3-S28p switch is involved for the initial eviction of EZH2-containing PRC2 (see H3S28 phosphorylation antagonizes Polycomb silencing section).

Classic PcG proteins function through repressing gene expression. The activation function of EZH1 raises several interesting questions. Is the activation function of EZH1 present in all cell types or only in some? Why do EZH1- and EZH2-containing PRC2 complexes exhibit such divergent functions? The structure of EZH1-containing PRC2 is crucial to explain the different activities of EZH1- and EZH2-containing PRC2.

H3K27M INHIBITS POLYCOMB REPRESSIVE COMPLEX 2 ACTIVITY AND LEADS TO PEDIATRIC GLIOBLASTOMA

In studies aiming to identify the causal mutations of pediatric glioblastomas (GBMs), researchers found frequent mutations in histone H3 coding genes. Interestingly, these different mutations result in amino acid substitutions at two specific positions of H3; K27M of H3.1 and H3.3; and G34R or G34V of H3.3 [101,102]. Up to 78% of pediatric diffuse intrinsic pontine gliomas (DIPGs) and 22% of nonbrain stem gliomas carry K27M, and gliomas that contain this mutation are exclusively of high grade (aggressive). H3K27 is the substrate of PRC2, leading to the question of whether H3K27M will affect H3K27 methylation. Indeed, the K27M mutation is shown to cause a global reduction of H3K27me2 and H3K27me3 in vivo, accompanied by a global increase of H3K27 acetylation [103–106]. What is interesting is that, in most patients, only one allele of histone H3 gene out of 32 is mutated. So H3K27M apparently acts in a dominant negative fashion.

Biochemical studies clearly show that H3K27M can inhibit PRC2 activity in vitro [104,105]. A peptide containing K27M is sufficient to inhibit PRC2. The IC_{50} of H3K27M peptide (67 nM) is at a similar range as that of GSK343 (27 nM), an inhibitor of EZH2 [105]. In another experiment, nucleosomal H3K27M inhibits PRC2 with $K_I = 2.1 \pm 0.9$ nM, whereas K_m for wild-type substrate is 67 nM. The inhibition appears to be competitive [107]. The strong inhibitory ability is in accordance with the fact that only a small fraction of H3K27M could cause the global decrease of H3K27me2/3 in patient cells. K27M might execute its inhibitory function through trapping PRC2 in vivo. Evidence shows that H3K27M peptide or nucleosomes can bind directly to EZH2 with higher affinity than wild-type nucleosomes [103–105], strongly suggesting that the sequestration of PRC2 by H3K27M causes decreased PRC2 activity and global loss of H3K27me2/3 (Fig. 10.5).

The structure of PRC2 with an H3K27M peptide reveals more details [32]. The mutant peptide indeed interacts with the EZH2 subunit and lies directly in the catalytic center. However, it is the H3R26 residue, but not the methionine residue, that occupies the lysine access channel where H3K27 should be located. The side chain of the methionine residue is not even visible in the crystal structure, arguing against the idea that methionine mimics methylated lysine. The clear part is that, arginine 26, a basic amino acid similar to but slightly larger than lysine, takes over the catalytic channel and plays an important role in "poisoning" PRC2. In support of this, mutation of R26 to alanine (H3R26AK27M) or asymmetric dimethylation (R26me2a) abrogated its inhibition of PRC2 activity [32,107]. It is unclear why only the K to M mutation (or K to I as shown in vitro) inhibits PRC2 activity, given that the side chain of methionine is not visible in crystal structure. Detailed biochemical and structural analysis are needed to answer this question.

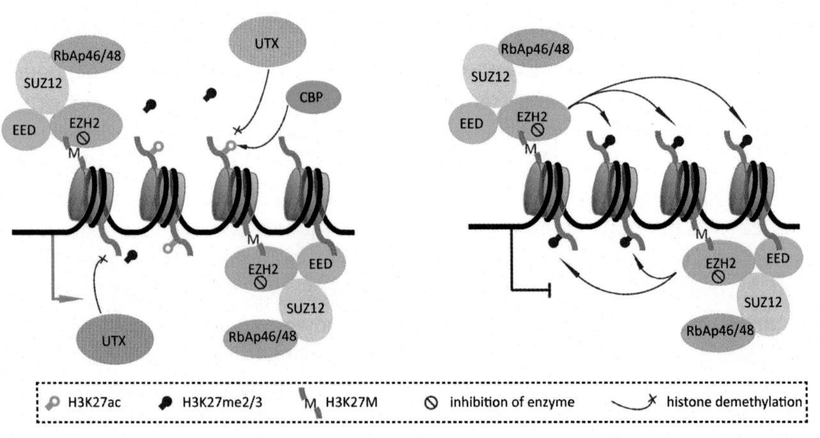

FIGURE 10.5 H3K27M sequesters PRC2 activity in pediatric glioblastomas. The heterozygous H3K27M mutation is frequently observed in pediatric glioblastomas. Histones carrying this mutation display higher affinity toward the PRC2 catalytic center. Therefore although only a small proportion of histone H3 is mutated in cancer cells, they can diminish PRC2 activity in a dominant negative manner through trapping PRC2. Reduced availability of PRC2 leads to a global decrease of H3K27me2/3 and increase of H3K27 acetylation. In contrast, some regions abnormally acquire H3K27me3 and are repressed. In this scenario, H3K27M might act as an initial recruiter of PRC2 on these regions.

Recently, the structure of human PRC2 together with an H3K27M peptide was reported. Different from the *Chaetomium thermophilum* PRC2 structure reported by Jiao et al. [32], the methionine 27 resided right in the lysine access channel of human PRC2. The difference may be due to either species specificity or different crystallization conditions. The new structure supports an inhibition model that is consistent with the known biochemical data [32a].

The highly specific and frequent mutation in H3K27 makes PRC2 a potential target for treating pediatric glioblastomas. However, several concerns should be carefully considered. First, H3K27M is not the only driver of pediatric glioblastomas, as overexpression of H3K27M in mice only leads to clusters of proliferating cells and does not induce gliomas, even in the absence of p53 [104]. A study reported the establishment of a pediatric glioblastoma model that was achieved by coexpressing H3.3K27M and PDGFRA D842V (a constitutively active form), accompanied by p53 loss in neural progenitor cells [108]. This finding suggests that multiple driving forces act together to promote glioblastoma, but not H3K27M alone. Second, the DNA methylation and H3K27me3 patterns are quite different between primary tumor samples and established tumor cell lines [103,105]. Third, despite the global decrease of H3K27me3, there are still hundreds of genes that display increased H3K27me3 levels. It is currently unknown how the escaping PRC2 is targeted to these regions. An interesting hypothesis is that H3K27M could act as a seed, triggering the spreading of H3K27me3 around it [109]. The genes that gain H3K27me3 are enriched for various cancer pathways [103], suggesting that the tumorigenesis promoted by H3K27M is a complex result of PRC2 dysfunction and not merely PRC2 inhibition. Therefore, the effect of H3K27M on the chromatin landscape and cell signaling is more complicated than expected. Thus, when studying GBM using a model system, it is important to establish standards to evaluate whether and to what extent the model cell line mimics in vivo GBM [105,110].

Attempts have been made to treat GBM by targeting histone modification pathways [110]. Inhibiting the H3K27me3 demethylase JMJD3 by GSKJ4 in K27M mutant brainstem glioma cell lines results in elevated H3K27me2/3 and inhibition of cell growth. GSKJ4 can also inhibit the growth of K27M brainstem glioma xenografts in vivo. In another study, a chemical screen that targeted epigenetic players was conducted in a GBM model cell line. The most effective hit in this screen was MI-2, an inhibitor of menin, which is a member of the trithorax family histone methyltransferase complex. MI-2 is also effective in inhibiting the growth of cancer cells derived from a patient carrying H3.3K27M positive DIPG. Therefore, epigenetic regulators could be considered potential therapeutic targets for treating pediatric GBM.

CONCLUSION

The noncatalytic subunits of PRC2 not only support the basal activity of PRC2 but also accept various inputs from chromatin, which are then translated into changes in EZH2 conformation and PRC2 activity. Depending the input properties, PRC2 activities are modulated in different directions. Repressive chromatin, such as that marked by H3K27me2/3, H2Aub, or with high density, can greatly stimulate PRC2. This facilitates the establishment and maintenance of H3K27me3, thus providing a self-reinforcing mechanism of heterochromatin. In contrast, active chromatin, characterized by open chromatin, H3K4me3 and H3K36me2/3 marks, could inhibit PRC2 activity, preventing the intrusion of PRC2 to active genes. The phosphorylation of the H3S28 residue counteracts Polycomb silencing through reducing the binding of PRC1 and PRC2 to H3K27me3. Thus, it is involved in modulating gene expression in response to external stimuli. PRC2 activity is also regulated by accessory components through multiple mechanisms, providing versatile ways to modulate PRC2 activity in various cells and biological processes. Thus, PRC2 is under tight regulation during development. Abnormal loss or gain of PRC2 activity is often linked to diseases such as cancer, and inhibitors of EZH2 are being tested in several clinical trials to treat different cancers.

LIST OF ACRONYMS AND ABBREVIATIONS

AR Androgen receptor
CRPC Castration-resistant prostate cancer
GBM Glioblastoma
H2Aub H2A K119 ubiquitination
H3K27me2/3 H3K27 di- and trimethylation
H3K27me3 H3K27 trimethylation
H3S28p H3S28 phosphorylation
JARID2-PRC2 PRC2 complex containing JARID2
PcG proteins Polycomb group proteins
PRC1 Polycomb repressive complex 1
PRC2 Polycomb repressive complex 2
PRE Polycomb response element
TSS Transcription start site

GLOSSARY

Accessory components of PRC2 Proteins that are not required for the basal activity of PRC2, and only contribute a subset of PRC2 in cell type specific manner. Including JARID2, AEBP2, PCL1/2/3, etc.

Active marker/Active modification Histone modifications that associate with actively transcribed genes, including H3K4me3, H3K36me2/3, histone acetylation, etc.

Allosteric activation Activation of an enzyme by an effector molecule that changes the conformation of the enzyme through binding to the noncatalytic site.

Repressive marker/Repressive modification Histone modifications that associate with repressed genes, including H3K27me2/3, H3K9me3, H4K20me3, H2A K119ub, etc.

ACKNOWLEDGMENTS

The lab of B.Z. is supported by the China National Science Foundation (Grants 31425013, 31530037, 31521002, 91419305), the Chinese Ministry of Science and Technology (Grant 2015CB856200), the Strategic Priority Research Program of the Chinese Academy of Sciences (Grant XDB08010103), and the Howard Hughes Medical Institute International Early Career Scientist Program.

REFERENCES

[1] Tie F, Furuyama T, Prasad-Sinha J, Jane E, Harte PJ. The *Drosophila* Polycomb Group proteins ESC and E(Z) are present in a complex containing the histone-binding protein p55 and the histone deacetylase RPD3. Development 2001;128:275–86.

[2] Cao R, Wang L, Wang H, Xia L, Erdjument-Bromage H, Tempst P, Jones RS, Zhang Y. Role of histone H3 lysine 27 methylation in Polycomb-group silencing. Science 2002;298:1039–43.

[3] Czermin B, Melfi R, McCabe D, Seitz V, Imhof A, Pirrotta V. *Drosophila* enhancer of Zeste/ESC complexes have a histone H3 methyltransferase activity that marks chromosomal Polycomb sites. Cell 2002;111:185–96.

[4] Kuzmichev A, Nishioka K, Erdjument-Bromage H, Tempst P, Reinberg D. Histone methyltransferase activity associated with a human multiprotein complex containing the Enhancer of Zeste protein. Genes Dev 2002;16:2893–905.

[5] Muller J, Hart CM, Francis NJ, Vargas ML, Sengupta A, Wild B, Miller EL, O'Connor MB, Kingston RE, Simon JA. Histone methyltransferase activity of a *Drosophila* Polycomb group repressor complex. Cell 2002;111:197–208.

[6] Cao R, Zhang Y. SUZ12 is required for both the histone methyltransferase activity and the silencing function of the EED-EZH2 complex. Mol Cell 2004;15:57–67.

[7] Peng JC, Valouev A, Swigut T, Zhang JM, Zhao YM, Sidow A, Wysocka J. Jarid2/Jumonji coordinates control of PRC2 enzymatic activity and target gene occupancy in pluripotent cells. Cell 2009;139:1290–302.

[8] Shen X, Kim W, Fujiwara Y, Simon MD, Liu Y, Mysliwiec MR, Yuan GC, Lee Y, Orkin SH. Jumonji modulates polycomb activity and self-renewal versus differentiation of stem cells. Cell 2009;139:1303–14.

[9] Landeira D, Sauer S, Poot R, Dvorkina M, Mazzarella L, Jorgensen HF, Pereira CF, Leleu M, Piccolo FM, Spivakov M, Brookes E, Pombo A, Fisher C, Skarnes WC, Snoek T, Bezstarosti K, Demmers J, Klose RJ, Casanova M, Tavares L, Brockdorff N, Merkenschlager M, Fisher AG. Jarid2 is a PRC2 component in embryonic stem cells required for multi-lineage differentiation and recruitment of PRC1 and RNA Polymerase II to developmental regulators. Nat Cell Biol 2010;12:618–24.

[10] Li G, Margueron R, Ku MC, Chambon P, Bernstein BE, Reinberg D. Jarid2 and PRC2, partners in regulating gene expression. Genes Dev 2010;24:368–80.

[11] Pasini D, Cloos PA, Walfridsson J, Olsson L, Bukowski JP, Johansen JV, Bak M, Tommerup N, Rappsilber J, Helin K. JARID2 regulates binding of the Polycomb repressive complex 2 to target genes in ES cells. Nature 2010;464:306–10.

[12] Nekrasov M, Klymenko T, Fraterman S, Papp B, Oktaba K, Kocher T, Cohen A, Stunnenberg HG, Wilm M, Muller J. Pcl-PRC2 is needed to generate high levels of H3-K27 trimethylation at Polycomb target genes. EMBO J 2007;26:4078–88.

[13] Cao R, Wang H, He J, Erdjument-Bromage H, Tempst P, Zhang Y. Role of hPHF1 in H3K27 methylation and Hox gene silencing. Mol Cell Biol 2008;28:1862–72.

[14] Sarma K, Margueron R, Ivanov A, Pirrotta V, Reinberg D. Ezh2 requires PHF1 to efficiently catalyze H3 lysine 27 trimethylation in vivo. Mol Cell Biol 2008;28:2718–31.

[15] Walker E, Chang WY, Hunkapiller J, Cagney G, Garcha K, Torchia J, Krogan NJ, Reiter JF, Stanford WL. Polycomb-like 2 associates with PRC2 and regulates transcriptional networks during mouse embryonic stem cell self-renewal and differentiation. Cell Stem Cell 2010;6:153–66.

[16] Ballare C, Lange M, Lapinaite A, Martin GM, Morey L, Pascual G, Liefke R, Simon B, Shi Y, Gozani O, Carlomagno T, Benitah SA, Di Croce L. Phf19 links methylated Lys36 of histone H3 to regulation of Polycomb activity. Nat Struct Mol Biol 2012;19:1257–65.

[17] Hunkapiller J, Shen Y, Diaz A, Cagney G, McCleary D, Ramalho-Santos M, Krogan N, Ren B, Song JS, Reiter JF. Polycomb-like 3 promotes polycomb repressive complex 2 binding to CpG islands and embryonic stem cell self-renewal. PLoS Genet 2012;8. e1002576.

[18] Casanova M, Preissner T, Cerase A, Poot R, Yamada D, Li XZ, Appanah R, Bezstarosti K, Demmers J, Koseki H, Brockdorff N. Polycomblike 2 facilitates the recruitment of PRC2 Polycomb group complexes to the inactive X chromosome and to target loci in embryonic stem cells. Development 2011;138:1471–82.

[19] Jones RS, Gelbart WM. The *Drosophila* Polycomb-group gene Enhancer of zeste contains a region with sequence similarity to trithorax. Mol Cell Biol 1993;13:6357–66.

[20] Smith BC, Denu JM. Chemical mechanisms of histone lysine and arginine modifications. Biochim Biophys Acta 2009;1789:45–57.

[21] Pasini D, Bracken AP, Jensen MR, Lazzerini Denchi E, Helin K. Suz12 is essential for mouse development and for EZH2 histone methyltransferase activity. EMBO J 2004;23:4061–71.

[22] Nekrasov M, Wild B, Muller J. Nucleosome binding and histone methyltransferase activity of *Drosophila* PRC2. EMBO Rep 2005;6:348–53.

[23] Ketel CS, Andersen EF, Vargas ML, Suh J, Strome S, Simon JA. Subunit contributions to histone methyltransferase activities of fly and worm polycomb group complexes. Mol Cell Biol 2005;25:6857–68.

[24] Cheng X, Collins RE, Zhang X. Structural and sequence motifs of protein (histone) methylation enzymes. Annu Rev Biophys Biomol Struct 2005;34:267–94.

[25] McCabe MT, Graves AP, Ganji G, Diaz E, Halsey WS, Jiang Y, Smitheman KN, Ott HM, Pappalardi MB, Allen KE, Chen SB, Della Pietra 3rd A, Dul E, Hughes AM, Gilbert SA, Thrall SH, Tummino PJ, Kruger RG, Brandt M, Schwartz B, Creasy CL. Mutation of A677 in histone methyltransferase EZH2 in human B-cell lymphoma promotes hypertrimethylation of histone H3 on lysine 27 (H3K27). Proc Natl Acad Sci USA 2012;109:2989–94.

[26] Sneeringer CJ, Scott MP, Kuntz KW, Knutson SK, Pollock RM, Richon VM, Copeland RA. Coordinated activities of wild-type plus mutant EZH2 drive tumor-associated hypertrimethylation of lysine 27 on histone H3 (H3K27) in human B-cell lymphomas. Proc Natl Acad Sci USA 2010;107:20980–5.

[27] Jung HR, Pasini D, Helin K, Jensen ON. Quantitative mass spectrometry of histones H3.2 and H3.3 in Suz12-deficient mouse embryonic stem cells reveals distinct, dynamic post-translational modifications at Lys-27 and Lys-36. Mol Cell Proteomics 2010;9:838–50.

[28] Voigt P, LeRoy G, Drury 3rd WJ, Zee BM, Son J, Beck DB, Young NL, Garcia BA, Reinberg D. Asymmetrically modified nucleosomes. Cell 2012;151:181–93.

[29] Yuan W, Xu M, Huang C, Liu N, Chen S, Zhu B. H3K36 methylation antagonizes PRC2-mediated H3K27 methylation. J Biol Chem 2011;286:7983–9.

[30] Young NL, DiMaggio PA, Plazas-Mayorca MD, Baliban RC, Floudas CA, Garcia BA. High throughput characterization of combinatorial histone codes. Mol Cell Proteomics 2009;8:2266–84.

[31] Yap DB, Chu J, Berg T, Schapira M, Cheng SWG, Moradian A, Morin RD, Mungall AJ, Meissner B, Boyle M, Marquez VE, Marra MA, Gascoyne RD, Humphries RK, Arrowsmith CH, Morin GB, Aparicio SAJR. Somatic mutations at EZH2 Y641 act dominantly through a mechanism of selectively altered PRC2 catalytic activity, to increase H3K27 trimethylation. Blood 2011;117:2451–9.

[32] Jiao L, Liu X. Structural basis of histone H3K27 trimethylation by an active polycomb repressive complex 2. Science 2015;350. aac4383.

[32a] Justin N, Zhang Y, Tarricone C, Martin SR, Chen S, Underwood E, De Marco V, Haire LF, Walker PA, Reinberg D, Wilson JR, Gamblin SJ. Structural basis of oncogenic histone H3K27M inhibition of human Polycomb Repressive Complex 2. Nat Commun 2016; 7:11316–26.

[33] Wu H, Zeng H, Dong A, Li F, He H, Senisterra G, Seitova A, Duan S, Brown PJ, Vedadi M, Arrowsmith CH, Schapira M. Structure of the catalytic domain of EZH2 reveals conformational plasticity in cofactor and substrate binding sites and explains oncogenic mutations. PLoS One 2013;8. e83737.

[34] Antonysamy S, Condon B, Druzina Z, Bonanno JB, Gheyi T, Zhang F, MacEwan I, Zhang A, Ashok S, Rodgers L, Russell M, Gately Luz J. Structural context of disease-associated mutations and putative mechanism of autoinhibition revealed by X-ray crystallographic analysis of the EZH2-SET domain. PLoS One 2013;8. e84147.

[35] Southall SM, Wong PS, Odho Z, Roe SM, Wilson JR. Structural basis for the requirement of additional factors for MLL1 SET domain activity and recognition of epigenetic marks. Mol Cell 2009;33:181–91.

[36] Xu K, Wu ZJ, Groner AC, He HH, Cai C, Lis RT, Wu X, Stack EC, Loda M, Liu T, Xu H, Cato L, Thornton JE, Gregory RI, Morrissey C, Vessella RL, Montironi R, Magi-Galluzzi C, Kantoff PW, Balk SP, Liu XS, Brown M. EZH2 oncogenic activity in castration-resistant prostate cancer cells is Polycomb-independent. Science 2012; 338:1465–9.

[37] Margueron R, Justin N, Ohno K, Sharpe ML, Son J, Drury WJ, Voigt P, Martin SR, Taylor WR, De Marco V, Pirrotta V, Reinberg D, Gamblin SJ. Role of the polycomb protein EED in the propagation of repressive histone marks. Nature 2009;461: 762–7.

[38] Ciferri C, Lander GC, Maiolica A, Herzog F, Aebersold R, Nogales E. Molecular architecture of human polycomb repressive complex 2. eLife 2012;1. e00005.

[39] Verreault A, Kaufman PD, Kobayashi R, Stillman B. Nucleosomal DNA regulates the core-histone-binding subunit of the human Hat1 acetyltransferase. Curr Biol 1998;8:96–108.

[40] Song JJ, Garlick JD, Kingston RE. Structural basis of histone H4 recognition by p55. Genes Dev 2008;22:1313–8.

[41] Nowak AJ, Alfieri C, Stirnimann CU, Rybin V, Baudin F, Ly-Hartig N, Lindner D, Muller CW. Chromatin-modifying complex component Nurf55/p55 associates with histones H3 and H4 and polycomb repressive complex 2 subunit Su(z)12 through partially overlapping binding sites. J Biol Chem 2011;286:23388–96.

[42] Schmitges FW, Prusty AB, Faty M, Stutzer A, Lingaraju GM, Aiwazian J, Sack R, Hess D, Li L, Zhou SL, Bunker RD, Wirth U, Bouwmeester T, Bauer A, Ly-Hartig N, Zhao KH, Chan HM, Gu J, Gut H, Fischle W, Muller J, Thoma NH. Histone methylation by PRC2 is inhibited by active chromatin marks. Mol Cell 2011;42:330–41.

[43] Lee ST, Li Z, Wu Z, Aau M, Guan P, Karuturi RK, Liou YC, Yu Q. Context-specific regulation of NF-kappaB target gene expression by EZH2 in breast cancers. Mol Cell 2011;43:798–810.

[44] Shi B, Liang J, Yang X, Wang Y, Zhao Y, Wu H, Sun L, Zhang Y, Chen Y, Li R, Zhang Y, Hong M, Shang Y. Integration of estrogen and Wnt signaling circuits by the polycomb group protein EZH2 in breast cancer cells. Mol Cell Biol 2007;27:5105–19.

[45] Varambally S, Dhanasekaran SM, Zhou M, Barrette TR, Kumar-Sinha C, Sanda MG, Ghosh D, Pienta KJ, Sewalt RGAB, Otte AP, Rubin MA, Chinnaiyan AM. The polycomb group protein EZH2 is involved in progression of prostate cancer. Nature 2002;419:624–9.

[46] Seruga B, Ocana A, Tannock IF. Drug resistance in metastatic castration-resistant prostate cancer. Nat Rev Clin Oncol 2011;8:12–23.

[47] Cha TL, Zhou BP, Xia W, Wu Y, Yang CC, Chen CT, Ping B, Otte AP, Hung MC. Akt-mediated phosphorylation of EZH2 suppresses methylation of lysine 27 in histone H3. Science 2005;310:306–10.

[48] Wu C, Jin X, Yang J, Yang Y, He Y, Ding L, Pan Y, Chen S, Jiang J, Huang H. Inhibition of EZH2 by chemo- and radiotherapy agents and small molecule inhibitors induces cell death in castration-resistant prostate cancer. Oncotarget 2016;7:3440–52.

[49] Barski A, Cuddapah S, Cui K, Roh TY, Schones DE, Wang Z, Wei G, Chepelev I, Zhao K. High-resolution profiling of histone methylations in the human genome. Cell 2007;129:823–37.

[50] Ferrari KJ, Scelfo A, Jammula S, Cuomo A, Barozzi I, Stutzer A, Fischle W, Bonaldi T, Pasini D. Polycomb-dependent H3K27me1 and H3K27me2 regulate active transcription and enhancer fidelity. Mol Cell 2014;53:49–62.

[51] Lee HG, Kahn TG, Simcox A, Schwartz YB, Pirrotta V. Genome-wide activities of Polycomb complexes control pervasive transcription. Genome Res 2015;25:1170–81.

[52] Wu H, Chen X, Xiong J, Li Y, Li H, Ding X, Liu S, Chen S, Gao S, Zhu B. Histone methyltransferase G9a contributes to H3K27 methylation in vivo. Cell Res 2011;21:365–7.

[53] Schwartz YB, Pirrotta V. Ruled by ubiquitylation: a new order for polycomb recruitment. Cell Rep 2014;8:321–5.

[54] Hansen KH, Bracken AP, Pasini D, Dietrich N, Gehani SS, Monrad A, Rappsilber J, Lerdrup M, Helin K. A model for transmission of the H3K27me3 epigenetic mark. Nat Cell Biol 2008;10:1291–300.

[55] Xu M, Wang W, Chen S, Zhu B. A model for mitotic inheritance of histone lysine methylation. EMBO Rep 2012;13:60–7.

[56] Scharf AN, Barth TK, Imhof A. Establishment of histone modifications after chromatin assembly. Nucleic Acids Res 2009;37:5032–40.

[57] Alabert C, Barth TK, Reveron-Gomez N, Sidoli S, Schmidt A, Jensen ON, Imhof A, Groth A. Two distinct modes for propagation of histone PTMs across the cell cycle. Genes Dev 2015;29:585–90.

[58] Huang C, Xu M, Zhu B. Epigenetic inheritance mediated by histone lysine methylation: maintaining transcriptional states without the precise restoration of marks? Philos Trans R Soc Lond B Biol Sci 2013;368:20110332.

[59] Martin C, Zhang Y. Mechanisms of epigenetic inheritance. Curr Opin Cell Biol 2007;19:266–72.

[60] Xu M, Long C, Chen X, Huang C, Chen S, Zhu B. Partitioning of histone H3-H4 tetramers during DNA replication-dependent chromatin assembly. Science 2010;328:94–8.

[61] Hall IM, Shankaranarayana GD, Noma K, Ayoub N, Cohen A, Grewal SI. Establishment and maintenance of a heterochromatin domain. Science 2002;297:2232–7.

[62] Liu N, Zhang Z, Wu H, Jiang Y, Meng L, Xiong J, Zhao Z, Zhou X, Li J, Li H, Zheng Y, Chen S, Cai T, Gao S, Zhu B. Recognition of H3K9 methylation by GLP is required for efficient establishment of H3K9 methylation, rapid target gene repression, and mouse viability. Genes Dev 2015;29:379–93.

[63] Al-Sady B, Madhani HD, Narlikar GJ. Division of labor between the chromodomains of HP1 and Suv39 methylase enables coordination of heterochromatin spread. Mol Cell 2013;51:80–91.

[64] Xu C, Bian CB, Yang W, Galka M, Hui OY, Chen C, Qiu W, Liu HD, Jones AE, MacKenzie F, Pan P, Li SSC, Wang HB, Min JR. Binding of different histone marks differentially regulates the activity and specificity of polycomb repressive complex 2 (PRC2). Proc Natl Acad Sci USA 2010;107:19266–71.

[65] Martin C, Cao R, Zhang Y. Substrate preferences of the EZH2 histone methyltransferase complex. J Biol Chem 2006;281:8365–70.

[66] Yuan W, Wu T, Fu H, Dai C, Wu H, Liu N, Li X, Xu M, Zhang Z, Niu T, Han Z, Chai J, Zhou XJ, Gao S, Zhu B. Dense chromatin activates Polycomb repressive complex 2 to regulate H3 lysine 27 methylation. Science 2012;337:971–5.

[67] Bhaumik SR, Smith E, Shilatifard A. Covalent modifications of histones during development and disease pathogenesis. Nat Struct Mol Biol 2007;14:1008–16.

[68] Blackledge NP, Farcas AM, Kondo T, King HW, McGouran JF, Hanssen LLP, Ito S, Cooper S, Kondo K, Koseki Y, Ishikura T, Long HK, Sheahan TW, Brockdorff N, Kessler BM, Koseki H, Klose RJ. Variant PRC1 complex-dependent H2A ubiquitylation drives PRC2 recruitment and polycomb domain formation. Cell 2014;157:1445–59.

[69] Cooper S, Dienstbier M, Hassan R, Schermelleh L, Sharif J, Blackledge NP, De Marco V, Elderkin S, Koseki H, Klose R, Heger A, Brockdorff N. Targeting polycomb to pericentric heterochromatin in embryonic stem cells reveals a role for H2AK119u1 in PRC2 recruitment. Cell Rep 2014;7:1456–70.

[70] Kalb R, Latwiel S, Baymaz HI, Jansen PW, Muller CW, Vermeulen M, Muller J. Histone H2A monoubiquitination promotes histone H3 methylation in Polycomb repression. Nat Struct Mol Biol 2014;21:569–71.

[71] Arrigoni R, Alam SL, Wamstad JA, Bardwell VJ, Sundquist WI, Schreiber-Agus N. The Polycomb-associated protein Rybp is a ubiquitin binding protein. FEBS Lett 2006;580:6233–41.

[72] Morey L, Aloia L, Cozzuto L, Benitah SA, Di Croce L. RYBP and Cbx7 define specific biological functions of polycomb complexes in mouse embryonic stem cells. Cell Rep 2013;3:60–9.

[73] Klymenko T, Muller J. The histone methyltransferases Trithorax and Ash1 prevent transcriptional silencing by Polycomb group proteins. EMBO Rep 2004;5:373–7.

[74] Schwartz YB, Kahn TG, Stenberg P, Ohno K, Bourgon R, Pirrotta V. Alternative epigenetic chromatin states of polycomb target genes. PLoS Genet 2010;6. e1000805.

[75] Voigt P, Tee WW, Reinberg D. A double take on bivalent promoters. Genes Dev 2013;27:1318–38.

[76] Sawicka A, Seiser C. Histone H3 phosphorylation - a versatile chromatin modification for different occasions. Biochimie 2012;94:2193–201.

[77] Gehani SS, Agrawal-Singh S, Dietrich N, Christophersen NS, Helin K, Hansen K. Polycomb group protein displacement and gene activation through MSK-dependent H3K27me3S28 phosphorylation. Mol Cell 2010;39:886–900.

[78] Lau PN, Cheung P. Histone code pathway involving H3 S28 phosphorylation and K27 acetylation activates transcription and antagonizes polycomb silencing. Proc Natl Acad Sci USA 2011;108:2801–6.

[79] Stojic L, Jasencakova Z, Prezioso C, Stutzer A, Bodega B, Pasini D, Klingberg R, Mozzetta C, Margueron R, Puri PL, Schwarzer D, Helin K, Fischle W, Orlando V. Chromatin regulated interchange between polycomb repressive complex 2 (PRC2)-Ezh2 and PRC2-Ezh1 complexes controls myogenin activation in skeletal muscle cells. Epigenet Chromatin 2011;4:16.

[80] Fischle W, Tseng BS, Dormann HL, Ueberheide BM, Garcia BA, Shabanowitz J, Hunt DF, Funabiki H, Allis CD. Regulation of HP1-chromatin binding by histone H3 methylation and phosphorylation. Nature 2005;438:1116–22.

[81] Hirota T, Lipp JJ, Toh BH, Peters JM. Histone H3 serine 10 phosphorylation by Aurora B causes HP1 dissociation from heterochromatin. Nature 2005;438:1176–80.

[82] Alekseyenko AA, Gorchakov AA, Kharchenko PV, Kuroda MI. Reciprocal interactions of human C10orf12 and C17orf96 with PRC2 revealed by BioTAP-XL cross-linking and affinity purification. Proc Natl Acad Sci USA 2014;111:2488–93.

[83] Shen X, Liu Y, Hsu YJ, Fujiwara Y, Kim J, Mao X, Yuan GC, Orkin SH. EZH1 mediates methylation on histone H3 lysine 27 and complements EZH2 in maintaining stem cell identity and executing pluripotency. Mol Cell 2008;32:491–502.

[84] Son J, Shen SS, Margueron R, Reinberg D. Nucleosome-binding activities within JARID2 and EZH1 regulate the function of PRC2 on chromatin. Genes Dev 2013;27:2663–77.

[85] Li XZ, Isono K, Yamada D, Endo TA, Endoh M, Shinga J, Mizutani-Koseki Y, Otte AP, Casanova M, Kitamura H, Kamijo T, Sharif J, Ohara O, Toyada T, Bernstein BE, Brockdorff N, Koseki H. Mammalian polycomb-like Pcl2/Mtf2 is a novel regulatory component of PRC2 that can differentially modulate polycomb activity both at the Hox gene cluster and at Cdkn2a genes. Mol Cell Biol 2011;31:351–64.

[86] Savla U, Benes J, Zhang J, Jones RS. Recruitment of *Drosophila* Polycomb-group proteins by Polycomblike, a component of a novel protein complex in larvae. Development 2008;135:813–7.

[87] Brien GL, Gambero G, O'Connell DJ, Jerman E, Turner SA, Egan CM, Dunne EJ, Jurgens MC, Wynne K, Piao LH, Lohan AJ, Ferguson N, Shi XB, Sinha KM, Loftus BJ, Cagney G, Bracken AP. Polycomb PHF19 binds H3K36me3 and recruits PRC2 and demethylase NO66 to embryonic stem cell genes during differentiation. Nat Struct Mol Biol 2012;19:1273–81.

[88] Musselman CA, Avvakumov N, Watanabe R, Abraham CG, Lalonde ME, Hong Z, Allen C, Roy S, Nunez JK, Nickoloff J, Kulesza CA, Yasui A, Cote J, Kutateladze TG. Molecular basis for H3K36me3 recognition by the Tudor domain of PHF1. Nat Struct Mol Biol 2012;19:1266–72.

[89] Cai L, Rothbart SB, Lu R, Xu BW, Chen WY, Tripathy A, Rockowitz S, Zheng DY, Patel DJ, Allis CD, Strahl BD, Song J, Wang GG. An H3K36 methylation-engaging Tudor motif of polycomb-like proteins mediates PRC2 complex targeting. Mol Cell 2013;49:571–82.

[90] Herz HM, Shilatifard A. The JARID2-PRC2 duality. Genes Dev 2010;24:857–61.

[91] Zhang Z, Jones A, Sun CW, Li C, Chang CW, Joo HY, Dai QA, Mysliwiec MR, Wu LC, Guo YH, Yang W, Liu KM, Pawlik KM, Erdjument-Bromage H, Tempst P, Lee Y, Min JR, Townes TM, Wang HB. PRC2 complexes with JARID2, MTF2, and esPRC2p48 in ES

cells to modulate ES cell pluripotency and somatic cell reprograming. Stem Cells 2011;29:229–40.

[92] Sanulli S, Justin N, Teissandier A, Ancelin K, Portoso M, Caron M, Michaud A, Lombard B, da Rocha ST, Offer J, Loew D, Servant N, Wassef M, Burlina F, Gamblin SJ, Heard E, Margueron R. Jarid2 methylation via the PRC2 complex regulates H3K27me3 deposition during cell differentiation. Mol Cell 2015;57:769–83.

[93] Laible G, Wolf A, Dorn R, Reuter G, Nislow C, Lebersorger A, Popkin D, Pillus L, Jenuwein T. Mammalian homologues of the Polycomb-group gene Enhancer of zeste mediate gene silencing in *Drosophila* heterochromatin and at *S. cerevisiae* telomeres. EMBO J 1997;16:3219–32.

[94] Ho L, Crabtree GR. An EZ mark to miss. Cell Stem Cell 2008;3:577–8.

[95] Margueron R, Li G, Sarma K, Blais A, Zavadil J, Woodcock CL, Dynlacht BD, Reinberg D. Ezh1 and Ezh2 maintain repressive chromatin through different mechanisms. Mol Cell 2008;32:503–18.

[96] Francis NJ, Kingston RE, Woodcock CL. Chromatin compaction by a polycomb group protein complex. Science 2004;306:1574–7.

[97] Mousavi K, Zare H, Wang AH, Sartorelli V. Polycomb protein Ezh1 promotes RNA polymerase II elongation. Mol Cell 2012;45:255–62.

[98] Riising EM, Helin K. A new role for the polycomb group protein Ezh1 in promoting transcription. Mol Cell 2012;45:145–6.

[99] Henriquez B, Bustos FJ, Aguilar R, Becerra A, Simon F, Montecino M, van Zundert B. Ezh1 and Ezh2 differentially regulate PSD-95 gene transcription in developing hippocampal neurons. Mol Cell Neurosci 2013;57:130–43.

[100] Margueron R, Reinberg D. The Polycomb complex PRC2 and its mark in life. Nature 2011;469:343–9.

[101] Schwartzentruber J, Korshunov A, Liu XY, Jones DT, Pfaff E, Jacob K, Sturm D, Fontebasso AM, Quang DA, Tonjes M, Hovestadt V, Albrecht S, Kool M, Nantel A, Konermann C, Lindroth A, Jager N, Rausch T, Ryzhova M, Korbel JO, Hielscher T, Hauser P, Garami M, Klekner A, Bognar L, Ebinger M, Schuhmann MU, Scheurlen W, Pekrun A, Fruhwald MC, Roggendorf W, Kramm C, Durken M, Atkinson J, Lepage P, Montpetit A, Zakrzewska M, Zakrzewski K, Liberski PP, Dong Z, Siegel P, Kulozik AE, Zapatka M, Guha A, Malkin D, Felsberg J, Reifenberger G, von Deimling A, Ichimura K, Collins VP, Witt H, Milde T, Witt O, Zhang C, Castelo-Branco P, Lichter P, Faury D, Tabori U, Plass C, Majewski J, Pfister SM, Jabado N. Driver mutations in histone H3.3 and chromatin remodelling genes in paediatric glioblastoma. Nature 2012;482:226–31.

[102] Wu G, Broniscer A, McEachron TA, Lu C, Paugh BS, Becksfort J, Qu C, Ding L, Huether R, Parker M, Zhang J, Gajjar A, Dyer MA, Mullighan CG, Gilbertson RJ, Mardis ER, Wilson RK, Downing JR, Ellison DW, Zhang J, Baker SJ, St. Jude Children's Research Hospital-Washington University Pediatric Cancer Genome Project. Somatic histone H3 alterations in pediatric diffuse intrinsic pontine gliomas and non-brainstem glioblastomas. Nat Genet 2012;44:251–3.

[103] Chan K-M, Fang D, Gan H, Hashizume R, Yu C, Schroeder M, Gupta N, Mueller S, James CD, Jenkins R, Sarkaria J, Zhang Z. The histone H3.3K27M mutation in pediatric glioma reprograms H3K27 methylation and gene expression. Genes Dev 2013;27:985–90.

[104] Lewis PW, Müller MM, Koletsky MS, Cordero F, Lin S, Banaszynski LA, Garcia BA, Muir TW, Becher OJ, Allis CD. Inhibition of PRC2 activity by a gain-of-function H3 mutation found in pediatric glioblastoma. Science 2013;340:857–61.

[105] Bender S, Tang Y, Lindroth AM, Hovestadt V, Jones DT, Kool M, Zapatka M, Northcott PA, Sturm D, Wang W, Radlwimmer B, Hojfeldt JW, Truffaux N, Castel D, Schubert S, Ryzhova M, Seker-Cin H, Gronych J, Johann PD, Stark S, Meyer J, Milde T,

Schuhmann M, Ebinger M, Monoranu CM, Ponnuswami A, Chen S, Jones C, Witt O, Collins VP, von Deimling A, Jabado N, Puget S, Grill J, Helin K, Korshunov A, Lichter P, Monje M, Plass C, Cho YJ, Pfister SM. Reduced H3K27me3 and DNA hypomethylation are major drivers of gene expression in K27M mutant pediatric high-grade gliomas. Cancer Cell 2013;24:660–72.

[106] Venneti S, Garimella MT, Sullivan LM, Martinez D, Huse JT, Heguy A, Santi M, Thompson CB, Judkins AR. Evaluation of histone 3 lysine 27 trimethylation (H3K27me3) and enhancer of Zest 2 (EZH2) in pediatric glial and glioneuronal tumors shows decreased H3K27me3 in H3F3A K27M mutant glioblastomas. Brain Pathol 2013;23:558–64.

[107] Brown ZZ, Muller MM, Jain SU, Allis CD, Lewis PW, Muir TW. Strategy for "detoxification" of a cancer-derived histone mutant based on mapping its interaction with the methyltransferase PRC2. J Am Chem Soc 2014;136:13498–501.

[108] Funato K, Major T, Lewis PW, Allis CD, Tabar V. Use of human embryonic stem cells to model pediatric gliomas with H3.3K27M histone mutation. Science 2014;346:1529–33.

[109] Lewis PW, Allis CD. Poisoning the "histone code" in pediatric gliomagenesis. Cell Cycle 2013;12:3241–2.

[110] Kallappagoudar S, Yadav R, Lowe B, Partridge J. Histone H3 mutations—a special role for H3.3 in tumorigenesis? Chromosoma 2015;124:177–89.

Chapter 11

Activating Mutations of the EZH2 Histone Methyltransferase in Cancer

R.G. Kruger, A.P. Graves, M.T. McCabe
GlaxoSmithKline, Collegeville, PA, United States

Chapter Outline

Introduction to Chromatin and EZH2 **259**
Amplification and Overexpression of EZH2 in Cancer **262**
Regulation of Normal B-Cell Differentiation by EZH2 **262**
Mutation and Biochemical Activity of EZH2 **264**
Discovery and Incidence of EZH2 Tyrosine 641 Mutations 264
Biochemical Activity of Y641 EZH2 Mutants 264
Discovery of Additional Gain-of-Function EZH2 Mutations **266**
Altered Substrate Specificity of A677G and A687V EZH2 Mutants 267
Structural Rationale for Altered Substrate Specificity in EZH2 Mutants **268**
Y641 Mutations Have Dual Effects on Substrate Preference 271
The A677G Mutation Optimizes Y641 Positioning for All Three Methylation Reactions 272
EZH2 A687 Coordinates a Water Molecule Required for Substrate Monomethylation 272
Cellular Activity of EZH2 Mutants **274**
Loss-of-Function EZH2 Mutations Commonly Occur in Myeloid Malignancies **275**
Discovery of EZH2 Inhibitors **275**
Mechanistic and Phenotypic Effects of EZH2 Inhibitors in Cancer Cells **278**
Conclusions **280**
List of Acronyms and Abbreviations **281**
References **282**

INTRODUCTION TO CHROMATIN AND EZH2

Every cell within the human body contains approximately 3.2 billion base pairs of DNA spread across 23 chromosome pairs and encodes 20,000–25,000 protein-coding genes. Despite the fact that every cell contains the same DNA sequence, different cell types can have very different morphologies and functions due in part to epigenetic regulation. As cells differentiate from

Polycomb Group Proteins. http://dx.doi.org/10.1016/B978-0-12-809737-3.00011-8

pluripotent stem cells, each commitment step is associated with many gene silencing and activation events which together dictate the transcriptome, and consequently the behavior, of daughter cells. These transcriptional regulation events are often associated with epigenetic processes such as histone modification (e.g., acetylation, methylation, phosphorylation, etc.) or DNA methylation. This epigenetic regulation is mediated primarily by large groups of proteins that are often referred to as readers, writers, erasers, and remodelers. The writers are enzymes that catalyze the transfer of posttranslational modifications onto DNA or histone substrates. The erasers remove the posttranslational modifications placed on DNA or histones by the writers. The reader proteins contain specialized protein domains that recognize the presence or absence of the posttranslational modifications. Lastly, the remodelers are generally ATP-dependent enzymes that can reposition or eject nucleosomes from chromatin. One such example of a writer enzyme, and the focus of this review, is EZH2.

The *Enhancer of zeste homolog 2 (EZH2)* gene encodes a SET domain-containing lysine methyltransferase that is typically found in association with a multiprotein complex referred to as the Polycomb repressive complex 2 (PRC2) (Fig. 11.1A). EZH2 is the only catalytic subunit of PRC2 and is believed to be primarily responsible for mediating methylation of histone H3 on lysine 27 (H3K27) [1–4] (Fig. 11.1B). When histones within a gene's promoter region are modified with dimethylation (me2) or trimethylation (me3) on H3K27, this is often associated with condensation and transcriptional repression of the local chromatin [1–6] (Fig. 11.1B). Enhancer of zeste (E(z)), the *Drosophila* ortholog of EZH2, was discovered because of its transcriptional regulation of the *white* (*w*) gene within the fly eye and for its significant effects on homeotic patterning [7,8]. In humans, PRC2 activity contributes to the maintenance of the self-renewal capacity of stem cells, both embryonic and tissue specific, and mediates the dynamic transcriptional regulation of genes required for proper development and differentiation [2,5,6,9].

In addition to EZH2, the core PRC2 complex also contains EED, SUZ12, AEBP2, and RbAp46/48 [2] (Fig. 11.1A). Although the functions of each component of PRC2 have not yet been fully characterized, a 3-member complex of EZH2, EED, and SUZ12 is minimally required for catalytic activity in vitro [3,4,10]. The RbAp46/48 and AEBP2 proteins, on the other hand, are thought to stimulate PRC2 activity by mediating interactions with chromatin [10]. In addition to the proteins within PRC2 itself, PRC2 also interacts with a wide array of proteins and noncoding RNAs within the cellular context. It is believed that these interactions function to modulate the catalytic activity of PRC2, substrate preference, and localization within the genome [11–14].

Over the past 10–15 years there has been significant interest in the biological activity of EZH2 both in normal and cancer cells. In particular, the finding that EZH2 is overexpressed in numerous cancers and that this

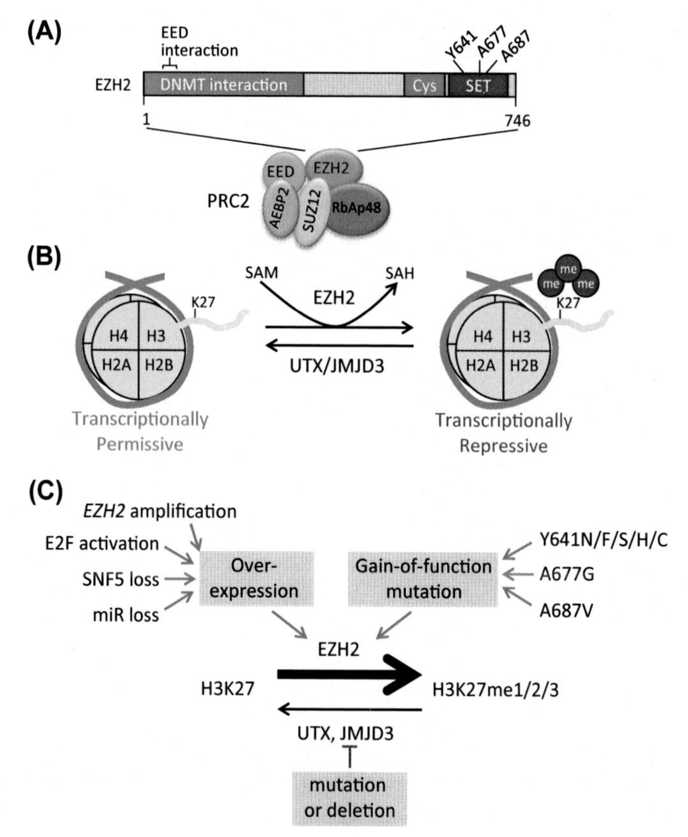

FIGURE 11.1 The EZH2 lysine methyltransferase maintains a transcriptionally repressive chromatin state and is dysregulated in cancer. (A) Domain structure of EZH2 including the EED and DNMT1 interaction domains, a cysteine-rich domain, and the catalytic SET domain. Amino acids whose mutation leads to gain-of-function phenotypes are indicated. Components of the Polycomb repressive complex 2 (PRC2) are also shown. (B) Schematic demonstrating methylation of the amino-terminal tail of histone H3 on lysine 27 (H3K27) within the context of a nucleosome. During this process a methyl group is transferred from SAM to the lysine generating SAH and methylated H3K27. Di- and trimethylation of H3K27 are associated with transcriptional repression. This process can be reversed by the lysine demethylases UTX and JMJD3. (C) The balance of methylated and unmethylated H3K27 is shifted toward increased H3K27me3 in many cancers. There can be a number of causes for this including, but not limited to, EZH2 overexpression, EZH2 gain-of-function mutation, and inactivation of UTX or JMJD3.

overexpression correlates with tumor aggressiveness, metastasis, and poor prognosis has generated significant interest in targeting EZH2 as a therapeutic strategy for cancer [15–18]. Furthermore, EZH2 was recently discovered to harbor somatic gain-of-function mutations in a subset of non-Hodgkin's lymphomas [19–22]. This chapter will discuss recent progress in understanding the role of EZH2 in human cancer with a particular focus on gain-of-function mutations, and review recent progress from both academic and industry groups working to develop methods for inhibiting EZH2.

AMPLIFICATION AND OVEREXPRESSION OF EZH2 IN CANCER

Overexpression of EZH2 was first documented in metastatic prostate cancer [17], but subsequent studies have revealed marked overexpression of EZH2 in the advanced stages of most solid tumors including breast, neuroendocrine lung, bladder, renal, skin, head and neck, liver, and others [15–18,23–28]. Consistent with EZH2 being overexpressed in aggressive and metastatic tumors, elevated expression of EZH2 also correlates with poor prognosis in patients. The mechanisms underlying this overexpression are quite diverse including amplification [29,30], activation of E2F transcription factors due to loss of *RB1* or *CDKN2A* [29], loss of the BAF chromatin remodeler complex member SNF5/INI1/SMARCB1 [31], and silencing of microRNAs (e.g., miR-101, -26a, and -214) that function to repress EZH2 transcripts [32–39] (Fig. 11.1C).

Interestingly, while EZH2 itself is dysregulated in a large fraction of tumors, other mechanisms also exist to disrupt levels and patterning of H3K27 methylation. Since methylation of H3K27 is reversible through the activity of histone demethylases such as UTX and JMJD3, H3K27 methylation levels are dictated by the relative rates of both methylation and demethylation reactions (Fig. 11.1C). In the case of EZH2 overexpression, increased EZH2-mediated methyltransferase activity increases H3K27 methylation. However, a similar effect can be achieved by decreasing the rate of H3K27 demethylation. In fact, the H3K27 demethylase UTX is inactivated by mutation or deletion at a relatively high frequency in a broad array of tumor types including transitional cell bladder carcinoma, esophageal squamous cell carcinoma, renal cell carcinoma, and multiple myeloma [40–42]. In renal cell carcinoma, H3K27me3 levels correlate with progression-free survival [43] and in esophageal squamous cell carcinoma disease severity and poor tumor differentiation correlate with H3K27me3 levels [44]. Recent work has also found that H3K27me3 levels are elevated in subgroup 4 medulloblastoma and detailed genomic analyses revealed that this is due to somatic mutation or deletion of *UTX* [45]. Thus, when considering the impact of dysregulated H3K27 methylation in cancer, it is necessary to consider a number of possible mechanisms that may impact this important histone modification.

REGULATION OF NORMAL B-CELL DIFFERENTIATION BY EZH2

While overexpression of EZH2 is prevalent in many solid tumor types, this is generally not the case in B-cell lymphomas. One explanation for this may be the high basal expression of EZH2 in normal proliferating B cells. EZH2 is normally highly expressed in pro-B cells and its expression progressively declines as cells transition into pre-B cells, then immature B cells, and eventually recirculating B cells [46] (Fig. 11.2). Within the context of the bone marrow, EZH2 expression is strictly required for differentiation of pro-B cells into pre-B cells and immature B cells as demonstrated by the accumulation of cells at the pro-B cell stage when EZH2 is genetically inactivated [46]. However, when EZH2 is inactivated after the pro-B cell stage, additional maturation steps occur normally suggesting that EZH2 functions primarily early in B-cell differentiation [46]. In contrast, when EZH2 is overexpressed in hematopoietic stem cells and evaluated in serial transplantation models, enhanced self-renewal capacity is observed indicating that EZH2 likely contributes to repopulating potential and prevents or delays replicative stress [47].

A mature B cell can be activated when it is exposed to an antigen and signals from T cells. These cells, termed centroblasts, divide rapidly within the dark zone of germinal centers (GCs) within lymph nodes. During this period of rapid expansion, GC B cells begin to express significant levels of EZH2 protein again [48,49]. These early GC B cells are thought to be the cell of origin for Burkitt's lymphomas and germinal center B cell (GCB) diffuse large B cell lymphomas (DLBCL) [50] (Fig. 11.2). Occasionally during the GC reaction, centroblasts exit the dark zone and enter the light zone where they cease division and become centrocytes. The cell of origin for follicular lymphomas (FLs) is believed to be a centrocyte (Fig. 11.2). Centrocytes interact with follicular dendritic cells and T cells within the light zone, and they can also differentiate into memory B cells or plasma cells [50]. Concomitant with the decreased proliferation that occurs in centrocytes, expression of EZH2 protein also decreases in these later GC B cells [48,49]. These findings clearly demonstrate that EZH2 expression is tightly correlated with both differentiation state and proliferation rate of normal B cells.

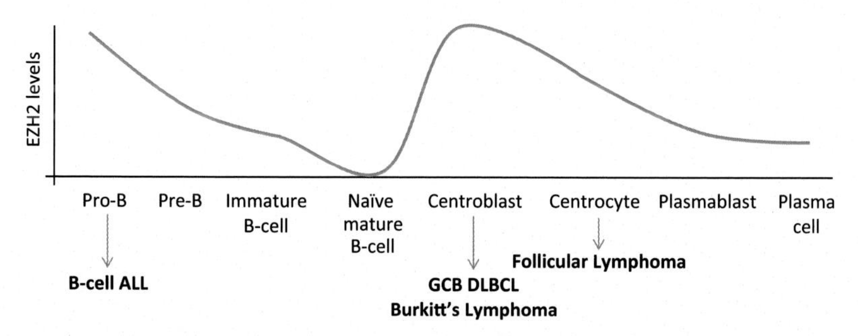

FIGURE 11.2 EZH2 is dynamically regulated throughout B-cell development. Graphical representation of the approximate levels of EZH2 protein at various stages of B-cell development. The cell of origin is also indicated for a number of tumor types, including some where EZH2 has been found to exhibit gain-of-function mutations.

MUTATION AND BIOCHEMICAL ACTIVITY OF EZH2

Discovery and Incidence of EZH2 Tyrosine 641 Mutations

Despite the lack of obvious overexpression of EZH2 in B-cell lymphomas, in 2009, Marco Marra and colleagues discovered recurrent somatic mutations of EZH2 in a subset of B-cell lymphomas [22]. Given the dynamic differential expression of EZH2 throughout multiple stages of B-cell development and maturation, it was intriguing that mutations were specifically found in GCB DLBCL and FL, but not in ABC DLBCL, mantle cell lymphoma, or a number of other lymphoma subtypes [22]. In a cohort of 83 GCB DLBCLs and 221 FLs, 18 (21.7%) and 16 (7.2%) cases, respectively, were found to harbor missense mutations at the EZH2 tyrosine 641 (Y641) residue (numbering based on EZH2 transcript NM_001203247) [22]. The mutations found included Y641 to histidine (H), asparagine (N), serine (S), phenylalanine (F), and cysteine (C) [22]. Many subsequent studies confirmed these results with mutation frequencies as high as 27.6% in FL [19,51–53].

In addition to GCB DLBCL and FLs, large-scale sequencing efforts have also identified a limited number of Y641 EZH2 mutations in select solid tumors. To date, three studies have reported mutation of EZH2 Y641 in skin cancers with mutations in one of eight cutaneous squamous cell carcinomas (cSCC) [54] and three of 268 melanomas [55,56]. The only other cancer reported to harbor EZH2 Y641 mutations is parathyroid cancer where 2 of 193 tumors harbored Y641N EZH2 mutations [57]. The basis for this tissue-specific mutation pattern is not understood.

Biochemical Activity of Y641 EZH2 Mutants

The Y641 residue of EZH2 is located within the protein's catalytic SET domain and is highly conserved across many species and lysine methyltransferases [21,22] (Figs. 11.1A and 11.3). This conservation implies

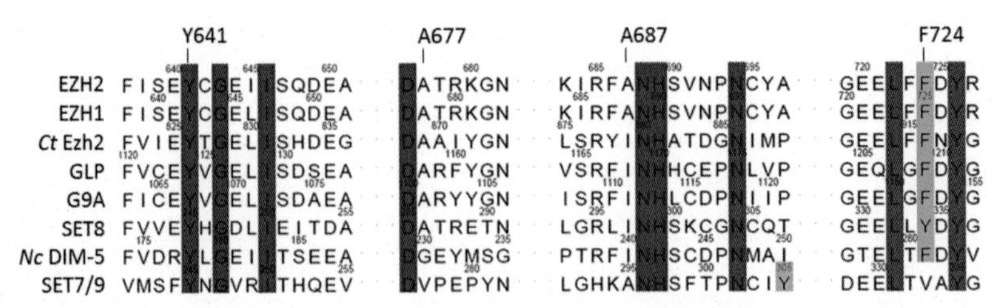

FIGURE 11.3 Amino acids involved in the coordination of the lysine substrate are highly conserved across homologs and orthologs. Alignment of protein sequences from human EZH2, human EZH1, *Chaetomium thermophilum* Ezh2, human GLP, human G9A, human SET8, *Neurospora crassa* DIM-5, and human SET7/9. Identical residues are highlighted in magenta, and the F/Y switch residue for each methyltransferase is highlighted in green. Although the F/Y switch residue of SET7/9 (Y305) is not aligned in the primary sequence, structurally it occupies similar space as the F/Y switch residues indicated for the homologous methlytransferases.

mechanistic or structural importance of the Y641 residue, and it was therefore not surprising when these EZH2 mutants were inactive in early biochemical assays [22]. Four of the Y641 mutant forms of EZH2 (Y641H/N/S/F) were expressed, copurified as 5-member PRC2 complexes, and found to be inactive in a biochemical methylation assay using a biotinylated peptide substrate composed of amino acids 21−44 of histone H3 [22]. However, given that: (1) the mutations appeared to be specific to Y641; (2) only a subset of possible Y641 mutations were identified (i.e., H/N/S/F/C); (3) mutations were always heterozygous leaving one wild-type (WT) allele intact; and (4) both alleles were always expressed [22], it was hypothesized by several groups that the mutations may lead to a more complex effect than a simple loss-of-function.

Indeed, detailed studies of the steady-state kinetics and substrate preferences of the EZH2 Y641 mutants revealed that these mutations were in fact not loss-of-function as previously reported, but rather they induced an altered substrate preference leading to an apparent change- or gain-of-function [58]. While cells containing WT EZH2 express detectable levels of mono-, di-, and trimethylation of H3K27, in the context of biochemical assays WT EZH2 exhibits a strong preference for unmethylated and monomethylated H3K27 substrates as measured by catalytic efficiency (k_{cat}/K_M) (H3K27me0:me1:me2 ratio = 9:6:1) [21,58] (Fig. 11.4). Consequently, WT EZH2 preferentially produces H3K27me1 and H3K27me2 and much less efficiently produces the transcriptionally repressive H3K27me3.

Tyrosine 641 EZH2 mutants, on the other hand, were found to act primarily on dimethylated H3K27 with little to no activity for unmodified or monomethylated H3K27 (H3K27me0:me1:me2 k_{cat}/K_M ratio = 1:2:13) [21,58,59] (Fig. 11.4). In fact, the Y641 mutants exhibit so little activity with an unmethylated substrate, that steady state kinetics could only be determined with this substrate using the most active Y641F EZH2 mutant, but even with this mutant a 10-fold decrease in catalytic efficiency was observed when comparing unmethylated versus dimethylated substrate [21,58] (Fig. 11.4). Since the initial studies of Y641 EZH2 mutants utilized unmodified histone

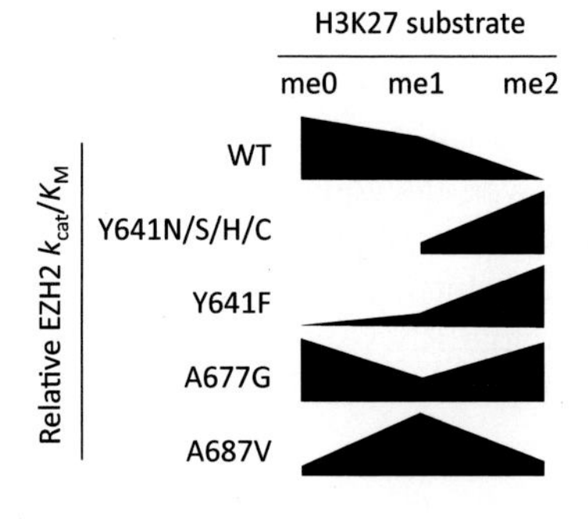

FIGURE 11.4 Gain-of-function mutations alter the normal substrate specificity of EZH2. An approximation of the catalytic activity (k_{cat}/K_M) of wild-type and mutant forms of EZH2 using H3K27me0, me1, or me2 as substrates.

peptide substrates where H3K27 is unmethylated, it is not surprising that these enzymes were found to be inactive as this is not a preferred substrate.

This differential substrate preference between WT and Y641 mutants combined with the observation that mutations were always heterozygous and both alleles were expressed led to a proposal that EZH2 Y641 mutant cancer cells required a coordination of WT and mutant activities to efficiently generate the abnormally elevated levels of fully methylated H3K27me3 product [58]. In fact, this was recently demonstrated to be the case by evaluating product generation via mass spectrometry in the presence of WT EZH2 alone, mutant EZH2 alone, or mixtures of the two [60]. When WT EZH2 was incubated with an unmethylated substrate, mono- and dimethylation were produced, but little H3K27me3 was detected. The Y641N/H/S/C EZH2 mutants alone produced no mono-, di-, or trimethylation of H3K27 when starting with an unmethylated substrate. On the other hand, when both WT EZH2 and either of Y641N/H/S/C EZH2 were incubated with unmethylated substrate, H3K27me3 was efficiently generated confirming that the two enzymes can work in coordination to efficiently trimethylate H3K27 [60]. Interestingly, the Y641F EZH2 mutant, the only Y641 mutant to have measurable activity with an unmethylated K27 substrate in earlier studies [21,58], was distinct from the other Y641 mutants in that it generated detectable amounts of H3K27me3 on its own [60].

DISCOVERY OF ADDITIONAL GAIN-OF-FUNCTION EZH2 MUTATIONS

Following the discovery of mutations at Y641 of EZH2, several groups began extensively sequencing B-cell lymphomas to establish the true incidence of Y641 EZH2 mutations and to determine whether there were additional gain-of-function mutations. While several additional mutated residues were identified [61–65], it was unclear which, if any, of these would alter the activity of EZH2. By quantifying the levels of H3K27me3 in >100 cancer cell lines, it was found that the majority of cell lines with the highest levels of H3K27me3 harbored Y641 EZH2 mutations [21]. However, the Pfeiffer DLBCL cell line exhibited some of the highest levels of H3K27me3, yet did not harbor a mutation at EZH2 Y641 [21]. Full-length sequencing of the *EZH2* gene revealed mutation of the A677 residue to a glycine (A677G) in this cell line [21]. Similar to EZH2 Y641, this residue is located within the catalytic SET domain and is most often conserved across many species and lysine methyltransferases [21] (Figs. 11.1A and 11.3). However, the equivalent residue in the *Neurospora crassa* DIM-5 methyltransferase is a glycine suggesting that this mutant may retain biochemical activity [21]. Interestingly, this mutation was previously identified in one sample in a large RNA-seq study of >100 non-Hodgkin lymphomas [62] and was subsequently identified in 1 of 41

DLBCLs [21] and 7 of 366 FLs [51] for an overall mutation frequency of approximately 1–2% of DLBCL and FLs.

A third EZH2 residue that exhibits recurrent somatic mutation in cancers is alanine 687 (A687). This residue resides within the EZH2 SET domain and lies immediately adjacent to the Asparagine-Histidine-Serine (NHS) motif which is [66] (Figs. 11.1A and 11.3). Interestingly, while the equivalent residues of human EZH1 and *Drosophila* E(z) are both alanine, this residue is often found as either a valine (31%) or isoleucine (35%) in other SET domain methyltransferases suggesting that this mutation may alter, rather than disrupt the enzyme's biochemical activity [66]. EZH2 A687 has been found to be mutated to a valine (A687V) in 1 of 127 non-Hodgkin lymphomas [62], 1 of 49 DLBCLs [67], 5 of 366 FLs [51], and 1 of 11 B-cell acute lymphoblastic leukemia (B-ALL) cell lines [66]. This mutation frequency is similar to A677 with approximately 1% of DLBCL and FL samples harboring the EZH2 A687V mutation.

While recurrent EZH2 mutations have thus far been reported primarily in GCB DLBCL and FL, the identification of potential gain-of-function mutations in B-cell ALL cell lines raises the question whether mutation of EZH2 may be an oncogenic strategy for other B-cell–derived cancers. As discussed previously, GCB DLBCL and FL arise from centroblasts and centrocytes, respectively [50,68]. However, common B-cell ALL arises from pre-pro-B cells or pro-B cells [69] (Fig. 11.2). Cells in these stages of B-cell differentiation have elevated EZH2 levels similar to B-cells undergoing the GC reaction [46,48]. While additional screening is required to determine the incidence of EZH2 mutations in B-cell ALL, it appears that activating mutations of EZH2 may preferentially occur in cancers derived from cell types that normally express relatively high levels of EZH2. This finding may indicate that silencing of EZH2 is required for progression to the next stage of differentiation while hyperactive forms of EZH2 impede further differentiation and may promote transformation in the presence of additional oncogenic mutations.

Altered Substrate Specificity of A677G and A687V EZH2 Mutants

Given the recurrent nature of the A677G and A687V EZH2 mutants and their location within the SET domain, both seemed to be likely candidates for change- or gain-of-function mutations similar to EZH2 Y641 mutations. Interestingly, while WT EZH2 displays decreasing catalytic efficiency with increased methylation of H3K27 and Y641 EZH2 mutants show increasing activity with increased methylation, A677G EZH2 displayed a distinct profile with all three methylation states being utilized with nearly equal efficiency (H3K27me0:me1:me2 k_{cat}/K_M ratio = 1.3:0.5:1) [21] (Fig. 11.4). When HeLa-purified nucleosomes were utilized as a substrate, the activity of A677G EZH2 was even greater than that of WT and Y641 mutant forms of EZH2 likely because of the ability of the A677G EZH2 complex to act upon a greater proportion of the heterogeneously modified HeLa histones [21]. Similar studies with A687V EZH2 revealed yet another distinctly altered substrate specificity whereby this enzyme exhibits reduced activity with an unmethylated substrate, increased activity with a monomethylated H3K27 substrate, and near WT activity with the dimethylated substrate (H3K27me0:me1:me2 k_{cat}/K_M ratio = 0.2:4.3:1) [70] (Fig. 11.4). Despite near normal catalytic efficiency of A687V EZH2 with a dimethylated substrate, recent work from Swalm et al. demonstrated that unlike WT EZH2, A687V EZH2 efficiently produces H3K27me3 over time in vitro. Thus, the effects of mutating EZH2 A687 to a valine may not be as simple as initially thought.

Considering that mutation of Y641, A677, or A687 affects the substrate specificity of EZH2 for histone H3 K27-containing peptides, it is reasonable to hypothesize that these mutations might also affect activity at other histone lysine residues. However, this has been shown not to be the case for all three mutated residues using a library of >600 peptides representing sequences within histones H2A, H2B, H3, or H4 and possessing up to five post-translational modifications, such as lysine and/or arginine methylation, lysine acetylation, or phosphorylation of serine, tyrosine, and/or threonine [21,66]. Therefore, these studies have demonstrated that each EZH2 mutation has its own unique effect upon the substrate specificity of EZH2 at H3K27; however, it does not appear that this leads to promiscuity at other histone modification sites beyond H3K27. More extensive studies will be needed to determine if mutated forms of EZH2 acquire any activity with nonhistone substrates.

STRUCTURAL RATIONALE FOR ALTERED SUBSTRATE SPECIFICITY IN EZH2 MUTANTS

Although the crystal structure of a functional human EZH2 complex remains elusive, recent advances have produced structures for a heavily truncated EZH2 SET domain [71,72] and a biochemically active PRC2 derived from *Chaetomium thermophilum* [73]. While these data have been informative, the utility of the truncated constructs has been somewhat limited because of the fact that they are not catalytically active and lack much of the binding pocket for *S*-adenosylmethionine (SAM), the methyl donor for the methylation reaction [71,72]. The *C. thermophilum* structure, on the other hand, is catalytically active and shares 37% identity and 57% similarity with human EZH2 within the catalytic SET domain. This is comparable to the 35% identity and 50% similarity for the SET domains of EZH2 and GLP, an H3K9 dimethyltransferase whose structure has previously been used to establish a homology model of EZH2 [21,59,66]. Herein, we have generated an updated homology model using both the crystal structures of *C. thermophilum* EZH2 (PDB ID 5CH1) [73] and GLP/EHMT1 bound to an H3K9me2 substrate (PDB ID 2RFI) [74] (Fig. 11.5).

These structures and models reveal that the EZH2 SET domain has two well-defined binding pockets that lie across from one another (Fig. 11.5A). One is more elongated and shallower and binds SAM while the other is narrower and deeper and binds the histone H3 lysine 27 side chain. Analysis of the homology model can provide potential explanations for the substrate preference of WT EZH2. Similar to other methyltransferases, such as the well-studied Set7/9, the terminal (ε) amine group of the unmethylated lysine substrate (H3K27me0) is positioned within the peptide pocket by two hydrogen bonds (Fig. 11.5A). One hydrogen bond is established between the ε-amine of the substrate and the hydroxyl group of the highly conserved Y641 residue [75,76], and the other is between the ε-amine of the substrate and an active site water molecule. This active site water molecule has been studied in other SET domain methyltransferases [75–77] and is positioned by A687 and I684 in

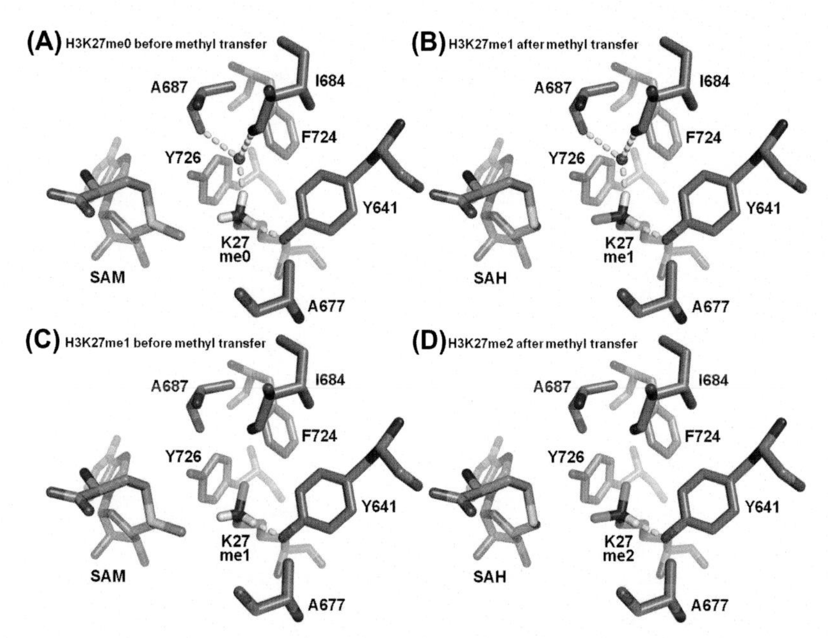

FIGURE 11.5 Mono- and dimethylation of histone H3K27 by wild-type (WT) EZH2. A homology model of EZH2 was generated using the crystal structures of *Chaetomium thermophilum* EZH2 (PDB ID 5CH1) [73] and human GLP/EHMT1 bound to an H3K9me2 substrate (PDB ID 2RFI) [74]. EZH2 residues are colored with gray carbons; SAM and SAH are colored with orange carbons; histone H3K27 is colored with cyan carbons; the conserved water molecule is represented as a *red sphere*; and hydrogen bonds are represented as *dashed yellow lines.* (A) WT EZH2 with SAM and H3K27me0 before the monomethylation reaction. (B) WT EZH2 with SAH and H3K27me1 after the monomethylation reaction. (C) WT EZH2 with SAM and H3K27me1 before the dimethylation reaction. (D) WT EZH2 with SAH and H3K27me2 after the dimethylation reaction. Figures were generated with the PyMOL Molecular Graphics System, Version 1.7.6.3 (Schrödinger, LLC).

EZH2. In enzymes where the F724 position of EZH2 is occupied by a tyrosine, the so-called phenylalanine–tyrosine switch position, the hydroxyl group of this tyrosine also participates in the coordination and stabilization of the water molecule [75–77].

After the lysine substrate is monomethylated, it must either be released and rebound, or it must rotate the methyl group away from the methyl transfer pore to align itself for a second methyl transfer reaction (Fig. 11.5B and C). When a monomethylated substrate is utilized by many SET domain enzymes, the Kme1 methyl group must rotate away from the methyl transfer pore so a second methyl group can be transferred. This requires that the active site water be displaced or relocated to make room for the larger substrate. At this point, the hydrogen bond between the Y641 hydroxyl group and the lysine ε-amine continues to orient the substrate for catalysis (Fig. 11.5C and D). However, with only ~2.9 Å between the hydroxyl oxygen of Y641 and the ε-amine group of H3K27me2, there appears to be very little room for the larger dimethylated lysine substrate to rotate or rebind in a position that properly orients the lysine substrate for a third methyl transfer reaction (Figs. 11.5D and 11.6A). Therefore, it appears that amino acid residues within the peptide-binding pocket

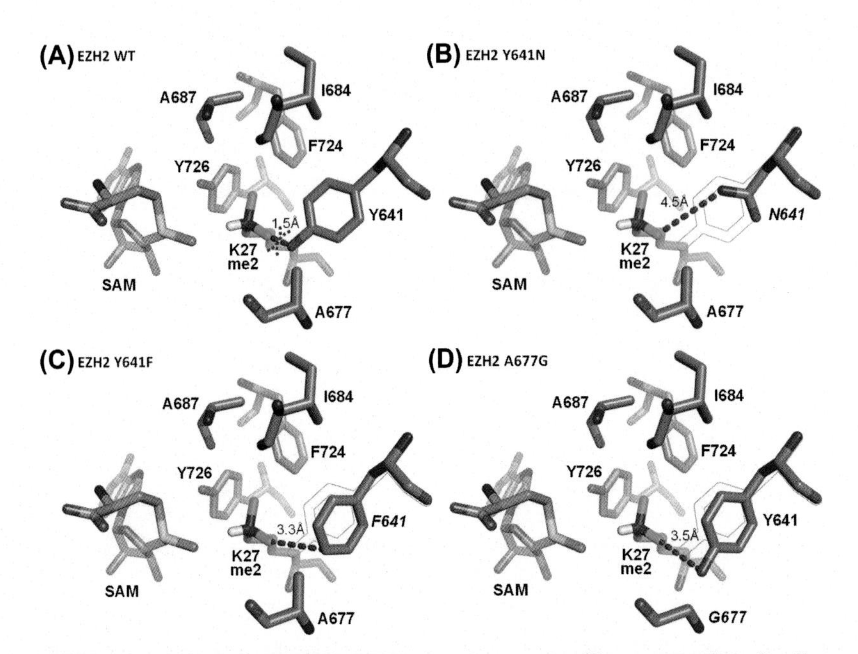

FIGURE 11.6 Y641X and A677G EZH2 mutants more readily accommodate the di- to trimethylation reaction. A homology model of EZH2 representing (A) WT EZH2, (B) Y641N EZH2, (C) Y641F EZH2, and (D) A677G EZH2 with SAM and H3K27me2. An outline of Y641 from WT EZH2 is overlaid in panels B, C, and D for reference. EZH2 residues are colored with gray carbons, SAM is colored with orange carbons, H3K27 is colored with cyan carbons, and distances between the lysine residue and the side chain of amino acid residue 641 are represented as *dashed magenta lines*. Figures were generated with the PyMOL Molecular Graphics System, Version 1.7.6.3 (Schrödinger, LLC).

constrain and orient the unmethylated and monomethylated substrates for optimal methyl transfer. However, while the Y641 residue of EZH2 participates in this process, it also sterically restricts activity with the larger dimethylated substrate. These observations correlate well with the WT enzyme's nine- and sixfold preferences for unmethylated and monomethylated K27 compared to the dimethylated form.

Y641 Mutations Have Dual Effects on Substrate Preference

If this dual function model for EZH2 Y641 is correct, then mutation of Y641 to a smaller residue, such as asparagine (Y641N), should provide additional room in the lysine tunnel to accommodate the di- to trimethylation reaction. A homology model of N641 EZH2 suggests that there is ~4.5 Å between the methyl of H3K27me2 and the N641 side chain compared to only 1.5 Å with Y641 in WT EZH2 (Fig. 11.6B). Importantly, while the pocket is now larger, the critical tyrosine hydroxyl to ε-amine hydrogen bond is lost, presumably decreasing the stability of the highly flexible unmodified or monomethylated lysine. To test this hypothesis, one can consider the range of mutations which change the enzyme's substrate preference. The Y641N, Y641S, and Y641C mutations significantly decrease the size of the residue at position 641 thereby enlarging the terminus of the peptide-binding pocket. The mutation of Y641 to a phenylalanine (Y641F), on the other hand, is relatively conservative with only loss of the tyrosine hydroxyl group (Fig. 11.6C). The Y641F EZH2 model suggests that the phenylalanine residue may adopt an alternative low-energy conformation in which it rotates away from the lysine tunnel toward the A677 residue positioned behind the 641 residue. This conformation likely provides sufficient space (H3K27me2 methyl to F641 side chain distance of at least 3.3 Å) for the dimethylated lysine to optimally orient for the third methyl transfer. However, loss of the stabilizing hydrogen bond that occurs between the tyrosine hydroxyl group and the lysine substrate is likely also responsible for the greatly reduced activity with unmodified lysines.

In addition to studying the effects of mutations within EZH2 itself, we can learn from the effects of amino acid substitutions in the SET domains of other methyltransferases as the SET domain is highly conserved across orthologs and homologs. For example, the effect of changing the EZH2 Y641 residue is predictable based on mutational studies from the human SET7/9 methyltransferase. SET7/9 normally monomethylates H3K4, however, when the SET7/9 Y245 residue (the equivalent of EZH2 Y641) is mutated to alanine, the mutant is no longer active with an H3K4me0 substrate [78]. Instead the Y245A SET7/9 mutant gains the ability to di- and trimethylate an H3K4me1 substrate [78]. Similarly, exchanging the Y641 equivalent of G9a (Y1067) with a phenylalanine converts the enzyme from an H3K9 mono- and dimethyltransferase to a trimethyltransferase [74]. Thus, numerous lines of evidence support this hypothesis that EZH2 Y641 both promotes monomethylation through substrate stabilization and obstructs trimethylation through steric hindrance.

The A677G Mutation Optimizes Y641 Positioning for All Three Methylation Reactions

While the EZH2 Y641 residue appears to line the peptide-binding pocket and interacts with the lysine substrate, the A677 residue does not interact directly with either H3K27 or SAM [21] (Fig. 11.6A). A677 is, however, in close proximity (~3.8 Å) to the hydroxyl group of Y641. Mutating A677 to the smaller glycine residue may permit Y641 to adopt an alternative conformation similar to that predicted for the phenylalanine of Y641F EZH2 (Fig. 11.6D). This conformation for Y641 could produce at least 3.5 Å between the hydroxyl oxygen of Y641 and the methyl of H3K27me2. This would be sufficient space to permit the lysine substrate to rotate into position for trimethylation. Importantly, since the A677G EZH2 mutant retains the tyrosine hydroxyl group at position 641, this would preserve the stabilizing interaction between Y641 and an H3K27me0 substrate. Therefore, retention of the Y641 residue combined with the ability of this residue to adopt an alternative conformation in the A677G EZH2 mutant may contribute to the mutant enzyme's unique ability to efficiently methylate H3K27me0, me1, and me2 [21].

Little work has been done to examine the role of A677 in EZH2 or other SET domain–containing enzymes, possibly because of the fact that EZH2 A677 doesn't immediately border the peptide-binding pocket. Review of the naturally occurring variation at this position across orthologs and homologs revealed that the structurally related SET domain methyltransferase DIM-5 from *N. crassa* also has a glycine at the equivalent position (Fig. 11.3). Interestingly, DIM-5 has been reported to perform mono-, di-, and trimethylation of its substrate H3K9 [79,80]. It appears likely that the alanine at residue 677 of EZH2, and likely equivalent residues in other SET domain methyltransferases, plays a critical role in the regulation of substrate specificity through indirect effects on residues lining the peptide-binding pocket.

EZH2 A687 Coordinates a Water Molecule Required for Substrate Monomethylation

The EZH2 A687 residue resides at the interface of the SAM and peptide-binding pockets nearly directly across from Y641 in the peptide-binding pocket [66] (Fig. 11.5). The side chain of A687 is buried from solvent and points away from the SAM pocket toward I684 and F724. The backbone carbonyl of A687, on the other hand, points toward the active site and likely forms H-bonds with both Y726 and the active site water molecule [66]. The Y726 residue appears to contribute to efficient binding of SAM/*S*-adenosylhomocysteine (SAH) and proper folding of the SET domain. However, the active site water molecule helps to properly position the lysine substrate (Fig. 11.5A). Consequently, A687 occupies an important and highly conserved position within the SET domain that may function in multiple roles to regulate binding of both SAM and lysine substrates.

In SET domain enzymes where the F/Y switch residue is a tyrosine, the active site water is coordinated by the equivalent residues of EZH2 A687, I684, the hydroxyl group of the F/Y switch tyrosine, and the lysine substrate itself. Similar to A677, not many studies have focused on the role of A687 or equivalent residues. However, extensive literature exists to explain the role of the F/Y switch residue in coordinating the active site water and regulating substrate preferences. In the SET7/9 methyltransferase, the F/Y switch residue is a tyrosine (Y305) and mutation to a phenylalanine increases production of Kme2 [75]. Similarly, when the F/Y switch residue of the SET8 monomethylase is mutated from Y334 to a phenylalanine, the mutant enzyme gains the ability to perform the dimethylation reaction [76]. Conversely, in G9A the F/Y switch residue is a phenylalanine (F1152) and mutation to a tyrosine reduces the production of Kme2 [81]. Since a tyrosine residue in the F/Y switch position contributes another H-bond with the water, displacement of the water molecule requires more energy and therefore decreased activity is observed with substrates that require the water to be displaced. In contrast, when the F/Y switch residue is a phenylalanine, this additional H-bond is not present and the water molecule is bound less tightly leading to the water being displaced more easily and increased ability to form Kme2 product. EZH2 contains a phenylalanine in the switch position which should facilitate dimethylation. However, it is proposed that the A687V mutation may result in subtle conformational changes in the surrounding residues (e.g., I684, F724, and Y726) that make up the lysine and SAM binding sites and therefore could affect the stability of the conserved water molecule and further facilitate the dimethylation reaction (Fig. 11.7A and B).

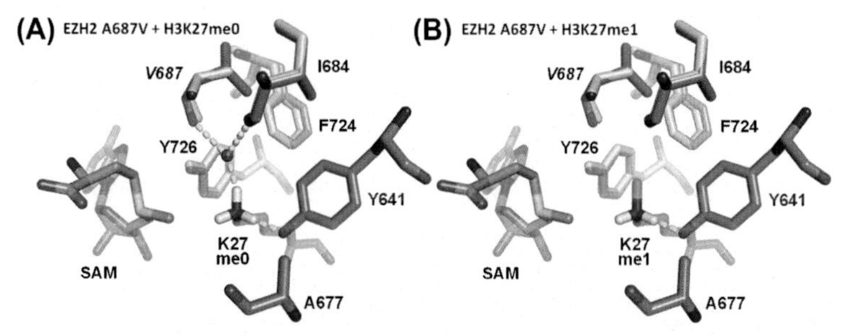

FIGURE 11.7 EZH2 A687 coordinates an active site water molecule required for the stabilization of the H3K27me0 substrate. A homology model of EZH2 overlaying the positions of amino acids in wild-type (WT) and A687V EZH2 with (A) H3K27me0 or (B) H3K27me1. A687V EZH2 residues are colored with gray carbons and WT EZH2 residues are colored with pink carbons. SAM is colored with orange carbons, H3K27 is colored with cyan carbons, the conserved water molecule is represented as a *red sphere*, and hydrogen bonds are represented as *dashed yellow lines*. Figures were generated with the PyMOL Molecular Graphics System, Version 1.7.6.3 (Schrödinger, LLC).

Although a crystal structure of full-length active EZH2 with substrate bound is not available, considering homology models and evolutionary conservation of specific residues permits us to generate plausible hypotheses for the roles and mechanisms associated with each of the commonly observed EZH2 SET domain mutations. It is clear that the coordination of the lysine substrate by EZH2 is complex and multifactorial, and nature has identified several opportunities to dysregulate proper substrate specificity to promote increased H3K27me3 in cancer.

CELLULAR ACTIVITY OF EZH2 MUTANTS

Considering the aberrant substrate specificities of Y641X, A677G, and A687V EZH2, it is not surprising that cell lines that harbor mutations in these residues almost uniformly exhibit elevated H3K27me3 [21,58–60,66]. In fact, it was in part through analysis of global H3K27me3 levels that the A677G EZH2 mutation was identified in the Pfeiffer DLBCL cell line [21]. Interestingly, while Y641F/N/S/H and A677G EZH2 mutants all promote hyper-trimethylation of H3K27, a concomitant marked reduction in global H3K27me2 is also observed in these cells suggesting that the mutant enzymes convert H3K27me2 into H3K27me3, consistent with the results of in vitro biochemical studies [21,58,60,66].

To explore the cellular consequences of EZH2 mutation, all of the known gain-of-function EZH2 mutants have been exogenously expressed in various cell lines. When either Y641F/N/S/H/C or A677G EZH2 was expressed in NIH3T3, MCF7, or HEK293T cells, H3K27me3 levels were markedly increased in as little as 3 days [21,52,59,66]. Also consistent with cell lines that naturally expressed EZH2 mutants, these engineered systems exhibited global reduction of H3K27me2 suggesting that this is a relatively early and likely a direct effect of mutant EZH2 activity [21,66]. Interestingly, when A687V EZH2 was exogenously expressed in MCF7 cells, it exhibited a significant and reproducible, albeit weaker, increase in H3K27me3 compared to Y641F/N/S/H/C and A677G EZH2 [66]. In contrast to the other mutants, A687V EZH2 showed little to no reduction in H3K27me2 levels. One possible explanation for this is that the elevated biochemical activity of A687V EZH2 with a monomethylated substrate helps to maintain H3K27me2 while increased activity with K27me2 promotes hypertrimethylation. This effect was also observed in a naturally occurring A687V EZH2 mutant cell line, SUP-B8, where H3K27me3 levels were elevated, but H3K27me2 levels were maintained at levels that were considerably higher than those of Y641F/N/S/H and A677G mutant cell lines.

These data support the notion that all reported Y641, A677, and A687 EZH2 mutants stimulate hypertrimethylation of H3K27 but vary in the extent to which they decrease H3K27me2. While both histone modifications correlate with reduced or absent gene expression [1], H3K27me2 appears to mark poised or lowly expressed genes [82] and may inactivate noncell type–specific enhancers [83]. Reinberg and colleagues have suggested that the conversion from H3K27me2 to H3K27me3 may result in more complete silencing of gene expression [82]. The transition from di- to trimethylation is normally stimulated by PHF1, a Polycomblike (Pcl) family member that interacts with a fraction of cellular PRC2 [82,84,85]. The existence of specific biological mechanisms for the differential regulation of H3K27me2 and H3K27me3 suggests that these modifications may indeed have distinct biological functions. Perhaps additional study of EZH2 mutant cell lines will provide further insights into the different functions of H3K27me2 and H3K27me3, but for now the relative contribution of H3K27me2 loss to lymphomagenesis remains unclear.

LOSS-OF-FUNCTION EZH2 MUTATIONS COMMONLY OCCUR IN MYELOID MALIGNANCIES

While gain-of-function mutations in EZH2 clearly play an important role in the formation of some B-cell malignancies, it has also become apparent that loss-of-function somatic mutations in EZH2 can be equally important in certain myeloid malignancies including myelodysplastic syndrome (MDS), primary myelofibrosis (PMF), chronic myelomonocytic leukemia (CMML), and T-ALL [20,41,64,86–90]. Unlike the EZH2 gain-of-function mutations which occur only at Y641, A677, and A687, EZH2 loss-of-function mutations occur throughout the gene sequence with no evidence of mutation hotspots. While the biological effects of loss-of-function mutations are not within the scope of this review, these mutations provide evidence that it is critical to maintain a proper balance of EZH2 function to maintain homeostasis. Tipping that balance in either direction can contribute equally to oncogenesis. While there are many differences between genetic inactivation and transient pharmacologic inhibition, this concept is important to consider as clinical trials evaluate the safety of using epigenetic modulatory agents in patients.

DISCOVERY OF EZH2 INHIBITORS

Extensive experimental and genetic data (e.g., gain-of-function mutations in DLBCL and FL) indicate that EZH2 may be a valuable target for cancer therapeutics. For example, many functional studies using siRNA or shRNA to knockdown EZH2 levels have demonstrated that EZH2 is required for cell growth, colony formation, migration, and/or invasion in multiple tumor types [2,10,15–17]. It is not surprising then that a large number of pharmaceutical companies, biotechs, and academic groups have been searching for drug-like EZH2 inhibitors for close to a decade.

The first tool reported as a pharmacologic inhibitor of EZH2 was 3-deazaneplanocin A (DZNep) (Fig. 11.8A). DZNep was reported to selectively

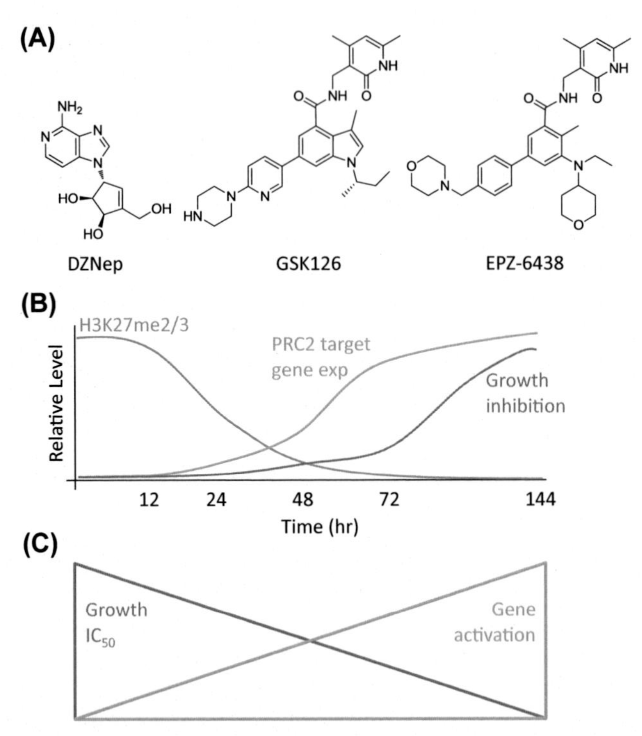

FIGURE 11.8 Discovery and activity of selective and potent small molecule EZH2 inhibitors. (A) Chemical structures of DZNep, GSK126, and EPZ-6438. DZNep is a SAH hydrolase inhibitor that has been reported to induce EZH2 degradation. GSK126 and EPZ-6438 are both direct EZH2 inhibitors that are currently being investigated in human clinical trials. (B) Temporal kinetics of the cellular effects induced by EZH2 inhibitors. Reduction of H3K27me2 and me3 levels begins as early as 24 h but requires approximately 3 days for maximal inhibition. Once H3K27me3 levels have been sufficiently reduced, a subset of repressed PRC2 target genes begin to be expressed between 24 and 72 h posttreatment, and this ultimately leads to growth inhibition through a G_1 arrest or cell death after 3–14+ days. (C) Schematic demonstrating the inverse relationship between growth IC_{50} value and number of gene activation events. In general, cell lines that are more sensitive to EZH2 inhibitor-induced growth inhibition (low gIC_{50} value) also exhibit the most gene expression changes in response to EZH2 inhibitor treatment.

inhibit the growth of cancer cells, but not normal cells, through a mechanism involving degradation of PRC2 components including EZH2, EED, and SUZ12 [91]. Unfortunately, DZNep is neither a selective nor direct inhibitor of EZH2, but rather it is a potent inhibitor of s-adenosylhomocysteine hydrolase (SAHH) [92]. While the mechanism whereby DZNep promotes the degradation of EZH2 is still not fully understood, inhibition of SAHH may promote an accumulation of SAH, the byproduct of SAM-dependent methylation reactions. Elevated SAH levels can then potentially lead to product inhibition of many SAM-dependent enzymes including, but not limited to EZH2. This was acknowledged by Tan et al. when they observed that only a minority of gene expression changes induced by DZNep overlapped with those induced by PRC2 siRNAs [91]. Additionally, subsequent studies demonstrated that DZNep was not selective for EZH2 and H3K27me3 but rather affected most histone methylation marks that were examined except trimethylation of H3K9 and H3K36 [93]. For these reasons, it is recommended that use of DZNep be limited to studies of SAHH inhibition as its many indirect downstream effects on epigenetic processes make it impossible to study the effects of EZH2 inhibition alone.

Fortunately, several potent and selective direct inhibitors of EZH2 catalytic activity have been identified over the past 6 years providing the scientific community with an array of tools to explore the effects of EZH2 inhibition. The first, and most numerous, group of EZH2 inhibitors includes several compounds that contain pyridones (Fig. 11.8A). These compounds were independently discovered by multiple groups and include GSK343, EPZ005687, EI-1, GSK503, CPI-360, CPI-169, GSK126, and EPZ-6438 [94–100]. They exhibit a range of potencies, but the most potent compounds have K_i values as low as 0.5 nM for WT and mutant EZH2 and inhibit 50% of global H3K27me3 levels in cells at concentrations as low as 7 nM. Importantly, when tested against a broad array of epigenetic and nonepigenetic enzymes, these inhibitors are highly selective for EZH2 with the sole exception of EZH1. However, despite the fact that the EZH2 and EZH1 SET domains are 96% identical, many of these inhibitors are still >150-fold selective for EZH2 versus EZH1.

A second group of EZH2 inhibitors containing a tetramethylpiperidinyl group has also been reported by Constellation Pharmaceuticals [101,102]. These compounds are potent biochemical inhibitors of both WT and mutant EZH2 but are less active in cells with EC_{50} values of 7 μM for reduction of global H3K27me3 [101]. While these compounds are less active than the pyridone series of EZH2 inhibitors, they provide a valuable second chemical series that may be further optimized to improve potency or incorporated into a hybrid molecule combining the best attributes of each series.

Using these inhibitors, the scaffolding and catalytic functions of EZH2 and PRC2 can now be separated and studied more easily. One of the challenges with using knockdown approaches to study the function of epigenetic and chromatin regulators is that in addition to their catalytic activities, many of these proteins participate in the formation of large protein complexes that interact with DNA, histones, other non-histone proteins, and noncoding RNAs. When using knockdown-based approaches, all of these functions can be disrupted. Small molecule inhibitors of catalytic activity, on the hand, often inhibit the enzyme's activity without affecting protein stability or protein:-protein interactions. In the case of PRC2, EED has been reported to interact with H3K27me3 through a WD40-repeat β-propeller domain [103], PRC2 interacts with HOTAIR and indirectly with LSD1 [14], and EZH2 may have other PRC2-independent [104] and noncatalytic roles [105,106] that are still poorly understood. Recently, a stabilized α-helix peptide disruptor of the EED:EZH2 interaction has been reported [105]. Through the use of the increasing number of molecular tools for EZH2 and PRC2, researchers can now readily explore the many functions of PRC2.

MECHANISTIC AND PHENOTYPIC EFFECTS OF EZH2 INHIBITORS IN CANCER CELLS

Considering the role of EZH2 in mediating H3K27me3, it comes as no surprise that treatment of cells with EZH2 inhibitors leads to a reduction in global H3K27me3 levels [66,95–98,101,102,107] detectable after 12–24 h and complete inhibition often takes 3 days or longer (Fig. 11.8B). In addition to H3K27me3, global H3K27me2 is also reduced upon EZH2 inhibitor treatment with potency similar to that for H3K27me3. In contrast, however, multiple groups have shown that H3K27me1 is not significantly depleted by EZH2 inhibition. Considering that EZH2 and EZH1 are the only known H3K27 methyltransferases, it is possible that H3K27me1 levels are maintained by EZH1. However, since most EZH2 inhibitors are less than 100-fold selective for EZH2 versus EZH1, EZH1 activity should be at least partially inhibited at higher doses. Thus, it may be necessary to develop more selective inhibitors for either EZH2 or EZH1 to more definitively address this question or perhaps there is another H3K27 methyltransferase that has yet to be discovered.

Interestingly, while EZH2 inhibitors decrease H3K27me2 and H3K27me3 similarly in nearly all cell lines tested, not all cell lines exhibit growth inhibition in response to this reduced histone methylation. When EZH2 WT and mutant DLBCL cell lines are examined using 6–14 day proliferation assays, the EZH2 mutant cell lines are significantly enriched among the most sensitive lines [95,97,100] suggesting that these cells are uniquely dependent on EZH2 activity for their growth and survival. However, while there was a clear enrichment for EZH2 mutant cell lines among the most sensitive cell lines, not all sensitive lines harbored EZH2 mutations and not all resistant lines were EZH2 WT indicating that additional factors beyond EZH2 mutation status must contribute to EZH2 inhibitor sensitivity. Importantly, not all cell lines that are sensitive to EZH2 inhibitors respond with the same phenotype. For example, the EZH2 mutant GCB DLBCL cell line KARPAS-422 exhibits a G_1 phase cell cycle arrest whereas the EZH2 mutant GCB DLBCL cell line Pfeiffer responds with a strong apoptotic response characterized by caspase-3 activation and increased sub-G_1 content [97].

The differential sensitivity of cell lines to EZH2 inhibitors may be explained, at least in part, by the transcriptional response of the cell lines to the inhibitors (Fig. 11.8B). Since H3K27me2 and H3K27me3 are both transcriptionally repressive histone modifications, it is expected that global reduction of these marks should lead to robust transcriptional activation.

Indeed this is sometimes the case, but surprisingly, this effect is not observed in all cell lines. Interestingly, when DLBCL cell lines are treated with concentrations of EZH2 inhibitor that deplete the majority of global H3K27me3, only sensitive cell lines respond with a transcriptional response [97]. For example, when cells were treated with GSK126 for 72 h, the most sensitive cell lines exhibited ~700–1000 differentially expressed genes whereas the most resistant exhibited <50 differentially expressed genes [97] (Fig. 11.8C). Consistent with H3K27me2 and H3K27me3 being repressive histone modifications, the majority of gene expression changes were upregulation. The upregulated genes were also found to be enriched in H3K27me3 prior to treatment further indicating that they are indeed derepressed PRC2 target genes. Since H3K27me3 levels are depleted in all cell lines, but gene reactivation is specific to only sensitive lines, this indicates that loss of H3K27me3 is not sufficient to activate gene expression on its own and that additional factors, such as redundant repressive marks (e.g., DNA methylation) or an absence of activating modifications (e.g., histone acetylation) or transcription factors, may be maintaining genes in a silenced state.

Consistent with the observation that cell lines exhibit different phenotypes in response to EZH2 inhibitors, the set of genes that is up or downregulated by EZH2 inhibitors is also cell line dependent. When a set of 10 GCB DLBCL cell lines was treated with GSK126, over 2500 microarray probe sets exhibited upregulation in at least one cell line; however, only 35 of these were reproducibly upregulated in at least four of the five sensitive EZH2 mutant lines [97]. While there was no single pathway or process significantly enriched among the overlapping set of upregulated genes, many of the cell lines exhibited enrichment of genes involved in cell cycle regulation and cell death. These findings are consistent with the phenotypes observed in cell growth assays and suggest that while aberrant Polycomb activity silences a unique set of genes in each cell line, the functional consequence of these events is to promote cell growth and survival.

When immunocompromised mice harboring xenografts of sensitive cell lines were treated with EZH2 inhibitors, global H3K27me3 levels were reduced in the tumors and a subset of genes exhibit transcriptional activation. The onset of these effects tended to be slower in vivo than in cell culture likely because of the effects of drug metabolism and excretion and perhaps the slower growth rate of cells in vivo. H3K27me3 loss was first observed after approximately 3 days and H3K27me3 levels continued to decline for over 1 week. Gene expression levels of reactivated genes tended to inversely correlate with H3K27me3 levels and therefore exhibited a similar time course as H3K27me3 reduction. These effects on H3K27me3 and gene expression resulted in tumor growth inhibition that often only became apparent after 7–10 days of treatment [97,100,107]. In some cases, tumor regression or shrinkage was also observed leading to complete loss of measurable tumors [97,107]. Following these mice for an additional 30+ days beyond the end of treatment with EZH2 inhibitors revealed no tumor regrowth suggesting that tumors had been eradicated from these mice [97,107].

Since EZH2 inhibitors have now been available commercially for several years, research groups have published accounts of EZH2 inhibitor activity in most major tumor types. Many of these data support what has been observed previously using RNAi-mediated knockdown approaches. However, the tumor type that has most convincingly been demonstrated to be sensitive to EZH2 inhibitors, other than DLBCL, is rhabdoid tumors. Rhabdoid tumors are a rare and aggressive pediatric cancer that are commonly diagnosed histologically by the presence of rhabdoid cells and/or genetically by the presence of inactivating mutations or deletions in the *SNF5/SMARCB1/INI1* gene. SNF5 is a component of the SWI/SNF nucleosome remodeling complex and was shown by Wilson et al. to exhibit an epigenetic antagonism with Polycomb [31]. These rhabdoid tumors which uniformly lack SNF5 protein exhibit markedly elevated expression of EZH2, and genetic deletion of SNF5 from CD8 T cells or MEFs leads to transcriptional upregulation of EZH2 [31]. These elevated PRC2 levels promote hypertrimethylation of H3K27 and consequently repression of Polycomb target genes. Knockdown of EZH2 leads to a reduced growth rate and induction of senescence in SNF5-deficient cells [31]. More impressively, while 100% of mice harboring *SNF5* deletion in the T-cell lineage develop CD8+ T-cell lymphomas and die within ~100 days, codeletion of *SNF5* and *EZH2* completely blocked tumor onset and 100% of mice survived beyond 150 days [31].

These data provided strong rationale to evaluate the activity of EZH2 inhibitors in this disease context. Similar to EZH2 mutant DLBCL, *SNF5*-deficient cell lines exhibited time-dependent growth inhibition and cell death marked by increased populations of G_1, sub-G_1, and TUNEL-positive cells [96]. This was associated with a transcriptional response that included activation of genes involved in neuronal differentiation and cell cycle inhibition and downregulation of hedgehog pathway genes, MYC, and EZH2 [96]. Lastly, EZH2 inhibitor treatment of mice harboring xenografts of G401 rhabdoid tumor cells resulted in global reduction of H3K27me3, transcriptional activation of several genes, and marked tumor growth inhibition and regression [96]. These studies highlight the complex interplay of chromatin-modifying and chromatin-remodeling pathways and further demonstrate that additional genetically-defined populations beyond EZH2 gain-of-function mutations may be sensitive to EZH2 inhibitors.

CONCLUSIONS

Based on the results of preclinical studies performed by many groups, several EZH2 inhibitors have now progressed into human clinical trials to evaluate the effectiveness of EZH2 inhibition in several tumor settings. These clinical compounds include EPZ-6438 from Epizyme, GSK2816126 from

GlaxoSmithKline, and CPI-1205 from Constellation Pharmaceuticals (clinicaltrials.gov). Current clinical trials are focused primarily on EZH2 WT and mutant GCB DLBCL and FL, multiple myeloma, synovial sarcoma, and rhabdoid tumors. While no manuscripts have been published on these trials yet, conference presentations have revealed encouraging data indicating antitumor activity in both EZH2 WT and mutant lymphomas as well as SNF5-deficient sarcomas. Additionally, no major safety concerns have been reported indicating that global inhibition of EZH2 activity may be well-tolerated.

While we await the final data from these phase I and II trials, preclinical research on the role of EZH2 in cancer continues to advance through the study of large genomic data sets, analysis of synthetic lethal screens, and continued study of EZH2 inhibitor activity. The discovery of EZH2 gain-of-function mutations kicked off an exciting avenue of research that has progressed rapidly from the first publication of EZH2 mutations in lymphoma in 2010 [22] to encouraging evidence presented in 2015 that this population is responding well to EZH2 inhibition in the clinic. While we now understand much about the frequency and distribution of EZH2 gain-of-function mutations in many different cancer settings, basic research has a long way to go to understand why EZH2 mutations selectively occur in only a very limited number of tumor types, how these hyperactive mutant enzymes modify the epigenome, and exactly how EZH2 inhibitors can be used to reverse these aberrant processes. We have scratched the surface of the role of EZH2 mutations in the context of B-cell lymphomas and now need to take a deeper dive into the basic and clinical implications of these important and intriguing mutations.

LIST OF ACRONYMS AND ABBREVIATIONS

A Alanine
ABC DLBCL Activated B-cell diffuse large B-cell lymphoma
BAF BRG1-or BRM-associated factor complex
B-ALL B-cell acute lymphoblastic leukemia
C Cysteine
CMML Chronic myelomonocytic leukemia
cSCC Cutaneous Squamous Cell Carcinoma
DIM-5 *Neurospora crassa* defective in methylation-5 lysine methyltransferase
DZNep 3-deazaneplanocin A, an inhibitor of SAHH
EED Embryonic ectoderm development, a protein component of PRC2
EZH1 Enhancer of *zeste* homolog 1, a SET domain H3K27 methyltransferase
EZH2 Enhancer of *zeste* homolog 2, a SET domain H3K27 methyltransferase
F Phenylalanine
FL Follicular lymphoma
G Glycine
G9A/GLP SET domain lysine methyltransferases responsible for H3K9me2
GC Germinal center
GCB DLBCL Germinal center B-cell diffuse large B-cell lymphoma
H Histidine

H3K27 Histone H3 lysine 27
I Isoleucine
JMJD3/KDM6B Jumonji D3 lysine demethylase
k_{cat} Catalytic rate constant
K_M Michaelis constant
MDS Myelodysplastic syndrome
me1, me2, me3 Mono-, di-, trimethylation
N Asparagine
PHF1 PHD Finger 1, Polycomb group protein
PMF Primary myelofibrosis
PRC2 Polycomb repressive complex 2
S Serine
SAH S-Adenosylhomocysteine
SAHH SAH hydrolase
SAM S-Adenosylmethionine
SET **S**u(var)3-9, **E**nhancer of zeste, and **T**rithorax lysine methyltransferase domain
SET7/9 SET domain lysine methyltransferase
SUZ12 Suppressor of *zeste* homolog 12, a protein component of PRC2
T-ALL T-cell acute lymphoblastic leukemia
UTX/KDM6A Ubiquitously Transcribed tetracopeptide repeat, X chromosome lysine demethylase
V Valine
WT Wild-Type
Y Tyrosine

REFERENCES

[1] Barski A, Cuddapah S, Cui K, Roh TY, Schones DE, Wang Z, et al. High-resolution profiling of histone methylations in the human genome. Cell 2007;129:823−37.
[2] Cao R, Wang L, Wang H, Xia L, Erdjument-Bromage H, Tempst P, et al. Role of histone H3 lysine 27 methylation in Polycomb-group silencing. Science 2002;298:1039−43.
[3] Kuzmichev A, Nishioka K, Erdjument-Bromage H, Tempst P, Reinberg D. Histone methyltransferase activity associated with a human multiprotein complex containing the Enhancer of Zeste protein. Genes Dev 2002;16:2893−905.
[4] Muller J, Hart CM, Francis NJ, Vargas ML, Sengupta A, Wild B, et al. Histone methyltransferase activity of a Drosophila Polycomb group repressor complex. Cell 2002;111:197−208.
[5] Bracken AP, Dietrich N, Pasini D, Hansen KH, Helin K. Genome-wide mapping of Polycomb target genes unravels their roles in cell fate transitions. Genes Dev 2006;20:1123−36.
[6] Kirmizis A, Bartley SM, Kuzmichev A, Margueron R, Reinberg D, Green R, et al. Silencing of human polycomb target genes is associated with methylation of histone H3 Lys 27. Genes Dev 2004;18:1592−605.
[7] Oguchi K, Takagi M, Tsuchida R, Taya Y, Ito E, Isoyama K, et al. Missense mutation and defective function of ATM in a childhood acute leukemia patient with MLL gene rearrangement. Blood 2003;101:3622−7.
[8] Wu CT, Jones RS, Lasko PF, Gelbart WM. Homeosis and the interaction of zeste and white in Drosophila. Mol Gen Genet 1989;218:559−64.
[9] Lee TI, Jenner RG, Boyer LA, Guenther MG, Levine SS, Kumar RM, et al. Control of developmental regulators by Polycomb in human embryonic stem cells. Cell 2006;125:301−13.

[10] Cao R, Zhang Y. SUZ12 is required for both the histone methyltransferase activity and the silencing function of the EED-EZH2 complex. Mol Cell 2004;15:57–67.

[11] Koh W, Park B, Lee S. A new kinetochore component CENP-W interacts with the polycomb-group protein EZH2 to promote gene silencing. Biochem Biophys Res Commun 2015;464:256–62.

[12] Mathiyalagan P, Okabe J, Chang L, Su Y, Du XJ, El-Osta A. The primary microRNA-208b interacts with Polycomb-group protein, Ezh2, to regulate gene expression in the heart. Nucleic Acids Res 2014;42:790–803.

[13] O'Connell S, Wang L, Robert S, Jones CA, Saint R, Jones RS. Polycomblike PHD fingers mediate conserved interaction with enhancer of zeste protein. J Biol Chem 2001;276:43065–73.

[14] Tsai MC, Manor O, Wan Y, Mosammaparast N, Wang JK, Lan F, et al. Long noncoding RNA as modular scaffold of histone modification complexes. Science 2010;329:689–93.

[15] Kleer CG, Cao Q, Varambally S, Shen R, Ota I, Tomlins SA, et al. EZH2 is a marker of aggressive breast cancer and promotes neoplastic transformation of breast epithelial cells. Proc Natl Acad Sci USA 2003;100:11606–11.

[16] Takawa M, Masuda K, Kunizaki M, Daigo Y, Takagi K, Iwai Y, et al. Validation of the histone methyltransferase EZH2 as a therapeutic target for various types of human cancer and as a prognostic marker. Cancer Sci 2011;102:1298–305.

[17] Varambally S, Dhanasekaran SM, Zhou M, Barrette TR, Kumar-Sinha C, Sanda MG, et al. The polycomb group protein EZH2 is involved in progression of prostate cancer. Nature 2002;419:624–9.

[18] Wagener N, Macher-Goeppinger S, Pritsch M, Husing J, Hoppe-Seyler K, Schirmacher P, et al. Enhancer of zeste homolog 2 (EZH2) expression is an independent prognostic factor in renal cell carcinoma. BMC Cancer 2010;10:524.

[19] Bodor C, O'Riain C, Wrench D, Matthews J, Iyengar S, Tayyib H, et al. EZH2 Y641 mutations in follicular lymphoma. Leukemia 2011;25:726–9.

[20] Ernst T, Chase AJ, Score J, Hidalgo-Curtis CE, Bryant C, Jones AV, et al. Inactivating mutations of the histone methyltransferase gene EZH2 in myeloid disorders. Nat Genet 2010;42:722–6.

[21] McCabe MT, Graves AP, Ganji G, Diaz E, Halsey WS, Jiang Y, et al. Mutation of A677 in histone methyltransferase EZH2 in human B-cell lymphoma promotes hyper-trimethylation of histone H3 on lysine 27 (H3K27). Proc Natl Acad Sci USA 2012;109: 2989–94.

[22] Morin RD, Johnson NA, Severson TM, Mungall AJ, An J, Goya R, et al. Somatic mutations altering EZH2 (Tyr641) in follicular and diffuse large B-cell lymphomas of germinal-center origin. Nat Genet 2010;42:181–5.

[23] Bachmann IM, Halvorsen OJ, Collett K, Stefansson IM, Straume O, Haukaas SA, et al. EZH2 expression is associated with high proliferation rate and aggressive tumor subgroups in cutaneous melanoma and cancers of the endometrium, prostate, and breast. J Clin Oncol 2006;24:268–73.

[24] Breuer RH, Snijders PJ, Smit EF, Sutedja TG, Sewalt RG, Otte AP, et al. Increased expression of the EZH2 polycomb group gene in BMI-1-positive neoplastic cells during bronchial carcinogenesis. Neoplasia 2004;6:736–43.

[25] Findeis-Hosey JJ, Huang J, Li F, Yang Q, McMahon LA, Xu H. High-grade neuroendocrine carcinomas of the lung highly express enhancer of zeste homolog 2, but carcinoids do not. Hum Pathol 2011;42:867–72.

[26] Sudo T, Utsunomiya T, Mimori K, Nagahara H, Ogawa K, Inoue H, et al. Clinicopathological significance of EZH2 mRNA expression in patients with hepatocellular carcinoma. Br J Cancer 2005;92:1754–8.

[27] Weikert S, Christoph F, Kollermann J, Muller M, Schrader M, Miller K, et al. Expression levels of the EZH2 polycomb transcriptional repressor correlate with aggressiveness and invasive potential of bladder carcinomas. Int J Mol Med 2005;16:349–53.

[28] Simon JA, Lange CA. Roles of the EZH2 histone methyltransferase in cancer epigenetics. Mutat Res 2008;647:21–9.

[29] Bracken AP, Pasini D, Capra M, Prosperini E, Colli E, Helin K. EZH2 is downstream of the pRB-E2F pathway, essential for proliferation and amplified in cancer. EMBO J 2003;22:5323–35.

[30] Saramaki OR, Tammela TL, Martikainen PM, Vessella RL, Visakorpi T. The gene for polycomb group protein enhancer of zeste homolog 2 (EZH2) is amplified in late-stage prostate cancer. Genes Chromosomes Cancer 2006;45:639–45.

[31] Wilson BG, Wang X, Shen X, McKenna ES, Lemieux ME, Cho YJ, et al. Epigenetic antagonism between polycomb and SWI/SNF complexes during oncogenic transformation. Cancer Cell 2010;18:316–28.

[32] Alajez NM, Shi W, Hui AB, Bruce J, Lenarduzzi M, Ito E, et al. Enhancer of Zeste homolog 2 (EZH2) is overexpressed in recurrent nasopharyngeal carcinoma and is regulated by miR-26a, miR-101, and miR-98. Cell Death Dis 2010;1:e85.

[33] Ciarapica R, Russo G, Verginelli F, Raimondi L, Donfrancesco A, Rota R, et al. Deregulated expression of miR-26a and Ezh2 in rhabdomyosarcoma. Cell Cycle 2009;8:172–5.

[34] Friedman JM, Liang G, Liu CC, Wolff EM, Tsai YC, Ye W, et al. The putative tumor suppressor microRNA-101 modulates the cancer epigenome by repressing the polycomb group protein EZH2. Cancer Res 2009;69:2623–9.

[35] Lu J, He ML, Wang L, Chen Y, Liu X, Dong Q, et al. MiR-26a inhibits cell growth and tumorigenesis of nasopharyngeal carcinoma through repression of EZH2. Cancer Res 2011;71:225–33.

[36] Sander S, Bullinger L, Klapproth K, Fiedler K, Kestler HA, Barth TF, et al. MYC stimulates EZH2 expression by repression of its negative regulator miR-26a. Blood 2008;112:4202–12.

[37] Smits M, Mir SE, Nilsson RJ, van der Stoop PM, Niers JM, Marquez VE, et al. Down-regulation of miR-101 in endothelial cells promotes blood vessel formation through reduced repression of EZH2. PLoS One 2011;6:e16282.

[38] Varambally S, Cao Q, Mani RS, Shankar S, Wang X, Ateeq B, et al. Genomic loss of microRNA-101 leads to overexpression of histone methyltransferase EZH2 in cancer. Science 2008;322:1695–9.

[39] Zhang B, Liu XX, He JR, Zhou CX, Guo M, He M, et al. Pathologically decreased miR-26a antagonizes apoptosis and facilitates carcinogenesis by targeting MTDH and EZH2 in breast cancer. Carcinogenesis 2011;32:2–9.

[40] Gui Y, Guo G, Huang Y, Hu X, Tang A, Gao S, et al. Frequent mutations of chromatin remodeling genes in transitional cell carcinoma of the bladder. Nat Genet 2011;43:875–8.

[41] Jankowska AM, Makishima H, Tiu RV, Szpurka H, Huang Y, Traina F, et al. Mutational spectrum analysis of chronic myelomonocytic leukemia includes genes associated with epigenetic regulation: UTX, EZH2, and DNMT3A. Blood 2011;118:3932–41.

[42] van Haaften G, Dalgliesh GL, Davies H, Chen L, Bignell G, Greenman C, et al. Somatic mutations of the histone H3K27 demethylase gene UTX in human cancer. Nat Genet 2009;41:521–3.

[43] Rogenhofer S, Kahl P, Mertens C, Hauser S, Hartmann W, Buttner R, et al. Global histone H3 lysine 27 (H3K27) methylation levels and their prognostic relevance in renal cell carcinoma. BJU Int 2012;109:459–65.

[44] Chen C, Zhao M, Yin N, He B, Wang B, Yuan Y, et al. Abnormal histone acetylation and methylation levels in esophageal squamous cell carcinomas. Cancer Invest 2011;29:548–56.

[45] Robinson G, Parker M, Kranenburg TA, Lu C, Chen X, Ding L, et al. Novel mutations target distinct subgroups of medulloblastoma. Nature 2012;488:43–8.

[46] Su IH, Basavaraj A, Krutchinsky AN, Hobert O, Ullrich A, Chait BT, et al. Ezh2 controls B cell development through histone H3 methylation and Igh rearrangement. Nat Immunol 2003;4:124–31.

[47] Kamminga LM, Bystrykh LV, de BA, Houwer S, Douma J, Weersing E, et al. The Polycomb group gene Ezh2 prevents hematopoietic stem cell exhaustion. Blood 2006;107:2170–9.

[48] Velichutina I, Shaknovich R, Geng H, Johnson NA, Gascoyne RD, Melnick AM, et al. EZH2-mediated epigenetic silencing in germinal center B cells contributes to proliferation and lymphomagenesis. Blood 2010;116:5247–55.

[49] van Galen JC, Dukers DF, Giroth C, Sewalt RG, Otte AP, Meijer CJ, et al. Distinct expression patterns of polycomb oncoproteins and their binding partners during the germinal center reaction. Eur J Immunol 2004;34:1870–81.

[50] Lenz G, Staudt LM. Aggressive lymphomas. N Engl J Med 2010;362:1417–29.

[51] Bodor C, Grossmann V, Popov N, Okosun J, O'Riain C, Tan K, et al. EZH2 mutations are frequent and represent an early event in follicular lymphoma. Blood 2013;122:3165–8.

[52] Ryan RJ, Nitta M, Borger D, Zukerberg LR, Ferry JA, Harris NL, et al. EZH2 codon 641 mutations are common in BCL2-rearranged germinal center B cell lymphomas. PLoS One 2011;6:e28585.

[53] Saieg MA, Geddie WR, Boerner SL, Bailey D, Crump M, da Cunha SG. EZH2 and CD79B mutational status over time in B-cell non-Hodgkin lymphomas detected by high-throughput sequencing using minimal samples. Cancer Cytopathol 2013;121:377–86.

[54] Durinck S, Ho C, Wang NJ, Liao W, Jakkula LR, Collisson EA, et al. Temporal dissection of tumorigenesis in primary cancers. Cancer Discov 2011;1:137–43.

[55] Hodis E, Watson IR, Kryukov GV, Arold ST, Imielinski M, Theurillat JP, et al. A landscape of driver mutations in melanoma. Cell 2012;150:251–63.

[56] Krauthammer M, Kong Y, Ha BH, Evans P, Bacchiocchi A, McCusker JP, et al. Exome sequencing identifies recurrent somatic RAC1 mutations in melanoma. Nat Genet 2012;44:1006–14.

[57] Cromer MK, Starker LF, Choi M, Udelsman R, Nelson-Williams C, Lifton RP, et al. Identification of somatic mutations in parathyroid tumors using whole-exome sequencing. J Clin Endocrinol Metab 2012;97:E1774–81.

[58] Sneeringer CJ, Scott MP, Kuntz KW, Knutson SK, Pollock RM, Richon VM, et al. Coordinated activities of wild-type plus mutant EZH2 drive tumor-associated hypertrimethylation of lysine 27 on histone H3 (H3K27) in human B-cell lymphomas. Proc Natl Acad Sci USA 2010;107:20980–5.

[59] Yap DB, Chu J, Berg T, Schapira M, Cheng SW, Moradian A, et al. Somatic mutations at EZH2 Y641 act dominantly through a mechanism of selectively altered PRC2 catalytic activity, to increase H3K27 trimethylation. Blood 2011;117:2451–9.

[60] Swalm BM, Knutson SK, Warholic NM, Jin L, Kuntz KW, Keilhack H, et al. Reaction coupling between wild-type and disease-associated mutant EZH2. ACS Chem Biol 2014;9:2459–64.

[61] Khan SN, Jankowska AM, Mahfouz R, Dunbar AJ, Sugimoto Y, Hosono N, et al. Multiple mechanisms deregulate EZH2 and histone H3 lysine 27 epigenetic changes in myeloid malignancies. Leukemia 2013;27:1301–9.

[62] Morin RD, Mendez-Lago M, Mungall AJ, Goya R, Mungall KL, Corbett RD, et al. Frequent mutation of histone-modifying genes in non-Hodgkin lymphoma. Nature 2011;476:298–303.

[63] Muto T, Sashida G, Oshima M, Wendt GR, Mochizuki-Kashio M, Nagata Y, et al. Concurrent loss of Ezh2 and Tet2 cooperates in the pathogenesis of myelodysplastic disorders. J Exp Med 2013;210:2627–39.

[64] Score J, Hidalgo-Curtis C, Jones AV, Winkelmann N, Skinner A, Ward D, et al. Inactivation of polycomb repressive complex 2 components in myeloproliferative and myelodysplastic/myeloproliferative neoplasms. Blood 2012;119:1208–13.

[65] Tatton-Brown K, Hanks S, Ruark E, Zachariou A, Duarte Sdel V, et al. Germline mutations in the oncogene EZH2 cause Weaver syndrome and increased human height. Oncotarget 2011;2:1127–33.

[66] Ott HM, Graves AP, Pappalardi MB, Huddleston M, Halsey WS, Hughes AM, et al. A687V EZH2 is a driver of histone H3 lysine 27 (H3K27) hypertrimethylation. Mol Cancer Ther 2014;13:3062–73.

[67] Lohr JG, Stojanov P, Lawrence MS, Auclair D, Chapuy B, Sougnez C, et al. Discovery and prioritization of somatic mutations in diffuse large B-cell lymphoma (DLBCL) by whole-exome sequencing. Proc Natl Acad Sci USA 2012;109:3879–84.

[68] Nogai H, Dorken B, Lenz G. Pathogenesis of non-Hodgkin's lymphoma. J Clin Oncol 2011;29:1803–11.

[69] Cobaleda C, Sanchez-Garcia I. B-cell acute lymphoblastic leukaemia: towards understanding its cellular origin. Bioessays 2009;31:600–9.

[70] Majer CR, Jin L, Scott MP, Knutson SK, Kuntz KW, Keilhack H, et al. A687V EZH2 is a gain-of-function mutation found in lymphoma patients. FEBS Lett 2012;586:3448–51.

[71] Antonysamy S, Condon B, Druzina Z, Bonanno JB, Gheyi T, Zhang F, et al. Structural context of disease-associated mutations and putative mechanism of autoinhibition revealed by X-ray crystallographic analysis of the EZH2-SET domain. PLoS One 2013;8:e84147.

[72] Wu H, Zeng H, Dong A, Li F, He H, Senisterra G, et al. Structure of the catalytic domain of EZH2 reveals conformational plasticity in cofactor and substrate binding sites and explains oncogenic mutations. PLoS One 2013;8:e83737.

[73] Jiao L, Liu X. Structural basis of histone H3K27 trimethylation by an active polycomb repressive complex 2. Science 2015;350:aac4383.

[74] Wu H, Min J, Lunin VV, Antoshenko T, Dombrovski L, Zeng H, et al. Structural biology of human H3K9 methyltransferases. PLoS One 2010;5:e8570.

[75] Del Rizzo PA, Couture JF, Dirk LM, Strunk BS, Roiko MS, Brunzelle JS, et al. SET7/9 catalytic mutants reveal the role of active site water molecules in lysine multiple methylation. J Biol Chem 2010;285:31849–58.

[76] Couture JF, Dirk LM, Brunzelle JS, Houtz RL, Trievel RC. Structural origins for the product specificity of SET domain protein methyltransferases. Proc Natl Acad Sci USA 2008;105:20659–64.

[77] Chu Y, Yao J, Guo H. QM/MM MD and free energy simulations of G9a-like protein (GLP) and its mutants: understanding the factors that determine the product specificity. PLoS One 2012;7:e37674.

[78] Xiao B, Jing C, Wilson JR, Walker PA, Vasisht N, Kelly G, et al. Structure and catalytic mechanism of the human histone methyltransferase SET7/9. Nature 2003;421:652–6.

[79] Tamaru H, Zhang X, McMillen D, Singh PB, Nakayama J, Grewal SI, et al. Trimethylated lysine 9 of histone H3 is a mark for DNA methylation in *Neurospora crassa*. Nat Genet 2003;34:75–9.

[80] Zhang X, Yang Z, Khan SI, Horton JR, Tamaru H, Selker EU, et al. Structural basis for the product specificity of histone lysine methyltransferases. Mol Cell 2003;12:177–85.

[81] Collins RE, Tachibana M, Tamaru H, Smith KM, Jia D, Zhang X, et al. In vitro and in vivo analyses of a Phe/Tyr switch controlling product specificity of histone lysine methyltransferases. J Biol Chem 2005;280:5563–70.

[82] Sarma K, Margueron R, Ivanov A, Pirrotta V, Reinberg D. Ezh2 requires PHF1 to efficiently catalyze H3 lysine 27 trimethylation in vivo. Mol Cell Biol 2008;28:2718–31.

[83] Ferrari KJ, Scelfo A, Jammula S, Cuomo A, Barozzi I, Stutzer A, et al. Polycomb-dependent H3K27me1 and H3K27me2 regulate active transcription and enhancer Fidelity. Mol Cell 2014;53:49–62.

[84] Nekrasov M, Klymenko T, Fraterman S, Papp B, Oktaba K, Kocher T, et al. Pcl-PRC2 is needed to generate high levels of H3-K27 trimethylation at Polycomb target genes. EMBO J 2007;26:4078–88.

[85] Cao R, Wang H, He J, Erdjument-Bromage H, Tempst P, Zhang Y. Role of hPHF1 in H3K27 methylation and Hox gene silencing. Mol Cell Biol 2008;28:1862–72.

[86] Abdel-Wahab O, Pardanani A, Patel J, Wadleigh M, Lasho T, Heguy A, et al. Concomitant analysis of EZH2 and ASXL1 mutations in myelofibrosis, chronic myelomonocytic leukemia and blast-phase myeloproliferative neoplasms. Leukemia 2011;25:1200–2.

[87] Ernst T, Pflug A, Rinke J, Ernst J, Bierbach U, Beck JF, et al. A somatic EZH2 mutation in childhood acute myeloid leukemia. Leukemia 2012;26:1701–3.

[88] Makishima H, Jankowska AM, Tiu RV, Szpurka H, Sugimoto Y, Hu Z, et al. Novel homo- and hemizygous mutations in EZH2 in myeloid malignancies. Leukemia 2010;24:1799–804.

[89] Ntziachristos P, Tsirigos A, Van VP, Nedjic J, Trimarchi T, Flaherty MS, et al. Genetic inactivation of the polycomb repressive complex 2 in T cell acute lymphoblastic leukemia. Nat Med 2012;18:298–301.

[90] Simon C, Chagraoui J, Krosl J, Gendron P, Wilhelm B, Lemieux S, et al. A key role for EZH2 and associated genes in mouse and human adult T-cell acute leukemia. Genes Dev 2012;26:651–6.

[91] Tan J, Yang X, Zhuang L, Jiang X, Chen W, Lee PL, et al. Pharmacologic disruption of Polycomb-repressive complex 2-mediated gene repression selectively induces apoptosis in cancer cells. Genes Dev 2007;21:1050–63.

[92] Glazer RI, Hartman KD, Knode MC, Richard MM, Chiang PK, Tseng CK, et al. 3-Deazaneplanocin: a new and potent inhibitor of S-adenosylhomocysteine hydrolase and its effects on human promyelocytic leukemia cell line HL-60. Biochem Biophys Res Commun 1986;135:688–94.

[93] Miranda TB, Cortez CC, Yoo CB, Liang G, Abe M, Kelly TK, et al. DZNep is a global histone methylation inhibitor that reactivates developmental genes not silenced by DNA methylation. Mol Cancer Ther 2009;8:1579–88.

[94] Beguelin W, Popovic R, Teater M, Jiang Y, Bunting KL, Rosen M, et al. EZH2 is required for germinal center formation and somatic EZH2 mutations promote lymphoid transformation. Cancer Cell 2013;23:677–92.

[95] Knutson SK, Wigle TJ, Warholic NM, Sneeringer CJ, Allain CJ, Klaus CR, et al. A selective inhibitor of EZH2 blocks H3K27 methylation and kills mutant lymphoma cells. Nat Chem Biol 2012;8:890–6.

[96] Knutson SK, Warholic NM, Wigle TJ, Klaus CR, Allain CJ, Raimondi A, et al. Durable tumor regression in genetically altered malignant rhabdoid tumors by inhibition of methyltransferase EZH2. Proc Natl Acad Sci USA 2013;110:7922–7.

[97] McCabe MT, Ott HM, Ganji G, Korenchuk S, Thompson C, Van Aller GS, et al. EZH2 inhibition as a therapeutic strategy for lymphoma with EZH2-activating mutations. Nature 2012;492:108–12.

[98] Qi W, Chan H, Teng L, Li L, Chuai S, Zhang R, et al. Selective inhibition of Ezh2 by a small molecule inhibitor blocks tumor cells proliferation. Proc Natl Acad Sci USA 2012;109:21360–5.

[99] Verma SK, Tian X, LaFrance LV, Duquenne C, Suarez DP, Newlander KA, et al. Identification of potent, selective, cell-active inhibitors of the histone lysine methyltransferase EZH2. ACS Med Chem Lett 2012;3:1091–6.

[100] Bradley WD, Arora S, Busby J, Balasubramanian S, Gehling VS, Nasveschuk CG, et al. EZH2 inhibitor efficacy in non-Hodgkin's lymphoma does not require suppression of H3K27 monomethylation. Chem Biol 2014;21:1463–75.

[101] Garapaty-Rao S, Nasveschuk C, Gagnon A, Chan EY, Sandy P, Busby J, et al. Identification of EZH2 and EZH1 small molecule inhibitors with selective impact on diffuse large B cell lymphoma cell growth. Chem Biol 2013;20:1329–39.

[102] Nasveschuk CG, Gagnon A, Garapaty-Rao S, Balasubramanian S, Campbell R, Lee C, et al. Discovery and optimization of tetramethylpiperidinyl benzamides as inhibitors of EZH2. ACS Med Chem Lett 2014;5:378–83.

[103] Margueron R, Justin N, Ohno K, Sharpe ML, Son J, Drury III WJ, et al. Role of the polycomb protein EED in the propagation of repressive histone marks. Nature 2009;461: 762–7.

[104] Xu K, Wu ZJ, Groner AC, He HH, Cai C, Lis RT, et al. EZH2 oncogenic activity in castration-resistant prostate cancer cells is Polycomb-independent. Science 2012;338: 1465–9.

[105] Kim W, Bird GH, Neff T, Guo G, Kerenyi MA, Walensky LD, et al. Targeted disruption of the EZH2-EED complex inhibits EZH2-dependent cancer. Nat Chem Biol 2013;9:643–50.

[106] Kim KH, Kim W, Howard TP, Vazquez F, Tsherniak A, Wu JN, et al. SWI/SNF-mutant cancers depend on catalytic and non-catalytic activity of EZH2. Nat Med 2015;21:1491–6.

[107] Knutson SK, Kawano S, Minoshima Y, Warholic NM, Huang KC, Xiao Y, et al. Selective inhibition of EZH2 by EPZ-6438 leads to potent antitumor activity in EZH2-mutant non-Hodgkin lymphoma. Mol Cancer Ther 2014;13:842–54.

Chapter 12

PcG Proteins in *Caenorhabditis elegans*

B. Tursun
Max-Delbrück-Center for Molecular Medicine, Berlin, Germany

Chapter Outline

Introduction **289**
PRC1 **292**
PRC1 in Other Species 292
No Obvious PRC1 in *Caenorhabditis elegans*? 294
SOR-1 and SAM Domain Containing SOP-2: Worm-Specific PRC1-Like Proteins? 295
PRC2 **296**
Identification of PRC2 Subunits in *Caenorhabditis elegans* 296
PRC2 Is Important for Germline Development and X Chromosome Repression 296
Transgenerational Inheritance of PRC2-Mediated Repression 298
H3K9me2 Compensates for the Loss of H3K27me3 on Sperm Chromosomes 299
Transmission of H3K27me3 in Dividing Cells During Embryonic Development 300
PRC2 Safeguards Germ Cells From Somatic Differentiation 301
PRC2 Restricts Plasticity of Embryonic Cells but is Dispensable for Somatic Differentiation in Worms 302
PRC2 Function in Somatic Cells of *Caenorhabditis elegans* 303
Antagonizing PRC2 Activity 304
Noncoding RNA-Mediated H3K27 Methylation 306
PcG Recruitment—Noncoding RNAs or PREs? 306
Conclusions **307**
List of Acronyms and Abbreviation **309**
References **309**

INTRODUCTION

While genetic screens using the fruit fly *Drosophila melanogaster* led to the initial identification of the first Polycomb genes [1–3], researchers using the nematode *Caenorhabditis elegans* as a genetic model organism revealed how epigenetic regulation by Polycomb proteins is transgenerationally inherited [4]. In general, studying Polycomb group (PcG) proteins in *C. elegans* can strengthen our understanding of the evolution and conservation of PcG

Polycomb Group Proteins. http://dx.doi.org/10.1016/B978-0-12-809737-3.00012-X

proteins. Furthermore, *C. elegans* as well as *D. melanogaster* lack CpG DNA methylation. In mammals, it is under debate as to whether there is cross talk between DNA methylation and PcG protein-mediated chromatin regulation [5–7] which can complicate the interpretation of observed PcG functions. Therefore, *C. elegans* and *D. melanogaster* provide an opportunity to reveal basic principles of PcG functions in a less convoluted context and can teach us how more intricate chromatin regulatory networks might have evolved.

C. elegans is the first multicellular organism whose genome was fully sequenced. Researchers using *C. elegans* as a genetic model organism pioneered highly conserved key biological findings. This is reflected by the awarding of Nobel Prizes to a number of *C. elegans* researchers such as Sydney Brenner, John Sulston, H. Robert Horvitz, Craig Mello, Andrew Fire, and Martin Chalfie (nobelprize.org). They have been honored for their findings concerning the genetics of organ development, apoptosis, RNA interference (RNAi), and for the first heterologous expression of GFP in an animal. Furthermore, research with *C. elegans* allowed several other key findings such as the first micro RNAs (miRNAs) *let-7* and *lin-4* [8,9], aging-regulating genes [10], and the axon guidance molecule Netrin, first identified in *C. elegans* in 1990 and named UNC-6 [11]. Such accomplishments highlight the power of *C. elegans* as a genetic model organism to reveal basic principles in biology, many of which are highly relevant for humans.

C. elegans has a short life cycle (2–3 weeks) and worm populations consist mainly of self-fertilizing hermaphrodites with males occurring less frequently (Fig. 12.1). One key feature of *C. elegans* is that forward genetic screens for mutants with a specific phenotype are straightforward. Such an approach led to the identification of the first PcG protein-related factors in *C. elegans* by screening for mutant animals that fail to have grandchildren (Fig. 12.2). Mutations in the corresponding genes caused a *maternal-effect sterile* (*mes*) phenotype [12]. These so-called *mes* genes encode factors that are maternally supplied and required for germline development in the offspring. Some of the identified *mes* genes turned out to encode protein components of the Polycomb repressive complex 2 (PRC2) in the nematode: MES-2, MES-3, and MES-6. Overall, the worm PRC2 complex resembles the Enhancer of zeste E(z) complex in flies and vertebrates with respect to composition and function such as mediating Lysine 27 trimethylation of Histone H3 (H3K27me3) thereby repressing gene expression [13,14]. PRC2 in *C. elegans* is mainly required for germline development such that animals without PRC2 are viable but lack a germline. Its activity in the germline is important for X chromosomal silencing and repression of somatic genes. Consequently, loss of PRC2 allows transcription factor (TF)-induced reprogramming of germ cells into somatic cells such as neuron or muscles.

In contrast, there is no canonical PRC1 complex in worms. However, PcG-related proteins such as MIG-32 and SPAT-3, which are functional homologs of human BMI-1 and RING2, mediate canonical PRC1 activities such as

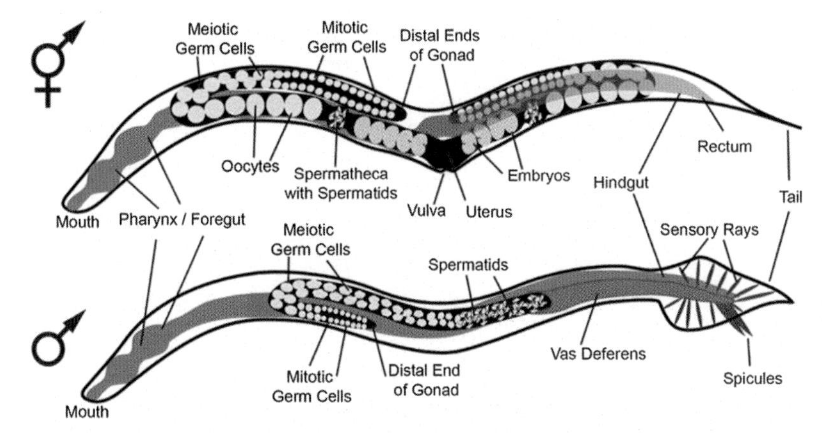

FIGURE 12.1 Adult *Caenorhabditis elegans* hermaphrodite and male. The majority of *C. elegans* nematode population consists of hermaphrodites (top) while males (bottom) occur at low frequency. The self-fertilizing hermaphrodite produces up to 300 offspring but can produce more than 1000 if additional sperm is provided by mating with males. Adult *C. elegans* nematodes are around 1 mm long and have a translucent body. Hermaphrodites have two gonad arms in which germ cells are being produced at the distal ends. The first 150 germ cells per gonad arm in the hermaphrodite germline mature to sperm and are pushed into the spermatheca by the following germ cells, which become oocytes. Oocytes passing the spermatheca are fertilized and give rise to embryos, which are laid through the uterus and vulva. Males mate with hermaphrodites by inserting their copulatory spicules, which are part of the specialized male tail, into the vulva of the hermaphrodite and ejaculating the sperm. They have one gonad arm which produces at the distal end germ cells that maturate to spermatids as they are pushed toward the vas deferens. Both hermaphrodites and males have five autosome pairs while hermaphrodites have two X chromosomes and males have only one X. The reproductive life cycle is about 3 days and the total life span is 2–3 weeks. *C. elegans* has a fixed number of cells of around 1000 somatic cells forming tissues such as the nervous system, muscles, epidermis, intestine, and others. *C. elegans* is a nonparasitic free-living nematode living in soil and can be cultured in the laboratory by feeding on *Escherichia coli.*

Histone H2A ubiquitylation [15] and appear to coregulate certain PRC2 targets in somatic cells [15]. An overview of PcG and PcG-like proteins in *C. elegans* is provided in Table 12.1.

When initially identified in *D. melanogaster*, one canonical set of target genes for repression by PcG proteins was homeotic (Hox) genes which determine body segmentation and subsequent organ development [2,16]. Recent studies in flies and mammals revealed that a large number of other target genes in different cellular and developmental contexts exist (reviewed in other chapters of this volume and in Refs. [16–19]).

This chapter provides an overview of PcG proteins and their functions in *C. elegans*, including recent findings concerning the transgenerational transmission of PRC2-mediated silencing and regulation of cellular plasticity by PRC2 in the nematode. An important aspect in the context of personalized epigenetics is the finding that the parental epigenetic memory established by

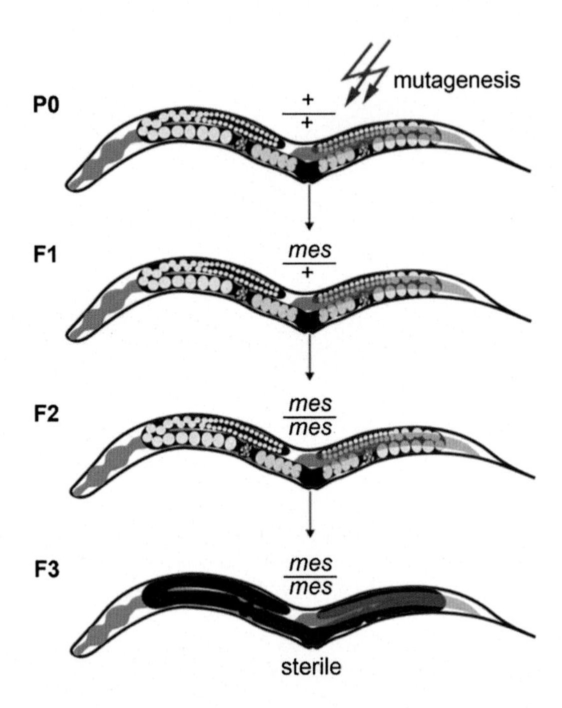

FIGURE 12.2 **Identification of PRC2 subunits encoded by *mes* genes.** The PRC2 subunits MES-2 (E(z)/EZH2), MES-3 (*Caenorhabditis elegans* specific), and MES-6 (ESC/EED) were discovered in a screen for mutant animals that fail to have grandchildren due to defects in their germline. Wild-type (+/+) P0 hermaphrodites were mutagenized giving rise to F1 progeny that are heterozygous for a *mes* mutation (*mes*/+). These F1 hermaphrodites produce F2 homozygous *mes* mutants (*mes/mes*) which develop normally with a functional germline because the maternally inherited load from the heterozygous (*mes*/+) F1 is compensating the lack of *mes* in the F2. The homozygous F2 (*mes/mes*) give rise to phenotypically healthy hermaphrodite progeny (F3); however, these F3 cannot develop a germline thus becoming sterile such that no further progeny can be generated [12,46].

PRC2 can be inherited through sperm and oocytes in *C. elegans*. Hence, the epigenetic signature of the developing animal is supplemented by the contribution from its parents, which could be an important facet of transgenerational influence on individual epigenetic signatures also in humans.

PRC1

PRC1 in Other Species

In mammals, PRC1 is a highly diverse multiprotein complex with several core components belonging to different protein families. Both mammalian and *D. melanogaster* PRC1 can include a subunit which recognizes Histone H3 trimethylated at Lysine 27 (H3K27me3) through a chromodomain such as the

TABLE 12.1 Polycomb Group-Like Proteins in *Caenorhabditis elegans*

C. elegans	Orthologs/Analogs		Characteristic protein domains	Molecular Function	Physiological Function	References
	Drosophila	*Human*				
PRC1-Like						
MIG-32	Lethal (3) 73Ah	BMI	RING	H2AK119 ubiquitylation	Male tail development; neuronal migration and process extension	[1]
SPAT-3	Ptip-PA	RING2	RING			[1]
SOP-2	Polyhomeotic-like	N/A	SAM	RNA binding	Hox gene regulation; tail development	[2,3]
SOR-1	N/A	N/A	N/A	N/A	Hox gene regulation; vulva development	[3]
PRC2						
MES-2	[E(z)]	EZH2	SET	H3K27 methylation	X chromosome silencing; germline development; germ cell safeguarding; male tail development	[4–9]
MES-3	N/A	N/A	N/A			
MES-6	Esc	EED	WD repeat			
LIN-53	Caf1-p55	CAF-1p48, RBBP4/7	Histone-binding; WD repeat			[7,10]
Antagonizing Factors and TrxG-Like Proteins						
MES-4	Dmel/Mes-4	NSD1	PHD, SET	H3K36 methylation	X silencing (indirectly) by restricting PRC2 activity; germline development; germ cell safeguarding	[4,7,8,11,12]
LIN-49	Br140-PB	BRD1	PHD/bromo	N/A	Hox gene regulation; hindgut development; male tail development; neuronal left/right asymmetry	[13–15]
LIN-59	ASH-1	ASH1L	SET/PHD	N/A	Hox gene regulation; hindgut development; male tail development	[13,15]

mammalian homolog of *D. melanogaster* Polycomb (Pc) termed Cbx [20,21]. Furthermore, mammalian and fly PRC1 complexes contain a Ring finger protein (RNF2 in mammals and RING in *Drosophila*) bearing ubiquitin ligase activity for the monoubiquitylation of Lysine 119 of histone H2A (H2AK119ubi) (reviewed by Refs. [17,22,23]). The prevailing canonical function of PRC1 as transcriptional silencing machinery is thought to be accomplished by interacting with trimethylated H3K27me3 and ubiquitylating H2A [17]. Interaction with PRC2-deposited H3K27me3 and the subsequent H2AK119 ubiquitylation have been thought to be successive and obligatory steps for PRC1-mediated silencing. However, recent studies suggest that more diverse PRC1-like complexes exist that can, for instance, contain proteins with histone demethylase activity and be recruited to target genomic loci independently of H3K27me3 [24—26]. Such PRC1 variants are known as the dRAF (RING associated factor) complex in flies or as BCOR (BCL6 corepressor) in humans that also contain the H3K36me2 demethylase KDM2 (KDMB2/FBXL10 in mammals). It is conceivable that such PRC1-like complexes silence gene expression via different modes [27—29]. A more comprehensive overview of PRC1 proteins in *D. melanogaster* and humans can be found in the other chapters of this volume and in Ref. [19].

No Obvious PRC1 in *Caenorhabditis elegans*?

PRC1-like multiprotein complexes seem to be absent in *C. elegans* based on amino acid sequence conservation. However, two proteins, MIG-32 and SPAT-3, are considered to be functional homologs of the human PRC1 subunits BMI-1 and RING2. Both, MIG-32 and SPAT-3 have been shown to be required for H2AK119 ubiquitylation in *C. elegans* [15]. While the mammalian counterparts were shown to have sufficient activity for ubiquitylation of H2AK119 simply by heterodimerizing with each other [30—32], it is not clear whether the *C. elegans* homologs bear such direct ubiquitylation activity. Physiologically, MIG-32 and SPAT-3 appear to be required for neuronal migration and extension of neuronal processes. However, defects in nervous system architecture [15] in *mig-32* or *spat-3* mutants are not due to misregulation of Hox genes arguing for noncanonical PRC1 gene targets distinct from Hox genes. Yet, male worms lacking the *mig-32* or *spat-3* genes show defects in the sensory rays of the male tail, which is an essential organ for mating. These defects resemble the phenotype of mutants for either *mes-2, mes-3,* or *mes-6* genes, which encode subunits of the *C. elegans* PRC2 complex [14,33—35]. In contrast to *mig-32* or *spat-3* mutants, PRC2 mutants show derepression of Hox genes such as *mab-5* or *egl-5* in the male tail [14]. Interestingly, an epistasis test by combining PRC2 and *mig-32* or *spat-3* mutants suggested that MIG-32 and SPAT-3 do coregulate target genes with PRC2 in the male tail. Though, none of these common target genes turned out to be Hox genes [15] this cooperation of MIG-32 and SPAT-3 with PRC2 in the male tail might reflect a more canonical pathway of PRC1-mediated silencing. Nevertheless, the fact that *mes-2/3/6* mutants are sterile due to germline defects, whereas *mig-32* and *spat-3* mutants are fertile [15], suggests that the worm PRC1-like proteins are functionally uncoupled from PRC2 in worms. Since there is also evidence in other species that PRC1 can function independently of PRC2 [24,29], the same observation in worms for MIG-32/SPAT-3 might reflect a more ancient mode of PRC1 activity.

Still, a PRC2-uncoupled function of a PRC1-like complex raises the unsolved question of how exactly MIG-32 and SPAT-3, and also PRC2 itself, are recruited to specific genetic loci, and how repression of transcription might be accomplished. It is possible that MIG-32 and SPAT-3 represent a PRC1-like complex and act similarly to their mammalian counterparts, BMI-1 and RING2 [30,31]. Interestingly, worms have a distant relative of the KDM2 demethylase which is termed JHDM-1 [36,37]. However, this putative KDM2-like protein is uncharacterized and misses some functional domains found in its fly and human counterparts. In general, it remains unclear how the PRC1-like complex containing MIG-32 and SPAT-3 exerts its function in worms and whether it might contain additional subunits.

SOR-1 and SAM Domain Containing SOP-2: Worm-Specific PRC1-Like Proteins?

Another group of hypothetically PcG-like proteins in *C. elegans* is the RNA-binding proteins SOP-2 (Suppressor of Pal-1) and SOR-1 (SOP-2 related) [38,39]. Though SOP-2 has a SAM (sterile alpha motif) domain, which is also found in the *Drosophila* PRC1 member Polyhomeotic [40–42], both have no orthologs in other organisms. Still, SOP-2 and SOR-1 have been implicated in the global regulation of Hox genes in worms such as *mab-5* and *ceh-13* which are the orthologs of *fushi tarazu* (*ftz*) and *labial* in flies, respectively [37,39,43]. Another Hox gene target of SOP-2/SOR-1 is *egl-5* which encodes the ortholog of *Drosophila* Abdominal-B and human Hox9-13 factors [37,38,44,45]. However, the exact mechanism of Hox gene regulation by SOP-2 and SOR-1 remains elusive. Both proteins lack H2AK119 ubiquitylation activity and, additionally, SOP-2 and SOR-1 appear to regulate Hox genes independently of PRC2 [39]. As mentioned before for MIG-32/SPAT-3, such independence from PRC2 might reflect a more ancient mode of PRC1-mediated gene repression. It is interesting to note that a mutation in the *sop-2* gene causes synthetic lethality as well as a defective vulva when combined with a *mes-6* (PRC2 subunit) mutation [38]. This observation could indicate that SOP-2/SOR-1 might cooperate with PRC2 in a context distinct from Hox gene regulation.

The lack of obvious or canonical PRC1-like proteins in worms can be interpreted to mean that *C. elegans* has no PRC1-like complexes. However, studies in *D. melanogaster* and mammals provide increasing evidence that PRC1/PRC1-like complexes are highly diverse, often missing canonical subunits, and appear to not always involve H2A ubiquitylation [19]. Considering such diversity, MIG-32/SPAT-3 and SOP-2/SOR-1 might represent a divergent form of PRC1 in worms. Nonetheless, future *C. elegans* studies might reveal even more PcG-like proteins with similar functions in regulating gene expression, as it has been described for PRC1-like complexes in other species.

PRC2

Identification of PRC2 Subunits in *Caenorhabditis elegans*

Pioneering work from the research group of Susan Strome led to the first identification of *C. elegans* PcG proteins forming the PRC2 complex in worms. By performing forward genetic screens Capowski and colleagues identified a set of genes termed *mes* which stands for *maternal-effect sterile* [12]. They were discovered in a screen for mutant animals that fail to have grandchildren due to defects in their germline. Products of the *mes* genes are maternally provided and are crucial for germline development in the offspring. Homozygous mutant *mes*$^{-/-}$ hermaphrodite worms derived from heterozygous *mes*$^{+/-}$ mothers give rise to phenotypically healthy hermaphrodite progeny. However, their progeny lose their germline during development thus becoming sterile such that no grandchildren progeny can be generated (Fig. 12.2) [12,46]. Interestingly, the F1 *mes*$^{-/-}$ males maintain their germline and remain fertile [47]. The reason for this exception has been revealed recently and will be discussed later.

Subsequent characterization of the *mes* genes revealed that they encode PcG proteins forming the *C. elegans* PRC2 complex [34,48]. The *C. elegans* PRC2 core proteins are the E(z)/EZH2 ortholog MES-2, the ESC/EED ortholog MES-6, and the worm-specific MES-3 subunit which has no homologs in species outside of the nematode phylum [13,34,46,48]. An obvious homolog of the canonical fly PRC2 subunit Su(Z)12 (mammalian SUZ12) seems to be absent in *C. elegans*. In mammals and flies, orthologs of the highly conserved *C. elegans* histone chaperone LIN-53 (Caf1p48/RBBP4/7 in humans, CAF1/p55/NURF55 in flies) have been shown to associate with PRC2 and are considered to be a core component of PRC2 [19,49–55]. Though a physical association of LIN-53 with MES-2/-3/-6 has not yet been investigated, a functional association of LIN-53 with PRC2 has been shown in the context of protecting germ cells from being reprogrammed into somatic cells [56,57]. It is therefore possible that, at least in the germline, *C. elegans* PRC2 might incorporate LIN-53 as a subunit.

PRC2 Is Important for Germline Development and X Chromosome Repression

The most important physiological role of PRC2 in *C. elegans* is to specify the epigenetic signature of the germline by regulating chromatin structure [47,58].

A predominant role of PRC2 in the germline of worms is also reflected by the expression pattern of the PRC2 subunits. Antibody staining in addition to a CRISPR/Cas9-based generated genomic MES-2:GFP reporter revealed that expression starts broad in most tissues in the embryo but is later restricted to the germline postdevelopmentally [34,48] (Fig. 12.3). The initial role of PRC2

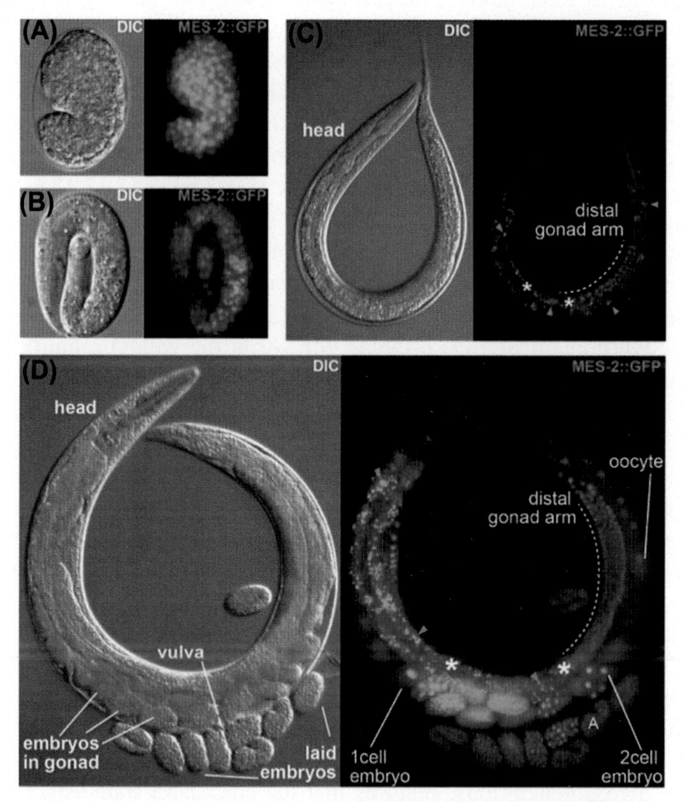

FIGURE 12.3 Expression of the PRC2 subunit MES-2 in the *Caenorhabditis elegans*. The panels show microscopic pictures of differential interference contrast (DIC) acquisitions and green fluorescent protein (GFP) signals derived from MES-2 fused to GFP. The endogenous *mes-2* gene locus was used to insert GFP-encoding DNA at the 3′ end before the stop codon of the *mes-2* gene. The resulting *C. elegans* strain has been named JH3203 [120]. (A) MES-2 is expressed ubiquitously in the early embryo (comma stage). (B) Embryo at the last stage before hatching (threefold stage) still shows high levels of ubiquitous MES-2 expression. (C) Hatched larva (L3 stage) shows mainly expression in the developing germline as indicated in the distal gonad arm. Some residual low-level expression is detectable in the head area. (D) Adult hermaphrodite, which are around 1 mm long, with MES-2 expression predominantly in the germline and in embryos that are either still in the gonad or have been extruded. 1-cell and 2-cell embryos are indicated that already have high levels of MES-2 due to the "maternal" load. In (C) and (D) the white asterisks indicate the distal ends of the gonad arms. Only one gonad arm is visible because the intestinal tract covers the other gonad arm. The *gray arrow*-heads indicate gut granules which show artificial green light signals due to autofluorescence.

in *C. elegans* was assigned to X chromosome silencing [58,59]. Hermaphrodite worms have two X chromosomes (XX) and males have one (XO). As in flies and humans, the ratio of gene expression from X versus autosomal chromosomes (A) needs to be adjusted since males have one X chromosome less (X:A ratio). Flies adjust the X:A ratio by a twofold increase of gene expression from the single X in males. In contrast, one entire X chromosome gets silenced in mammalian females while *C. elegans* hermaphrodites have both X chromosomes downregulated uniformly (reviewed in Ref. [60]). The repression of X-linked genes in the soma requires the dosage compensation complex (DCC) in *C. elegans*. But DCC does not seem to contribute to X chromosome silencing in the germline [60]. Instead, PRC2 is the main germline X chromosome silencing machinery, repressing gene expression through the deposition of the silencing mark H3K27me3 [4,33,47,61] and is therefore crucial for germline development in hermaphrodite worms. Interestingly, a contribution of PRC2 to X chromosomal silencing has also been shown in mammalian X chromosome inactivation [62–64] and could thus be an evolutionary conserved function of PRC2.

The SET domain of E(z) ortholog MES-2 is the catalytic entity for the methylation of H3K27 [33,60]. Mutants for *mes-2*, *mes-3*, or *mes-6* show accumulation of active histone modifications on their X chromosomes accompanied by an increase in active RNA polymerase II, leading to the upregulation of X-linked genes in the germline [33,58,65]. As mentioned before, it is important to note that these effects are not observed in homozygous $mes^{-/-}$ mutants which are derived from heterozygous $mes^{+/-}$ mothers. Rather, gene silencing is lost in the subsequent F1 $mes^{-/-}$ progeny, leading to a decay of the primordial germ cells, thereby failing to give rise to a germline (Fig. 12.2). Remarkably, apart from the missing germline resulting in sterility, the F1 $mes^{-/-}$ show no other obvious defects [12,34]. Hence, PRC2 is dispensable in somatic cells for survival but is required to modulate gene expression as mentioned earlier, for instance, in the *C. elegans* mail tail.

The transgenerational effect described above, that is, mutant phenotype retention across generations, suggests that the chromatin state established by PRC2 is inherited and maintained in the germ cells of wild-type worms.

Transgenerational Inheritance of PRC2-Mediated Repression

Besides its role in X chromosomal silencing, PRC2 is also essential in the germline for repressing somatic genes on autosomes. How this epigenetic memory is transmitted to the next generation and, furthermore, how dividing cells in embryos inherit and maintain this epigenetic memory were revealed only recently by Gaydos and colleagues [4]. By following the H3K27me3 transmission in PRC2 mutant backgrounds, they observed that the H3K27me3 marks could be transmitted to the next generation through oocytes as well as sperm. In contrast, the PRC2 transcripts are provided solely through oocytes to the zygote [4]. Moreover, X chromosomes that are inherited from sperm (XO) retain their repressed state, while an X chromosome inherited from the oocyte loses its transcriptional repression before fertilization. It is important to note that, although X chromosomes in wild-type worms are globally repressed in the germline, they lose repression during regular oocyte maturation in hermaphrodites [60]. Hence, PRC2 mutant males (XO germline) that inherited their X chromosome from the oocyte become sterile while PRC2 mutant males with the X chromosome derived from sperm are fertile [4,47]. These observations raise the question of how male (XO) germlines can maintain X chromosome silencing independently of H3K27me3 deposition by PRC2? An alternative repression mechanism must exist which participates in maintaining repression of sperm-inherited X chromosomes. In general, the fact that the epigenetic memory can be inherited through sperm and oocytes into the embryo is an important aspect in the context of personalized epigenetics. The epigenetic signature of the developing animal is likely to be influenced by the transmitted parental epigenetic memory.

H3K9me2 Compensates for the Loss of H3K27me3 on Sperm Chromosomes

In addition to H3K27me3, there are other histone modifications associated with repressed chromatin. The repressive heterochromatin mark H3K9me2 is mediated by the Histone methyltransferases (HMTs) MET-2, the homolog of mammalian SETDB1 [66], and SET-25, a distant homolog of the fly Su(var) 3–9 [37]. Sperm-inherited X chromosomes that lack H3K27me3 make use of the repression mediated by MET-2 and SET-25. Methylation of H3K9 was shown to be sufficient for maintaining the repressed state of the sperm-inherited X chromosome thereby ensuring proper development of the male germline [4]. In contrast, when both repressive marks, H3K27me3 and H3K9me2, are absent, males cannot generate germ cells and become sterile. Remarkably, both repressive marks H3K27me3 and H3K9me2, but not MES proteins, can be transmitted to the embryo via sperm. Consequently, when wild-type sperm fertilizes a PRC2 mutant oocyte, which globally lacks H3K27me3 on all chromosomes, the resulting one-cell embryo also lacks PRC2 (Fig. 12.4). Nevertheless, in this scenario, all sperm-derived chromosomes of the PRC2-lacking embryo carry H3K27me3 and H3K9me2 while maternal chromosomes are devoid of these marks [4]. Interestingly, global H3K9 methylation is one of the hallmarks of inactivated mammalian X chromosomes [67]. In *C. elegans* the repression of genes by H3K9 methylation on X chromosomes seems to be redundant with the activity of PRC2. While H3K9me2 is sufficient to maintain the repressed state of sperm X chromosomes, the lack of H3K9me in SET-25 and MET-2 mutants does not result in

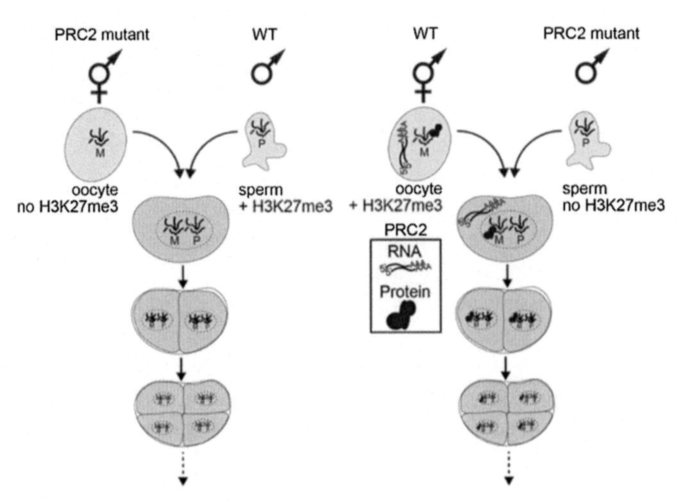

FIGURE 12.4 Inheritance of the epigenetic memory in *Caenorhabditis elegans*. H3K27me3 marks generated by PRC2 are inherited through sperm and oocytes but they persist on chromosomes in a gamete-of-origin manner during several rounds of divisions in early embryos. This mode of inheritance could be demonstrated in the embryos that are derived from oocytes lacking PRC2 and H3K27me3 fertilized by sperm carrying H3K27me3 (left side) [4]. Sperm cannot transmit PRC2-encoding transcripts and protein to the embryo but oocytes can do as shown in the reciprocal scenario on the right. Interestingly, the restriction of H3K27me3 to the gamete-of-origin persists during the early embryo development even in the presence of PRC2 and does not spread to all other chromosomes in the dividing cells. Later during development H3K27me3 spreads to all chromosomes [4].

loss of X chromosome repression and has milder effects on male fertility when PRC2 is present [4,68].

Transmission of H3K27me3 in Dividing Cells During Embryonic Development

Tracking the transmission of H3K27me3 on sperm-derived chromosomes in embryos which lack PRC2, revealed that the sperm-inherited H3K27me3 was detectable on chromatin throughout approximately four rounds of cell division. While it is unclear whether the sperm-inherited chromosomes might carry PRC2 components, the H3K27me3 appeared to persist only on the chromosomes originating from sperm and did not seem to spread to all other chromosomes in the dividing cells (Fig. 12.4). In the reciprocal combination of PRC2 mutant sperm lacking H3K27me3 with wild-type oocytes that provide H3K27me3-carrying chromosomes together with PRC2 transcripts, H3K27me3 is maintained at high levels. However, the H3K27me3 marks remained restricted to the oocyte-derived chromosomes in developing embryos although de novo deposition of H3K27me3 is possible due to the presence of PRC2 [4]. In contrast, H3K9me3 marks appear rapidly on all chromosomes when SET-25 and MET-2 were present [4]. Only later, during postembryonic development when the germline begins to expand and mature, did H3K27me3 marks spread to all chromosomes in the germ cells leading to the reestablishment of the H3K27me3 repressive signature on all germline chromosomes.

In summary, H3K27me3 marks generated by PRC2 are inherited through sperm and oocytes but they persist on chromosomes in a gamete-of-origin manner during several rounds of division in early embryos. The fact that this restriction persists, even in the presence of PRC2, could be due to a similar mode of PRC2 action as in mammals. That is, preexisting H3K27me3 marks may recruit PRC2 via the EED (MES-6 in *C. elegans*) subunit thereby propagating methylation of neighboring H3K27 [69]. Such a conserved feature of PRC2 in *C. elegans* would explain the observed gamete-of-origin mode of H3K27me3 propagation during cell divisions in the early embryo.

In general, it is conceivable that mammals also inherit sperm-derived H3K27me3 which are present on mammalian sperm chromatin [70]. The transmission of such marks or "epigenetic memory" could play a crucial role during embryonic development and may be key to maintaining totipotency of the germline across several generations in mammals.

PRC2 Safeguards Germ Cells From Somatic Differentiation

Apart from the crucial role in germline development, PRC2 has also been shown to be essential for safeguarding germ cells from being reprogrammed into somatic cells in *C. elegans* [56,71]. Germ cells can be reprogrammed into muscle-like cells or specific neurons when PRC2 subunits are depleted by RNAi and a muscle or neuronal fate-inducing TF is overexpressed [56]. Preceding this discovery, the histone chaperone LIN-53 (ortholog of human Caf1p48/RBBP4/7 and *D. melanogaster* CAF1/p55/NURF55) was attributed with the same safeguarding function for germ cells [57]. As mentioned earlier, the human and fly orthologs of LIN-53 are associated with PRC2 [19,49–55]. This property of LIN-53 orthologs in other species suggested that, in worms, LIN-53 might associate with PRC2 in the germline to safeguard germ cells from differentiation into somatic cells. Indeed, reprogramming phenotypes of germ cells are identical when either LIN-53 or a PRC2 subunit is depleted and both depletion backgrounds result in the global loss of H3K27me3 in germ cells [56]. Such functional overlap strongly suggests that LIN-53 depletion affects PRC2 function in the worm germline and further highlights the requirement for repression of germline chromosomes by H3K27me3 to protect totipotency of germ cells. Interestingly, *D. melanogaster* mutants that lack the PRC2 subunit E(z) show expression of somatic cell fate markers in the germ cells of adult testes [72]. Remarkably, this seems to be due to a non–cell-autonomous role of E(z) in the somatic gonadal cells, which appears to act via the epidermal growth factor signaling pathway to prevent expression of somatic genes in germ cells [72]. A similar observation has been made following the transcriptional profiling of dissected germlines from mutant worms lacking PRC2. A number of genes that are specific for somatic cells are activated in the germline of *mes*$^{-/-}$ mutants which, in contrast to flies, are due to cell autonomous effects in the germ cells [65]. Notably, germ cells with upregulated somatic genes do not convert into somatic cells. Their direct reprogramming into neurons or muscle cells requires the forced expression of a fate-inducing TF as described earlier [56,57]. It would be interesting to see whether germ cells in flies that lack PRC2 and show derepression of somatic gene expression are also permissive for the direct reprogramming into somatic cells.

PRC2 Restricts Plasticity of Embryonic Cells but is Dispensable for Somatic Differentiation in Worms

A non–cell-autonomous role of PRC2 in safeguarding cellular fates could also exist in *C. elegans*. It has been shown that worm embryos which entirely lack PRC2, either by using a *mes-2* mutant background or by RNAi, exhibit increased plasticity of the somatic cell lineages [73]. This increase in plasticity primarily affects the time window during which the embryonic cells could be converted to other lineage identities. This was shown by the forced expression of TFs, which can induce somatic differentiation, such as the muscle fate-inducing MyoD homolog HLH-1 or the endodermal fate-inducing GATA factor END-1 [73–75]. It was shown earlier that over-expression of such TFs, even in wild-type embryos, is potent enough to convert different lineages to other fates. However, embryonic cells show an intrinsic permissiveness for TF overexpression-induced lineage conversion only during early embryonic development (reviewed in Ref. [71]). In contrast, depletion of PRC2 allows the forced lineage conversion of somatic cells also during late embryonic stages [73]. The extended time window for the plasticity of mutant embryonic cells is likely due to loss of PRC2-mediated gene silencing indicating that PRC2 restricts plasticity during early differentiation. Interestingly, a functional analogy exists in mammalian tissues with respect to the protection of cell fates by PRC2 in *C. elegans*. It has been shown that depletion of PRC2 subunits facilitates the reprogramming of mammalian somatic cells into induced pluripotent stem cells [76,77]. Furthermore, depletion of the histone chaperone complex CAF-1, which contains the LIN-53 ortholog Caf1p48/RBBP4/7, enhances reprogramming of mouse fibroblasts [78]. Such analogies in antagonizing either direct reprogramming in germ cells of *C. elegans* or iPCS reprogramming in mammalian tissues could reflect conserved functions of PRC2 and LIN-53 with respect to safeguarding cellular fates.

PRC2 Function in Somatic Cells of *Caenorhabditis elegans*

The extension of the plasticity time window in PRC2-depleted worms only applies to the embryonic stage of developing animals. Postembryonically developing somatic cells were refractory to the induction of ectopic fates irrespective of PRC2 absence [73]. These observations suggest that, as long as somatic cell lineages are not challenged during development, PRC2 is dispensable. Indeed, lack of PRC2 does not impair proper differentiation of the somatic cells in *C. elegans* suggesting that mechanisms independent of PRC2-mediated repression must ensure development and maintenance of somatic cell fate in *C. elegans* [4,12,14,58]. This is in stark contrast to other organisms such as flies or vertebrates where the lack of PRC2 leads to severe developmental defects and diseases such as cancer (reviewed in Refs. [22,69,79–81]).

PRC2's predominant role in germline development and maintenance in *C. elegans* is also reflected by the PRC2 expression pattern in adult worms. The subunits MES-6 and MES-2 are detected mainly in the germline of adults (see figure and Refs. [14,34,48]). Expression levels of the PRC2 subunits in the embryonic somatic cell lineages gradually decline during postembryonic development and become undetectable in the soma by the time animals reach adulthood (Fig. 12.3) [14]. The fact that homozygous *mes* mutants give rise to progeny with proper somatic development [14,34,48,82] emphasizes the notion that, as long as worms are grown under normal conditions, PRC2 is needed only for developing and maintaining the germline. A protective role of PRC2 in somatic cells of developing embryos is only apparent when the embryonic cells are challenged by overexpression of fate-inducing TFs [73].

Nevertheless, though worms lacking PRC2 do not display any obvious defects in the development of somatic cells, subtle perturbation of gene expression in specific somatic cells has been noted in PRC2 mutants. As briefly described above, PRC2 depletion in worms affects expression of the Hox genes *mab-5* and *egl-5* in the male tail [14]. Absence of one of the MES-2/3/6 proteins leads to defects in the development of the sensory organs in the male tail (termed rays) which could be attributed to the derepression of Hox target genes [14]. Although *C. elegans* males of PRC2 mutants remain fertile and do not show any obvious health issues, the identification of Hox gene derepression in somatic cells of PRC2 mutants suggests that worms retain some aspects of the PRC2-mediated Hox gene regulation shown in *D. melanogaster* and vertebrates (reviewed in Refs. [83,84]). It is possible that tissue-specific characterization of transcriptomes in *mes* mutants could reveal additional derepressed genes in the soma despite the fact that loss of such gene repression is not causing obvious somatic phenotypes in PRC2 mutants.

Antagonizing PRC2 Activity

The histone lysine methyltransferase activity of the *C. elegans* PRC2 resides in the SET domain of the MES-2 subunit. Besides MES-2, another SET domain containing MES protein was identified in the initial screen for grandchildless mutants denoted as MES-4 [12]. MES-4 has a SET domain [13,33,34] and represents the worm ortholog of the mammalian nuclear SET domain (NSD) containing proteins such as NSD1 [85]. Mutant *mes-4* worms show the same *mes* phenotype as *mes-2/3/6* mutants; however, MES-4 does not associate with MES-2/3/6 [58,65]. Intriguingly, its SET domain has been shown to deposit the histone mark H3K36me3 which is usually associated with transcriptionally active chromatin loci [58,85]. The MES-4 activity of catalyzing the H3K36me3 mark as opposed to a silencing mark (H3K27me3) by the MES-2/3/6 complex (PRC2) appears to bare a dichotomy since the *mes-4* and PRC2 mutants all show the same *mes* phenotype. Moreover, loss of either MES-4 or MES-2/3/6 leads to the desilencing of X chromosomes in the germline [58,65,82] and, in addition, to permissiveness for direct reprogramming of germ cells into somatic cells as described previously [56].

A number of key observations regarding MES-4 localization on chromosomes in combination with sophisticated genetics shed light on this conundrum. In early embryos and the germline, MES-4 can be detected along all five autosomes while being mostly absent from X chromosomes. However, PRC2 depletion leads to extensive binding of MES-4 on X chromosomes [58,82]. The mutually exclusive binding pattern of MES-4 and PRC2 on germline chromosomes raised the possibility that the activity of MES-4 and the placing of H3K36me3 marks might repel PRC2. Indeed, such antagonism could be confirmed by transcript profiling of dissected germlines and genome-wide chromatin immunoprecipitations (ChIP) using H3K27me3 or H3K36me3 antibodies [65]. It turned out that MES-4 excludes PRC2 from autosomal genes and thereby leads to concentrated PRC2 activity on specific chromatin loci. Focusing PRC2 is important to prevent spreading of H3K27me3 on autosomes, which would otherwise lead to silencing of genes that need to be active. Importantly, the ectopic spreading of PRC2 activity leads to a loss of concentrated repression on other loci [56,65]. Hence, in *mes-4* mutants PRC2 loses its focused activity and ubiquitously spreads on chromosomes leading to lower levels of repression on X chromosomes and autosomal loci. Decreased levels of H3K27me3 at loci that need high levels of H3K27 methylation for proper repression become ectopically activated. Thereby, *mes-4* mutants mimic the phenotype of lost PRC2 activity in *mes-2/3/6 mutants*. The proposed model of antagonistic H3K27me3 versus H3K36me3 is further supported by observations in other species. For example, it has been shown in *D. melanogaster* that both modifications have a mutually exclusive distribution in the genome and can even prevent the placing of other modifications on the same histone tail [86–89].

Another observation concerning the MES-4 versus MES-2/3/6 antagonism is that the PRC2 proteins appear to be dispersed in the nucleoplasm of embryonic cells rather than showing a clear chromatin-bound pattern like MES-4 [34,48,82]. The reason for this disparity in nuclear localization is unknown. The delocalized PRC2 pattern of expression may reflect a characteristic mode of action for PRC2 in *C. elegans*. That is, PRC2 might act in a more "kiss-and-go" manner for the deposition of H3K27me3 on chromatin.

Another group of proteins that have been described to counteract PcG-mediated gene repression in flies and mammals are Trithorax group (TrxG) proteins [18,19,90–93]. However, it is not known whether such antagonism exists in *C. elegans*. The only well-described TrxG protein in *C. elegans* so far is ASH-2, the ortholog of *Drosophila* Ash2, which is a member of the conserved H3K4 trimethylation (H3K4me3) complex. Interestingly, a recent study in *D. melanogaster* suggests that the antagonistic relationship of PcGs and TrxG is required for life span regulation in flies [94]. In this context, the ASH-2-containing trithorax complex in *C. elegans* has been implicated in life span regulation though it is unclear whether *C. elegans* PRC2 has a role in this context as well [95]. The life span-regulating function of the ASH-2 complex takes place in the germline, the tissue in which *C. elegans* PRC2 plays its predominant role in chromatin silencing. Hence, it is possible that PRC2 activity might be indirectly implicated in ASH-2 trithorax complex-mediated regulation of life span.

As described earlier, mutants lacking PRC2 subunits (MES-2, MES-3, or MES-6) as well as the putative *C. elegans* PRC1-like proteins (MIG-32, SPAT-3, SOP-2, or SOR-1) show developmental defects in the male tail [14,39]. A similar defect has been observed for mutants of two other genes: *lin-49* and *lin-59*, which encode PHD/bromodomain and SET domain-containing proteins, respectively [96]. Both have been suggested to function similar to TrxG proteins as in *Drosophila* [96]. The defects in the male tail in *lin-49* and *lin-59* mutants could be attributed to the misregulation of Hox genes [14,39,96], though it is not clear, whether this misregulation reflects an antagonistic relationship of LIN-49 and LIN-59 versus PRC2. Moreover, the PHD/bromodomain-containing LIN-49 has been ascribed activating functions independent of a putative antagonistic role to PcGs during hindgut development and the regulation of left/right asymmetry in the nervous system [97–99]. Also the SET domain-containing LIN-59, which has been suggested to be related to the *Drosophila* TrxG protein ASH1, is required for proper hindgut development like LIN-49 [37,99,100]. Therefore, it remains unclear whether an interplay of LIN-49 and LIN-59 with PRC2 exists during gene expression regulation.

In general, a number of poorly characterized SET domain–containing proteins with predicted H3K4 HMT activity are present in *C. elegans* and, therefore, it is possible that additional factors with similar antagonistic functions as TrxG proteins in other species might exist in worms.

Noncoding RNA-Mediated H3K27 Methylation

The role of noncoding RNAs (ncRNA) in recruiting PcG complexes to specific genomic loci has been documented in mammals and *D. melanogaster.* One example is the long ncRNA HOTAIR which is the antisense transcript of the HOXC gene cluster. HOTAIR recruits PRC2 to Hox gene clusters thereby promoting H3K27 methylation of the respective loci [101,102]. Another example is the long ncRNA *Xist* which recruits PRC2 in the context of X chromosome inactivation [103,104]. While such long ncRNA-mediated recruitment of PcG proteins have not been discovered in *C. elegans* so far, a recent study reports that small RNAs might trigger H3K27 methylation in nematodes [105]. Findings from this study suggest that both exogenously and endogenously provided small RNAs with complementary sequence to target genomic loci can trigger H3K27 methylation in embryos. This process depends on the nuclear RNAi defective pathway in *C. elegans* which includes Argonaute proteins and is required for siRNA transport, inheritance of RNAi as well as H3K9 methylation [106,107]. Interestingly, these small RNA-induced H3K27me3 appear to be transmitted to progeny as described for endogenous H3K27me3 [4,105]. These discoveries connect the RNA interference (RNAi) pathway to chromatin silencing and are reminiscent of the RNA-induced transcriptional silencing in the fission yeast *Schizosaccharomyces pombe* which leads to gene silencing by H3K9 methylation [108]. An interesting observation is that the nematode-specific PRC2 subunit MES-3 is not involved [105]. This could suggest that a minimal MES-2/MES-6 complex is sufficient for small RNA-mediated H3K27me3. Yet, the mechanism of methylation triggering and recruitment of MES-2/MES-6 remains elusive.

PcG Recruitment—Noncoding RNAs or PREs?

The interconnection of small RNAs and histone methylation through PcGs has been reported in other species such as the protozoan *Tetrahymena thermophila* and mammalian cells [109–112]. In *T. thermophila*, the E(z) analog EZL1, which methylates H3K27 but also H3K9, appears to be targeted to chromatin by a small RNA-involving component of the RNAi machinery such as Argonaute proteins [109]. In human T cells and mouse ES cells, short RNAs originating from the PRC2-repressed genetic loci form a stem-loop structure resembling the binding sites of PRC2 in the long ncRNA *Xist* [103,113]. These short ncRNAS ranging in size from 50–200 nucleotides bind PRC2 through the SUZ12 subunit and may play a role in the association of PRC2 with its target genes [113]. Additionally, the Argonaute protein AGO1 can associate with RNA polymerase II in human HEK293T cells together with EZH2. This association might trigger H3K27 methylation at target genes of PcG silencing

[112]. While these findings insinuate that small RNAs might play a role in PcG targeting, the mechanisms or a clear proof for the requirement of small RNAs for PcG-mediated silencing in mammalian cells is lacking. As mentioned above, the finding in *C. elegans* that small RNAs and members of the RNAi machinery can trigger heritable H3K27me3 could indicate an involvement of small ncRNAs in PRC2 recruitment in nematodes [105].

While there is no evidence in *D. melanogaster* regarding the implication of small RNAs in PcG-mediated silencing, long ncRNAs have been ascribed a role in regulating PcG target gene regulation. In flies, PcGs are recruited to their target genes through specific DNA sequences termed Polycomb response elements (PREs) [114,115]. It has been suggested that long ncRNAs derived from transcribed PRE loci can associate with PcG proteins and might mediate either activation or silencing of PcG target genes [116]. However, no obvious PREs could be identified in other organisms [114,117]. Nevertheless, the potential recruitment of PcG proteins by small RNAs in *C. elegans* and protozoans or even vertebrates might reflect one important facet of how PcGs find their specific target loci. However, while there is accumulating evidence that both small and long ncRNAs may be involved in PcG recruitment, it remains unclear how universal such a mechanism could be.

CONCLUSIONS

While canonical PcG proteins of the PRC1 complex are absent in *C. elegans*, the identification of the PcG proteins MES-2, MES-3, and MES-6 [12] revealed that the PRC2 is highly conserved in nematodes. Distantly PRC1-related proteins such as MIG-32 and SPAT-3 provide some PRC-1-like functions during homeotic gene regulation and ubiquitylation of histone H2A. The lack of a canonical PRC1 complex suggests that PRC2-mediated silencing can be uncoupled from PRC1, at least in the *C. elegans* germline. The predominant role of PRC2 in *C. elegans* is the repression of somatic genes and X chromosome silencing in the germline. Importantly, accurate genomic distribution of the H3K27me3 marks in germ cells depends on methylated H3K36 deposited by the methylase MES-4 [65], the mouse NSD1 homolog [85,118]. This finding corroborates the notion derived from *D. melanogaster* studies that methylated H3K36 might restrict ectopic spreading of PRC2-mediated H3K27me3 [119]. Such a mechanism could reflect a more general antagonistic relationship between H3K27me3 and H3K36me rather than a mutually exclusive one. The importance of this antagonism in *C. elegans* is highlighted by the fact that spreading of H3K27me3 upon loss of MES-4 leads to reduced, or in other words, "diluted" H3K27me3 levels at repressed genomic loci in the germline [65]. Consequently, this compromised repression makes germ cells permissive to being reprogrammed into neurons or muscle-like cells [56]. It is conceivable that such a "dilution effect" due to ectopic H3K27me3 spreading is a general phenomenon also in other organisms explaining why seemingly antagonistic mechanisms can result in the same phenotype as seen for the loss of either MES-2/3/6 or MES-4 in the *C. elegans* germline.

More recent studies on PRC2 in *C. elegans* revealed a much-discussed possibility of transgenerational and cellular inheritance of PRC2-mediated silencing. First, the parental epigenetic memory established by PRC2 can be inherited through sperm and oocytes in *C. elegans*. Second, the H3K27me3 marks are propagated on chromosomes in a gamete-of-origin fashion during embryogenesis. Such mode of cellular transmission implies that modified histones are passaged during cell division in *cis* and only locally [4]. These insights gained by investigating *C. elegans* could provide an important aspect in the context of personalized epigenetics for humans. It is possible that the epigenetic signature of individuals is significantly influenced by transgenerational inheritance also in humans. Therefore, it would be interesting to see whether PcG-mediated silencing might indeed be inherited in humans as well.

The open question in *C. elegans* remains how exactly H3K27me3 is distributed to all chromosomes in the developing germline of animals. Worms that start out embryonic development with chromosomes carrying H3K27me3 exclusively on paternal or maternal chromosomes acquire H3K27me3 marks on all chromosomes de novo later in the germline [4]. Precisely, how PRC2 might be recruited to specific sites to silence somatic genes on autosomal chromosomes and more generally on the X chromosome, remains speculative. Since the primordial germ cells turn on their transcriptional activity during postembryonic development, TFs in conjunction with the distribution of antagonizing H3K36me patterns might limit PRC2 activity to specific domains. In this context, PREs as in *D. melanogaster* have not been identified in *C. elegans*. Worms might rely on a more basic and context-dependent scheme of recruiting PRC2 to specific loci. Whether, for instance, TFs or initial silencing mechanisms need to precede PRC2 recruitment requires further investigation. Interestingly, the discovery of small RNAs binding to MES-2/MES-6 which appears to trigger sequence-specific H3K27 methylation by PRC2 could explain, to some extent, PcG recruitment in *C. elegans* [105]. While such a role for small ncRNAs in PcG recruitment has been suggested previously in other species, whether small RNAs could indeed direct specificity of PcG-mediated silencing is a subject to debate.

Future "basic research" using model organisms such as *Drosophila* and *C. elegans* could provide significant insight into these open questions with respect to PcG recruitment and silencing specificity. *C. elegans* and *D. melanogaster* allow straightforward genetic screens and the opportunity to test the physiological relevance of proposed mechanisms in vivo. Cunningly designed future genetic screens with both model organisms in sensitized genetic backgrounds could help to identify additional players in the regulation of PcG specificity.

LIST OF ACRONYMS AND ABBREVIATION

Ash1 (Absent, small, or homeotic)-like (*Drosophila*)
CAF Chromatin assembly factor
C. elegans *Caenorhabditis elegans*
D. melanogaster *Drosophila melanogaster*
DIC Differential interference contrast
EED Embryonic ectoderm development
EZH2 Enhancer of zeste homolog 2
E(z) Enhancer of zeste
F1 First filial generation
GFP Green fluorescent protein
H2AK119 Histone H2A lysine residue 119 ubiquitylated
H3K4me Histone H3 lysine residue four methylated
H3K9me Histone H3 lysine residue nine methylated
H3K9ac Histone H3 lysine residue nine acetylated
H3K27me2 Histone H3 lysine residue 27 trimethylated
H3K27me3 Histone H3 lysine residue 27 trimethylated
Hox Homeobox
HMT Histone methyltransferase
MES Maternal-effect sterile
NSD Nuclear SET domain
NURF Nucleosome-remodeling factor
PHD Plant homeodomain
P0 Parental generation
PcG Polycomb group
PRC Polycomb repressive complex
RAF RING associated factor
RBBP Retinoblastoma binding protein
SET Su(var)3-9 + Enhancer of zeste + Trithorax
SUZ12 Suppressor of zeste 12 homolog
TrxG Trithorax group

REFERENCES

[1] Struhl G. A gene product required for correct initiation of segmental determination in Drosophila. Nature 1981;293:36–41.

[2] Lewis EB. A gene complex controlling segmentation in Drosophila. Nature 1978;276:565–70.

[3] Duncan IM. Polycomblike: a gene that appears to be required for the normal expression of the bithorax and antennapedia gene complexes of *Drosophila melanogaster*. Genetics 1982;102:49–70.

[4] Gaydos LJ, Wang W, Strome S. Gene repression. H3K27me and PRC2 transmit a memory of repression across generations and during development. Science 2014;345:1515–8.

[5] Viré E, Brenner C, Deplus R, Blanchon L, Fraga M, Didelot C, et al. The Polycomb group protein EZH2 directly controls DNA methylation. Nature 2006;439:871–4.

[6] Reddington JP, Sproul D, Meehan RR. DNA methylation reprogramming in cancer: does it act by re-configuring the binding landscape of Polycomb repressive complexes? BioEssays 2013;36:134–40.

[7] Rose NR, Klose RJ. Understanding the relationship between DNA methylation and histone lysine methylation. Biochim Biophys Acta 2014;1839:1362–72.
[8] Wightman B, Ha I, Ruvkun G. Posttranscriptional regulation of the heterochronic gene lin-14 by lin-4 mediates temporal pattern formation in *C. elegans*. Cell 1993;75:855–62.
[9] Lee RC, Feinbaum RL, Ambros V. The *C. elegans* heterochronic gene lin-4 encodes small RNAs with antisense complementarity to lin-14. Cell 1993;75:843–54.
[10] Kenyon CJ. The genetics of ageing. Nature 2010;464:504–12.
[11] Hedgecock EM, Culotti JG, Hall DH. The unc-5, unc-6, and unc-40 genes guide circumferential migrations of pioneer axons and mesodermal cells on the epidermis in *C. elegans*. Neuron 1990;4:61–85.
[12] Capowski E, Martin P, Garvin C, Strome S. Identification of grandchildless loci whose products are required for normal germ-line development in the nematode *Caenorhabditis elegans*. Genetics 1991;129:1061–72.
[13] Xu L, Fong Y, Strome S. The *Caenorhabditis elegans* maternal-effect sterile proteins, MES-2, MES-3, and MES-6, are associated in a complex in embryos. Proc Natl Acad Sci USA 2001;98:5061–6.
[14] Ross J, Zarkower D. Polycomb group regulation of Hox gene expression in *C. elegans*. Dev Cell 2003;185:523–35.
[15] Karakuzu O, Wang DP, Cameron S. MIG-32 and SPAT-3A are PRC1 homologs that control neuronal migration in *Caenorhabditis elegans*. Development 2009;136:943–53.
[16] Margueron R, Reinberg D. The Polycomb complex PRC2 and its mark in life. Nature 2011;469:343–9.
[17] Simon JA, Kingston RE. Mechanisms of Polycomb gene silencing: knowns and unknowns. Nat Rev Mol Cell Biol 2009:1–12.
[18] Di Croce L, Helin K. Transcriptional regulation by Polycomb group proteins. Nat Struct Mol Biol 2013;20:1147–55.
[19] Schwartz YB, Pirrotta V. A new world of Polycombs: unexpected partnerships and emerging functions. Nat Rev Genet 2013;14:853–64.
[20] Levine SS, Weiss A, Erdjument-Bromage H, Shao Z, Tempst P, Kingston RE. The core of the polycomb repressive complex is compositionally and functionally conserved in flies and humans. Mol Cell Biol 2002;22:6070–8.
[21] Gao Z, Zhang J, Bonasio R, Strino F, Sawai A, Parisi F, et al. Pcgf homologs, CBX proteins, and RYBP define functionally distinct PRC1 family complexes. Mol Cell 2012;45:344–56.
[22] Su S, Zhang M, Li L, Wu M. Polycomb group genes as the key regulators in gene silencing. Wuhan Univ J Nat Sci 2014;19:1–7.
[23] del Prete S, Mikulski P, Schubert D, Gaudin V. One, two, three: polycomb proteins hit all dimensions of gene regulation. Genes 2015;6:520–42.
[24] Tavares L, Dimitrova E, Oxley D, Webster J, Poot R, Demmers J, et al. RYBP-PRC1 complexes mediate H2A ubiquitylation at polycomb target sites independently of PRC2 and H3K27me3. Cell 2012;148:664–78.
[25] Leeb M, Pasini D, Novatchkova M, Jaritz M, Helin K, Wutz A. Polycomb complexes act redundantly to repress genomic repeats and genes. Genes Dev 2010;24:265–76.
[26] Wu X, Johansen JV, Helin K. Fbxl10/Kdm2b recruits polycomb repressive complex 1 to CpG islands and regulates H2A ubiquitylation. Mol Cell 2013;49:1134–46.
[27] Sanchez-Pulido L, Devos D, Sung ZR, Calonje M. RAWUL: A new ubiquitin-like domain in PRC1 Ring finger proteins that unveils putative plant and worm PRC1 orthologs. BMC Genomics 2008;9:308.

[28] Gearhart MD, Corcoran CM, Wamstad JA, Bardwell VJ. Polycomb group and SCF ubiquitin ligases are found in a novel BCOR complex that is recruited to BCL6 targets. Mol Cell Biol 2006;26:6880–9.

[29] Lagarou A, Mohd-Sarip A, Moshkin YM, Chalkley GE, Bezstarosti K, Demmers JAA, et al. dKDM2 couples histone H2A ubiquitylation to histone H3 demethylation during Polycomb group silencing. Genes Dev 2008;22:2799–810.

[30] Cao R, Tsukada Y-I, Zhang Y. Role of Bmi-1 and Ring1A in H2A ubiquitylation and Hox gene silencing. Mol Cell 2005;20:845–54.

[31] Buchwald G, van der Stoop P, Weichenrieder O, Perrakis A, van Lohuizen M, Sixma TK. Structure and E3-ligase activity of the Ring-Ring complex of polycomb proteins Bmi1 and Ring1b. EMBO J 2006;25:2465–74.

[32] Li Z, Cao R, Wang M, Myers MP, Zhang Y, Xu R-M. Structure of a Bmi-1-Ring1B polycomb group ubiquitin ligase complex. J Biol Chem 2006;281:20643–9.

[33] Bender LB, Cao R, Zhang Y, Strome S. The mes-2/mes-3/mes-6 complex and regulation of histone H3 methylation in *C. elegans*. Curr Biol 2004;14:1639–43.

[34] Holdeman R, Nehrt S, Strome S. MES-2, a maternal protein essential for viability of the germline in *Caenorhabditis elegans*, is homologous to a Drosophila Polycomb group protein. Development 1998;125:2457–67.

[35] Fong AP, Yao Z, Zhong JW, Cao Y, Ruzzo WL, Gentleman RC, et al. Genetic and epigenetic determinants of neurogenesis and myogenesis. Dev Cell 2012;22:721–35.

[36] Zheng Y, Hsu F-N, Xu W, Xie X-J, Ren X, Gao X, et al. A developmental genetic analysis of the lysine demethylase KDM2 mutations in *Drosophila melanogaster*. Mech Dev 2014;133:36–53.

[37] Shaye DD, Greenwald I. OrthoList: a compendium of *C. elegans* genes with human orthologs. PLoS ONE 2011;6:e20085.

[38] Zhang T, Sun Y, Tian E, Deng H, Zhang Y, Luo X, et al. RNA-binding proteins SOP-2 and SOR-1 form a novel PcG-like complex in *C. elegans*. Development 2006;133:1023–33.

[39] Zhang H, Azevedo RBR, Lints R, Doyle C, Teng Y, Haber D, et al. Global regulation of Hox gene expression in *C. elegans* by a SAM domain protein. Dev Cell 2003;4:903–15.

[40] Robinson AK, Leal BZ, Chadwell LV, Wang R, Ilangovan U, Kaur Y, et al. The growth-suppressive function of the polycomb group protein polyhomeotic is mediated by polymerization of its sterile Alpha Motif (SAM) domain. J Biol Chem 2012;287:8702–13.

[41] Peterson AJ, Kyba M, Bornemann D, Morgan K, Brock HW, Simon J. A domain shared by the Polycomb group proteins Scm and ph mediates heterotypic and homotypic interactions. Mol Cell Biol 1997;17:6683–92.

[42] Kim CA, Gingery M, Pilpa RM, Bowie JU. The SAM domain of polyhomeotic forms a helical polymer. Nat Struct Biol 2002;9:453–7.

[43] Waring DA, Kenyon C. Regulation of cellular responsiveness to inductive signals in the developing *C. elegans* nervous system. Nature 1991;350:712–5.

[44] Wang BB, Müller-Immergluck MM, Austin J, Robinson NT, Chisholm A, Kenyon C. A homeotic gene cluster patterns the anteroposterior body axis of *C. elegans*. Cell 1993;74:29–42.

[45] Schaller D, Wittmann C, Spicher A, Müller F, Tobler H. Cloning and analysis of three new homeobox genes from the nematode *Caenorhabditis elegans*. Nucleic Acids Res 1990;18:2033–6.

[46] Paulsen JE, Capowski EE, Strome S. Phenotypic and molecular analysis of mes-3, a maternal-effect gene required for proliferation and viability of the germ line in *C. elegans*. Genetics 1995;141:1383–98.

[47] Garvin C, Holdeman R, Strome S. The phenotype of mes-2, mes-3, mes-4 and mes-6, maternal-effect genes required for survival of the germline in *Caenorhabditis elegans*, is sensitive to chromosome dosage. Genetics 1998;148:167–85.

[48] Korf I, Fan Y, Strome S. The Polycomb group in *Caenorhabditis elegans* and maternal control of germline development. Development 1998;125:2469–78.

[49] Tie F, Furuyama T, Prasad-Sinha J, Jane E, Harte PJ. The Drosophila Polycomb Group proteins ESC and E(Z) are present in a complex containing the histone-binding protein p55 and the histone deacetylase RPD3. Development 2001;128:275–86.

[50] Müller J, Hart CM, Francis NJ, Vargas ML, Sengupta A, Wild B, et al. Histone methyltransferase activity of a Drosophila Polycomb group repressor complex. Cell 2002;111:197–208.

[51] Czermin B, Melfi R, McCabe D, Seitz V, Imhof A, Pirrotta V. Drosophila enhancer of Zeste/ESC complexes have a histone H3 methyltransferase activity that marks chromosomal Polycomb sites. Cell 2002;111:185–96.

[52] Nowak AJ, Alfieri C, Stirnimann CU, Rybin V, Baudin F, Ly-Hartig N, et al. Chromatin-modifying complex component Nurf55/p55 associates with histones H3 and H4 and polycomb repressive complex 2 subunit Su(z)12 through partially overlapping binding sites. J Biol Chem 2011;286:23388–96.

[53] Cao R, Wang L, Wang H, Xia L, Erdjument-Bromage H, Tempst P, et al. Role of histone H3 lysine 27 methylation in Polycomb-group silencing. Science 2002;298:1039–43.

[54] Kuzmichev A, Margueron R, Vaquero A, Preissner TS, Scher M, Kirmizis A, et al. Composition and histone substrates of polycomb repressive group complexes change during cellular differentiation. Proc Natl Acad Sci USA 2005;102:1859–64.

[55] Ciferri C, Lander GC, Maiolica A, Herzog F, Aebersold R, Nogales E. Molecular architecture of human polycomb repressive complex 2. eLife 2012;1:e00005.

[56] Patel T, Tursun B, Rahe DP, Hobert O. Removal of polycomb repressive complex 2 makes *C. elegans* germ cells susceptible to direct conversion into specific somatic cell types. Cell Rep 2012;2:1–9.

[57] Tursun B, Patel T, Kratsios P, Hobert O. Direct conversion of *C. elegans* germ cells into specific neuron types. Science 2011;331:304–8.

[58] Fong Y, Bender L, Wang W, Strome S. Regulation of the different chromatin states of autosomes and X chromosomes in the germ line of *C. elegans*. Science 2002;296:2235–8.

[59] Kelly W, Schaner C, Dernburg A, Lee M, Kim S, Villeneuve A, et al. X-chromosome silencing in the germline of *C. elegans*. Development 2002;129:479.

[60] Strome S, Kelly WG, Ercan S, Lieb JD. Regulation of the X chromosomes in *Caenorhabditis elegans*. Cold Spring Harb Perspect Biol 2014;6:a018366.

[61] Seydoux G, Strome S. Launching the germline in *Caenorhabditis elegans*: regulation of gene expression in early germ cells. Development 1999;126:3275–83.

[62] Mak W, Baxter J, Silva J, Newall AE, Otte AP, Brockdorff N. Mitotically stable association of polycomb group proteins eed and enx1 with the inactive X chromosome in trophoblast stem cells. Curr Biol 2002;12:1016–20.

[63] Silva J, Mak W, Zvetkova I, Appanah R, Nesterova TB, Webster Z, et al. Establishment of histone h3 methylation on the inactive X chromosome requires transient recruitment of Eed-Enx1 polycomb group complexes. Dev Cell 2003;4:481–95.

[64] Pinter SF, Sadreyev RI, Yildirim E, Jeon Y, Ohsumi TK, Borowsky M, et al. Spreading of X chromosome inactivation via a hierarchy of defined Polycomb stations. Genome Res 2012;22:1864–76.

[65] Gaydos LJ, Rechtsteiner A, Egelhofer TA, Carroll CR, Strome S. Antagonism between MES-4 and polycomb repressive complex 2 promotes appropriate gene expression in *C. elegans* germ cells. Cell Rep 2012;2:1169–77.

[66] Towbin BD, González-Aguilera C, Sack R, Gaidatzis D, Kalck V, Meister P, et al. Step-wise methylation of histone H3K9 positions heterochromatin at the nuclear periphery. Cell 2012;150:934–47.

[67] Heard E, Rougeulle C, Arnaud D, Avner P, Allis CD, Spector DL. Methylation of histone H3 at Lys-9 is an early mark on the X chromosome during X inactivation. Cell 2001;107:727–38.

[68] Andersen EC, Shimko TC, Crissman JR, Ghosh R, Bloom JS, Seidel HS, et al. A powerful new quantitative genetics platform, combining *Caenorhabditis elegans* high-throughput fitness assays with a large collection of recombinant strains. G3 2015;5:911–20.

[69] Margueron R, Justin N, Ohno K, Sharpe ML, Son J, Drury III WJ, et al. Role of the polycomb protein EED in the propagation of repressive histone marks. Nature 2009;461:762–7.

[70] Hammoud SS, Nix DA, Zhang H, Purwar J, Carrell DT, Cairns BR. Distinctive chromatin in human sperm packages genes for embryo development. Nature 2009;460:473–8.

[71] Tursun B. Cellular reprogramming processes in *Drosophila* and *C. elegans*. Curr Opin Genet Dev 2012;22:475–84.

[72] Eun SH, Shi Z, Cui K, Zhao K, Chen X. A non–cell autonomous role of E(z) to prevent germ cells from turning on a somatic cell marker. Science 2014;343:1513–6.

[73] Yuzyuk T, Fakhouri THI, Kiefer J, Mango SE. The polycomb complex protein mes-2/E(z) promotes the transition from developmental plasticity to differentiation in *C. elegans* embryos. Dev Cell 2009;16:699–710.

[74] Fukushige T, Hawkins MG, McGhee JD. The GATA-factor elt-2 is essential for formation of the *Caenorhabditis elegans* intestine. Dev Biol 1998;198:286–302.

[75] Fukushige T, Krause M. The myogenic potency of HLH-1 reveals wide-spread developmental plasticity in early *C. elegans* embryos. Development 2005;132:1795–805.

[76] Onder TT, Kara N, Cherry A, Sinha AU, Zhu N, Bernt KM, et al. Chromatin-modifying enzymes as modulators of reprogramming. Nature 2012:1–7.

[77] Fragola G, Germain P-L, Laise P, Cuomo A, Blasimme A, Gross F, et al. Cell reprogramming requires silencing of a core subset of polycomb targets. PLoS Genet 2013;9:e1003292.

[78] Cheloufi S, Elling U, Hopfgartner B, Jung YL, Murn J, Ninova M, et al. The histone chaperone CAF-1 safeguards somatic cell identity. Nature 2015;528:218–24.

[79] Conway E, Healy E, Bracken AP. PRC2 mediated H3K27 methylations in cellular identity and cancer. Curr Opin Cell Biol 2015;37:42–8.

[80] Aloia L, Di Stefano B, Di Croce L. Polycomb complexes in stem cells and embryonic development. Development 2013;140:2525–34.

[81] Laugesen A, Helin K. Chromatin repressive complexes in stem cells, development, and cancer. Cell Stem Cell 2014;14:735–51.

[82] Strome S. Specification of the germ line. WormBook 2005:1–10.

[83] Schuettengruber B, Ganapathi M, Leblanc B, Portoso M, Jaschek R, Tolhuis B, et al. Functional anatomy of polycomb and trithorax chromatin landscapes in Drosophila embryos. PLoS Biol 2009;7:e13.

[84] Ringrose L, Paro R. Polycomb/trithorax response elements and epigenetic memory of cell identity. Development 2007;134:223–32.

[85] Rechtsteiner A, Ercan S, Takasaki T, Phippen TM, Egelhofer TA, Wang W, et al. The histone H3K36 methyltransferase MES-4 acts epigenetically to transmit the memory of germline gene expression to progeny. PLoS Genet 2010;6:e1001091.
[86] Liu T, Rechtsteiner A, Egelhofer TA, Vielle A, Latorre I, Cheung M-S, et al. Broad chromosomal domains of histone modification patterns in *C. elegans*. Genome Res 2011;21:227–36.
[87] Kharchenko PV, Alekseyenko AA, Schwartz YB, Minoda A, Riddle NC, Ernst J, et al. Comprehensive analysis of the chromatin landscape in *Drosophila melanogaster*. Nature 2010;471:480–5.
[88] Schmitges FW, Prusty AB, Faty M, Stützer A, Lingaraju GM, Aiwazian J, et al. Histone methylation by PRC2 is inhibited by active chromatin marks. Mol Cell 2011;42:330–41.
[89] Yuan W, Xu M, Huang C, Liu N, Chen S, Zhu B. H3K36 methylation antagonizes PRC2-mediated H3K27 methylation. J Biol Chem 2011;286:7983–9.
[90] Ringrose L. Polycomb, trithorax and the decision to differentiate. BioEssays 2006;28:330–4.
[91] Papp B, Müller J. Histone trimethylation and the maintenance of transcriptional ON and OFF states by trxG and PcG proteins. Genes Dev 2006;20:2041.
[92] Schuettengruber B, Cavalli G. Recruitment of polycomb group complexes and their role in the dynamic regulation of cell fate choice. Development 2009;136:3531.
[93] Grimaud C, Nègre N, Cavalli G. From genetics to epigenetics: the tale of Polycomb group and trithorax group genes. Chromosome Research 2006;14:363–75.
[94] Siebold AP, Banerjee R, Tie F, Kiss DL, Moskowitz J, Harte PJ. Polycomb repressive complex 2 and trithorax modulate Drosophila longevity and stress resistance. Proc Natl Acad Sci USA 2010;107:169–74.
[95] Greer EL, Maures TJ, Hauswirth AG, Green EM, Leeman DS, Maro GS, et al. Members of the H3K4 trimethylation complex regulate lifespan in a germline-dependent manner in *C. elegans*. Nature 2010;466:1–7.
[96] Chamberlin HM, Thomas JH. The bromodomain protein LIN-49 and trithorax-related protein LIN-59 affect development and gene expression in *Caenorhabditis elegans*. Development 2000;127:713–23.
[97] Chang S, Johnston RJ, Hobert O. A transcriptional regulatory cascade that controls left/right asymmetry in chemosensory neurons of *C. elegans*. Genes Dev 2003;17:2123–37.
[98] Sarin S, O'Meara MM, Flowers EB, Antonio C, Poole RJ, Didiano D, et al. Genetic screens for *Caenorhabditis elegans* mutants defective in left/right asymmetric neuronal fate specification. Genetics 2007;176:2109–30.
[99] Chamberlin HM, Brown KB, Sternberg PW, Thomas JH. Characterization of seven genes affecting *Caenorhabditis elegans* hindgut development. Genetics 1999;153:731–42.
[100] Tripoulas N, LaJeunesse D, Gildea J, Shearn A. The Drosophila ash1 gene product, which is localized at specific sites on polytene chromosomes, contains a SET domain and a PHD finger. Genetics 1996;143:913–28.
[101] Tsai M-C, Manor O, Wan Y, Mosammaparast N, Wang JK, Lan F, et al. Long noncoding RNA as modular scaffold of histone modification complexes. Science 2010;329:689–93.
[102] Rinn JL, Kertesz M, Wang JK, Squazzo SL, Xu X, Brugmann SA, et al. Functional demarcation of active and silent chromatin domains in human HOX loci by noncoding RNAs. Cell 2007;129:1311–23.
[103] Zhao J, Sun BK, Erwin JA, Song J-J, Lee JT. Polycomb proteins targeted by a short repeat RNA to the mouse X chromosome. Science 2008;322:750–6.

[104] Kanduri C, Whitehead J, Mohammad F. The long and the short of it: RNA-directed chromatin asymmetry in mammalian X-chromosome inactivation. FEBS Lett 2009;583:857–64.

[105] Mao H, Zhu C, Zong D, Weng C, Yang X, Huang H, et al. The Nrde pathway mediates small-RNA-directed histone H3 lysine 27 trimethylation in *Caenorhabditis elegans*. Curr Biol 2015;25:2398–403.

[106] Guang S, Bochner AF, Pavelec DM, Burkhart KB, Harding S, Lachowiec J, et al. An argonaute transports siRNAs from the cytoplasm to the nucleus. Science 2008;321:537–41.

[107] Guang S, Bochner AF, Burkhart KB, Burton N, Pavelec DM, Kennedy S. Small regulatory RNAs inhibit RNA polymerase II during the elongation phase of transcription. Nature 2010;465:1097–101.

[108] Verdel A, Jia S, Gerber S, Sugiyama T, Gygi S, Grewal SIS, et al. RNAi-mediated targeting of heterochromatin by the RITS complex. Science 2004;303:672–6.

[109] Liu Y, Taverna SD, Muratore TL, Shabanowitz J, Hunt DF, Allis CD. RNAi-dependent H3K27 methylation is required for heterochromatin formation and DNA elimination in *Tetrahymena*. Genes Dev 2007;21:1530–45.

[110] Weinberg MS, Villeneuve LM, Ehsani A, Amarzguioui M, Aagaard L, Chen Z-X, et al. The antisense strand of small interfering RNAs directs histone methylation and transcriptional gene silencing in human cells. RNA 2006;12:256–62.

[111] Ting AH, Schuebel KE, Herman JG, Baylin SB. Short double-stranded RNA induces transcriptional gene silencing in human cancer cells in the absence of DNA methylation. Nat Genet 2005;37:906–10.

[112] Kim DH, Villeneuve LM, Morris KV, Rossi JJ. Argonaute-1 directs siRNA-mediated transcriptional gene silencing in human cells. Nat Struct Mol Biol 2006;13:793–7.

[113] Kanhere A, Viiri K, Araújo CC, Rasaiyaah J, Bouwman RD, Whyte WA, et al. Short RNAs are transcribed from repressed polycomb target genes and interact with polycomb repressive Complex-2. Mol Cell 2010;38:675–88.

[114] Bauer M, Trupke J, Ringrose L. The quest for mammalian Polycomb response elements: Are we there yet? Chromosoma 2015:1–26.

[115] Kassis JA, Brown JL. Polycomb group response elements in Drosophila and vertebrates. 1st ed. Elsevier Inc.; 2013.

[116] Hekimoglu B, Ringrose L. Non-coding RNAs in polycomb/trithorax regulation. RNA Biol 2009;6:129–37.

[117] Okulski H, Druck B, Bhalerao S, Ringrose L. Quantitative analysis of polycomb response elements (PREs) at identical genomic locations distinguishes contributions of PRE sequence and genomic environment. Epigenet Chromatin 2011;4:4.

[118] Furuhashi H, Takasaki T, Rechtsteiner A, Li T, Kimura H, Checchi PM, et al. Trans-generational epigenetic regulation of *C. elegans* primordial germ cells. Epigenet Chromatin 2010;3:15.

[119] Klymenko T, Müller J. The histone methyltransferases Trithorax and Ash1 prevent transcriptional silencing by Polycomb group proteins. EMBO Rep 2004;5:373–7.

[120] Paix A, Wang Y, Smith HE, Lee C-YS, Calidas D, Lu T, et al. Scalable and versatile genome editing using linear DNAs with microhomology to Cas9 Sites in *Caenorhabditis elegans*. Genetics 2014;198:1347–56.

Chapter 13

Global Functions of PRC2 Complexes

V. Pirrotta
Rutgers University, Piscataway, NJ, United States

Chapter Outline

Introduction	317
Targeted Silencing Functions	318
Global Functions of PRC2	319
Genomic Distribution of H3K27 Methylation	321
Role of Global H3K27 Methylation	323
The Role of UTX: H3K27 Demethylation or Not?	324
H3K27 Acetylation	326
Roaming Activities	328
The Accessibility Hypothesis	329
Recruitment of PRC2 by a PRC1 Type of Complex	332
Does H2A Ubiquitylation Play a Role in Global PRC2 Activity?	333
Polycomb Repressive Activities	336
Evolutionary Aspects of PRC2 Function	338
References	342

INTRODUCTION

The classical view of Polycomb repressive mechanisms involves the interplay of two types of Polycomb Group (PcG) complex: PRC1 and PRC2. Of these, PRC2 has primarily the enzymatic function of methylating histone H3 at lysine 27. PRC1 includes the canonical Polycomb (Pc) component (mammalian CBX), which contains a chromodomain able to recognize the H3K27me3 modification. The discovery of a whole family of PRC1-related complexes, some containing and some lacking a chromodomain component, and possessing histone H2AK119 ubiquityl transferase activity, has now complicated the story in various ways and has raised the possibility of other kinds of interplay between PRC1-type and PRC2 complexes in addition to the chromodomain-mediated recognition of the H3K27 methylation mark.

In this chapter I will distinguish between classical PcG silencing of target genes and a more global transcriptional interference. I will argue that PcG complexes in general have global functions, opportunistic functions, as well as

Polycomb Group Proteins. http://dx.doi.org/10.1016/B978-0-12-809737-3.00013-1

specifically targeted functions although the lines separating these three aspects may be less than clear-cut at individual sites. Some of the ideas discussed here were presented in an earlier form [1]. As PcG complexes, I will consider the two main varieties: the very diverse PRC1 type and the more well-defined but still diverse PRC2. The core feature of the PRC1 complexes is the presence of a RING E3 ubiquityl transferase heterodimer. A further distinction may be made between canonical PRC1 complexes, containing a chromodomain subunit (Polycomb in *Drosophila*, CBX in mammals), and able to recognize methylated histone H3K27, and noncanonical or variant PRC1 complexes, lacking a chromodomain subunit. The core of the PRC2 complex includes the catalytic subunit E(z) or its homologs, the highly conserved Esc/EED component, and the Su(z)12 component or its homologs. The last of these is absent in nematodes and microorganisms but present in flies, mammals, and plants. Additional components such as AEBP2, JARID2, and PCL appear to be variable and substoichiometric, but may play important roles in stabilizing the complex and targeting it to specific sites.

An important question that is not often discussed concerns the extent to which these complexes exist as distinct free species in the nucleus, before binding to chromatin, the degree to which they might be dynamically interconvertible, and the relative importance of their targeted versus untargeted activities. Biochemists study structures that can be characterized and purified or even crystallized, but to do so requires concentrations of the components not normally occurring in vivo. Complexes are often isolated after overexpressing multiple components in the same cells to allow complex formation, presumably cotranslationally, but quite possibly driven by high concentrations and by the presence of abundant chaperone proteins. The degree to which complex formation may be driven by multiple interactions at the chromatin target site has not been explored. One argument for the free existence of at least enzymatically active stable cores of the PcG complexes is that the histone modifications that they produce on genomic chromatin can be found widely spread without a concomitant presence of stably bound complexes. This suggests that a substantial and important aspect of PcG activities in the nucleus may be a sort of "hit-and-run" transient and nonspecific association with chromatin, with a residence time that is at least long enough to allow the histone modification but not sufficiently long to allow detection by the commonly used chromatin crosslinking and immunoprecipitation technologies (ChIP, chromatin immunoprecipitation).

TARGETED SILENCING FUNCTIONS

With a few notable exceptions, both PRC1 and PRC2 complexes must be recruited to a target gene to establish the classical silenced state. The presence of the recruiting element to which PRC1 and PRC2 complexes are stably bound defines the set of PcG target genes in the genome. PRC2 activity at

Polycomb target genes is easily visualized by the formation of well-defined domains of tens of kilobases in which the level of H3K27me3 is strikingly elevated relative to the genomic average or to surrounding chromatin. These domains can range from a few to hundreds of Kb and include sites where PRC2 can be detected by chromatin immunoprecipitation techniques. In *Drosophila*, these binding sites are localized, usually in an interval of 1 kb, and correspond to Polycomb response elements or PREs, where PRC1 components are also found. Both PRC1 and PRC2 are thought to be recruited by the cooperative action of specific DNA-binding proteins (see Chapter 5 in this volume). In mammalian genomes, H3K27me3 domains may be focused on CpG islands or may spread over larger distances, particularly in differentiated tissues.

In the *Drosophila* genome, only a handful of sites bind PRC1 but not PRC2 and contain no H3K27me3 [2]. One of these sites, the locus of the two *ph* genes that encode the Ph components of PRC1, has been studied in greater detail [3–5]. This locus is particularly interesting because it cannot be said to be silenced since the Ph proteins are needed for PcG silencing itself but it is nevertheless auto- and downregulated, presumably to maintain a steady but not excessive level of the Ph proteins. A similar situation obtains at the *Psc-Su(z)2* locus encoding two other paralogous PRC1 components, although in this case both PRC1 and PRC2 are present and the entire locus is marked by H3K27me3 [5,6]. It is clear therefore that the two loci remain at least partly active, though PcG-regulated, but the lack of outright silencing cannot be attributed to the presence or absence of PRC2 or H3K27me3. In mammalian cells also not all loci containing PRC2 or H3K27me3 also bind PRC1 [7], but the details have not been analyzed. Using knockout mutations of RING1B (RING2) or EED, it has been shown, however, that while a large set of genes is regulated by both PRC1 and PRC2, another set requires only PRC1 and a third set only PRC2 [8].

GLOBAL FUNCTIONS OF PRC2

Over the past 15 years, studies of PRC2 function have been focused on these Polycomb target sites and on the H3K27me3 histone modification associated with them. In this chapter, we are concerned with a different aspect of PcG complexes: their untargeted activities on chromatin. These have been largely neglected partly for historical reasons and partly for their lack of connection with specific gene functions. Quantitatively, however, they represent the major activities of both PRC1-type and PRC2 complexes. The distribution of the three degrees of H3K27 methylation reveals in fact a vastly wider role of the PRC2 complex in the genome. In particular, H3K27me2 is both virtually ubiquitous and extremely abundant.

Knockout experiments have shown that the PRC2 complex is responsible for all the di- and trimethylation, as well as most of the monomethylation of

H3K27 in the genome [9,10]. The residual PRC2-independent monomethylation of H3K27 is likely due to some unknown histone methyltransferase, though not to the H3K9 methylases G9A or GLP, which, in vitro, have some methylation activity at this position [10]. *In vitro* studies have shown that the catalytic efficiency of PRC2 decreases with the addition of methyl groups to the target so that H3K27me0 > H3K27me1 > H3K27me2 in the ratio 9:6:1 [11]. These values were obtained using a recombinant PRC2 complex including AEBP2 and an H3K27 peptide containing residues 21−44. The exact values may therefore differ depending on the presence of additional cofactors and on a nucleosomal rather than peptide substrate. However, the data are consistent with the idea that trimethylation is a significantly more difficult step than mono- or dimethylation.

Experiments with SILAC (stable isotope labeling with amino acids in cell culture) and mass spectrometry show that, upon DNA replication, the old histones retain their posttranscriptional modifications while newly deposited histones acquire their modifications gradually during the cell cycle [12]. The abundance of H3K27me1 on new histones shortly after deposition suggests that the methylation is not processive but stepwise and argues against the idea that the restoration of H3K27me3 is due to PRC2 associated with the replication fork. Nevertheless, H3K27me1 and H3K27me2 on new histones reach their steady-state values within one cell cycle. In contrast, H3K27 trimethylation is a very slow process, requiring up to three cell cycles to bring newly deposited H3 to the genomic average level. Overall, therefore, while monomethylation may be restored by PRC2 associated with the replication fork, as has been proposed [13], H3K27me2 constitutes the substrate from which most H3K27me3 is produced by addition of one methyl group at a slower rate [13,14]. These rates are, however, genomic averages. They do not tell us if particular regions are more efficiently targeted than others. PRC2 has been reported to associate with the replication fork [13] and, in *Drosophila*, it was found to remain associated with PREs during replication [15,16]. It is likely, therefore, that sites of stable PRC2 binding recover full H3K27 methylation faster than the genomic average. In addition, PRC2 exhibits a product feedback mechanism: the Esc/EED subunit recognizes previously di- or trimethylated regions, and this interaction stimulates the catalytic activity of the PRC2 complex [17]. Another mechanism that promotes catalytic activity is the recognition of densely packed nucleosomes through the Su(var)12 subunit [18]. Denser nucleosome packing tends to be found in transcriptionally silent regions. As a result of these feedback effects, Polycomb target genes that were previously di- or trimethylated, and therefore likely to contain a higher nucleosome density, are better targets for trimethylation and may in fact be remethylated more rapidly than would otherwise be expected.

Globally, however, the slow rate of trimethylation suggests that H3K27me3 domains may not be very densely trimethylated at any one time. This is reflected by the relative steady-state abundance of the three levels of modification, evaluated by mass spectrometry, which shows that 50–70% of all histone H3 is H3K27me2, 5% is H3K27me1, and 5–10% is H3K27me3 [10,19,20] (Table 13.1). Thus, seen quantitatively, the most important activity of PRC2 is the dimethylation of most nucleosomal H3K27. If we add to these percentages ~2% for H3K27 acetylation, it is evident that a relatively small fraction of histone H3 remains unmodified at lysine 27 at any one time, estimated as 16% [10]. This implies that H3K27 is a very tightly regulated position of histone H3 and that it is likely to play a particularly important role in chromatin transactions.

GENOMIC DISTRIBUTION OF H3K27 METHYLATION

The surprisingly high level of H3K27 dimethylation raises the urgent question of its distribution and relationship to that of H3K27me1 and H3K27me3. The latter has been abundantly reported in many different kinds of cells and organisms because of its connection to Polycomb silencing. More recently, however, attention has turned to the distribution of the other two degrees of H3K27 methylation. The three methylation states were compared in mouse embryonic stem cells [10] and in *Drosophila* cultured cells [21]. The distributions of these three methylation states follow a very similar pattern in the two organisms (Fig. 13.1). In contrast to the specific enrichment of H3K27me3 at silenced genes, H3K27me1 is found only at transcriptionally active genes, while H3K27me2 is enriched at all intergenic and

TABLE 13.1 Abundances of H3K27 Modifications

Modification	% of H3
H3K27me0	16
H3K27me1	5
H3K27me2	60–70
H3K27me3	5–10
H3K27ac	2

Abundances are expressed as % of total histone H3.
Data are taken from Ferrari KJ, Scelfo A, Jammula S, Cuomo A, Barozzi I, Stützer A, et al. Polycomb-dependent H3K27me1 and H3K27me2 regulate active transcription and enhancer fidelity. Mol Cell 2014;53:49–62; Jung HR, Pasini D, Helin K, Jensen ON. Quantitative mass spectrometry of histones H3.2 and H3.3 in Suz12-deficient mouse embryonic stem cells reveals distinct, dynamic post-translational modifications at Lys-27 and Lys-36. Mol Cell Proteomics 2010;9:838–50; Voigt P, LeRoy G, Drury WJI, Zee BM, Son J, Beck DB, et al. Asymmetrically modified nucleosomes. Cell 2012;151:181–93.

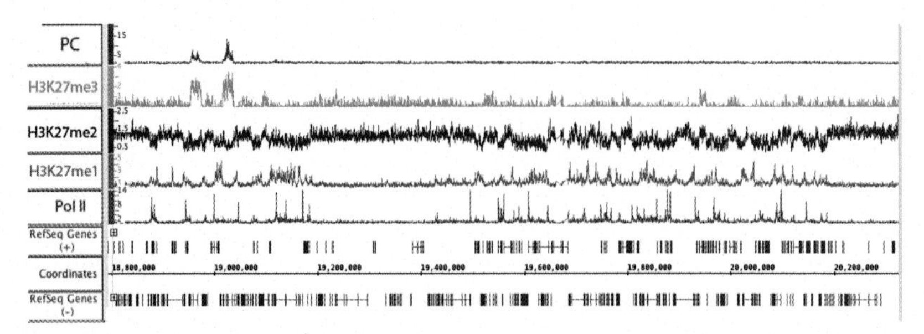

FIGURE 13.1 H3K27 methylation profiles. Chromatin immunoprecipitation sequencing profiles of a representative region of the *Drosophila* genome, showing the distribution of H3K27me1, H3K27me2, and H3K27me3. Also shown is the distribution of Polycomb (Pc) and RNA Pol II. While the major peaks of H3K27me3 are associated with Polycomb-repressed genes, significant H3K27me3 is present also in other regions. H3K27me2 is found abundantly in all regions not actively transcribed. H3K27me1 is found at transcriptionally active genes. *Data from Lee H-G, Kahn TG, Simcox A, Schwartz YB, Pirrotta V. Genome-wide activities of Polycomb complexes control pervasive transcription. Genome Res 2015;25:1170–81.*

nontranscribed regions. Both studies confirmed that H3K27me3 enrichment is found only in regions of stable binding of PRC2 and that H3K27me1- and H3K27me2-containing regions have no detectable bound PRC2. Relatively efficient H3K27 trimethylation occurs in regions that already have preexistent methylation and densely packed nucleosomes, as discussed above. It may also require the involvement of additional PRC2 cofactors: PCL, JARID2, AEBP2, possibly lncRNAs, necessary either to facilitate stable binding or stimulate catalytic activity, or both.

Since the SILAC experiments show that H3K27 methylation is acquired slowly and sequentially by newly deposited histones [12], we must conclude that at least dimethylation does not take place in association with the replication fork but builds up gradually by a hit-and-run type of mechanism that likely targets the entire genome. In this view, the hit-and-run mechanism dimethylates efficiently but rarely reaches the trimethylated state. Nevertheless, the genomic ChIP sequencing (ChIP-seq) experiments revealed the presence of a low but significant level of H3K27me3 that accompanies H3K27me2 in most transcriptionally inactive regions [21]. Comparison with a conditional inactivation of E(z) or with a *Su(z)12* knockout mutant cell line shows that this is not simply a background noise level but represents a real trimethylating activity that is apparently achievable by the hit-and-run mechanism. Although very low, this untargeted trimethylation is very widely distributed and probably constitutes a substantial part of the total genomic H3K27me3. It is very likely, however, that this fraction of the total H3K27me3 is reached with much slower kinetics than that associated with PRC2 target genes. As a result, data for the overall H3K27me3 kinetics must be interpreted with caution since they may reflect the significant contribution of the untargeted H3K27me3 with different kinetics than the trimethylation of PRC2 target genes.

The relationship between mono- and dimethylation is essentially complementary. H3K27me1 is found primarily in active transcription units, where H3K27me2 is low. Mapping of the H3K27me2 and H3K27me1 distribution in a metagene representation shows that the level of the former decreases in proportion to the level of expression while the level of the latter increases with transcriptional activity [21]. In agreement with this, H3K27me1 was strongly correlated with the level of H3K36me3 [10]. In *Drosophila*, at least, these effects are also closely correlated with the distribution of the H3K27 demethylase Utx, whose binding in transcription units increases with the level of transcriptional activity [21].

What might be the role of H3K27me1 in transcribed regions? Since the Utx demethylase is recruited to these regions and is responsible for the demethylation of H3K27me2, the simplest interpretation is that H3K27me1 is an intermediate in the demethylation. In vitro, in fact, H3K27me1 is a poorer substrate for Utx than H3K27me2 [22,23]. However, the observations are also compatible with the idea that H3K27me1 results from remethylation activity of PRC2 [10] that is partially inhibited by the H3K36me3 produced in transcribed regions by the Set2 methyltransferase. In vitro experiments have shown that nucleosomal substrates containing H3K36me3 inhibit PRC2-dependent dimethylation but not monomethylation of H3K27 [24,25]. Overall, it seems likely that, although associated with transcriptional activity, H3K27me1 does not have a positive effect on transcription, consistently with its effect in blocking H3K27 acetylation (see later discussion).

ROLE OF GLOBAL H3K27 METHYLATION

What might be the function of global H3K27 dimethylation? Is it just an intermediate step to ensure that the trimethylated state can be more easily reached at the sites to be repressed by stably bound PRC2? Using a temperature-sensitive E(z) allele, Lee et al. [21] found that loss of E(z) function caused a global increase in transcriptional activity. Not surprisingly, this increase was most pronounced in genes that had previously less or no transcriptional activity, and therefore higher levels of H3K27me2. Genes with low steady-state levels of transcripts increased those levels 20- to 50-fold but even very highly transcribed genes increased 2- to 4-fold. Normally silent intergenic regions acquired significant levels of transcripts. This suggests that the loss of H3K27me2 in transcribed regions is not due to lower PRC2 activity caused by transcription-associated chromatin changes but rather the reverse: H3K27me2 must be removed to allow transcriptional activity. The effect on highly transcribed genes indicates that, even at these sites where little H3K27me2 is detectable in the steady state, H3K27 methylation is nevertheless important in governing the efficiency of transcription.

The effect of H3K27 methylation is not limited to transcriptionally active regions. The fact that loss of PRC2 function causes transcriptional activity to appear at inactive genes or intergenic regions shows that loss of H3K27 methylation renders chromatin accessible to RNA polymerase and the transcriptional apparatus. That this is attributable to the ubiquitous H3K27me2 is supported by the fact that RNAi knockdown of Utx, the only known H3K27 demethylase in *Drosophila*, strongly suppresses the appearance of transcriptional activity in intergenic regions [21]. This result is particularly instructive: inactivation of PRC2 increases transcription locally, in the case of Utx about fivefold; RNAi knockdown of Utx reduces the fivefold higher level of Utx mRNA to the same level as before PRC2 inactivation. In other words, the decrease in PRC2 activity reduces H3K27me2, which in turn increases transcription of Utx and results in greater demethylation of H3K27me2.

THE ROLE OF UTX: H3K27 DEMETHYLATION OR NOT?

Two main factors control the level of H3K27 methylation in active genes: inhibition of PRC2 by H3K36me3, which is deposited by Setd2, and demethylation by H3K27 demethylases. To this may be added the direct competition by H3K27 acetylation and, in very active genes, nucleosome eviction and turnover. All these factors are associated with transcriptional activity. Mammalian genomes contain two distinct H3K27 demethylase genes: JMJD3 and Utx, both containing JumonjiC domains. Both enzymes can act on all three grades of methylation when assayed on histone peptides but are less efficient and demethylate H3K27me1 very poorly on nucleosomes, suggesting that additional cofactors or histone modifications may be required for efficient demethylation in vivo [22,26–29]. Both JMJD3 and Utx are needed in mammals for proper differentiation, but different cell lineages apparently specifically require either one or the other [22,30]. Knockout experiments in mouse indicate that Utx is more important in early development while JMJD3 comes into play at later stages. An important complication is due to the fact that the Utx gene resides in a region of the X chromosome that escapes X inactivation. A homologous gene, *Uty*, is present on the Y chromosome. *Uty* is functional and produces an abundant protein that is, however, devoid of demethylase activity in vitro ([22], 4029). UTY is probably not without other functions, as shown by the fact that female mouse embryos lacking UTX die in early development, while male embryos lacking UTX but retaining UTY survive to term [31–34]. Importantly, even when both UTX and JMJD3 demethylase activities are lacking, male embryos can still remove H3K27me3 from Polycomb-repressed genes during tissue differentiation. It appears then that UTY can supply a critical function for the removal of H3K27me3 although it lacks demethylase activity in vitro.

In *Drosophila*, no JMJD3 homolog is encoded in the genome and Utx is the only known H3K27 demethylase, thus facilitating the genetic analysis of its function. In flies, genetic analysis of a *Utx* deletion mutant, lacking the entire catalytic domain shows that even when both maternal and zygotic demethylase activities are absent, embryonic development can be completed and death ensues only in the larval stages, associated with partial loss of *Hox* gene expression [35]. Like mammalian UTX, *Drosophila* Utx is found at actively transcribed genes [21], but, while in mammals UTX is specifically enriched in the promoter region, in flies it colocalizes with elongating RNA polymerase and coimmunoprecipitates with it [23]. The association of Utx with transcriptional activity is not limited to Polycomb target genes but extends to all active genes, confirming that its role is not just to control H3K27me3 but is needed to remove the ubiquitous H3K27me2. The Utx H3K27 demethylase is in fact present in transcription units in direct proportion to their transcriptional activity, and this presence is reflected by a similar proportional loss of H3K27me2. This leaves us with a conundrum: how could Utx contribute to the removal of H3K27 methylation even in the absence of H3K27 demethylase activity and what might be the nonenzymatic role of Utx?

A clue to the solution came from the discovery that both JDMD3 and Utx are integrally implicated in a number of chromatin-opening activities. They are involved with chromatin-remodeling complexes recruited by transcription factors, an activity that does not require their demethylase function [36]. Both JDMD3 and Utx interact with a BRG1-containing remodeling complex and are recruited by transcription factors to promoter and enhancer target regions, rendering them more accessible to restriction enzymes. Activation of some target genes is independent of the H3K27 demethylase activity, but interaction with the remodeling complex requires a protein domain close to or overlapping with the jumonji catalytic domain.

Utx is also an integral component of the Trr/MLL3/4 complex, responsible for H3K4 monomethylation in mammals and in flies [37]. The phenotype of *Drosophila trr* mutants resembles that of *Utx* mutants. Both are associated with reduced levels of H3K4me1 but normal levels of H3K4me2,3. *trr* mutant tissues were found to display an overgrowth phenotype [38] similar to that reported for *Utx* by Herz et al. [39], but this is disputed by Copur and Müller [35], who observed no overgrowth in parallel experiments. Additional links between Utx and TRR/MLL3/4 function are provided by H3K27 acetylation. Loss of *trr* causes not only a global decrease of H3K4me1, its enzymatic product, but also a global decrease of H3K27ac [37]. This suggests that the Tr/MLL3,4 complex has additional, demethylase-independent ways to remove H3K27 methylation and promote H3K27 acetylation. As discussed below, Utx interacts directly with the CREB-binding protein (CBP) histone acetyltransferase that targets H3K27 (Fig. 13.2). The importance of the activities of Utx and Trr/MLL3,4 in controlling access to regulatory regions is underlined by the fact that they are the targets of mutations identified in a variety of cancers [40–42]. Interestingly, recent work also shows that overexpression of H3K27 demethylases in *Caenorhabditis elegans* induces the

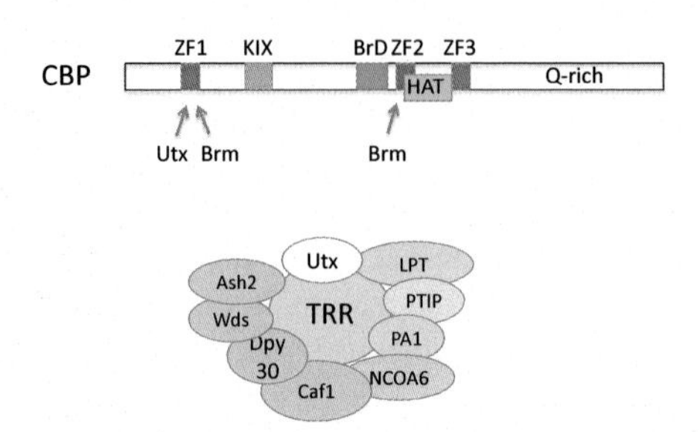

FIGURE 13.2 The Utx demethylase is a component of the CREB-binding protein acetyltransferase (CBP) and the Trr/MLL3,4 complexes. The diagram above shows the domains of the CBP protein. The three zinc fingers (ZF) and the KIX domain are binding sites for many transcription factors. BrD indicates the bromodomain and HAT the acetyltransferase catalytic domain. The binding sites of Utx and Brm are taken from reference ([56], 6806). Below is shown a schematic composition of the Trr complex, homologous to the mammalian MLL3,4 complexes. The Ash2, Wds, Dpy30, and Caf1 proteins are also components of the related Set1 and Trx complexes. *The diagram is drawn after Mohan M, Herz H-M, Smith ER, Zhang Y, Jackson J, Washburn MP, et al. The COMPASS family of H3K4 methylases in* Drosophila. *Mol Cell Biol 2011;31:4310–18.*

unfolded protein response and a constellation of activities that results in life span extension [43,44]. In particular, overexpression of Jmjd-1.2 alone, which demethylates H3K27me2 but not H3K27me3, is sufficient for this effect. Consistently, PRC2 mutations in *Drosophila* enhance longevity and resistance to stress [45].

H3K27 ACETYLATION

Why the H3K27 methylation mark, whether dimethyl or trimethyl, is repressive is a question to which we will have to return. It is clear, however, that it would act as a direct antagonist of activating functions if these require H3K27 acetylation, a mark specifically associated with active chromatin. H3K27ac is specifically enriched at the 5′ ends of transcriptionally active genes [46]. Although it is known to turn over rapidly, it normally is thought to accelerate transcription by enhancing the binding of transcriptional activators and by promoting transition of RNA polymerase from initiation to elongation [47]. Significantly, H3K27ac, together with H3 K4me1, forms a chromatin signature associated with activated enhancer regions [48–50].

Tie et al. [51] showed that, in *Drosophila*, H3K27 acetylation is specifically produced by CBP, is associated with transcriptionally active genes, and antagonizes Polycomb repression. These authors were the first to point out that, at many sites, H3K27ac and H3K27me3 alternated, depending on whether the associated gene was active or Polycomb-repressed. They proposed that the two histone marks are mutually antagonistic. A similar antagonistic relationship between H3K27 methylation and acetylation catalyzed by CBP or its close homolog p300 was shown in mammalian cells [52]. CBP/p300-dependent acetylation is a hallmark of active enhancers often in association with H3K4me1 [48]. Most transcription factors are unable to bind to their cognate sites on a nucleosomal template. However, when a binding site is located in a more unprotected region at the edge of a nucleosome, it is more likely to become transiently accessible. In addition some transcription factors, called pioneer factors, are able to bind to nucleosomal DNA [53]. Activation of an enhancer is thought to follow a series of steps in which initial binding of factors recruits chromatin modifying and remodeling activities resulting in displacement of nucleosomes and access to additional DNA-binding sites (reviewed in Ref. [54]).

In mammals, CBP and p300 are powerful acetyltransferases that target in particular histone H3K27, H3K18, and H4K8 [52,55] but also acetylate many transcription factors. Just as important, CBP/p300 also contains interactions sites for over 400 other proteins, many of which are DNA-binding transcription factors. These activators can therefore recruit CBP/p300 to their binding sites in enhancer or promoter regions. Enhancer opening and activation are likely to ensue from pioneer factor binding or even from the transient binding of a transcription factor to its target site followed by the recruitment of CBP/p300 to help open the region and make it accessible to other enhancer factors or transcriptional activators needed for enhancer function. The association between CBP and chromatin-remodeling activity necessary to open chromatin is supported by the fact that it binds the BRM component of the SWI/SNF nucleosome remodeling complex, as well as the Utx H3K27 demethylase [56]. The enhancer opening process, therefore, both remodels nucleosomes and makes them available for H3K27 acetylation. That the acetylation itself has an instructive role in inducing transcriptional activity is shown by the fact that the catalytic domain of p300 can activate transcription when artificially targeted to promoters or enhancers [57].

Less well understood is the mechanistic connection between the CBP/p300-mediated chromatin acetylation and opening and the H3K4 methylation that is associated with it at enhancers as well as at promoter regions. H3K4me1 at enhancers generally precedes detectable H3K27ac and enhancer activation, although it is not necessarily clear that transient H3K27 acetylation is not also involved. Nevertheless, H3K4me1 is often found at poised enhancers that await additional signals to become fully activated [50,58].

As we have seen, transcriptionally inactive regions are heavily dimethylated at H3K27. To acetylate, and therefore to open a previously inactive region, the preexisting H3K27me2 must first be removed by the Utx H3K27 demethylase or by nucleosome turnover. A connection between the H3K27 acetylation and the H3K4 methylation activities is revealed by the fact that both CBP and Trr/MLL3,4 bind Utx and therefore bring along with them the prerequisite demethylation activity. However, CBP has been reported to bind not only Utx but also the BRM remodeling ATPase protein [56], suggesting that both demethylation and nucleosome turnover can contribute to the removal of H3K27 methylation.

Pengelly et al. [59] showed that the histone mutation H3K27R causes derepression of homeotic genes in *Drosophila* imaginal discs. Since this mutation prevents both methylation and acetylation, they argued that H3K27 acetylation is not necessary for activation but only to antagonize methylation. This may be a premature conclusion. Many PcG-repressed genes in flies as in mammals contain loaded but stalled RNA polymerase. In their experiments, Pengelly et al. induced their mutant clones in imaginal discs, at a time when they would already have loaded the RNA Polymerase on the promoters of the homeotic genes [60] and would no longer require H3K27 acetylation. Nevertheless, the questions of how PcG mechanisms repress and whether H3K27 methylation has an intrinsic repressive role remain unsettled.

ROAMING ACTIVITIES

Although chromatin modifying activities such as H3K27 methylation, H3K27 acetylation, or H3K4 monomethylation are often viewed as targeted functions involving enhancers or promoters of specific target genes, several lines of evidence suggest that they begin as untargeted roaming or hit-and-run activities. The first involves a general argument: to find specific sites where they can bind for longer times, all such activities must be able to sample all genomic chromatin. The question is then whether enzymatically productive encounters with nucleosomes can occur frequently in the absence of stable binding. As discussed earlier, for H3K27 methylation this frequency is high and results in high levels of H3K27me2 at all sites where methylation is not locally inhibited, blocked, or removed. The presence of H3K27 methylation naturally blocks acetylation at this position, but evidence for a continuous process of surveillance and opportunistic activity is provided by the fact that when H3K27 methylation is impaired, the level of H3K27 acetylation increases globally [10,21,51,52,61]. This increase is not just localized to target genes or even to enhancers and promoters but can be detected over transcriptionally inactive genes and intergenic regions.

Loss of H3K27 methylation also increases global levels of H3K4me1 [21] both in transcribed regions and in normally nontranscribed genes and intergenic regions. At sites where the enzymatic activities bind more stably, the effects are much more significant. Poised enhancer sites are premarked with H3K4me1 but not activated and therefore lack H3K27ac and possess instead H3K27me2. Loss of PRC2 activity resulted in strong H3K27 acetylation at these poised enhancer sites and consequent transcriptional activation of the associated genes [10]. The conclusion proposed was that global H3K27me2 serves as a barrier to inappropriate activation of tissue-specific enhancers. Conversely, inhibition of CBP/p300 caused loss of H3K27ac at thousands of enhancers and its replacement with H3K27me2 [10].

THE ACCESSIBILITY HYPOTHESIS

Taken together, these results indicate that H3K27 methylation and H3K27 acetylation, as well as, in some way, H3K4 methylation, are in a constant, genome-wide antagonism, and the corresponding complexes engage in a sort of global chromatin surveillance to exploit any transient gap in the modification produced by the antagonist to control access to genomic DNA. Specific control of different parts of the genome requires the ability to read the DNA sequence by specific DNA-binding proteins. For the majority of DNA-binding proteins, access to the nucleotide sequence is prevented by the wrapping of the DNA around the histone core of nucleosomes or, worse, the folding of a string of nucleosomes into higher order chromatin structures. Nucleosome-remodeling machines can open chromatin regions and make DNA available for binding, but the sequence specificity can only be provided by surveying the entire genome. This can be accomplished by a roaming process that transiently opens every sequence, giving it a brief window of accessibility. We know that opening enhancers and promoter regions involve H3K27 acetylation and H3K4 methylation. The results discussed earlier indicate that these are the acute phases of a more global survey process that is antagonized by a roaming PRC2-dependent H3K27 methylation activity (Fig. 13.3). The global distribution and high level of saturation of the H3K27me2 mark are important to limit the accessibility of genomic DNA to a variety of agents: aggressive chemicals, adventurous RNA polymerase, DNA-binding proteins of all sorts and, not least, it has been argued, to invasive foreign DNA [62].

To maintain a dynamic equilibrium, activities that remove both the activating and the repressive histone marks must also be involved. We have already seen that both the CBP/p300 acetylase and the Trr/MLL3,4 H3K4 methylase are associated with the Utx H3K27 demethylase. The Rpd3 histone deacetylase has been reported to be associated with PRC2 [63,64], but it might also constitute an independent roaming activity as it does in budding yeast [65,66]. Several enzymes are known to remove H3K4me1,2, but it is not known whether they have a roaming, untargeted component.

The result of these roaming activities is to provide a transient opportunity for DNA-binding factors to find and bind to their recognition sequences. Pioneer factors [67] able to bind to nucleosomal DNA also contribute to this process, whose ultimate objective is to secure local chromatin remodeling and the longer-term binding of CBP/p300 and Trr/MLL, thus guaranteeing a more stable chromatin opening and binding of the sequence-specific factors (see Ref. [1] for an earlier version of this argument). The chromatin opening process is ensured by the recruitment of nucleosome remodelers through direct association with CBP/p300, binding to acetylated histones mediated by the bromodomains contained in the remodeling complexes, and not least through the lower stability of modified nucleosomes (reviewed in Ref. [54]). In fact, an important role of the H3K3me1 mark is to recruit nucleosome remodelers such

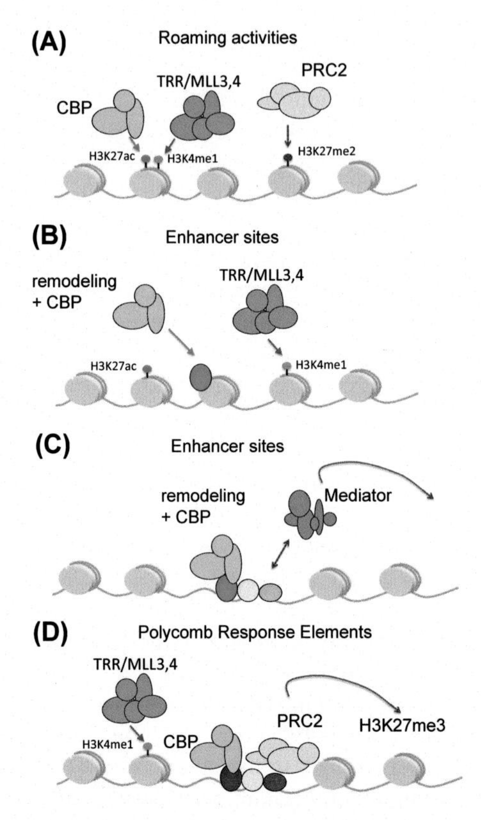

FIGURE 13.3 **Roaming, hit-and-run activities.** (A) Roaming activities. Genomic chromatin is constantly but randomly visited by several contrasting activities, some promoting and some inhibiting opening. A key chromatin-opening activity is H3K27 acetylation by CREB-binding protein acetyltransferase (CBP). This is reversed by a roaming deacetylase activity that normally limits the lifetime of the acetylated state. A second important chromatin-opening activity is the H3K4 monomethylation carried out by the TRR/MLL3,4 complex, which includes the H3K27 demethylase. These activities are antagonized by a roaming PRC2 that normally generates a high level of H3K27me2. The H3K27 methylation clearly blocks the acetylation, but it is not clear how it also antagonizes the H3K4 methylation, unless at least transient acetylation is required for H3K4 methylation. These contrasting activities produce an equilibrium dominated by H3K27me2 and hence predominantly repressive. (B) At enhancer sites, DNA-binding factors such as pioneer factors can bind to nucleosomal DNA, recruit more stable association with CBP and probably TRR/MLL3,4 and associated chromatin-remodeling activities. These remodel or displace nucleosomes and allow binding of additional enhancer factors. (C) These factors can now promote transcription, usually through a Mediator complex. (D) A similar process occurs at Polycomb response elements, where the DNA-binding factors recruit instead PRC1 and PRC2 to generate domains of H3K27me3 and H2AK118ub to generate stable repression of associated genes.

as TIP60 that catalyze the exchange of histone H2A by H2A.Z (or H2Av in *Drosophila*) [68]. The nucleosome remodeling that displaces nucleosomes is counteracted by de novo, replication-independent deposition of nucleosomes, a process that utilizes histone H3.3 [69,70]. Nucleosomes with the combination H2A.Z and H3.3 have been shown to be unusually labile [71,72], accounting for the rapid turnover of the nucleosomes bordering the binding sites of regulatory factors [69,70]. Evidence from some specific enhancers that have been analyzed in detail suggests that H3K4me1 occurs before H3K27 acetylation and can be found at "poised" enhancers that are not yet activated and still occupied by H3K27me3 [49,50]. We cannot exclude, however, that H3K27 acetylation at these sites occurs transiently, allowing the H3K4 monomethylation before the availability of factors that induce stable binding of the CBP/p300 acetylase.

This series of events would be involved at all sites that require binding to DNA of sequence-specific factors, including enhancer and promoter factors but also, in principle, site-specific recombination factors, insulator proteins or similar architectural proteins, and repressive factors or complexes. In fact, *Drosophila* PREs, the sites that recruit stable binding of both PRC1 and PRC2 are also DNase hypersensitive sites [73,74] associated with high turnover of histone H3.3 [70], flanked by nucleosomes marked with H3K4me1 [75]. In addition, PREs of Polycomb-repressed genes are the strongest CBP-binding sites in the *Drosophila* genome [76]. While this might seem paradoxical, given the commonly held view that Polycomb repression involves chromatin compaction, it is consistent with the idea that all sites that involve DNA-binding factors require access to the DNA and therefore continuous chromatin-opening activities. PRE regions are not associated with H3K27ac or other histone acetylation mark, and it is not known whether CBP catalytic activity is necessary for PRE function but if any acetylation takes place, it would very likely turn over very rapidly. Schematically, this view of genomic accessibility could be summarized as shown in Fig. 13.3.

The evidence presented in the preceding sections indicates that both opening and closing activities are global, pervasive, and in constant antagonism such that a decrease on one side rapidly results to an increase on the other. It is upon this background that we must envisage the occurrence of enhancer activation, promoter activation, Polycomb repression, and all the variety of phenomena that specifically control individual genes or chromatin regions. Opportunistic binding of DNA-binding factors or the colonization of specific sites by pioneer factors creates the opportunity for specific recruitment of the H3K27 acetylase function and the H3K4 monomethylase function. The more stable binding of these functions permits the action of the H3K27 demethylase, the nucleosome remodeling, the histone turnover and replacement by the H2A.Z and H3.3 variants, and the stable opening of a region corresponding to one or several nucleosomes. What is done with these open regions depends on the nature of the DNA-binding proteins. In the case of enhancer factors or promoter factors, the result might be the association with transcriptional apparatus. The opened region would in any case be sensitive to binding of RNA polymerase and result in the production of transcripts. These might be unscheduled and short and constitute part of the pervasive transcriptional noise in the nucleus. Or they may be functional RNA molecules such as enhancer-associated eRNAs or transcripts associated with promoter regions. Some, corresponding to a constellation of appropriate factors, sequence features and coding regions, would be gene transcripts and eventually encode proteins. On the other hand, the DNA-binding factors might be of the sort associated with *Drosophila* PREs, which cooperatively recruit Polycomb complexes PRC1 and PRC2. The result would then be enhanced H3K27 methylation highly enriched in H3K27me3, as well as inhibition of transcriptional activity. The repressive state does not preclude the binding of the CBP or the H3K4 monomethylation, both of which are found at repressed PREs. In mammalian genomes, the place of the PRE is taken by CpG islands which tend to be nucleosome free and can act either as Polycomb recruiters or as promoter regions.

RECRUITMENT OF PRC2 BY A PRC1 TYPE OF COMPLEX

In *Drosophila* most Polycomb target genes recruit the PRC1 and PRC2 complexes through specific sequence elements that assemble cooperatively recruiting platforms, the PREs (see Chapter 5). In mammalian genomes the major mechanism that targets stable binding of the repressive complexes appears much more general. It relies on and potentially targets all CpG islands and, since some 70% of protein-coding genes have CpG island promoters, a large fraction of the genes are possible targets for Polycomb repression. The Polycomb recruiting mechanism takes advantage of features that are also important for promoter activity. CpG islands tend to be protected against DNA methylation and relatively depleted of nucleosomes. Proteins containing CXXC domains can bind to CpG-rich regions [77]. One such protein is the KDM2B histone H3K36 demethylase, a component of a variant PRC1 complex called PRC1.1 [78–80]. DNA methylation antagonizes this process, probably by interfering with the binding of the KDM2B to CpG-rich regions. Transcriptional activity also inhibits PRC1.1 recruitment, but the reasons for this are not clear.

In a recently proposed scenario, the recruitment of PRC1.1 is followed by that of PRC2 [81,82]. Evidence for this comes from reconstruction experiments in which PRC1.1 or components thereof were tethered to a target sequence that normally binds neither PRC1 nor PRC2. The remarkable results were that PRC2 recruitment requires only the RING-PCGF heterodimer or, more precisely, a catalytically active fragment of this complex able to produce histone H2AK119 ubiquitylation. Mutations of the catalytic domain prevent PRC2 recruitment, suggesting that PRC2 recognizes and binds to

H2AK119ub. To complete the argument, Kalb et al. [83] showed that in flies and mammals H2AK119ub binds AEBP2 and, to a lesser extent, JARID2, two substoichiometric components of PRC2. This supports the conclusion that H2AK119ub can recruit PRC2 complexes containing these components. Therefore, in this view, variant PRC1 complexes recruit PRC2, which in turn trimethylate H3K27 and allow the recruitment of the canonical PRC1 complex containing the chromodomain component (PC or CBX). It should be noted, however, that although Wu et al. [80] confirmed that a tethered KDM2B recruits the PRC1.1 complex, they did not detect recruitment of EZH2, as predicted by this model. This significant difference may be due to the type of cell used: mouse embryonic stem cells in one study ([81], 8249), but the human embryonic kidney 293FT transformed cell line in the other study ([80], 7308). Alternatively, other features may limit the applicability of this model such that we cannot determine how widely used it might be to recruit PRC2 complexes.

In fact, KDM2B is enriched at unmethylated CpG islands throughout the genome whether or not they bind PRC1 [78], implying that not only DNA methylation but also other chromatin features modulate the ability to carry out the recruitment of Polycomb repression. This argument was extended by Klose et al. [84] to argue that recruitment of Polycomb complexes is a constantly ongoing process that is antagonized and modulated by other chromatin features, among which are those associated with transcriptional activity. A consequence of this model is that stable binding of PRC2 and consequent H3K27 trimethylation require prior transcriptional downregulation. Since, as we have seen, low transcriptional activity results in H3K27 dimethylation by untargeted, roaming PRC2, the target of Polycomb repression is already primed for H3K27 trimethylation. This is consistent with SILAC and mass spectrometry experiments showing that H3K27me3 derives primarily from H3K27me2. Remarkably, the arrest of transcriptional elongation by inhibitors of RNA polymerase II results in the recruitment of PRC1.1 and binding of PRC2 [85]. According to this model, then, loss of transcription in principle would allow recruitment of PRC2 at any CpG-rich promoter region.

DOES H2A UBIQUITYLATION PLAY A ROLE IN GLOBAL PRC2 ACTIVITY?

Given the link proposed between H2A ubiquitylation and PRC2, is there a connection between the widely spread H3K27 dimethylation activity of PRC2 and the genomic distribution of H2AK119ub? Similarly, might the repressive activity of the global H3K27me2 be mediated by a PRC1 complex? Although the antagonism between H3K27 methylation and H3K27 acetylation could be sufficient to account for the interference between H3K27me2 and transcriptional activity, we might also wonder if PRC1 complexes might be at least transiently involved in the repression. There are then two questions concerning the possible involvement of a PRC1 complex in the global action of PRC2. One involves the possible role of H2A ubiquitylation in facilitating the hit-and-run action of PRC2. The other concerns the transient involvement of a canonical PRC1 complex recruited by the global H3K27 methylation.

To take the second question first, it is often assumed that the PRC1 complex provides the repressive activity while the PRC2-dependent H3K27 trimethylation confers the epigenetic memory. The two are linked by the ability of the canonical PRC1 complex to recognize the H3K27me3 mark through the chromodomain of PC or the CBX homologs. In mammals this model poses some problems because most of the CBX chromodomains bind to H3K9me3 as well or better than to H3K27me3 [86]. In *Drosophila*, however, the specificity of the PC chromodomain is well able to distinguish the two. The chromodomain of PC has also a significant affinity for H3K27me2, though five times weaker than for H3K27me3 [87,88]. In addition, a low but measurable level of H3K27me3 accompanies H3K27me2 [21]. Together, these methylations raise the possibility that a canonical PRC1 complex might be transiently attracted to visit dimethylated genomic regions and contribute to repression. PRC1 repressive action has been broadly assumed, with little direct evidence, to result from some chromatin compaction that prevents access to transcription factors. However, regardless of its validity at Polycomb target genes, this model or any model involving direct repressive action by PRC1 cannot be applied to regions of H3K27me2, where the association would only be highly transient. A transient presence could achieve significant repressive effects only if it left behind a stable chromatin modification. A PRC1 complex can, in principle, ubiquitylate histone H2AK119, a modification that has been reported to interfere with transcriptional activity [89]. Is there a broad genomic distribution of H2AK119ub to account for a global repressive activity? In mammalian cells this distribution has not been well characterized but, in *Drosophila*, H2AK118A ubiquitylation, corresponding to mammalian H2AK119ub, appears to be in fact very broadly distributed (Fig. 13.4), although at steady-state levels much lower than those of H3K27me2 [21].

It has been reported that, in vitro, canonical PRC1 complexes in mammals and in flies have but little ubiquityl transferase activity and that most of the H2AK119ub is produced by variant PRC1 complexes lacking the chromodomain component [79,90]. In *Drosophila*, H2AK118ub (corresponding to mammalian H2AK119ub) was attributed to a variant PRC1 complex called dRAF [90], which was reported to include the KDM2 homolog, RING/Sce and PSC but not PC. More recent results indicate that the canonical *Drosophila* PRC1 does contribute a substantial amount of H2AK118ub, prominent peaks of which are found at most but not all Polycomb target genes [21]. Interestingly, the entire bithorax complex and the *Antennapedia* gene, but not the remaining homeotic genes or other PcG targets, are entirely lacking H2AK118ub, though fully repressed in these cells (Fig. 13.4). The absence of H2AK118ub at these genes appears to be due to the PR-DUB complex, whose

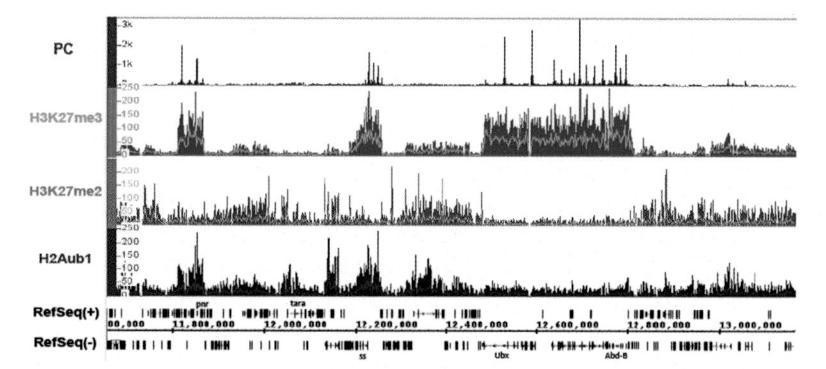

FIGURE 13.4 Distribution of H2A ubiquitylation. The figure compares the distribution of H2AK118ub with those of H3K27me2 and H3K27me3. The distribution of Polycomb (Pc), representing Polycomb-repressed genes, is also shown. A large block of H3K27me3 corresponds to the repressed bithorax complex. Note that little H2A ubiquitylation is present in this domain, while two other Polycomb-repressed sites have prominent peaks of H2AK118ub. *Data from Lee H-G, Kahn TG, Simcox A, Schwartz YB, Pirrotta V. Genome-wide activities of Polycomb complexes control pervasive transcription. Genome Res 2015;25:1170–81.*

deubiquitylating activity is in fact required for effective repression of these homeotic genes [91,119]. At least at these genes, therefore, H2AK118ub is not only not required but also actually interferes with Polycomb repression.

Knockdown experiments show that most of the H2AK118ub in the *Drosophila* genome is not attributable to dRAF, or at any rate not to a complex containing KDM2, but to a variant PRC1-like complex lacking PC or PSC but containing L(3)73Ah [21]. This is a protein required for viability but whose loss of function does not cause typical homeotic derepression [92]. It is now identified as a close homolog of mammalian PCGF3 and therefore a likely RING/Sce partner in the PRC1 catalytic core, although this has not yet been directly demonstrated. Although this variant PRC1 complex has not yet been biochemically characterized, its activity is very broadly distributed rather than associated with stable PRC1-binding sites. Knockdown of L(3)73Ah eliminates the broadly distributed H2AK118ub but has only a small effect on the H2AK118ub associated with PRC1-binding sites, indicating a division of labor between activities at targeted PRC1 sites and genome-wide variant PRC1 activities. The broad H2AK118ub distribution is enriched in transcriptionally silent and intergenic regions, generally the same regions enriched in H3K27me2. However, knockdown of PC, or KDM2, or L(3)73Ah had only minor effects on the levels of global H3K27me2 and knockdown of L(3)73Ah had only slight effects on global transcriptional repression [21]. It is unlikely, therefore, that H2A ubiquitylation plays a major role in maintaining the global levels of H3K27me2, either as a recruiter of a repressive PRC2 activity or as the product of PRC1 recruited by H3K27 methylation, at least in *Drosophila* cultured cells.

Further recent results have raised additional questions about a possible role of H2A ubiquitylation as a recruiter of PRC2. In mouse embryonic stem cells, mutations in the catalytic domain of Ring1b (RING2) that eliminate its H2A ubiquitylation activity do not prevent binding of PRC1 complexes to target genes or what is reported as their chromatin compaction [93,94]. Repression of target genes also occurs independently of H2A ubiquitylation but less efficiently, leading to loss of embryonic stem cells [94]. In mouse embryos, these mutations allow early embryonic development and confirm that the repressive role of PRC1 is independent of its catalytic activity [95]. In *Drosophila*, mutations in histones H2A and H2Av that preclude their ubiquitylation do not interfere with full Polycomb repression of the homeotic genes [96]. This in itself is not surprising, given the fact that the major homeotic genes lack H2AK118ub [21], and does not exclude the possibility that other Polycomb targets may in fact require H2AK118ub. In another experiment, the RING/Sce protein was mutated in the catalytic domain and unable to ubiquitylate. In embryos lacking zygotic RING function, most tissue patterning is normal, but small groups of cells do show morphogenetic defects [96], suggesting the possibility that the stability of the Polycomb-repressed state is reduced. Finally, in mouse intestinal stem cells, knockout of Ring1,2 eliminates H2AK119ub but has no effect on H3K27me3 [97].

In conclusion, H2A ubiquitylation is not required for the recruitment of PRC2, for Polycomb repression of at least the major target genes or for transcriptional repression at sites normally enriched for H2A ubiquitylation. Loss of H2AK118/119 ubiquitylation is ultimately lethal in flies and mammals, not because of inability to establish Polycomb repression but perhaps due to less effective maintenance of the repressed state.

POLYCOMB REPRESSIVE ACTIVITIES

The results discussed in the preceding sections leave the classical picture of Polycomb mechanisms somewhat in disarray not only with respect to the recruiting mechanism and the coordination between PRC1 and PRC2 but to the very nature of the repressive activity.

While the evidence points to a connection between PRC2 and H2AK119ub, it is unlikely that this constitutes a general recruiting mode for PRC2. It should rather be viewed as a possible mechanism in certain situations, perhaps in embryonic stem cells or in the early stages of development. However, the conditions that make this mode feasible are not at all clear at this point. Also in question is the role of H2AK119ub in repression. It seems clear that the presence of H2AK119ub is insufficient for transcriptional repression. Also clear is that the catalytic activity of the RING protein is not required for repressive function or for what is described as a chromatin-condensing activity. The evidence for such a condensing activity comes from chromatin imaging experiments in which the distances between different parts of *Hox* gene clusters decrease when the *Hox* genes are repressed compared to when they are active [93,95,98]. These results tell us about the folding or looping of the chromatin fiber but not necessarily about a compaction at the level of the nucleosomes that would prevent access of transcription factors. The compaction model is difficult to apply in *Drosophila*, where the binding sites of PRC1 and PRC2 are often distant from the promoters of the target genes. What is clear is that the repressed state does not prevent access of factors or chromatin complexes to the target genes since the PRC1- and PRC2-binding regions are themselves the sites of brisk turnover of histone H3.3, Polycomb proteins, and other chromatin-remodeling activities [70,74]. An alternative view that seems increasingly plausible is that a large part of the repressive activity of Polycomb complexes is due to the PRC2-dependent H3K27 methylation and its interference with chromatin opening, enhancer accessibility, and promoter function. The difference between chromatin domains containing H3K27me2 and those containing H3K27me3 may lie in the ability of H3K27me3 to enlist the help of chromodomain-containing complexes; however, they may be recruited to target genes, for the surveillance and efficient maintenance of the epigenetic mark and its predominant role over antagonistic activities at enhancers and promoters.

The repressive effect of H3K27me3 is not a new activity but rather the same repressive activity exerted by H3K27me2 genome-wide, only more specifically focused by the collaboration with the chromodomain-containing PRC1 complexes, without which the repressive state would be vulnerable to the normal competitive activities discussed in earlier sections. The canonical PRC1 could well bring additional features useful for generating a more stable and better coordinated repressive activity. For example, the canonical PRC1 might be important to help rapid remethylation repressed after the passage of the replication fork. Features of the Polycomb and the PH components would favor interactions with other Polycomb-repressed domains in the genome and the formation of Polycomb "bodies" or repressive compartments in the nucleus where the concentration of Polycomb components is locally high. PRC1 complexes have one other activity that could play an important role in repression: in vitro, they can antagonize chromatin remodeling by a mechanism that has remained unclear [99]. While this activity might contribute to repression of specific target genes, it is unlikely to be responsible for the global suppression of transcription associated with H3K27me2, since this does not involve stably bound PRC1. Instead, it seems increasingly likely that this global repression and perhaps also much of targeted gene repression is attributable to the mutual antagonism between methylation and acetylation of H3K27 and the chromatin remodeling and H3K4 methylation activities associated with H3K27ac.

There are no conclusive experimental data to favor this view of Polycomb repression over the more traditional view, which assigned the bulk of the repressive activities to PRC1 complexes but while repression by PRC1 is largely conjectural, repression through PRC2 seems more solidly founded on the antagonism between H3K27 methylation and H3K27 acetylation. In the next section, a brief survey of what is known of the role of Polycomb complexes in phylogeny may help to put these ideas in perspective.

EVOLUTIONARY ASPECTS OF PRC2 FUNCTION

Given the global nature of the role attributed to PRC2 in the preceding discussion, it might be expected that it would be evolutionarily very ancient. Although early studies suggested that Polycomb mechanisms might be associated with multicellularity and the need to specify differential gene expression in different cell lineages, it is now clear that PRC2 complexes containing at least the two most essential components, Esc/EED and E(z), are ancient and found in unicellular organisms belonging to the fungal and protist kingdoms as well as in multicellular animals and plants (Table 13.2). Less well conserved is the Su(z)12 component, which in metazoans is also essential for PRC2 function. It is not entirely clear whether Su(z)12 was ancestral but lost in various branches or whether it was added to a more ancient E(z) + Esc/EED core in a common ancestor of plants and animals. But how is H3K27 methylation used? Given the global activity of PRC2 and its importance for keeping chromatin accessibility and thus pervasive transcriptional activity under control, it might be wondered if this was the basic and ancestral function of H3K27 methylation. A few examples follow.

In *Chlamydomonas*, which possesses well conserved E(z) and ESC homologs but not Su(z)12, the genome contains H3K27me1 and H3K27me2 but not the trimethylated form [100]. Knockdown of the E(z) component was found to increase expression of multicopy transgenes, which are normally transcriptionally silenced. Expression of multicopy retrotransposons also

TABLE 13.2 Phylogenetic Conservation of PcG Proteins

Organism	E(z)	Esc	Su(z)12	RING	Pc
Chlamydomonas	+	+	−	NA	−
Neurospora	+	+	+	NA	−
Cryptococcus	+	+	−	NA	−
Arabidopsis	+	+	+	+	−
Caenorhabditis elegans	+	+	−	+	−
Drosophila	+	+	+	+	+
Mammals	+	+	+	+	+

increased, but no evidence was presented for derepression of single copy genes. The results indicate that the role of PRC2 in *Chlamydomonas* is generally repressive and may serve a surveillance function [100] but too little is known to conclude anything more.

In *Neurospora*, the conserved PRC2 complex, including a Su(z)12 homolog, lays down H3K27 methylation predominantly near the ends of the chromosomes and not overlapping with the H3K9me3 distribution, which is mainly in the proximal regions of chromosomes [101]. Surprisingly, the patterns of H3K27me2 and H3K27me3 are nearly identical and together they cover about 7% of the genome, including hundreds of genes that are normally kept silent. The exclusion of H3K27me2/me3 from centric regions is due to the binding of HP1 to H3K9me3, which antagonizes PRC2. Loss of HP1 causes the redistribution of H3K27 methylation, which colonizes the areas occupied by H3K9me3 at the expense of the telomeric-proximal regions [102].

In the yeast *Cryptococcus neoformans*, a telomeric distribution of H3K27me3 (H3K27me2 was not determined) similar to that in *Neurospora* is produced by PRC2 [103]. The *Cryptococcus* PRC2 complex contains conserved E(z), Esc/EED, and Caf1/RBBP4 homologs but lacks the Su(z)12 component. Instead, it includes other proteins, in particular Ccc1, a chromodomain-containing protein that binds H3K27me3. Loss of Ccc1 causes a redistribution of H3K27me3 similar to that observed in *Neurospora*: HP1 mutants: H3K27me3 is largely lost at telomeric regions and accumulates in centric heterochromatic regions that contain H3K9me2. This redistribution appears to be driven by H3K9 methylation because loss of the H3K9 methyltransferase eliminates both H3K9me2 and the redistributed H3K27me3. The Esc/EED component of PRC2, with its affinity for methylated H3K9, may be responsible at least in part for targeting H3K27 methylation to heterochromatin.

In neither of these two unicellular organisms is there evidence of specific recruitment of PRC2, and the pattern of activity seems more related to the global type of activity seen in *Drosophila* and mammalian cells, modulated by specific attractions or inhibitions such as product recognition and H3K9-related antagonism. Unfortunately, the distribution of the PRC2 components themselves was not determined in *Neurospora* or *Cryptococcus*, nor was the relationship between H3K27me2 and H3K27me3 after redistribution. We do not know therefore if these organisms display stable binding sites of PRC2 as opposed to a purely hit-and-run mode of action.

In plants, a well-conserved PRC2 complex is involved in the repression of specific sets of genes. In fact, plants have evolved more than one type of PRC2 complex containing different E(z) homologs to regulate expression and parental imprinting of embryonic genes, vernalization genes, flowering, and morphogenetic genes (for review, see Ref. [104]). Plant PRC2 complexes contain the core E(z), Esc/EED, and Su(z)12 homologs, whose tissue-specific variants in some way determine the specific gene targets, apparently by association with specific targeting components, which also help to broaden the domain of H3K27me3. The broadening is required for stable silencing, as has also been argued for the *Drosophila* Polycomb complexes [105,106].

Much less conserved phylogenetically is a recognizable PRC1 complex that operates in partnership with PRC2. The unicellular organisms that possess a PRC2 complex lack a Polycomb homolog although they retain an HP1 homolog and, in the case of *Cryptococcus*, include in the PRC2 complex itself a chromodomain protein that recognizes H3K27me3. It is unclear, therefore, whether they require a PRC1 partner to effect transcriptional repression. Nor is it clear whether the core RING proteins and the H2A ubiquitylation function are present. More information is available for plants. The catalytic core of a PRC1 complex, including a RING protein and PCGF partners, is present in *Arabidopsis* and is implicated in regulation of at least some of the PRC2 target genes, but a chromodomain-containing Polycomb homolog has not been identified. Plants have instead an H3K27me3-recognition component, LHP1, that is not homologous to metazoan Polycomb proteins. It contains a chromodomain and a chromo shadow domain structurally related to those of HP1 although it is associated with euchromatin, H3K27me3, and PRC2-repressed genes. Recent results indicate that LHP1 and the RING components bind to the PRC2 complex. This would provide PRC2 with a product recognition module that would help to target it to sites previously H3K27-methylated [107] and would be functionally similar to the product recognition component in the *C. neoformans* PRC2 complex and to the methyl lysine recognition feature of the ESC/EED component in higher metazoans. Whether this PRC1-like complex makes additional contributions to repression and whether the ubiquityl transferase activity targets histone H2A remain unknown, but these results indicate that plants have developed a somewhat different strategy to achieve stable silencing of specific genes (for reviews, see Refs. [108,109]).

Finally I will consider the case of the nematode *C. elegans*, a lower metazoan and an intensively studied model organism. *C. elegans* has conserved the E(z) and Esc/EED core PRC2 components but lacks a Su(z)12 component. The PRC2 function in *C. elegans* was first discovered for its role in suppressing expression of the X chromosome and of somatic genes in the germ line (see Chapter 11 of this volume). PRC2 function is primarily required for the maintenance of the germ line, and loss of H3K27 methylation produces only subtle somatic phenotypes [110–112]. A canonical PRC1 complex with a chromodomain-containing Polycomb homolog is absent, but RING and PCGF homologs are present and genetic evidence suggests that they play a role in regulating at least some of the somatic genes targeted by PRC2 [113]. However, the major function of PRC2 in repressing X chromosome genes and somatic genes in the germ line does not involve PRC1 components and can, to some extent, be substituted by an H3K9 methylation function [112]. This feature is supported by the substantial overlap in *C. elegans* between H3K27me3 and H3K9me3 distributions and the observation that silencing of multicopy transgenes requires both histone modifications [114,115]. Such an overlap in distribution and function is not observed in flies or mammals. No information is available for H3K27me2 or ubiquitylated H2A distributions. However, H3K27me1 is localized to active transcription units [116] as in flies and mammals, and the specific enrichment of H3K27me3 on the X chromosome of germ line cells is caused by high levels of H3K36me3 on autosomes, which are known to antagonize PRC2 [117]. In conclusion, there is much to suggest that the role of PRC2 in *C. elegans* resembles the global functions of PRC2 in flies and mammals rather than the targeted repression of specific genes.

The sense of the preceding discussion is that, while the core components of the PRC2 complex are phylogenetically ancient, there is little evidence for a specific targeting of Polycomb repression in lower eukaryotes, or even in lower metazoans such as *C. elegans*. It is more difficult to determine how much of this is due to loss of components in individual phylogenetic branches, but it seems clear that all three core PRC2 components were present in the last common ancestor of plants and animals and the mode of action of the complex was of a whole-genome surveying type, modified by positive and negative chromatin modifications. The much more heterogeneous ancestry of PRC1 components raises questions about the history of the partnership between PRC1 and PRC2. There is no evidence for the connection between the two in unicellular eukaryotes either in terms of a chromodomain-containing Polycomb homolog or of a relationship between H3K27 methylation and H2A ubiquitylation, but more data may change this picture.

While we wait for more evidence, it is possible to argue that an ancient PRC2 complex, whose primary task was to survey the genome and modulate accessibility to the underlying DNA, has retained this function in animals, handled primarily by H3K27 dimethylation. In higher eukaryotes it also acquired a more precise targeting mechanism and a link to a partner PRC1 complex although it is not clear whether the latter occurred independently in plants and animals. The ability to bind stably, through various cofactors, possibly including H2Aub modules, allows PRC2 to produce higher local levels of H3K27me3, and the evolutionary advent of a chromodomain component in PRC1 promotes the action of PRC2 in the surrounding chromatin, producing a broader and stabler domain of H3K27 trimethylation.

Finally, it might be asked why just H3K27? Why should this lysine be the critical key to opening or shutting the chromatin door? Is it just an evolutionary accident? There is no answer at present. In fact, however, H3K27 is probably not the only possible controlling site. There is evidence that H3K9 methylation can play a similar role but tends to be reserved for heterochromatin and therefore kept more under control in euchromatic regions. Nevertheless, when HP1 is absent, H3K9 methylation in euchromatin increases to proportions similar to those of H3K27 methylation. But this, and the whole complex question of heterochromatic silencing, is another story.

REFERENCES

[1] Pirrotta V. The necessity of chromatin: a view in perspective. In: Allis CD, Caparros M-L, Jenuwein T, Reinberg D, Lachner M, editors. Epigenetics. 2nd ed. Cold Spring Harbor Laboratory Press; 2014. p. 927–43.

[2] Rastelli L, Chan CS, Pirrotta V. Related chromosome binding sites for zeste, suppressors of zeste and Polycomb group proteins in *Drosophila* and their dependence on Enhancer of Zeste function. EMBO J 1993;12:1513–22.

[3] Fauvarque M-O, Dura J-M. *Polyhomeotic* regulatory sequences induce developmental regulator-dependent variegation and targeted P-element insertions in *Drosophila*. Genes Dev 1993;7:1508–20.

[4] Bloyer S, Cavalli G, Brock HW, Dura J-M. Identification and characterization of *polyhomeotic* PREs and TREs. Dev Biol 2003;261:426–42.

[5] Schwartz YB, Kahn TG, Nix DA, Li X-Y, Bourgon R, Biggin M, et al. Genome-wide analysis of Polycomb targets in *Drosophila melanogaster*. Nat Genet 2006;38:700–5.

[6] Park SY, Schwartz YB, Kahn TG, Asker D, Pirrotta V. Regulation of polycomb group genes *Psc* and *Su(z)2* in *Drosophila melanogaster*. Mech Dev 2012;2012(128):536–47.

[7] Boyer LA, Plath K, Zeitlinger J, Brambrink T, Medeiros LA, Lee TI, et al. Polycomb complexes repress developmental regulators in murine embryonic stem cells. Nature 2006;441:349–53.

[8] Leeb M, Pasini D, Novatchkova M, Jaritz M, Helin K, Wutz A. Polycomb complexes act redundantly to repress genomic repeats and genes. Genes Dev 2010;24:265–76.

[9] Montgomery ND, Yee D, Chen A, Kalantry S, Chamberlain SJ, Otte AP, et al. The murine Polycomb group protein Eed is required for global histone H3 lysine-27 methylation. Curr Biol 2005;15:942–7.

[10] Ferrari KJ, Scelfo A, Jammula S, Cuomo A, Barozzi I, Stützer A, et al. Polycomb-dependent H3K27me1 and H3K27me2 regulate active transcription and enhancer fidelity. Mol Cell 2014;53:49–62.

[11] McCabe MT, Graves AP, Ganji G, Diaz E, Halsey WS, Jiang Y, et al. Mutation of A677 in histone methyltransferase EZH2 in human B-cell lymphoma promotes hypertrimethylation of histone H3 on lysine 27 (H3K27). Proc Natl Acad Sci USA 2012;109:2989–94.

[12] Alabert C, Barth TK, Reverón-Gómez N, Sidoli S, Schmidt A, Jensen ON, et al. Two distinct modes for propagation of histone PTMs across the cell cycle. Genes Dev 2015;29:585–90.

[13] Hansen KH, Bracken AP, Pasini D, Dietrich N, Gehani SS, Monrad A, et al. A model for transmission of the H3K27me3 epigenetic mark. Nat Cell Biol 2008;10:1291–300.

[14] Zee BM, Levin RS, Xu B, LeRoy G, Wingreen NS, Garcia BA. In vivo residue-specific histone methylation dynamics. J Biol Chem 2010;285:3341–50.

[15] Petruk S, Sedkov Y, Johnston DM, Hodgson JW, Black KL, Kovermann SK, et al. TrxG and PcG proteins but not methylated histones remain associated with DNA through replication. Cell 2012;150:922–33.

[16] Petruk S, Black KL, Kovermann SK, Brock HW, Mazo A. Stepwise histone modifications are mediated by multiple enzymes that rapidly associate with nascent DNA during replication. Nat Commun 2013;4:2841.

[17] Margueron R, Justin N, Ohno K, Sharpe ML, Son J, Drury III WJ, et al. Role of the polycomb protein EED in the propagation of repressive histone marks. Nature 2009;461:762–7.

[18] Yuan W, Wu T, Fu H, Dai C, Wu H, Liu N, et al. Dense chromatin activates polycomb repressive complex 2 to regulate H3 lysine 27 methylation. Science 2012;337:971–5.
[19] Jung HR, Pasini D, Helin K, Jensen ON. Quantitative mass spectrometry of histones H3.2 and H3.3 in Suz12-deficient mouse embryonic stem cells reveals distinct, dynamic post-translational modifications at Lys-27 and Lys-36. Mol Cell Proteomics 2010;9:838–50.
[20] Voigt P, LeRoy G, Drury WJI, Zee BM, Son J, Beck DB, et al. Asymmetrically modified nucleosomes. Cell 2012;151:181–93.
[21] Lee H-G, Kahn TG, Simcox A, Schwartz YB, Pirrotta V. Genome-wide activities of Polycomb complexes control pervasive transcription. Genome Res 2015;25:1170–81.
[22] Lan F, Bayliss PE, Rinn JL, Whetstine JR, Wang JK, Chen S, et al. A histone H3 lysine 27 demethylase regulates animal posterior development. Nature 2007;449:689–94.
[23] Smith ER, Lee MG, Winter B, Droz NM, Eissenberg JC, Shiekhattar R, et al. *Drosophila* UTX is a histone H3 Lys27 demethylase that colocalizes with the elongating form of RNA polymerase II. Mol Cell Biol 2008;28:1041–6.
[24] Schmitges FW, Prusty AB, Faty M, Stützer A, Lingaraju GM, Aiwazian J, et al. Histone methylation by PRC2 is inhibited by active chromatin marks. Mol Cell 2011;42:330–41.
[25] Yuan W, Xu M, Huang C, Liu N, Chen S, Zhu B. H3K36 methylation antagonizes PRC2-mediated H3K27 methylation. J Biol Chem 2011;286:7983–9.
[26] Hong S, Cho Y-W, Yu L-R, Yu H, Veenstra TD, Ge K. Identification of JmjC domain-containing UTX and JMJD3 as histone H3 lysine 27 demethylases. Proc Natl Acad Sci USA 2007;104:18439–44.
[27] De Santa F, Totaro MG, Prosperini E, Notarbartolo S, Testa G, Natoli G. The histone H3 lysine-27 demethylase Jmjd3 links inflammation to inhibition of Polycomb-mediated gene silencing. Cell 2007;130:1083–94.
[28] Lee MG, Villa R, Trojer P, Norman J, Yan K-P, Reinberg D, et al. Demethylation of H3K27 regulates Polycomb recruitment and H2A ubiquitination. Science 2007;318:447–50.
[29] Agger K, Cloos PAC, Christensen J, Pasini D, Rose S, Rappsilber J, et al. UTX and JMJD3 are histone H3K27 demethylases involved in HOX gene regulation and development. Nature 2007;449:731–4.
[30] Park DH, Hong SJ, Salinas RD, Liu SJ, Sun SW, Sgualdino J, et al. Activation of neuronal gene expression by the JMJD3 demethylase is required for postnatal and adult brain neurogenesis. Cell Rep 2014;8:1290–9.
[31] Welstead GG, Creyghton MP, Bilodeau S, Cheng AW, Markoulaki S, Young RA, et al. X-linked H3K27me3 demethylase Utx is required for embryonic development in a sex-specific manner. Proc Natl Acad Sci USA 2012;109:13004–9.
[32] Shpargel KB, Sengoku T, Yokoyama S, Magnuson T. UTX and UTY demonstrate histone demethylase-independent function in mouse embryonic development. PLoS Genet 2012;8:e1002964.
[33] Shpargel KB, Starmer J, Yee D, Pohlers M, Magnuson T. KDM6 demethylase independent loss of histone H3 lysine 27 trimethylation during early embryonic development. PLoS Genet 2014;10:e1004507.
[34] Wang C, Lee J-E, Cho Y-W, Xiao Y, Jin Q, Liu C, et al. UTX regulates mesoderm differentiation of embryonic stem cells independent of H3K27 demethylase activity. Proc Natl Acad Sci USA 2012;109:15324–9.
[35] Copur Ö, Jäckle H. The histone H3-K27 demethylase Utx regulates HOX gene expression in *Drosophila* in a temporally restricted manner. Development 2013;140:3478–85.

[36] Miller SA, Mohn SE, Weinmann AS. Jmjd3 and UTX play a demethylase-independent role in chromatin remodeling to regulate T-box family member-dependent gene expression. Mol Cell 2010;40:594–605.
[37] Herz H-M, Mohan M, Garruss AS, Liang K, Takahashi Y-h, Mickey K, et al. Enhancer-associated H3K4 monomethylation by trithorax-related, the *Drosophila* homolog of mammalian Mll3/Mll4. Genes Dev 2012;26:2604–20.
[38] Kanda H, Nguyen A, Chen L, Okano H, Hariharan IK. The *Drosophila* ortholog of MLL3 and MLL4, trithorax related, functions as a negative regulator of tissue growth. Mol Cell Biol 2013;33:1702–10.
[39] Herz H-M, Madden LD, Chen Z, Bolduc C, Buff E, Gupta R, et al. The H3K27me3 demethylase dUTX is a suppressor of notch- and Rb-dependent tumors in Drosophila. Mol Cell Biol 2010;30:2485–97.
[40] Morin RD, Mendez-Lago M, Mungall AJ, Goya R, Mungall KL, Corbett RD, et al. Frequent mutation of histone-modifying genes in non-Hodgkin lymphoma. Nature 2011;476:298–303.
[41] van Haaften G, Dalgliesh GL, Davies H, Chen L, Bignell G, Greenman C, et al. Somatic mutations of the histone H3K27 demethylase gene UTX in human cancer. Nat Genet 2009;41:521–3.
[42] Parsons DW, Li M, Zhang X, Jones S, Leary RJ, Lin JC-H, et al. The genetic landscape of the childhood cancer medulloblastoma. Science 2011;331:435–9.
[43] Merkwirth C, Jovaisaite V, Durieux J, Matilainen O, Jordan SD, Quiros PM, et al. Two conserved histone demethylases regulate mitochondrial stress-induced longevity. Cell 2016;165:1209–23. http://dx.doi.org/10.1016/j.cell.2016.04.012.
[44] Tian Y, Garcia G, Bian Q, Steffen KK, Joe L, Wolff S, et al. Mitochondrial stress induces chromatin reorganization to promote longevity and UPR^{mt}. Cell 2016;165:1197–208.
[45] Siebold AP, Banerjee R, Tie F, Kiss DL, Moskowitz J, Harte PJ. Polycomb repressive complex 2 and trithorax modulate *Drosophila* longevity and stress resistance. Proc Natl Acad Sci USA 2010;107:169–74.
[46] Wang Z, Zang C, Rosenfeld JA, Schones DE, Barski A, Cuddapah S, et al. Combinatorial patterns of histone acetylations and methylations in the human genome. Nat Genet 2008;40:897–903.
[47] Stasevich TJ, Hayashi-Takanaka Y, Sato Y, Maehara K, Ohkawa Y, Sakata-Sogawa K, et al. Regulation of RNA polymerase II activation by histone acetylation in single living cells. Nature 2014;516:272–5.
[48] Heintzman ND, Stuart RK, Hon G, Fu YT, Ching CW, Hawkins RD, et al. Distinctive and predictive chromatin signatures of transcriptional promoters and enhancers in the human genome. Nat Genet 2007;39:311–8.
[49] Creyghton MP, Cheng AW, Welstead GG, Kooistra T, Carey BW, Steine EJ, et al. Histone H3K27ac separates active from poised enhancers and predicts developmental state. Proc Natl Acad Sci USA 2010;107:21931–6.
[50] Rada-Iglesias A, Bajpai R, Swigut T, Brugmann SA, Flynn RA, Wysocka J. A unique chromatin signature uncovers early developmental enhancers in humans. Nature 2011;470:279–83.
[51] Tie F, Banerjee R, Stratton CA, Prasad-Sinha J, Stepanik V, Zlobin A, et al. CBP-mediated acetylation of histone H3 lysine 27 antagonizes *Drosophila* Polycomb silencing. Development 2009;136:3131–41.

[52] Pasini D, Malatesta M, Jung HR, Walfridsson J, Willer A, Olsson L, et al. Characterization of an antagonistic switch between histone H3 lysine 27 methylation and acetylation in the transcriptional regulation of Polycomb group target genes. Nucl Acids Res 2010;38:4958–69.

[53] Sekiya T, Muthurajan UM, Luger K, Tulin AV, Zaret KS. Nucleosome-binding affinity as a primary determinant of the nuclear mobility of the pioneer transcription factor FoxA. Genes Dev 2009;23:804–9.

[54] Calo E, Wysocka J. Modification of enhancer chromatin: what, how, and why? Mol Cell 2013;49:825–37.

[55] Feller C, Forné I, Imhof A, Becker PB. Global and specific responses of the histone acetylome to systematic perturbation. Mol Cell 2015;57:559–71.

[56] Tie F, Banerjee R, Conrad PA, Scacheri PC, Harte PJ. Histone demethylase UTX and chromatin remodeler BRM bind directly to CBP and modulate acetylation of histone H3 lysine 27. Mol Cell Biol 2012;32:2323–34.

[57] Hilton IB, D'Ippolito AM, Vockley CM, Thakore PI, Crawford GE, Reddy TE, et al. Epigenome editing by a CRISPR-Cas9-based acetyltransferase activates genes from promoters and enhancers. Nat Biotech 2015;33:510–7. http://dx.doi.org/10.1038/nbt.3199.

[58] Zentner GE, Scacheri PC. The chromatin fingerprint of gene enhancer elements. J Biol Chem 2012;287:30888–96.

[59] Pengelly AR, Copur Ö, Jäckle H, Herzig A, Müller J. A histone mutant reproduces the phenotype caused by loss of histone-modifying factor polycomb. Science 2013;339:698–9.

[60] Chopra VS, Hong J-W, Levine M. Regulation of Hox gene activity by transcriptional elongation in *Drosophila*. Curr Biol 2009;19:688–93.

[61] van Galen P, Viny AD, Ram O, Ryan RJH, Cotton MJ, Donohue L, et al. A multiplexed system for quantitative comparisons of chromatin landscapes. Mol Cell 2016;61:170–80.

[62] Madhani HD. The frustrated gene: origins of eukaryotic gene expression. Cell 2013;155:744–9.

[63] Czermin B, Melfi R, McCabe D, Seitz V, Imhof A, Pirrotta V. *Drosophila* enhancer of Zeste/ESC complexes have a histone H3 methyltransferase activity that marks chromosomal Polycomb sites. Cell 2002;111:185–96.

[64] Kuzmichev A, Nishioka K, Erdjument-Bromage H, Tempst P, Reinberg D. Histone methyltransferase activity associated with a human multiprotein complex containing the Enhancer of Zeste protein. Genes Dev 2002;16:2893–905.

[65] Suka N, Suka Y, Carmen AA, Wu J, Grunstein M. Highly specific antibodies determine histone acetylation site usage in yeast heterochromatin and euchromatin. Mol Cell 2001;8:473–9.

[66] Kurdistani SK, Robyr D, Tavazoie S, Grunstein M. Genome-wide binding map of the histone deacetylase Rpd3 in yeast. Nat Genet 2002;31:248–54.

[67] Zaret KS, Carroll JS. Pioneer transcription factors: establishing competence for gene expression. Genes Dev 2011;25:2227–41.

[68] Jeong KW, Kim K, Situ AJ, Ulmer TS, An W, Stallcup MR. Recognition of enhancer element-specific histone methylation by TIP60 in transcriptional activation. Nat Struct Mol Biol 2011;18:1358–65.

[69] Ahmad K, Henikoff S. The histone variant H3.3 marks active chromatin by replication-independent nucleosome assembly. Mol Cell 2001;9:1191–200.

[70] Mito Y, Henikoff JG, Henikoff S. Histone replacement marks the boundaries of cis-regulatory domains. Science 2007;315:1408–11.

[71] Jin C, Felsenfeld G. Nucleosome stability mediated by histone variants H3.3 and H2A.Z. Genes Dev 2007;21:1519–29.

[72] Jin C, Zang C, Wei G, Cui K, Peng W, Zhao K, et al. H3.3/H2A.Z double variant-containing nucleosomes mark 'nucleosome-free regions' of active promoters and other regulatory regions. Nat Genet 2009;41:941–5.

[73] Mishra RK, Mihaly J, Barges S, Spierer A, Karch F, Hagstrom K, et al. The iab-7 Polycomb response element maps to a nucleosome-free region of chromatin and requires both GAGA and pleiohomeotic for silencing activity. Mol Cell Biol 2001;21:1311–8.

[74] Dellino GI, Schwartz YB, Farkas G, McCabe D, Elgin SCR, Pirrotta V. Polycomb silencing blocks transcription initiation. Mol Cell 2004;13:887–93.

[75] Kharchenko PV, Alekseyenko AA, Schwartz YB, Minoda A, Riddle NC, Ernst J, et al. Comprehensive analysis of the chromatin landscape in *Drosophila melanogaster.* Nature 2011;471:480–5.

[76] Philip P, Boija A, Vaid R, Churcher A, Meyers D, Cole P, et al. CBP binding outside of promoters and enhancers in *Drosophila melanogaster.* Epigenetics Chromatin 2015;8:48.

[77] Shin Voo K, Carlone DL, Jacobsen BM, Flodin A, Skalnik DG. Cloning of a mammalian transcriptional activator that binds unmethylated CpG motifs and shares a CXXC domain with DNA methyltransferase, human trithorax, and methyl-CpG binding domain protein 1. Mol Cell Biol 2000;20:2108–21.

[78] Farcas AM, Blackledge NP, Sudbery I, Long HK, McGouran JF, Rose NR, et al. KDM2B links the polycomb repressive complex 1 (PRC1) to recognition of CpG islands. eLife Sci 2012;1:e00205.

[79] Gao Z, Zhang J, Bonasio R, Strino F, Sawai A, Parisi F, et al. Pcgf homologs, CBX proteins, and RYBP define functionally distinct PRC1 family complexes. Mol Cell 2012;45:344–56.

[80] Wu X, Johansen JV, Helin K. Fbxl10/Kdm2b recruits polycomb repressive complex 1 to CpG islands and regulates H2A ubiquitylation. Mol Cell 2013;49:1134–46.

[81] Blackledge NP, Farcas AM, Kondo T, King HW, McGouran JF, Hanssen LLP, et al. Variant PRC1 complex-dependent H2A ubiquitylation drives PRC2 recruitment and Polycomb domain formation. Cell 2014;157:1445–59.

[82] Cooper S, Dienstbier M, Hassan R, Schermelleh L, Sharif J, Blackledge NP, et al. Targeting polycomb to pericentric heterochromatin in embryonic stem cells reveals a role for H2AK119u1 in PRC2 recruitment. Cell Rep 2014;7:1456–70.

[83] Kalb R, Latwiel S, Baymaz HI, Jansen PWTC, Müller CW, Vermeulen M, et al. Histone H2A monoubiquitination promotes histone H3 methylation in Polycomb repression. Nat Struct Mol Biol 2014;21:569–71.

[84] Klose RJ, Cooper S, Farcas AM, Blackledge NP, Brockdorff N. Chromatin sampling – an emerging perspective on targeting polycomb repressor proteins. PLoS Genet 2013;9:e1003717.

[85] Riising EM, Comet I, Leblanc B, Wu X, Johansen JV, Helin K. Gene silencing triggers polycomb repressive complex 2 recruitment to CpG islands genome wide. Mol Cell 2014;55:347–60.

[86] Bernstein E, Duncan EM, Masui O, Gil J, Heard E, Allis CD. Mouse polycomb proteins bind differentially to methylated histone H3 and RNA and are enriched in facultative heterochromatin. Mol Cell Biol 2006;26:2560–9.

[87] Fischle W, Wang Y, Jacobs SA, Kim Y, Allis CD, Khorasanizadeh S. Molecular basis for the discrimination of repressive methyl-lysine marks in histone H3 by Polycomb and HP1 chromodomains. Genes Dev 2003;17:1870−81.

[88] Min J, Zhang Y, Xu R-M. Structural basis for specific binding of Polycomb chromodomain to histone H3 methylated to Lys 27. Genes Dev 2003;17:1823−8.

[89] Zhou W, Zhu P, Wang JK, Pascual G, Ohgi KA, Lozach J, et al. Histone H2A monoubiquitination represses transcription by inhibiting RNA polymerase II transcriptional elongation. Mol Cell 2008;29:69−80.

[90] Lagarou A, Mohd-Sarip A, Moshkin YM, Chalkley GE, Bezstarosti K, Demmers JAA, et al. dKDM2 couples histone H2A ubiquitylation to histone H3 demethylation during Polycomb group silencing. Genes Dev 2008;22:2799−810.

[91] Scheuermann JC, de Ayala Alonso AG, Oktaba K, Ly-Hartig N, McGinty RK, Fraterman S, et al. Histone H2A deubiquitinase activity of the Polycomb repressive complex PR-DUB. Nature 2010;465:243−7.

[92] Irminger-Finger I, Nöthiger R. The *Drosophila melanogaster* gene *lethal(3)73Ah* encodes a ring finger protein homologous to the oncoproteins MEL-18 and BMI-1. Gene 1995;163:203−8.

[93] Eskeland R, Leeb M, Grimes GR, Kress C, Boyle S, Sproul D, et al. Ring1B compacts chromatin structure and represses gene expression independent of histone ubiquitination. Mol Cell 2010;38:452−64.

[94] Endoh M, Endo TA, Endoh T, Isono K, Sharif J, Ohara O, et al. Histone H2A monoubiquitination is a crucial step to mediate PRC1-dependent repression of developmental genes to maintain ES cell identity. PLoS Genet 2012;8:e1002774.

[95] Illingworth RS, Moffat M, Mann AR, Read D, Hunter CJ, Pradeepa MM, et al. The E3 ubiquitin ligase activity of RING1B is not essential for early mouse development. Genes Dev 2015;29:1897−902.

[96] Pengelly AR, Kalb R, Finkl K, Müller J. Transcriptional repression by PRC1 in the absence of H2A monoubiquitylation. Genes Dev 2015;29:1487−92.

[97] Chiacchiera F, Rossi A, Jammula S, Piunti A, Scelfo A, Ordóñez-Morán P, et al. Polycomb complex PRC1 preserves intestinal stem cell identity by sustaining Wnt/β-catenin transcriptional activity. Cell Stem Cell 2016;18:91−103.

[98] Williamson I, Eskeland R, Lettice LA, Hill AE, Boyle S, Grimes GR, et al. Anterior-posterior differences in HoxD chromatin topology in limb development. Development 2012;139:3157−67.

[99] Shao Z, Raible F, Mollaaghababa R, Guyon JR, Wu CT, Bender W, et al. Stabilization of chromatin structure by PRC1, a Polycomb complex. Cell 1999;98:37−46.

[100] Shaver S, Casas-Mollano JA, Cerny RL, Cerutti H. Origin of the polycomb repressive complex 2 and gene silencing by an E(z) homolog in the unicellular alga Chlamydomonas. Epigenetics 2010;5:301−12.

[101] Jamieson K, Rountree MR, Lewis ZA, Stajich JE, Selker EU. Regional control of histone H3 lysine 27 methylation in Neurospora. Proc Natl Acad Sci USA 2013;110:6027−32.

[102] Jamieson K, Wiles ET, McNaught KJ, Sidoli S, Leggett N, Shao Y, et al. Loss of HP1 causes depletion of H3K27me3 from facultative heterochromatin and gain of H3K27me2 at constitutive heterochromatin. Genome Res 2016;26:97−107.

[103] Dumesic PA, Homer CM, Moresco JJ, Pack LR, Shanle EK, Coyle SM, et al. Product binding enforces the genomic specificity of a yeast Polycomb repressive complex. Cell 2015;160:204−18.

[104] Grossniklaus U, Paro R. Transcriptional silencing by Polycomb-group proteins. In: Allis CD, Caparros M-L, Jenuwein T, Reinberg D, Lachner M, editors. Epigenetics. 2nd ed. Cold Spring Harbor Laboratory Press; 2014. p. 463–506.

[105] De Lucia F, Crevillen P, Jones AME, Greb T, Dean C. A PHD-Polycomb repressive complex 2 triggers the epigenetic silencing of FLC during vernalization. Proc Natl Acad Sci USA 2008;105:16831–6.

[106] Kahn TG, Schwartz YB, Dellino GI, Pirrotta V. Polycomb complexes and the propagation of the methylation mark at the *Drosophila Ubx* gene. J Biol Chem 2006;281:29064–75.

[107] Derkacheva M, Steinbach Y, Wildhaber T, Mozgova I, Mahrez W, Nanni P, et al. *Arabidopsis* MSI1 connects LHP1 to PRC2 complexes. EMBO J 2013;32:2073–85.

[108] Mozgova I, Köhler C, Hennig L. Keeping the gate closed: functions of the polycomb repressive complex PRC2 in development. Plant J 2015;83:121–32.

[109] Merini W, Calonje M. PRC1 is taking the lead in PcG repression. Plant J 2015;83:110–20.

[110] Ross JM, Zarkower D. Polycomb group regulation of *Hox* gene expression in *C. elegans*. Dev Cell 2003;4:891–901.

[111] Korf I, Fan Y, Strome S. The Polycomb group in *Caenorhabditis elegans* and maternal control of germline development. Development 1998;125:2469–78.

[112] Gaydos LJ, Wang W, Strome S. H3K27me and PRC2 transmit a memory of repression across generations and during development. Science 2014;345:1515–8.

[113] Karakuzu O, Wang DP, Cameron S. MIG-32 and SPAT-3A are PRC1 homologs that control neuronal migration in *Caenorhabditis elegans*. Development 2009;136:943–53.

[114] Ho JWK, Jung YL, Liu T, Alver BH, Lee S, Ikegami K, et al. Comparative analysis of metazoan chromatin organization. Nature 2014;512:449–52.

[115] Towbin BD, González-Aguilera C, Sack R, Gaidatzis D, Kalck V, Meister P, et al. Step-wise methylation of histone H3K9 positions heterochromatin at the nuclear periphery. Cell 2012;150:934–47.

[116] Gerstein MB, Lu ZJ, Van Nostrand EL, Cheng C, Arshinoff BI, Liu T, et al. Integrative analysis of the *Caenorhabditis elegans* genome by the modENCODE project. Science 2010;2012(330):1775–87.

[117] Gaydos LJ, Rechtsteiner A, Egelhofer TA, Carroll CR, Strome S. Antagonism between MES-4 and polycomb repressive complex 2 promotes appropriate gene expression in *C. elegans* germ cells. Cell Rep 2012;5:1169–77.

[118] Mohan M, Herz H-M, Smith ER, Zhang Y, Jackson J, Washburn MP, et al. The COMPASS family of H3K4 methylases in *Drosophila*. Mol Cell Biol 2011;31:4310–8.

[119] Kahn TG, Dorafshan E, Schultheis D, Zare A, Stenberg P, Reim I, et al. Interdependence of PRC1 and PRC2 for recruitment to Polycomb Response Elements. Nucleic Acids Res 2016. http://dx.doi.org/10.1093/nar/gkw701.

Index

'*Note:* Page numbers followed by "f" indicate figures and "t" indicate tables.'

A

Abdominal-B (Abd-B), 16–17, 139f, 295
Adipocyte enhancer-binding protein 2 (AEBP2), 168–169, 207–208, 207f, 242–243, 242f
A677G
  A687V altered substrate specificity, 267–268
  mutation, 270f, 272
Amino acid sequences, 35–36
Antagonizing activity, 304–305
Antennapedia complex (ANT-C), 135
Arabidopsis thaliana, 137–138, 195–196
Arginine, 44–45

B

B-cell differentiation, 262–263, 263f
BCL6 corepressor (BCOR), 10–11
Bithorax complex (BX-C), 33–34, 113, 135
Black chromatin, 136–137

C

Caenorhabditis elegans, 3, 196–197
  defined, 289–290
  homeotic (Hox) genes, 291
  life cycle, 290
  Polycomb group-like proteins, 293t
  PRC1, 294–295
    other species, 292–294
  PRC2
    antagonizing activity, 304–305
    embryonic cells, restricts plasticity of, 302
    embryonic development, 300–301
    germline development, 296–298, 297f
    H3K9me2, 299–300, 300f
    H3K27me3, 299–301, 300f
    noncoding RNA-mediated H3K27 methylation, 306
    PcG recruitment, 306–307
    PREs, 306–307
    somatic cells, 303
    somatic differentiation, safeguards germ cells from, 301–302
    sperm chromosomes, 299–300
    subunits identification, 296
    transgenerational inheritance, 298–299
    X chromosome repression, 296–298, 297f
    XO germline, 298–299
  SOR-1 and SAM domain, 295–296
Canonical Polycomb repressive complexes 1 (cPRC1 complex)
  background, 57–58
  biochemical compositions, 58, 58f
  biological importance, 65–72
  chromatin and mechanism, 59, 60f
  H2A monoubiquitination, 65–72, 69t–71t
  mechanism, 59–63
  perspectives, 72–74
  posttranslational modifications (PTMs), 57–58
Castration-resistant prostate cancer (CRPC), 229
CBX, 2
CBX7 chemical probes, 48–50, 49f
Cdkn2a gene, 66–67
Cellular activity, 274–275
Cellular memory modules (CMMs), 114–115
Chaetomium thermophilum, 170–171, 200, 227, 268, 269f
Chemical probes, 48–50
ChIP sequencing (ChIP-seq), 322–323
Chlamydomonas reinhardtii, 196–197, 338–339
Chromatin, 199–202, 199f, 234–236, 235f, 259–262, 261f
Chromatin Assembly Factor 1 (CAF1), 228
Chromatin immunoprecipitations (ChIP), 112–113, 304
Chromatin immunoprecipitation sequencing (ChIP-seq), 68–72
Chromobox 2 (CBX2), 41

Chromobox (CBX)
  domain, 166
  proteins, 7
Chromodomain
  CBX7 chemical probes, 48–50, 49f
  heterochromatin-associated protein one (HP1), 45
  H3K27me3, 35–36, 36f, 37t
    mammalian Polycomb homologs, 41–43, 43f
    posttranslational modifications, 44–45
    structural basis, 38–41, 39f
  noncanonical partners, 46–48
  noncoding RNA, 46–48
  overview, 33–35
  putative nonhistone targets, 45–46
  yeast and plant, Polycomb homologs from, 50–51
Chronic myelomonocytic leukemia (CMML), 275
Clasp theory, 42–43
Clustered regularly interspaced short palindromic repeat (CRISPR)/Cas9 genome, 147–148, 296–298
CREB-binding protein (CBP), 115, 325–327, 326f, 330f
Cryptococcus neoformans, 196–197, 339

D

3-Deazaneplanocin A (DZNep), 275–277, 276f
Deubiquitinase (Dub) enzymes, 85
3D gene networks, 147
Dissociation constant (K_d), 35–36, 41
Dorsal switch protein 1 (Dsp1), 117
Dosage compensation complex (DCC), 296–298
dPC protein, 66
dRing-associated factors (dRAF), 132–133
Drosophila, 1, 33–36, 41, 46–47, 51, 58f, 59, 62–63, 65–66, 68–72, 112–114, 114f
  Nurf55, 173–174, 173f
  reporter genes, 22–23
  transgenic cell lines, 35

E

Ectoderm development protein (EED), 135, 166–167, 171–173, 172f, 176–177, 176f, 233–234
Embryonic cells, restricts plasticity of, 302
Embryonic development, 300–301
Embryonic stem cells (ESCs), 135
End helix (EH), 14–16, 61
Enhancer of zeste homolog 1 (EZH1), 204–206, 229–230, 246–247
Enhancer of zeste homolog 2 (EZH2), 177–179, 204–206, 229–230, 246–247
  additional gain-of-function, 266–268
  A677G and A687V altered substrate specificity, 267–268
  A677G mutation, 270f, 272
  altered substrate specificity, structural rationale for, 268–274, 269f
  amplification and overexpression, 262
  cellular activity, 274–275
  chromatin, 259–262, 261f
  first SANT (SANT1) domain, 169–170
  inhibitors, 275–278, 276f
  mechanistic and phenotypic effects, 276f, 278–280
  myeloid malignancies, 275
  normal B-cell differentiation, 262–263, 263f
  substrate monomethylation, A687 for, 272–274, 273f
  Y641 mutations, 264–266, 265f, 270f, 271
ESC-E(Z) complex, 192–193
EZH1. *See* Enhancer of zeste homolog 1 (EZH1)
EZH2. *See* Enhancer of zeste homolog 2 (EZH2)

F

Follicular lymphomas (FLs), 263

G

Germinal center B cell (GCB), 263
Germinal centers (GCs), 263
Germline development, 296–298, 297f
Green chromatin, 136–137

H

H2A K119 ubiquitination, 237–239, 238f
H2A ubiquitylation, 1–2
Heterochromatin-associated protein one (HP1), 34
High mobility group (HMG), 117
Histone demethylase lysine demethylase 2B (KDM2B), 9

Histone methyltransferases (HMTs), 193–195, 299–300
Histone phosphorylation, 240
H3K9me2, 299–300, 300f
H3K27me1, 202–204
H3K27me2, 202–204
H3K27me3, 2–3, 61–62, 202–204, 299–301, 300f
  activation mechanism, 180–181, 181f
  repressive domains, 136
H3K27 methylation
  accessibility hypothesis, 329–332, 330f
  acetylation, 326–328
  genomic distribution, 321–323, 322f
  roaming activities, 328
  role, 323–324
  UTX, 324–326
H3K27M inhibition mechanism, 179–180, 180f
Homeotic (HOX) genes, 6, 65–66
Homology domain 1 (HD1) domain, 17
HOTAIR, 149, 306
HOTTIP, 149
HOX genes, 61–62, 336–337
H3S28 phosphorylation (H3S28p), 236
Human CBX7, 47–48
Hydrophobic clasp, 36f, 42–43, 43f, 46

I

Immunocompromised mice harboring xenografts, 279–280
Ink4a transcript, 66–67
Internal tandem duplications (ITDs), 12–14, 13t

J

JARID2, 209–211, 244–245

K

KDM2B, 63–64, 237, 238f

L

Like Heterochromatin Protein 1 (LHP1), 51
Linker histone H1.4, 46
long noncoding RNAs (lncRNA), 168
Long-range interactions, 145
Loop domains, 141–142

M

Malignant brain tumor (MBT), 7, 38
mouse embryonic stem cells (mESC), 167–168
MTF2, 243–244
Multiple Polycomb-binding sites, 61
Myelodysplastic syndrome (MDS), 275
Myeloid malignancies, 275

N

Negative staining electron microscopy (EM), 168–169
Neurospora crassa, 266–267
Noncanonical Polycomb repressive complexes 1 (ncPRC1 complex)
  biological importance, 65–72
  H2A monoubiquitination, 65–72, 69t–71t
  mechanism, 63–65, 65f
  perspectives, 72–74
Noncoding RNA, 46–48
Noncoding RNA-mediated H3K27 methylation, 306
Nuclear SET domain (NSD), 304
Nucleosome remodeling deacetylase complex (NuRD), 228
Nucleosome-Remodeling Factor 55 (NURF-55), 132–133
Nucleosome remodeling factor complex (NURF), 228

O

O-linked b-NAcetylglucosamine (O-GlcNAc), 20–21

P

Paramecium tetraurelia, 196–197
PCGF2, 58, 66–67
PCGF4, 58, 66–67
PcG foci, 136
PCGF1 ubiquitin fold discriminator (PUFD), 11
PcG/H3K27me3-repressive domains, 136–137
PcG-mediated silencing, 193–195
PcG RING Finger (PCGF) protein, 6
PCL, 208–209, 243–244
PHF1, 243–244
PHF19, 243–244
Pho-repressive complex (PHO-RC), 115, 132–133

Ph SAM Polymerization, 17–18
midloop/helix (ML), 61
Pleiohomeotic (Pho) repression complex (PhoRC), 22
Polar clasp, 36f, 44–45
Polar fingers, 42–43
Polycomb bodies, 61
Polycomb group (PcG) proteins, 112
biochemical activities, 133f
cancer, three-dimensional genomics in, 151
chromatin, 132–134, 133f
chromatin compaction, 137–138
dynamic multilooped three-dimensional structures, target loci form, 138–141, 139f
long-range chromosomal interactions, 144–148
noncoding RNA, 148–151, 150f
overview, 131–132
Polycomb domains, 135–137
Polycomb-repressed domains, 141–144
three-dimensional gene networks, 144–148
three-dimensional organization, 148–151
topologically associating domains (TADs), 141–144, 142f
Polycomb repressive complex 1 (PRC1), 1–3, 81–82, 132–133
Bmi1, 89
Bmi1-Ring1B
E2 enzymes, 87t–88t, 91–92
heterodimer, 89–91, 90t
Bmi1-Ring1B-UbcH5c, 87t–88t, 93–94
H2A ubiquitination, 99–102
Ph SAM polymerization, 16–17
dependent repression, 22–23
function, 16–17
H2Aub activity, 21–22
polymer regulation, 20–21
SAM-polymerized PRC1 stoichiometry, 18–20
Scm SAM-dependent repression, 22–23
subnuclear organization, 17–18
Polycomb group (PcG), 6–7, 7t, 8f
Polycomb group SAMS, 14–23, 15f, 15t
PRC1 ubiquitin ligase module, 94–96, 95f
RAWUL, 7–14
BCOR PUFD, 12–14, 13t
complex structures, 10f
heterodimerization, 11–12, 12f
internal tandem duplications (ITDs), 12–14
Polycomb group SAM, 14–23
structural basis, 9–11
Ring1, 89
Ring E3 ligase, 85–86
ubiquitin
conjugation, 82–85, 82f
deconjugation, 82–85, 82f
transfer, mechanism, 96–99
uH2A, 102–104
Polycomb repressive complex 2 (PRC2), 1–3, 132–133
accessory components, 241–247
AEBP2, 242–243, 242f
EZH1, 246–247
JARID2, 244–245
PCL proteins, 243–244
activity, 226–227
adipocyte enhancer-binding protein 2 (AEBP2), 168–169
Arabidopsis thaliana, 195–196
Caenorhabditis elegans, 196–197
Chlamydomonas reinhardtii, 196–197
chromatin, 199–202, 199f, 234–236, 235f
cofactors, 206–212
AEBP2, 207–208, 207f
identified cofactors, 211–212
JARID2, 209–211
PCL, 208–209
conservation and divergence, 193, 194f
core complex, 197–206
Cryptococcus neoformans, 196–197
C. thermophilum, 200
domain architecture, 167f
Drosophila Nurf55, 173–174, 173f
EED, 171–173, 172f, 176–177, 176f, 228, 233–234
electron microscopy studies, 169–171, 169f
enzymatic activity, 197–200, 199f, 225–230
ESC-E(Z) complex, 192–193
evolutionary aspects, 338–341, 338t
evolutionary conservation, 194f, 195–197
Ezh2, 177–179
EZH1/EZH2, 204–206, 229–230, 246–247
functions, 319–321, 321t
glioblastomas (GBMs), 247–249, 248f
H2A K119 ubiquitination, 237–239, 238f
H2A ubiquitylation, 333–336, 335f
histone methyltransferase activity, 193–195

histone modifications, 236–241
H3K27me1, 202–204
H3K27me2, 202–204
H3K27me3, 202–204
H3K27me3 activation mechanism, 180–181, 181f
H3K4me2/3 and H3K36me2/3, 239–240
H3K27 methylation, 230–232
accessibility hypothesis, 329–332, 330f
acetylation, 326–328
genomic distribution, 321–323, 322f
propagation, 233–234
roaming activities, 328
role, 323–324
UTX, 324–326
H3K27M inhibition mechanism, 179–180, 180f, 247–249, 248f
HMTase, 200–201
H3S28 phosphorylation antagonizes Polycomb silencing, 240–241
negative staining electron microscopy (EM), 168–169
overview, 165–169, 317–318
Paramecium tetraurelia, 196–197
PcG-mediated silencing, 193–195
Polycomb-like proteins (PCL), 168
Polycomb machinery, 212–214
Polycomb repressive activities, 336–338
PRC1 recruitment, 332–333
Protein Data Bank (PDB), 171
RbAp48, 173–174, 173f
structure, 226f, 227–228
SUZ12, 201
Suz12^{VEFS}, 177
targeted silencing functions, 318–319
ternary Ezh2–EED–Suz12 complex, 174–179, 175f–176f
Tetrahymena thermophila, 196–197
transcriptional regulation, 214–216
Ultrabithorax (Ubx), 194–195
X-ray crystallography studies, 171–179
Polycomb-repressive deubiquitinase (PR-DUB), 132–133
Polycomb response elements (PREs), 133–134, 231–232
cellular memory modules, 114–115
cooperative recruitment, DNA platforms for, 118–121, 119f
Drosophila, 112–114, 114f
mammals, 121–122
overview, 111–112
pairing-sensitive silencing, 114f
Polycomb group protein complexes, 112
sequence-specific DNA-binding proteins, 115–117, 116t
Polyhomeotic homolog 1 (PHC1), 9
Posttranslational modifications (PTMs), 34, 44–45
PRC2–AEBP2 complex, electron microscopy, 169f, 170
Primary myelofibrosis (PMF), 275
Primitive neuroectodermal tumors of the central nervous system (CNS-PNET), 12–14
Protein acetyltransferase KAT5, 45
Protein code, 45–46
Protein Data Bank (PDB), 171

R

RAWUL. *See* RING finger and WD40-associated ubiquitin-like (RAWUL)
RbAp48, 173–174, 173f
Really interesting new gene (RING) E3 ligases, 82–83
Repressive chromatin hub, 138–140
Ring1a, 67–68
Ring1b, 67–68, 73, 336
RING1B YY1–Binding Protein (RYBP), 9–10
RING domain proteins, 83–85
RING finger and WD40-associated ubiquitin-like (RAWUL), 5–6
BCOR PUFD, 12–14, 13t
complex structures, 10f
heterodimerization, 11–12, 12f
internal tandem duplications (ITDs), 12–14
Polycomb group SAM, 14–23
structural basis, 9–11
RNA interference (RNAi) pathway, 145, 306

S

S-adenosyl-L-homocysteine (SAH), 226
Schizosaccharomyces pombe, 35, 306
Second SANT domain (SANT2), 169–170
Sequence-specific DNA-binding adaptor proteins, 115–117, 116t
Sex comb on midleg (SCM), 132–133
Somatic cells, 303
Somatic differentiation, safeguards germ cells from, 301–302

Sperm chromosomes, 299–300
Sterile alpha motif (SAM), 5–6, 61, 137–138
Stimulation-responsive motif (SRM), 234
Stochastic optical reconstruction microscopy (STORM), 18, 144
Substrate monomethylation, A687 for, 272–274, 273f
SUZ12, 201
Suz12^{VEFS}, 177

T

Tetrahymena thermophila, 196–197, 306–307
Transcriptional regulation, 214–216
Transcription start site (TSS), 114, 135, 230–231
Transgenerational inheritance, 298–299
Translocation ets leukemia (TEL), 14–16
Trithorax group (TrxG), 166, 182, 305
Trithorax-like gene, 116
Trithorax (Trx) protein, 115
Tyrosine 641 mutations, 264

U

Ubiquitin
  conjugation, 82–85, 82f
  deconjugation, 82–85, 82f
  transfer, mechanism, 96–99
uH2A, 102–104
Ultrabithorax (Ubx), 113–114, 194–195
UNC3866, 49

X

X chromosome inactivation, 149
X chromosome repression, 296–298, 297f
XO germline, 298–299
X-ray crystallography studies, 171–179

Y

Y641 mutations, 264–266, 265f, 270f, 271
YY1-associated factor (YAF) proteins, 9–10